D1384185

COMPUTATIONAL AUDITORY SCENE ANALYSIS

IEEE Press
445 Hoes Lane
Piscataway, NJ 08854

COMPUTATIONAL AUDITORY SCENE ANALYSIS

Principles, Algorithms, and Applications

Edited by

DeLiang Wang
The Ohio State University

Guy J. Brown
University of Sheffield

IEEE PRESS

A JOHN WILEY & SONS, INC., PUBLICATION

Published by John Wiley & Sons, Inc., Hoboken, New Jersey.
Published simultaneously in Canada.

Library of Congress Cataloging-in-Publication Data is available.

ISBN-13 978-0-471-74109-1
ISBN-10 0-471-74109-4

Printed in the United States of America.

10 9 8 7 6 5 4 3 2 1

Contents

Foreword

This book provides exactly what is needed by a young scientist or engineer who wants to get started in the exciting new field of computational auditory scene analysis. In describing what the book is about, it is helpful to start by considering a particular difficult problem that a human faces in everyday listening. A number of sounds are present at the same time, but the listener is interested in only one of them. How can the desired sound be isolated from the mixture?

When you ask people how they can listen to only one of the sounds in a mixture, a typical listener might answer, "I just listen to one and try not to be distracted by the others." Notice that this answer presupposes that a number of separate sounds have already been discerned and the only problem is to listen to just one of them. It illustrates the limitations of introspection, because the scientist knows facts about sound that our typical listener does not. The first is that although each sound-producing event in the environment radiates its own pattern of pressure waves into the air, these individual sound-wave patterns are summed together on each of the two eardrums of the listener. Consequently, each eardrum vibrates in a pattern that is the sum of all the wave patterns that reach it. Unfortunately, this summed waveform does not have written on it how many sounds have been added together to create it. There are an infinite number of possible ways in which simpler sounds could have been added together to make this sum. So the listener's brain has to solve the difficult problem of deriving the actual waveforms that have contributed to the mixture—a process known as auditory scene analysis (ASA).

Years of research have shown that human listeners solve the ASA problem by exploiting the regularities of the world. Let me give an example. There are many periodic sounds in the world, among them the sounds of musical instruments, buzzes of insects, animal vocalizations, and many of the sounds present in human speech. For example, a vowel, produced on a certain pitch, is a complex tone that can be analyzed into a set of harmonics—pure tones whose frequencies are multiples of a fundamental frequency. When two vowels are heard at the same time, the ear encounters a set of components that consists of a mixture of the two sets of harmonics. This fact about the regularity of harmonics is a characteristic that does not yet have anything to do with human hearing. However, the human auditory system, in analyzing the incoming mixture, can look for sets of components that relate to different fundamentals. Suppose it finds that a majority of the incoming components can be assigned to two subsets, one whose fundamental is 110 Hz (110, 220, 330, 440, . . .) and a second whose fundamental is 160 Hz (160, 320, 480, . . .), how many distinct sounds should it conclude that it is hearing? The obvious answer is two. It is extremely unlikely that the frequency components present in a single sound or in several concurrent sounds would line up into exactly two sets of harmonics.

This use of the regularity found in harmonic sounds is an example of exploiting the regularities of the world—inferring the number of acoustic sources from the properties of the acoustic input, under the assumption that the input is a mixture of natural sounds. However, not all sounds are harmonic; so we have to use other properties of a mixture that arise because it is the sum of a number of distinct natural sounds, probably separated in space. Over the years, we have found out what many of these properties are and have studied how they are used by human listeners.

Our success in doing so has raised a number of interesting questions:

- Do animals other than humans perform ASA on auditory inputs, and, if so, do they exploit the same environmental regularities as humans do, or are there other regularities specific to the animals' own environments?
- What is the neurological basis of ASA?
- Do very young infants already perform ASA as adults do, or must this be learned?
- How do composers take advantage of the listener's ASA to control whether musical sounds will blend with, or stand out from, one another?
- When mixtures involve speech sounds, are they subject to the same ASA processes as mixtures of nonspeech sounds?
- Can the use of sound in data displays (sonification) profit from knowledge about human ASA?
- How can auditory prostheses help their users to segregate sounds?

The challenge of answering these questions has attracted the attention of researchers in a number of disciplines, and promising starts have already been made.

There is another question of tremendous practical importance: Is it possible to program computers to perform ASA? There have been many interesting approaches to this problem, and the evolving research field has been named "computational auditory scene analysis" (CASA). Solving this problem would have valuable applications. Computer programs for speech recognition tend to run into serious trouble when other sounds accompany the target speech. Solving the problem of mixtures would allow speech-recognition programs to operate in complex acoustic environments so that they could, for instance, interact verbally with a user, either face to face or by telephone, and respond to commercial orders, database queries, or other messages. Research on CASA can also serve theoretical purposes. For example, if computer systems utilize methods similar to those proposed as explanations of human ASA, their success or failure can serve as a test of the adequacy of these explanations, and can raise new questions for research on humans and other animals.

Many scientific meetings have been held on the subject of CASA, bringing together researchers from different institutions and countries. The present volume has been preceded by other books on the subject, but these have been reports of the research meetings on CASA, providing accounts of specific pieces of research or individual approaches to this problem. The present book is unique. Although written by leading researchers in the field, it is designed as a textbook, giving overall accounts of the different approaches to CASA with only as much detail as needed for clarity of exposition. The reader's access to the field as a whole is greatly facilitated by the comprehensiveness of the chapters and by the inclusion of extensive bibliographies (in the neighborhood of 100 references per chapter). The broad coverage plus the abundant references provide everything that is needed to rapidly come up to speed on CASA research. The editors and the other authors are to be congratulated on putting together a volume that should prove to be extremely useful in opening up this field to a large number of students and researchers.

ALBERT S. BREGMAN

December 2005

Preface

The acoustic signal reaching our ears comprises sound waves from multiple sources and their reflections from surfaces in the environment. One of the most remarkable achievements of human perception is the ability to disentangle this extremely complex mixture and group the sound components that originate from the same event. This perceptual accomplishment, often taken for granted since it is such a common experience, may not be truly appreciated until one undertakes the effort to construct a machine system that matches human performance.

In 1990, Albert Bregman published his seminal book, *Auditory Scene Analysis,* which was the first to systematically explain the principles underlying the perception of complex acoustic mixtures. Decades of psychophysical research, and Bregman's book in particular, have motivated an emerging field of study called *computational auditory scene analysis* (CASA), which aims to develop machine systems that achieve sound-source separation by exploiting perceptual principles. The CASA approach, which may be characterized as the reverse engineering of human auditory scene analysis, stands in sharp contrast with traditional approaches to sound separation, such as speech enhancement, spatial filtering using microphone arrays, and independent component analysis. On the other hand, the approach has been heavily influenced by David Marr's pioneering investigation of computational vision as embodied in his 1982 book *Vision.*

Interest in CASA, and sound separation in general, has grown substantially in recent years. This has been partly fueled by the widespread realization that automatic speech and speaker recognition, which focuses on single-source signals, breaks down in real (i.e., noisy) acoustic environments. Also, the problem of environmental noise is regarded as the greatest obstacle in hearing aid design. However, although there is now a substantial body of CASA work, there is no single book that gives a comprehensive introduction to the field. This book was conceived to fill this void, with the intention of presenting fundamental concepts and state-of-the-art techniques within a coherent framework. Although there is a long way to go to meet the ultimate goals of CASA, we believe that the time is ripe to review the current state of CASA research, and to define the challenges ahead.

We hope that the result of this effort is an account that is authoritative and coherent, but also widely accessible. The book is intended for graduate students and researchers who are interested in sound separation, robust automatic speech recognition, music processing, and the neural and perceptual modeling of auditory scene analysis. The text is intended to be largely self-contained, but assumes some basic knowledge of signal processing. Psychoacoustical concepts are briefly explained when they are introduced; Bregman's book is, of course, recommended for those who wish to study the psychology of hearing in more detail.

The book has a companion website that includes software for generating some of the demonstrations in this book, as well as other standard CASA tools. The site also includes test corpora, sound demonstrations, and links to other useful information. The URL of the website is http://www.casabook.org.

ORGANIZATION OF THE BOOK

Chapter 1 offers a high-level introduction to CASA, defining the problem and the scope, and explaining a number of basic concepts and representations. It also provides a brief history of CASA development. The remainder of the book focuses on specific topics in CASA. Chapter 2 deals with multipitch tracking, which is a fundamental problem for CASA systems that address the segregation of pitched sounds, such as voiced speech. Two different, but interrelated approaches to source separation are then addressed: feature-based segregation (Chapter 3) and model-based segregation (Chapter 4). The following two chapters deal with topics in binaural CASA systems, namely sound localization (Chapter 5) and localization-based grouping (Chapter 6). We then consider the issue of reverberation (Chapter 7), which presents a significant problem for the use of CASA in real acoustic environments. Two applications of CASA technology are considered in the subsequent chapters. Chapter 8 introduces the emerging field of music scene description, which uses CASA-like processing to extract salient information from musical signals. Chapter 9 reviews the relationship between CASA and automatic speech recognition, which is a key application area for CASA technology. Finally, Chapter 10 considers CASA systems that have a strong biological foundation, and discusses the insights that have come from computational models of neural and perceptual phenomena concerning auditory scene analysis.

ACKNOWLEDGMENTS

First of all, we thank the authors for giving so much of their time to this collective project, and for accepting the nagging of the editors with good humor. We are grateful to Prof. Albert Bregman for his generous foreword. Thanks are due to Jeanne Audino of the IEEE Press and Lisa Morano Van Horn of Wiley. Special thanks to Stuart Wrigley for assistance with proofreading. We also want to express our gratitude to Dr. Willard Larkin of AFOSR for his support to the field, and for funding several related workshops in which many of the authors participated.

DeLiang Wang
Guy J. Brown

Columbus, Ohio
Sheffield, United Kingdom
July 2006

Contributors

Jon Barker
Department of Computer Science
University of Sheffield
Sheffield, United Kingdom

Guy J. Brown
Department of Computer Science
University of Sheffield
Sheffield, United Kingdom

Alain de Cheveigné
Audition—Psychophysique, Modélisation, Neurosciences
Universite Paris 5, École Normale Supérieure
Paris, France

Daniel P. W. Ellis
Department of Electrical Engineering
Columbia University
New York, New York

Albert S. Feng
Beckman Institute
University of Illinois at Urbana-Champaign
Urbana, Illinois

Masataka Goto
National Institute of Advanced Industrial Science and Technology (AIST)
Tsukuba, Ibaraki, Japan

Douglas L. Jones
Department of Electrical and Computer Engineering
University of Illinois at Urbana-Champaign
Urbana, Illinois

Kalle J. Palomäki
Laboratory of Computer and Information Science
Helsinki University of Technology
Helsinki, Finland

Richard M. Stern
Department of Electrical and Computer Engineering and Language Technologies Institute
Carnegie Mellon University
Pittsburgh, Pennsylvania

DeLiang Wang
Department of Computer Science and Engineering and Center for Cognitive Science
Ohio State University
Columbus, Ohio

Acronyms

ACF	Autocorrelation function
AGC	Automatic gain control
AIC	Akaike information criterion
ALI	Attentional leaky integrator
AM	Amplitude modulation
AMS	Amplitude modulation spectrum
ART	Adaptive resonance theory
ASA	Auditory scene analysis
ASR	Automatic speech recognition
ASW	Apparent source width
BIC	Bayesian information criteria
BMLD	Binaural masking level difference
CASA	Computational auditory scene analysis
CMN	Cepstral mean normalization
CN	Cochlear nucleus
CD	Characteristic delay
CF	Characteristic frequency
CVC	Consonant–vowel–consonant
DC	Direct current
DCT	Discrete cosine transform
DDF	Double difference function
DFT	Discrete Fourier transform
DRR	Direct sound-to-reverberation ratio
EC	Equalization-cancellation
EE	Excitatory–excitatory
EEG	Electroencephalogram
EI	Excitatory–inhibitory
EM	Expectation-maximization
ERB	Equivalent rectangular bandwidth

ERP	Event-related potential
***F*0**	Fundamental frequency
Δ*F*0	Difference in fundamental frequency
FFT	Fast Fourier transform
FIR	Finite impulse response
FM	Frequency modulation
GMM	Gaussian mixture model
HMM	Hidden Markov model
HRIR	Head-related impulse response
HRTF	Head-related transfer function
IBI	Interbeat interval
IC	Interaural coherence
ICA	Independent component analysis
IF	Instantaneous frequency
IID	Interaural intensity difference
IMTF	Inverse modulation transfer function
IOI	Interonset interval
ITD	Interaural time difference
LEGION	Locally excitatory, globally inhibitory network
LEV	Listener envelopment
LP	Linear prediction
MAA	Minimum audible angle
MAP	Maximum a posteriori
MCMC	Markov chain Monte Carlo
MFCC	Mel-frequency cepstral coefficient
MIDI	Musical instrument digital interface
MINT	Multiple input–output theorem
MLD	Masking level difference
MMN	Mismatch negativity
MPEG	Moving picture experts group
MTF	Modulation transfer function
MUSIC	Multiple signal classification
OPTIMA	Organized processing toward intelligent music scene analysis
ORN	Object-related negativity
PB	Phonetically balanced
PDF	Probability density function
PLP	Perceptual linear prediction
PreFEst	Predominant *F*0 estimation
QBH	Query by humming
RASTA-PLP	Relative spectral–perceptual linear prediction
RefraiD	Refrain detection
RIR	Room impulse response
RMS	Root mean square
SACF	Summary autocorrelation function
SID	Speaker identification

SNR	Signal-to-noise ratio
SPIN	Speech perception in noise test
SRT	Speech reception threshold
STFT	Short-time Fourier transform
STI	Speech transmission index
T-F	Time-frequency
TRT	Tone repetition time
VLSI	Very large-scale integration

CHAPTER 1

Fundamentals of Computational Auditory Scene Analysis

DELIANG WANG and GUY J. BROWN

Listen. What do you hear? Perhaps a passing car, distant voices, the hum of a computer or music playing quietly in the background. Regardless, it is likely that you can identify more than one source of sound. This illustrates a property of human hearing that is so familiar we usually take it for granted: we are able to distinguish individual sound sources from a complex mixture of sounds reaching our ears. A striking example of this perceptual feat is given by Helmholtz [53] in his 1863 masterpiece on audition:

> In the interior of a ball-room . . . we have a number of musical instruments in action, speaking men and women, rustling garments, gliding feet, clinking glasses, and so on . . . a tumbled entanglement [that is] complicated beyond conception. And yet . . . the ear is able to distinguish all the separate constituent parts of this confused whole. ([53], pp. 26–27)

A particularly important aspect of this problem, which is apparent from Helmholtz's description, is the perception of speech in such complex acoustic environments. Writing in the 1950s, Cherry [24, 25] termed this the "cocktail party problem":

> One of our most important faculties is our ability to listen to, and follow, one speaker in the presence of others . . . we may call it "the cocktail party problem." No machine has yet been constructed to do just that. ([25], p. 280)

Cherry's assessment of machine performance is as pertinent today as it was half a century ago. How indeed might one construct a device that can segregate the mixture of sounds emanating from a cocktail party, or from any other acoustic environment, with a performance that matches that of human listeners? This challenging problem is the topic of this book.

Insight into the processes underlying the perceptual separation of sound sources

Computational Auditory Scene Analysis. Edited by DeLiang Wang and Guy J. Brown

has come from several decades of psychophysical research. In 1990, Bregman [13] published a coherent account of this work. He contends that there are many similarities between audition and vision. When we view a visual scene, the edges, textures and colors are analyzed and interpreted as perceptual wholes (e.g., a face or a chair). Similarly, Bregman argues that the sound reaching our ears is subjected to an *auditory scene analysis* (ASA). Conceptually, this may be regarded as a two-stage process. In the first stage (which has been called *segmentation*), the acoustic input is decomposed into a collection of local time-frequency regions (which we term "segments"). The second stage (*grouping*) combines segments that are likely to have arisen from the same environmental source into a perceptual structure, which is termed a *stream*. Bregman's account therefore makes an important distinction between the acoustic source (such as the plucking of a guitar string) and the corresponding perceptual stream (the mental representation of a guitar being played).

By providing a coherent framework for understanding the perceptual organization of sound, Bregman's book has stimulated much interest in computational studies of ASA. Such studies are motivated in part by the demand for practical sound separation systems, which have many applications including noise-robust automatic speech recognition, hearing prostheses, and automatic music transcription. This emerging field has become known as *computational auditory scene analysis* (CASA). This book aims to provide a comprehensive account of the state-of-the-art in CASA, in terms of the underlying principles, the algorithms and system architectures that are employed, and the likely applications of this new technology.

In this introductory chapter, we review general principles of auditory scene analysis in listeners and machines. Section 1.1 briefly describes the structure and function of the human auditory system, and then discusses the auditory organization of simple stimuli (such as sequences of tones). We then consider the performance of human listeners in cocktail party scenarios, in which speech is contaminated by noise or other voices, and review the grouping principles that are thought to underlie ASA. Section 1.2 defines CASA and its goal in an attempt to provide a coherent theme for the field. The section also discusses potential applications of CASA technology. Section 1.3 introduces some fundamental concepts of CASA systems, including system architecture and auditory-motivated signal representations. Section 1.4 discusses how to properly evaluate a CASA system. Section 1.5 summarizes other major sound separation approaches, and contrasts them with CASA. Finally, we end the chapter with a historical account of CASA development, and a brief review of the following chapters of this book.

1.1 HUMAN AUDITORY SCENE ANALYSIS

1.1.1 Structure and Function of the Auditory System

Before considering ASA in human listeners, it is useful to have some appreciation of the structure and function of the early stages of auditory processing. The review that follows is necessarily brief; the reader is referred to standard texts such as [108, 68] for further details.

The auditory periphery is a complex transducer that converts sound vibrations into action potentials ("spikes") in the auditory nerve. Broadly, the periphery can be divided into three areas: the outer, middle, and inner ear. The outer ear consists of the external ear (pinna), the ear canal (meatus), and the eardrum (tympanic membrane); the ear canal connects the external ear with the eardrum. The pinnae play a role in spatial hearing, by imposing spectral characteristics on sound that depend on its direction of incidence with respect to the head. Sound traveling down the ear canal causes the eardrum to vibrate. These vibrations are transmitted to the fluid-filled cochlea by the middle ear, in which three tiny bones (ossicles) form a lever, the action of which matches the impedance of air to that of the cochlear fluids.

The cochlea in the inner ear is the organ of hearing. It consists of a coiled, fluid-filled tube that is divided along its length by two membranes, Reissner's membrane and the basilar membrane. The basilar membrane varies in mass and stiffness along its length, so that different regions of the membrane vibrate at different resonant frequencies. As a result, the basilar membrane exhibits a complex pattern of motion in response to sound-induced movement of the cochlear fluids. More specifically, the response of the basilar membrane to a sinusoidal stimulus is a traveling wave that moves along the length of the cochlea. Importantly, the traveling wave oscillates at the frequency of the stimulus and, therefore, provides a timing code for stimulus frequency. Additionally, the traveling wave reaches its peak amplitude at a location where the stimulus frequency matches the resonant frequency of the basilar membrane. Hence, stimulus frequency is also represented by a place code.

The movements of the basilar membrane are transduced into neural activity by inner hair cells. The hairs at the tops of these cells are displaced when the basilar membrane moves up and down. As a result, the inner hair cells are activated, and this leads to the initiation of action potentials in the spiral ganglion cells whose axons make up the auditory nerve. It should be noted that the auditory nerve encodes a half-wave rectified version of the stimulus, because action potentials are only initiated by the movement of hairs in one direction.

Auditory nerve responses exhibit a number of important properties. They respond preferentially to certain frequencies, so that the frequency selectivity originating at the basilar membrane is maintained. If the stimulus frequency is low, the fibers respond to individual vibrations of the stimulating waveform, exhibiting the so-called "phase locking" phenomenon. Auditory nerve fibers fire spontaneously in the absence of a stimulus. The firing rate of a fiber is related to stimulus level by a sigmoidal function, so that the neural response exhibits level compression and saturation effects. Also, auditory nerve fibers adapt to a steady stimulus; they respond vigorously at stimulus onset, and then the firing rate declines to a steady state. Finally, there is a refractory period for a brief time after the offset of the stimulus, during which the firing rate falls below the spontaneous level.

Beyond the auditory periphery, the pathway leading to the auditory cortex passes through four neural structures: cochlear nucleus, superior olive, inferior colliculus, and medial geniculate nucleus. Neurons throughout these higher centers of the auditory pathway and the auditory cortex appear to be tuned to particular stimulus features, such as periodicity, sound intensity, amplitude and frequency modulation, and interaural time and intensity differences.

Although it is clear that human listeners are able to solve the ASA problem from eardrum vibrations in response to sound waves, relatively little is known at the physiological level about the way in which the higher auditory centers perform ASA. However, insights have also come from perceptual experiments; it is these that we consider next.

1.1.2 Perceptual Organization of Simple Stimuli

Much of the evidence for ASA has come from perceptual experiments with simple stimuli. A typical example is the sequence of alternating high- and low-frequency tones shown in Fig. 1.1, due to Noorden [126]. When this sequence is presented to listeners at a slow rate (e.g., when the time between tone onsets is 150 ms or so), the perceptual organization of the tones depends upon the frequency difference between them. If the frequency difference is small (less than 4 semitones or so; see Fig. 1.1, panel **A**) then listeners perceive a single stream consisting of a sequence of high- and low-pitched tones with a galloping rhythm. However, when the high and low tones are separated in frequency by more than 12 semitones or so (panel **B**), the galloping rhythm is no longer heard. Instead, the sequence segregates into two streams, each of which consists of tones with the same pitch that are equally spaced in time. This phenomenon is known as *streaming*. At intermediate frequency separations, listeners may hear either organization, and can switch between one or two streams with conscious effort.

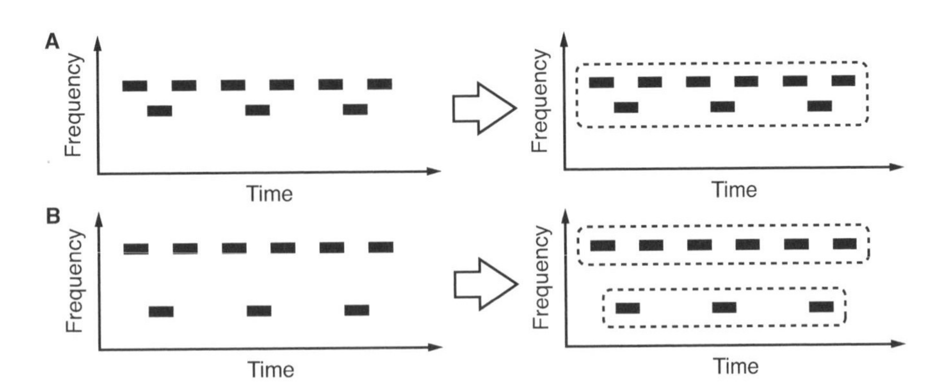

Figure 1.1 Perceptual organization of alternating tone sequences. (**A**) A tone sequence in which there is a small difference in frequency between the high and low tones is perceived as a perceptual whole (i.e., a single stream). (**B**) A larger difference in frequency causes the high and low tones to segregate into different streams.

The streaming phenomenon is also dependent on the rate of presentation; at faster rates, the range of ambiguity is reduced, and streaming can be induced by relatively small differences in frequency. Indeed, when the rate of presentation is fast and there is a large frequency separation between the high and low tones, listeners cannot hear a single stream, even with conscious effort. Bregman [13] argues that this effect is due to the existence of automatic, *primitive* processes of auditory organization. We will expand on this important point in Section 1.1.4. However, before doing so we return to the issue of "cocktail party" listening, in which the stimulus reaching the ears is considerably more complex than the tone sequences discussed above.

1.1.3 Perceptual Segregation of Speech from Other Sounds

Although Helmholtz marveled at the ability of human hearing in cluttered acoustic environments, he was also very much aware of its limitations. Specifically, he noted that in order for a specific sound to be heard out from the whole, it must not be "too much overpowered by the mere loudness of the others" ([53], p. 25). In essence, Helmholtz's comment refers to the phenomenon of masking, in which the threshold of audibility for one sound is raised by the presence of another sound (which is called the masker). So how well can the auditory system segregate, and hence recognize, speech in the presence of masking sounds?

One way to quantify speech intelligibility in noise is the speech reception threshold (SRT), which is the signal-to-noise ratio (SNR) measured in decibels (dB) required for a 50% intelligibility score. Steeneken [122] has investigated the intelligibility of different kinds of speech material when presented in a background of speech-shaped noise (i.e., stationary noise with a spectrum similar to the long-term speech spectrum). His data are shown in Fig. 1.2, and demonstrate that the speech intelligibility score varies with the speech material used. Nonsense consonant-vowel-consonant (CVC) syllables are the most susceptible to noise because they lack contextual cues, having a SRT of about –4 dB. The most intelligible speech materials are spoken digits and letters of the alphabet, which have a SRT of about –10 dB. To put these figures into context, Bronkhorst and Plomp [14] report that for "cocktail party"-like situations in which all voices are equally loud, speech is intelligible for normal-hearing listeners even when as many as six interfering talkers are present. Assuming that the individual utterances are uncorrelated, this situation amounts to a SNR of about –7.8 dB.

The nature of the interference also has a strong influence on the SRT. Miller [93] reviewed the masking of speech by a variety of tones, broadband noises, and other voices. Subjects were tested for their word intelligibility scores, and the results are shown in Fig. 1.3. In general, tones are less effective maskers than broadband noises. For example, speech is intelligible even when corrupted by a complex tone glide that is 20 dB more intense (pure tones are even weaker maskers). Broadband noise is the most effective speech masker, and the corresponding SRT is about 2 dB. When the masker consists of other voices, the SRT varies within a considerable range depending on how many talkers are present. As shown in Fig. 1.3, the SRT is

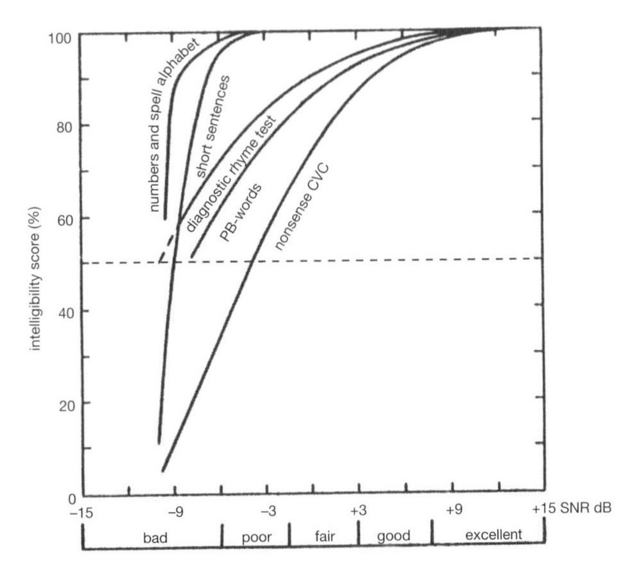

Figure 1.2 Speech intelligibility score with respect to SNR for different speech materials (adapted with permission from [122]). The 50% intelligibility score is indicated by the dashed line. A diagnostic rhyme test determines an intelligibility score from a limited number of alternative words that differ on a single articulatory feature. The interference is speech-shaped noise. "PB" stands for "phonetically balanced."

about −10 dB for single-voice interference and quickly rises to −2 dB for two-voice interference. The SRT stays about the same (around −1 dB) when the masker contains four or more voices.

The relative ease of speech segregation with a single competing talker is usually attributed to the ability of listeners to "glimpse" the target voice during gaps in the masker [94, 27, 2]. More specifically, speech energy is sparsely distributed in the time-frequency plane, and, hence, gaps occur in the spectrum of the masker during which listeners can obtain an uncorrupted estimate of the spectrum of the target speech signal. When the interference consists of multiple voices or broadband noise, the opportunity for glimpsing is much reduced. However, even with a single interfering voice, speech intelligibility is worse when the competing talker is of the same sex than when it is of a different sex, and it is worst when the same talker is used both as the target and interference [22]. Although glimpsing is likely to play a role, it may not adequately explain the large intelligibility differences under such conditions [121]. Additionally, auditory scene analysis may contribute to a glimpsing process by identifying time-frequency regions that are dominated by a single sound source.

It is worth noting that the curves shown in Figs. 1.2 and 1.3 are steep in the region of the SRT, so that even a small gain in SNR leads to an appreciable increase in intelligibility (e.g., a SNR gain of 1 dB near to the SRT leads to an increase in in-

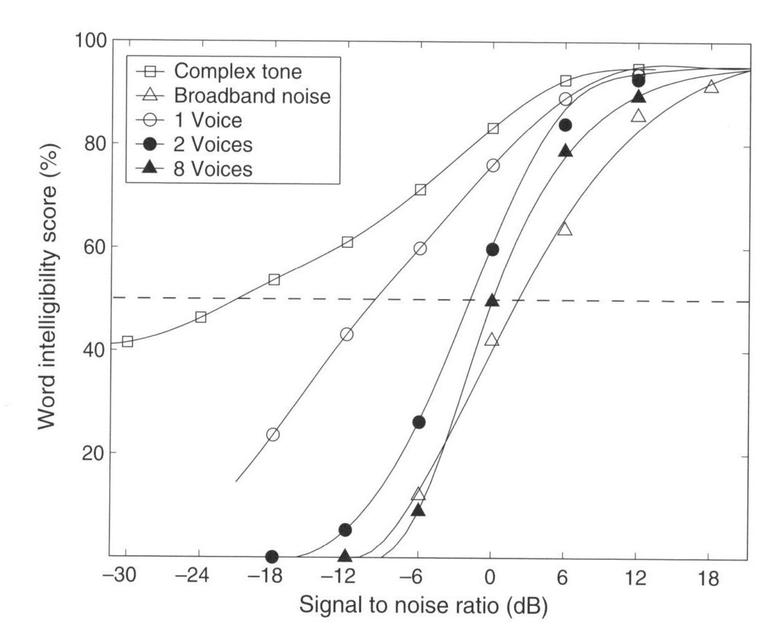

Figure 1.3 Speech intelligibility score with respect to SNR for different types of interference (redrawn from [93]). The level of the target speech is fixed at 95 dB. The 50% intelligibility is indicated by the dashed line. The pitch frequency of the complex tone rises slowly and drops suddenly in a periodic manner. The broadband noise covers the spectrum from 20 Hz to 4 kHz. For speech interference, data are shown for 1, 2, and 8 competing speakers.

telligibility of 5–10%, depending on the interference material). From a different perspective, it has been estimated that hearing-impaired listeners typically need a SNR that is 5–10 dB higher than that of normal listeners in order to achieve the same intelligibility score [110, 111, 1]. This suggests that consistent improvements in SNR of just a few decibels can yield significant improvements in speech intelligibility.

All of the data discussed above concern mixtures of speech and noise that are created in the laboratory and presented to listeners over a single audio channel. However, in natural "cocktail party" environments, listeners receive an input from both ears and must also contend with room reverberation. What is the effect of these factors?

When target speech and noise intrusions are presented from different spatial locations, binaural hearing can lead to a significant reduction in the SRT [14, 38]. At least two mechanisms appear to contribute to this "binaural advantage." First, when sounds are presented at different locations, the SNR at one ear will be higher than the SNR at the other. Hence, listeners may gain some benefit by selectively attending to the ear in which the SNR is greatest. Second, interaural interaction enables location-based organization of the auditory scene, a topic discussed in Chapters 5 and 6.

Although speech perception in quiet is robust to room reverberation [97, 7], speech perception in noise is sensitive to reverberation effects. The studies by Plomp [109] and Culling et al. [30] demonstrate a significant increase in the SRT when speech is presented together with an interfering sound in reverberant conditions. For example, Culling et al. report a 5 dB increase in the SRT for naturally intonated speech presented with a reverberation time of 400 ms, compared to anechoic conditions (monotone speech fares a little better). The same study also finds that the benefit of spatial separation disappears in reverberant conditions. The effects of reverberation on speech perception and auditory grouping cues are addressed in Chapter 7.

1.1.4 Perceptual Mechanisms

What are the perceptual mechanisms that underlie auditory organization in human listeners? Bregman's book, *Auditory Scene Analysis,* gives the most systematic and comprehensive treatise on this question to date [13]. As explained in the introduction to the book, Bregman's conceptual framework for ASA amounts to an analysis–synthesis process; the acoustic scene is decomposed into a collection of segments, which are subsequently grouped to form coherent streams. Such grouping processes may operate both simultaneously (in which concurrent segments are integrated) or sequentially (in which segments are grouped across time). Bregman also makes a further distinction between *primitive* grouping and *schema-based* grouping. The former is regarded as an innate and bottom-up process, which relies on the intrinsic structure of environmental sounds. Schema-based grouping refers to the fact that auditory features belonging to the same learned pattern, such as a syllable, tend to be bound together. It is therefore a top-down process, and based on prior knowledge.

According to Bregman, primitive grouping is governed by mechanisms that are analogous to those proposed by the Gestalt psychologists in relation to visual perception (for a review, see [102]). If we view the acoustic signal in the form of a time-frequency scene (i.e., like a spectrogram—see Figs. 1.1 and 1.4), then the major primitive grouping principles can be summarized as follows (see among others [126, 51, 13, 31, 64, 141, 8, 96]).

- **Proximity in frequency and time.** The closer acoustic components are in frequency, the greater is the tendency to group them into the same stream. This is apparent from the alternating tone sequences shown in Fig. 1.1. Similarly, acoustic components tend to be perceptually grouped if they are close in time; this explains the effect of presentation rate on the streaming phenomenon, as discussed in Section 1.1.2.
- **Periodicity.** A set of acoustic components that are harmonically related (i.e., have frequencies that are integer multiples of the same fundamental frequen-

cy) tend to be grouped into the same stream. This is a simultaneous grouping principle.

- **Continuous or smooth transition.** Frequency components tend to be grouped into the same stream if they form a continuous trajectory (the temporal continuity principle) or a discontinuous but smooth trajectory. Similarly, smooth changes in pitch contour, intensity, spatial location, and spectrum tend to be interpreted as the continuation of an existing sound, whereas abrupt changes signify the onset of a new sound source.
- **Onset and offset.** Listeners tend to group frequency components into the same stream if they have the same onset time or, to a lesser extent, the same offset time. This is a kind of simultaneous organization.
- **Amplitude and frequency modulation.** Frequency components that exhibit the same temporal modulation tend to be grouped together into the same stream. This simultaneous grouping principle applies to both amplitude modulation (AM) and frequency modulation (FM).
- **Rhythm.** A sequence of rhythmically related tones tends to be integrated into the same stream. This is a sequential organization cue that applies to events that are separated in time.
- **Common spatial location.** Concurrent sounds that originate from the same location in space tend to be grouped. Also, as noted above, sequential integration is promoted by smoothly varying location for the case of moving sound sources. However, the fact that listeners can separate sounds that originate from the same location, and can segregate acoustic mixtures when listening monaurally, suggests that spatial location may play a secondary role to spectral cues.

Since these organizational principles have largely been derived from the study of simple laboratory stimuli, we should ask whether they are also applicable to complex, real-world sounds such as speech and music. Figure 1.4 illustrates some of the grouping cues that are present in a speech utterance. Both broadband and narrowband spectrograms are displayed in the figure. Prominent features in this example include the temporal continuity of harmonics and formants, harmonicity of voiced speech, onset and offset synchrony, and coherent amplitude modulation. Despite the complexity of the speech signal, the figure suggests that many primitive grouping cues are applicable to the perceptual organization of natural speech. Further discussion of the role of ASA in speech perception can be found in Chapter 9.

Similarly, it is well known that music is subject to auditory organization. Indeed, this fact was exploited by some Baroque composers, who used sequences of rapidly alternating high and low notes, like those shown in panel **B** of Fig. 1.1, to give the impression that a single instrument was playing two melodic lines (so-called "virtual polyphony"; see [13], p. 464). A further example is the perception of "interleaved melodies" [36, 37]. In this paradigm, the notes of two melodies are interleaved in time, so that adjacent tones are from different melodies. If the pitch ranges of the two melodies do not overlap, the interleaved sequence is perceptually segregated

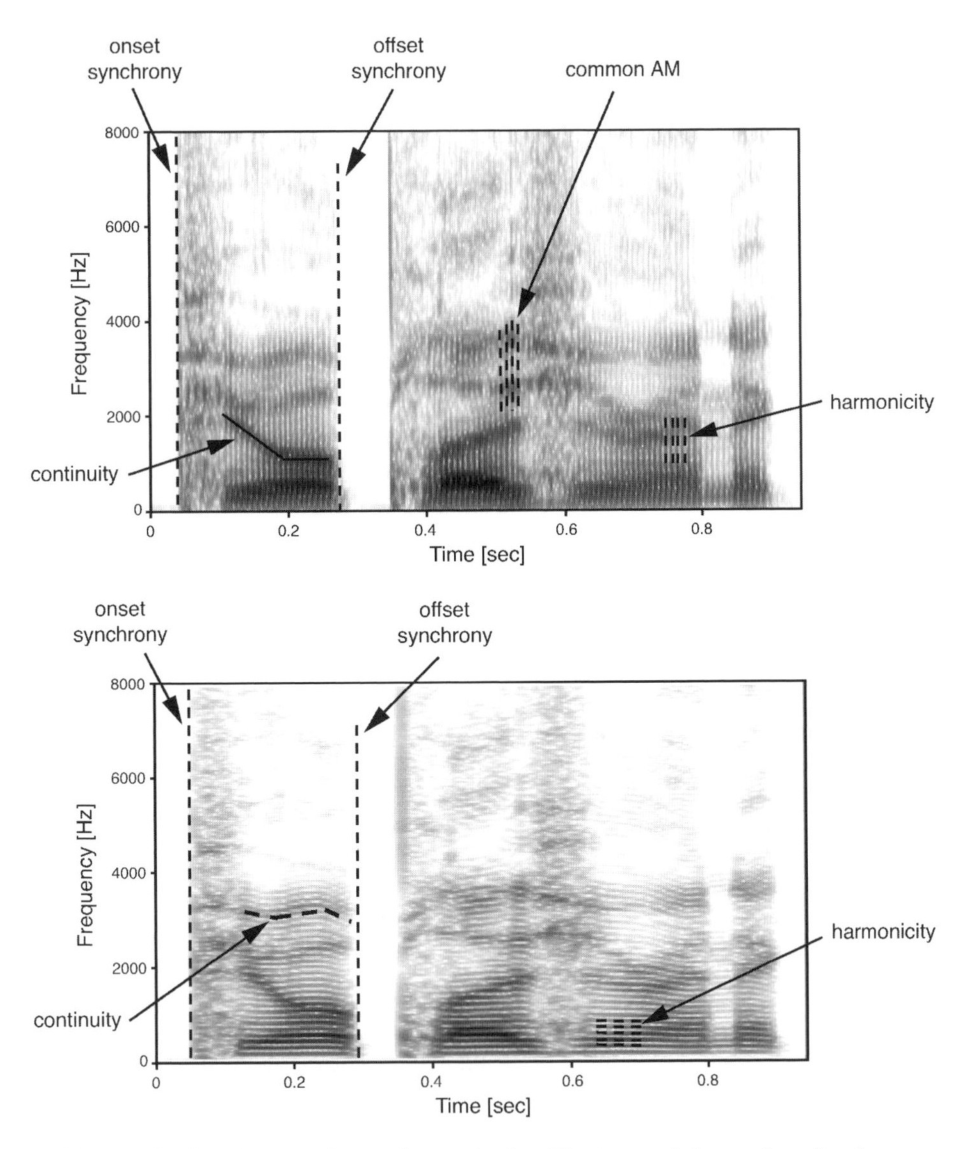

Figure 1.4 Grouping cues in speech organization. The top panel shows a broadband spectrogram of the utterance "pure pleasure." The cues shown in the figure are temporal continuity of a formant, onset synchrony of a stop consonant, offset synchrony at the end of the word "pure," and common AM and harmonicity corresponding to glottal pulses. The bottom panel shows a narrowband spectrogram of the same utterance. The cues shown are temporal continuity of a single harmonic, onset and offset synchrony, and harmonicity.

into two streams, which correspond to the two melodies. However, if the two melodies have overlapping pitch ranges, listeners are unable to segregate them. Similar effects are found if the notes of the two melodies are distinguished by timbre, rate of attack and decay, or duration [52].

1.2 COMPUTATIONAL AUDITORY SCENE ANALYSIS (CASA)

1.2.1 What Is CASA?

Broadly speaking, CASA may be defined as the study of auditory scene analysis by computational means. To follow the earlier quote from Cherry, one may define the CASA problem as the challenge of constructing a machine system that achieves human performance in ASA. This definition is entirely functional; it makes no reference to underlying mechanisms. In other words, it makes no commitment as to whether CASA should employ the perceptual and neural mechanisms known to be used by the human auditory system.

One way to make CASA more biologically relevant is to limit the scope of investigation to monaural (one-microphone) or binaural (two-microphone) input, because the information available to the auditory system for solving the ASA problem is the sound received at the two ears. Without some constraint on the number and placement of microphones, the CASA problem could be totally circumvented; imagine, say, a solution that deploys a dedicated microphone for each sound source. As discussed in Section 1.5, it is well understood that a microphone array can be constructed to perform beamforming and spatial filtering. If the target source originates from a distinct direction, beamforming is able to segregate the target sound. However, such a solution is based on a principle that cannot be applied in monaural or binaural conditions, and it only works in constrained environments. On the basis of these considerations, we propose a more specific definition of CASA: *It is the field of computational study that aims to achieve human performance in ASA by using one or two microphone recordings of the acoustic scene.*

As noted above, the notion of CASA does not imply that the human auditory system should be slavishly modeled. However, many workers adopt an approach that is to some extent based on the principles of processing in the human auditory system. As a result, the term CASA has come to be associated with a perceptually motivated approach that is distinct from other approaches to sound separation (see Section 1.5). In practice, the influence of perceptual studies on CASA research varies considerably. Indeed, the following chapters of this book illustrate that CASA systems range from models that explicitly simulate perceptual and physiological data, to sound separation systems that are related to perception only at a very abstract level.

It is sometimes said that the CASA field lacks a common theory. Comparison is often made with automatic speech recognition (ASR), in which the goal is clear and the evaluation metric is common. Although it is true that there is no consensus on the criteria for evaluating CASA systems (see Section 1.4.1), we should bear in mind that the problem of CASA is considerably broader than the ASR problem. Given the interdisciplinary nature of the field and the diverse range of motivations and applications, it should not be surprising that there are different opinions on the essence of CASA and it will likely remain so in the foreseeable future. One could say the same of computational vision, which has been developed by many more workers for a much longer time. In our opinion, it is healthy to have diversity in the early stages of development of this new field.

1.2.2 What Is the Goal of CASA?

In his influential treatise on computational vision, Marr [85] makes a compelling argument for separating three levels of analysis in order to understand complex information processing systems. The first level—computational theory—is concerned with analyzing the goal of computation and the general processing strategy. The second level—representation and algorithm—is concerned with the representation of the input and output, and the algorithm that transforms from the input representation to the output representation. The third level—implementation—is concerned with how to physically realize the representation and algorithm. Marr states that "Each of the three levels of description will have its place in the eventual understanding of perceptual information processing" (p. 25).

Marr's framework suggests a key question—What is the goal of CASA? In order to answer this, we should, as suggested above, put ASA in the broader context of auditory perception and ask what purpose perception serves. From an information processing perspective, which is shared by human and machine perception, Gibson [45] considers perceptual systems to be ways of seeking and extracting information about the environment from the sensory input. More specifically, the purpose of vision, according to Marr [85], is to produce a visual description of the environment for the viewer. By analogy with this, we may state that the purpose of audition is to produce an auditory description of the environment for the listener. Two related points are worth noting here. First, perception is a process private to the perceiver, despite the fact that the physical environment may be shared by different perceivers. Second, what perception gives us at a given time is *a* description of the environment, not *the* description. That the perceiver constructs only a partial picture of the environment is well illustrated by the perceptual phenomenon of change blindness: a change to an image during a flicker or other interruption often goes undetected, despite the fact that the change is easily seen once the attention of the viewer is directed to it (for example, see [112]). Change blindness is attributed to the limited capacity of attention.

Considering auditory scene analysis more specifically, Bregman [13] states that the goal of ASA is to produce separate streams from the auditory input, such that each stream represents a single sound source in the acoustic environment. To extrapolate from this, we may state that the goal of CASA is to computationally extract individual streams from one or two recordings of an acoustic scene. Note that here and elsewhere in this volume we use the term "stream" to refer both to the perceptual representation of a sound source, and to the representation of a sound source in computer memory.

An analysis of the CASA problem at the computational theory level has led Wang [132] to propose that the goal of CASA should be to estimate an ideal time-frequency (T-F) mask. Consider a time-frequency representation such as the spectrogram shown in Fig. 1.4, in which the frequency axis and time axis are divided into discrete units. An ideal T-F mask is then a binary matrix, whose value is one for a T-F unit in which the target energy is stronger than the total interference energy, and is zero otherwise. Further discussion of this approach is given in Section 1.3.5.

1.2.3 Why CASA?

In addition to the scientific challenge of constructing a "cocktail party processor," research in CASA has a number of important applications, several of which are listed below.

- **Robust automatic speech and speaker recognition.** Much progress has been made in automatic speech and speaker recognition in recent years. However, the performance of these systems degrades rapidly in the presence of acoustic interference, and is much poorer than human performance [81, 59]. Arguably, the current predicament of recognition systems in real environments is largely due to a preoccupation with the recognition of clean speech. This stands in contrast to work in computer vision, in which the central role of scene (image) analysis was recognized from the beginning [85, 43]. CASA offers to redress this imbalance by providing a perspective in which speech is regarded as just one of many sound sources in a complex acoustic environment.
- **Hearing prostheses.** About 10% of the population suffers from hearing loss, and for most of them hearing aids are the primary means of alleviating the associated deficits. Listeners with hearing loss often have difficulty in understanding speech in noisy environments (i.e., their ability to perform "cocktail party" processing is reduced). In such cases, hearing aids that provide amplification are of little help, since they amplify both the speech and the noise. Noise robustness is regarded as the greatest obstacle in hearing aid design [35, 76], and it is a problem for which CASA could provide a solution.
- **Automatic music transcription.** Automatic music transcription aims to derive a symbolic score (usually in a note-based form) from a musical audio recording. This is a challenging technical problem in its own right, and a solution to the problem also offers the potential for notating ethnic music that has no written form. The problem is similar in some respects to automatic speech recognition, but with the added complication that music contains concurrent instruments that must be segregated before they can be individually transcribed. CASA may be able to provide such a segregation (e.g., see [17, 48]). More broadly, CASA can also contribute to solving the problem of music scene description, which is addressed in Chapter 8.

- **Audio information retrieval.** A huge volume of audio recordings are available in private archives and via the Internet, and a key research problem is how to search the audio content efficiently. Because recordings generally contain mixtures of sound sources, the separation of such mixtures is often a prerequisite for retrieving audio information from sound files.
- **Auditory scene reconstruction.** Once an acoustic mixture has been segregated, an auditory scene can be reconstructed in which acoustic sources are placed at arbitrary locations. For example, an interface may be designed to selectively modify desired acoustic events, or to present different sounds at perceptually distinct locations for enhanced human listening.
- **Contribution to hearing science.** CASA can make a contribution to hearing science, by suggesting computational mechanisms that help to explain how the auditory system solves the ASA problem.

1.3 BASICS OF CASA SYSTEMS

In this section, we introduce some of the fundamental ideas associated with the design and implementation of CASA systems. First, we discuss the issue of system architecture—the sequence of processing blocks that comprise a CASA system, and the manner in which information is passed between them. We then discuss some auditory-motivated signal representations (e.g., the correlogram and cross-correlogram), which have frequently been used in CASA systems. The notion of a time-frequency mask is then introduced, and, finally, we discuss strategies for resynthesizing audio waveforms of segregated sound sources.

1.3.1 System Architecture

Figure 1.5 shows a representative architecture of CASA systems, which broadly follows Bregman's conceptual framework for ASA. In this architecture, a digitally recorded acoustic mixture first undergoes peripheral analysis, giving a time-fre-

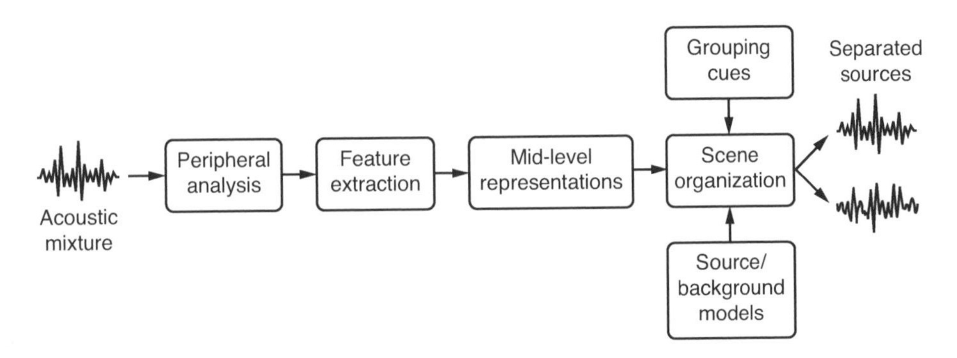

Figure 1.5 System architecture of a typical CASA system.

quency representation of auditory activity (e.g., a cochleagram; see Section 1.3.2). Acoustic features are then extracted, such as periodicity, onsets, offsets, amplitude modulation, and frequency modulation. Mid-level representations, such as segments or other intermediate descriptions, are then formed using these features. Scene organization takes place on the basis of primitive grouping cues and trained models of individual sound sources (and the background), producing separate streams. Finally, an audio waveform is resynthesized from a separated stream, so that segregation performance can be assessed by listening tests or by metrics that compare the segregated waveform with a ground truth.

1.3.2 Cochleagram

The first stage of processing in a CASA system usually derives a time-frequency representation of the acoustic input. This is often achieved by using a computer model of the peripheral auditory system, in order to provide a frequency analysis that is consistent with the known properties of human frequency selectivity. Computer models of the auditory periphery broadly follow the structure described in Section 1.1.1, with processing components that simulate the outer/middle ear, cochlear frequency selectivity, and transduction by hair cells. Notable integrated models of this type include those of Lyon [82], Seneff [117], Deng and Geisler [34], Patterson et al. [104], and Zhang et al. [142].

When modeling cochlear frequency selectivity, it is usual to adopt a functional approach rather than use a detailed model of basilar membrane mechanics. Generally, the basilar membrane is modeled either as a transmission line (i.e., a cascade of filter sections) [82] or as a filterbank, in which each filter (or "channel") models the frequency response associated with a particular point on the basilar membrane. Here, we focus on a widely adopted example of the latter approach, which is based on the gammatone filter.

The gammatone filter was popularized by Johannsma [63] as a model for the impulse response function of auditory nerve fibers, as estimated by the reverse correlation of spike patterns (see also [42, 32]). The gammatone is a bandpass filter, whose impulse response $g_{f_c}(t)$ is the product of a gamma function and a tone (hence "gammatone"):

$$g_{f_c}(t) = t^{N-1} \exp[-2\pi t b(f_c)] \cos(2\pi f_c t + \phi) u(t) \tag{1.1}$$

Here, N is the filter order, f_c is the filter center frequency (in Hz), ϕ is the phase, and $u(t)$ is the unit step function (i.e., $u(t) = 1$ for $t \geq 0$, and 0 otherwise). The function $b(f_c)$ determines the bandwidth for a given center frequency. For $f_c/b(f_c)$ sufficiently large, a good approximation to the frequency response of the gammatone is given by [55]

$$G(f) \approx \left[1 + \frac{j(f-f_c)}{b(f_c)}\right]^{-N} \qquad (0 < f < \infty) \tag{1.2}$$

It can therefore be seen that the filter is symmetric about f_c on a linear frequency scale. Patterson et al. [105] show that for $N = 4$, the gammatone filter gives an excellent fit to experimentally derived estimates of human auditory filter shape.

The bandwidth of the gammatone filter is usually set according to measurements of the equivalent rectangular bandwidth (ERB) of human auditory filters [46]. The ERB of a filter is defined as the bandwidth of an ideal rectangular filter that has the same peak gain, and which passes the same total power for a white-noise input. For auditory filters, the ERB may be regarded as a measure of critical bandwidth [96], and a good match to human data is given by

$$\mathrm{ERB}(f) = 24.7 + 0.108f \tag{1.3}$$

Furthermore, it is usual to assume that the filter center frequencies are distributed over frequency in proportion to their bandwidths, which for fourth-order filters is given by [105]

$$b(f) = 1.019\ \mathrm{ERB}(f) \tag{1.4}$$

In this regard, it is often convenient to distribute the filter center frequencies on the so-called "ERB-rate" scale. This is a warped frequency scale, similar to the critical-band scale of the human auditory system, on which filter center frequencies are uniformly spaced according to their ERB bandwidth. The ERB-rate scale is an approximately logarithmic function relating frequency to the number of ERBs, $E(f)$, and is given by

$$E(f) = 21.4\ \log_{10}(0.00437f + 1) \tag{1.5}$$

Figure 1.6 shows the impulse responses and frequency responses for eight gammatone filters that are equally spaced on the ERB-rate scale. Note that at low frequencies, the filters have narrow bandwidths, and are more closely spaced in frequency (panel **B**). In the figure, the filters have been normalized to have the same peak gain at their center frequency. In practice, however, the gains can be set in order to simulate the resonances due to the outer and middle ears, which boost sound energy in the 2–4 kHz range. Alternatively, the outer/middle ear transfer function can be imposed on the input signal, prior to frequency analysis, by a simple linear filter.

Panel **A** of Fig. 1.6 also illustrates that the impulse responses of the low-frequency filters peak at a later time than those of the high-frequency filters, due to their narrow bandwidths. If across-frequency comparisons are to be made in a CASA system (e.g., in order to detect event onsets in different frequency channels) then it may be convenient to phase-compensate the gammatone filterbank, so that the peaks of the impulse responses are aligned. Holdsworth et al. [55] show that this can be achieved by introducing a time lead $t_c = (N - 1)/2\pi b(f)$ to the output of each filter, in order to align the peaks of their envelopes. A further phase correction $\phi = -2\pi f_c t_c$ is needed to align the peak of the envelope of each impulse response with the peak of its fine structure, giving the phase-compensated filter

$$\tilde{g}f_c(t) = (t + t_c)^{N-1} \exp[-2\pi b(f_c)(t + t_c)] \cos(2\pi f_c t) u_c(t) \quad (1.6)$$

where $u_c(t) = 1$ for $t \geq -t_c$, and 0 otherwise. Note that the resulting filter is non-causal.

Efficient digital implementations of the gammatone filter have been proposed by Holdsworth et al. [55], Cooke [28], and Slaney [118]. The output of the filter may be regarded as a measure of basilar membrane displacement, which is subjected to further processing in order to derive a simulation of auditory nerve activity. For reasons of computational efficiency, most CASA systems use a representation of firing rate in the auditory nerve, rather than a spike-based representation. Most simply, this can be obtained by half-wave rectification of the filterbank outputs, followed by a static nonlinearity (such as a square root). A more sophisticated approach is to model adaptation using an automatic gain control (AGC) [117], or to use a model of hair cell transduction (see [54] for a review).

The model of hair cell transduction proposed by Meddis [87, 88, 91] is often paired with the gammatone filterbank, because it represents a good compromise between accuracy and computational efficiency. The model is based on the assumption that three reservoirs of transmitter substance exist within the hair cell, and that transmitter is released in proportion to the degree of basilar membrane displacement. The amount of transmitter released is equated with the probability of a spike being generated in the associated auditory nerve fiber. The Meddis model replicates many of the characteristics of auditory nerve responses discussed above, including rectification, compression, spontaneous firing, saturation effects and adaptation.

Panel **A** of Fig. 1.7 shows the response of the Meddis hair cell model to a pure tone stimulus, where the basilar membrane displacement has been provided by a

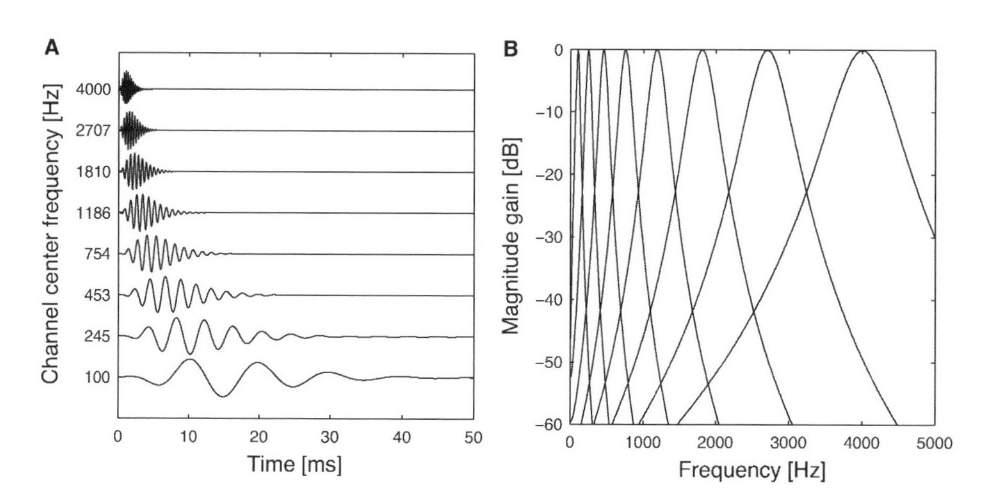

Figure 1.6 Gammatone filters. (**A**) Impulse responses for eight gammatone filters, with center frequencies equally spaced between 100 Hz and 4 kHz on the ERB-rate scale of Glasberg and Moore [46]. (**B**) Frequency responses of the filters shown in panel **A**.

gammatone filter channel centered on the tone frequency. Panel **B** of the figure shows the response of a complete cochlear model (gammatone filterbank and Meddis hair cell model) to three periods of a steady-state vowel. Within each frequency channel, the response can be interpreted as the instantaneous firing rate within an auditory nerve fiber. The resulting time-frequency representation is often termed a *neural activity pattern* or *cochleagram*. A cochleagram may also refer to a simpler form of cochlear representation: filtering by an auditory filterbank followed by some form of nonlinear rectification.

For visual representation of long signals, the waveform display used in panel **B** of Fig. 1.7 contains too detailed information. A representation that is more like a conventional spectrogram can be obtained by smoothing the time series associated with each frequency channel, downsampling, and then mapping the resulting values to color or a gray scale. An example of a cochleagram computed in this way is shown in Fig. 1.8, for the utterance "they enjoy it when I audition" spoken by a male talker. The figure was produced by smoothing the output from the Meddis hair cell model with a Hanning window of width 20 ms, and sampling the output at 5 ms intervals. A conventional spectrogram is also shown in the figure, which was also produced using Hanning-windowed frames of width 20 ms and a frame interval of 5 ms. A number of differences between the two representations can be noted. First, the representation of low frequencies is expanded in the cochleagram, because of the ERB-rate spacing of the filter center frequencies. Second, the first formant seen in the spectrogram is resolved into harmonics in the cochleagram, because auditory filters with low center frequencies have a narrow

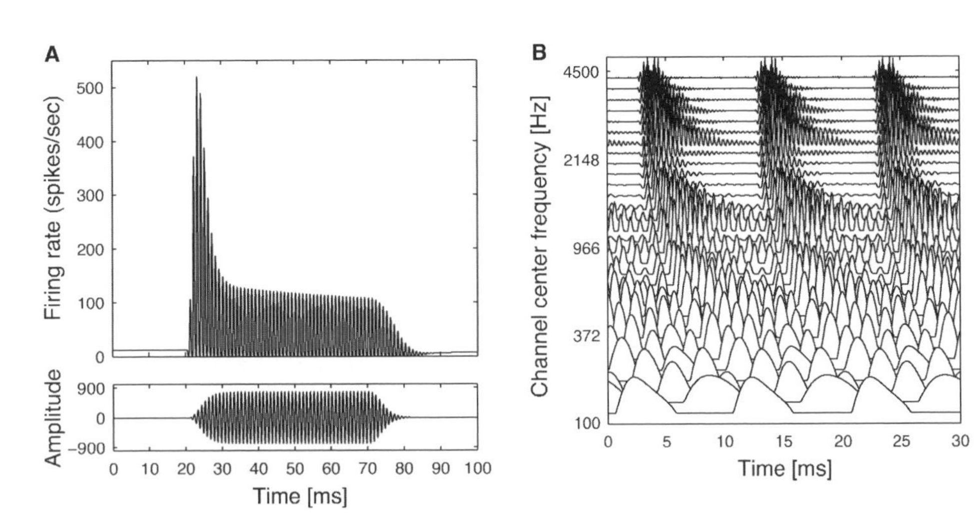

Figure 1.7 Simulated auditory nerve activity. (**A**) The lower panel shows the response of a gammatone filter with a center frequency of 500 Hz to a pure tone (also with a frequency of 500 Hz). The upper panel shows the corresponding output of the Meddis hair cell model. (**B**) Neural activity pattern (computed using a gammatone filterbank and Meddis hair cell model) for the steady-state vowel /er/ with a fundamental frequency of 100 Hz.

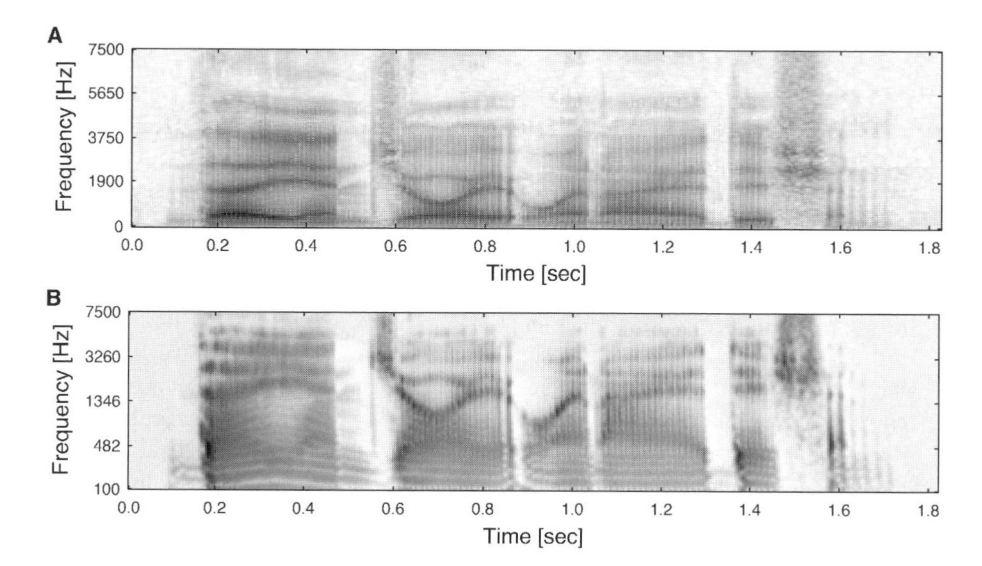

Figure 1.8 Spectrogram (**A**) and cochleagram (**B**) for the utterance "they enjoy it when I audition" spoken by a male talker. Dark pixels indicate regions of high energy, and light pixels indicate regions of low energy.

bandwidth. Finally, adaptation in the hair cell model causes acoustic onsets to be emphasized in the cochleagram.

1.3.3 Correlogram

The autocorrelation-based theory of pitch perception proposed by Licklider [77] has formed the basis for many computational models of fundamental frequency ($F0$) estimation and $F0$-based sound separation. The first computational implementation of Licklider's scheme was demonstrated by Richard Lyon, and later published by Lyon [84] and Weintraub [135, 136]. These authors referred to it as an "auto-coincidence" or "coincidence" representation. Subsequently, Slaney and Lyon [119] introduced the term *correlogram*.

The correlogram is usually computed in the time domain, by autocorrelating the simulated auditory nerve firing activity at the output of each cochlear filter channel:

$$acf(n, c, \tau) = \sum_{k=0}^{K-1} a(n-k, c)a(n-k-\tau, c)h(k) \tag{1.7}$$

Here, $a(n, c)$ represents the simulated auditory nerve response for frequency channel c at discrete time n, and τ is the time lag. The autocorrelation is computed over a window of length K samples, which is shaped by a window function h (which is typically chosen to be a Hanning, exponential, or rectangular shape). For greater efficiency, the *correlogram* can also be computed in the frequency domain using the discrete Fourier transform (DFT) and its inverse transform (IDFT), using the relation

$$acf(\mathbf{x}_{n,c}) = \mathrm{IDFT}(|\mathrm{DFT}(\mathbf{x}_{n,c})|^p) \tag{1.8}$$

where $\mathbf{x}_{n,c}$ is a windowed section of the simulated auditory nerve activity obtained from frequency channel c, starting at time n. In Eq. 1.8, the use of a parameter p allows for a generalized autocorrelation. Setting $p = 2$ gives a conventional autocorrelation function, but smaller values of p may be advantageous because they yield sharper peaks in the resulting function [124].

Panel **A** of Fig. 1.9 shows a correlogram for a vowel /er/ with $F0$ = 100 Hz, which is derived from the neural activity pattern shown in Fig. 1.7. The figure shows one time frame of the three-dimensional volume defined by $acf(n, c, \tau)$, in which frequency channel and autocorrelation lag are plotted on orthogonal axes. It should be noted that the squaring inherent in the autocorrelation function means that $acf(n, c, \tau)$ requires twice as much dynamic range as $a(n, c)$; in the figure, this has been addressed by normalizing each channel to a maximum value of one.

The correlogram is important as a model of pitch perception because it unifies two schools of pitch theories: the place theories (which emphasize the role of resolved harmonics) and the temporal theories (which emphasize the role of unresolved harmonics [77, 96]. Figure 1.9 demonstrates that the correlogram combines

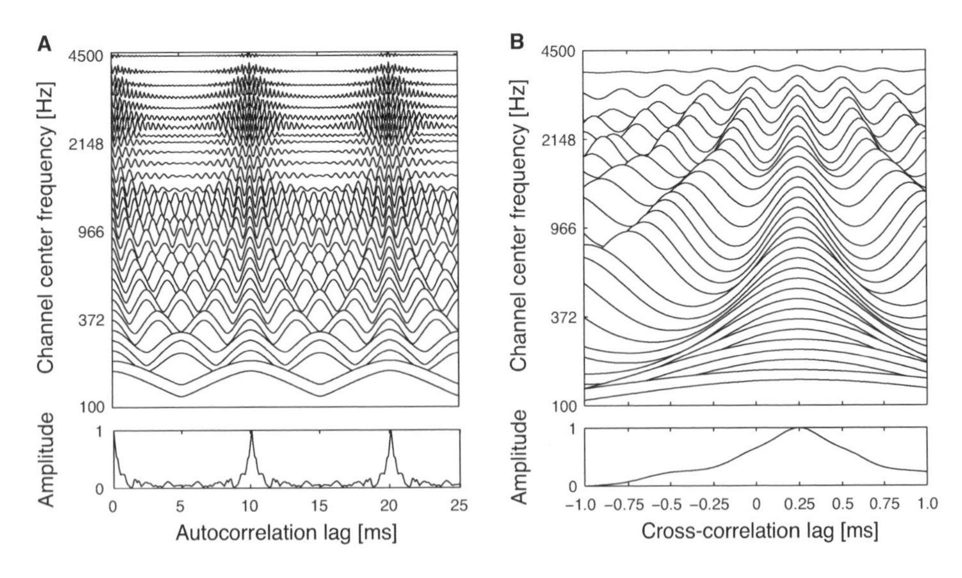

Figure 1.9 Correlation representations. (**A**) Correlogram of the steady-state vowel /er/ with a fundamental frequency of 100 Hz (upper panel) and summary correlogram (lower panel). (**B**) Cross-correlogram of the steady-state vowel /er/, presented to a binaural auditory model with an interaural time difference of 0.25 ms (upper panel) and summary cross-correlogram (lower panel).

pitch information from both resolved and unresolved harmonic regions. Consider the lowest frequency channels in the figure, which are excited by the $F0$ of the vowel (100 Hz). They exhibit a peak at the period of the $F0$ (10 ms), but a peak also occurs at 20 ms because the autocorrelation of a periodic signal is itself a periodic function. Similarly, channels driven by the second harmonic exhibit a peak at the period of 200 Hz (5 ms), but also at 10, 15, 20, 25 ms, and so on. Hence, a peak at 10 ms occurs in all correlogram channels that are excited by a resolved harmonic. At high frequencies, auditory filters are wider and a number of harmonics interact within the same filter. This results in amplitude modulation ("beating") at the rate of the $F0$. Hence, the high-frequency channels of the correlogram also exhibit a peak at the period of the $F0$, and the correlogram for a periodic sound displays a characteristic "spine" that is centered on the fundamental period (and its multiples).

This $F0$-related structure can be emphasized by pooling the information in the correlogram across frequency. The resulting *summary correlogram* is given by

$$sacf(n, \tau) = \sum_{c} acf(n, c, \tau) \tag{1.9}$$

The summary correlogram is shown below the correlogram in panel **A** of Fig. 1.9, and exhibits two strong peaks. The peak that occurs at the shortest lag (10 ms) corresponds to the fundamental period. Meddis and Hewitt [89, 90] have demonstrated that the position of peaks in the summary correlogram corresponds closely to perceived pitch. Furthermore, concurrent sources with different $F0$s give rise to multiple peaks in the summary correlogram, which can be used as the basis for multipitch tracking and $F0$-based sound separation algorithms (see Chapters 2 and 3).

1.3.4 Cross-Correlogram

As with the correlogram, the basic idea underpinning the *cross-correlogram* was described in the literature some 30 years before a computational implementation was attempted. Specifically, the cross-correlogram is motivated by the work of Jeffress [62], who describes a mechanism by which human listeners might localize sound using interaural time differences (ITD). Jeffress proposes a neural circuit in which firing activity that arises from the same critical band of each ear travels along a pair of delay lines. When the net delay corresponds to the interaural time difference, activity in the two delay lines coincides and is detected by a coincidence unit. This mechanism is equivalent to an interaural cross-correlation of the form

$$ccf(n, c, \tau) = \sum_{k=0}^{M-1} a_L(n-k, c)\, a_R(n-k-\tau, c)\, h(k) \tag{1.10}$$

where $a_L(n, c)$ and $a_R(n, c)$ represent the simulated auditory nerve response at discrete time n and frequency channel c, for the left and right ears, respectively. The parameter τ is the lag, and the cross-correlation is computed over a window h of size M samples (again, a Hanning, exponential, or rectangular window is often used). Cross-correlogram representations of this form have been used in CASA algorithms by Lyon [83] and many others.

Panel **B** of Fig. 1.9 shows a cross-correlogram for a steady-state vowel /er/, which has been presented to a binaural auditory model with an interaural time difference of 0.25 ms. As with the correlogram, the cross-correlogram is actually a three-dimensional volume defined by $ccf(n, c, \tau)$ and the figure shows a slice through this volume taken at a particular time instant. Again, each channel of the cross-correlogram has been normalized to a maximum value of one. The cross-correlogram has a spine centered on the ITD of the stimulus, which is surrounded by "sidelobes" due to harmonic components and filterbank resonances. Changes in the ITD of the stimulus simply cause this pattern to be shifted to the left or the right.

The spinal structure in the cross-correlogram can be emphasized by pooling over frequency, in a manner analogous to that described by Eq. 1.9. The resulting *summary cross-correlogram* is shown beneath the cross-correlogram in Fig. 1.9, and has a peak at 0.25 ms, which corresponds to the ITD of the stimulus. If the input signal consists of a mixture of sound sources that originate from different locations, then the summary cross-correlogram will generally contain multiple peaks that can be used as the basis for localizing and separating each constituent source (see Chapters 5 and 6).

It should be noted that in the form described above, the cross-correlogram is only sensitive to ITD; interaural intensity differences (IIDs) do not cause a change in the response pattern. Subsequent studies have introduced IID sensitivity into the dual delay-line model by using an inhibitory input from the contralateral ear [79]. Additionally, various enhancements of the basic cross-correlogram scheme have been proposed that allow it to explain aspects of the precedence effect [80] and binaural signal detection [26]. Finally, the notion of an interaural cross-correlation mechanism is broadly supported by physiological studies, which have revealed systematic arrangement of ITD-sensitive neurons in the auditory midbrain (e.g., [140]).

1.3.5 Time-Frequency Masks

Many CASA systems achieve source segregation by computing a *mask* to weight a T-F representation of the acoustic input such as a cochleagram. The mask applies a weight to each T-F unit, such that spectrotemporal regions that are dominated by the target source are emphasized, and regions that are dominated by other sources are suppressed. The values in the mask may be binary or real-valued; in the latter case, the mask value may be interpreted as the ratio of target energy to mixture energy (as in a Wiener filter) or the probability that the T-F unit "belongs" to the target source. A time-frequency weighting of this kind was first employed in the binaural source separation algorithm described by Lyon [83], and has subsequently been adopted by other workers [136, 15, 133, 115].

Perceptually, the use of binary T-F masks is motivated by the phenomenon of masking in auditory perception, in which a sound is rendered inaudible by a louder sound within a critical band (for a review, see [96]). Additionally, different lines of computational consideration have converged on the use of binary masks as described in the following text.

First, Joujine et al. [65] and Roweis [116] have noted that a speech signal is sparsely distributed in a high-resolution T-F representation and, as a result, different speech utterances tend not to overlap in individual T-F units. This observation leads to the property of orthogonality between different speech utterances. In this case, binary masks are sufficient for decomposing sound mixtures into their constituent sources. The orthogonality assumption holds well for mixtures of speech and other sparsely distributed signals (e.g., complex tones), but is not valid for speech babble or other broadband intrusions.

Second, the notion of a time-frequency mask is central to the missing data approach to ASR proposed by Cooke et al. [29]. In this approach, the mask indicates whether each acoustic feature should be regarded as reliable or unreliable evidence of a target speech source, so that it can be treated appropriately during recognition. Furthermore, Cooke et al. introduce the notion of an a priori mask, which is used to assess the upper limit on the performance of a missing data ASR system that uses binary masks. The a priori mask is so called because it is computed using speech and noise signals before mixing. Specifically, Cooke et al. employ a priori masks in which reliable time-frequency regions are assumed to be those for which the mixture energy lies within 3 dB of the energy in the premixing speech signal.

Third, Wang and colleagues [57, 113, 58] have suggested an ideal binary mask as a computational goal of CASA (see Section 1.2.2). Specifically, denote the target energy in a T-F unit as $s(t,f)$ and the interference energy as $n(t,f)$. The ideal binary mask is given by

$$m(t,f) = \begin{cases} 1 & \text{if } s(t,f) - n(t,f) > \theta \\ 0 & \text{otherwise} \end{cases} \tag{1.11}$$

where t and f index the time and frequency dimensions, respectively. The parameter θ is typically chosen to be 0, corresponding to a 0 dB criterion. Figure 1.10 shows an ideal time-frequency mask for a mixture of speech and a telephone sound, together with cochleagrams for the mixture and clean signals. Figure 1.5, panel **E**, illustrates the result of applying the ideal mask to the mixture; it is clear from the figure that the masked mixture is much closer to the clean target than the mixture itself.

Support for this notion comes from the observation that target sound reconstructed from an ideal binary mask is of high perceptual quality [115, 113]. The ideal binary mask also provides a very effective front end for automatic speech recognition in noise [29, 113]. Recent experiments with human listeners have found that ideal masking leads to substantial improvements of speech intelligibility; in particular, the ideal binary mask defined where $\theta = -6$ dB appears to be most effective for human speech intelligibility [113, 23, 21].

1.3.6 Resynthesis

A number of CASA systems include a resynthesis pathway, which allows an audio waveform to be reconstructed from a group of segments that correspond to an indi-

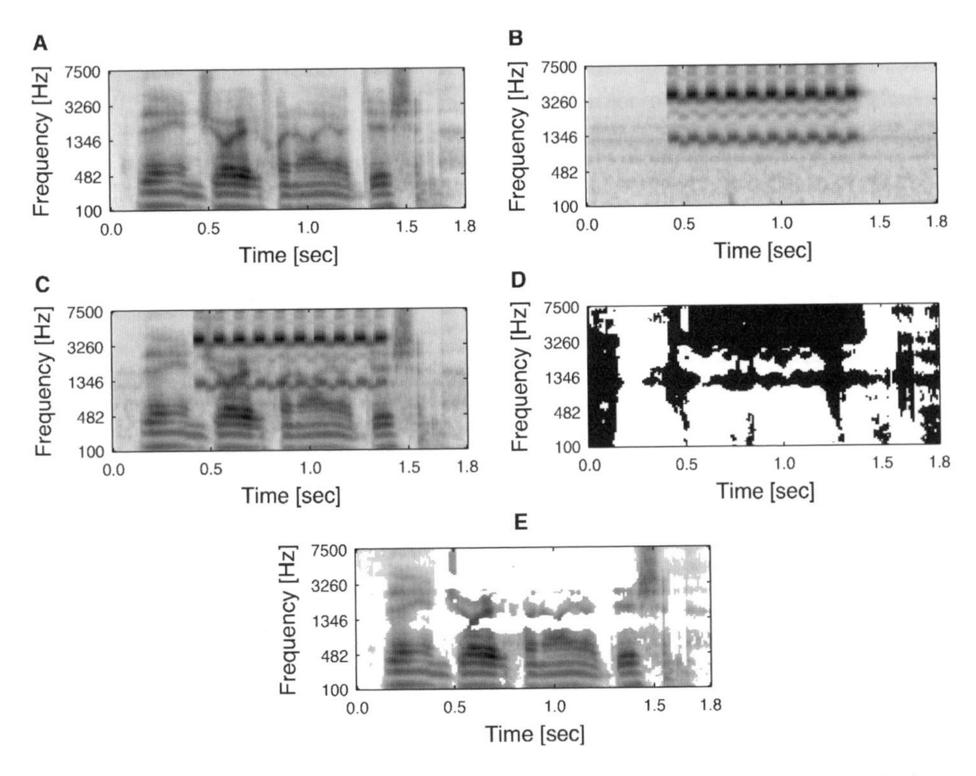

Figure 1.10 Binary time-frequency masks. (**A**) Cochleagram of the utterance "they enjoy it when I audition" spoken by a male talker. (**B**) Cochleagram of a telephone ring (this is the signal denoted "n6" in Cooke's [28] corpus of noise intrusions). (**C**) Cochleagram of a mixture of speech and telephone ring, with SNR = 0 dB. (**D**) Ideal binary mask, in which white regions indicate time-frequency elements that are dominated by the speech. (**E**) Result of applying the ideal binary mask to the cochleagram of the mixture; note the similarity with the cochleagram of the original speech (panel **A**).

vidual sound source. This allows the quality of segregation to be assessed by listening tests, or by measuring the signal-to-noise ratio before and after processing. Resynthesis is usually achieved by inverting a time-frequency representation (for example, see Cooke [28]). Other auditory-motivated representations can also be inverted, including the correlogram [120].

In systems that use T-F masks as described above, it is straightforward to resynthesize an audio signal for the target sound source from the output of an auditory filterbank, such as the gammatone filterbank (see Section 1.3.2). This can be achieved using an approach introduced by Weintraub [136] (see also [16]). First, across-channel phase shifts in the filterbank output are removed by reversing the response of each filter in time, passing the reversed response back through the filter, and then time-reversing the filtered response again. Alternatively, the phase-corrected form of the gammatone filter can be used (Eq. 1.6). The phase-corrected output from each filter channel is then divided into time frames by windowing with a raised cosine, with a frame size that matches the original decomposition into T-F units. The energy in each T-F unit is then weighted by the corresponding T-F mask value (which may be binary or real-valued between 0 and 1). The weighted responses are then summed across all frequency channels to yield a reconstructed audio waveform. High-quality resynthesis has been obtained in this way for speech and other signal types.

1.4 CASA EVALUATION

Quantitative evaluation of CASA systems is critical for gauging the progress made in the field. Systematic evaluation also serves to guide the development of a model. The evaluation process typically consists of choosing an evaluation criterion or metric, and selecting an evaluation corpus that is representative of the application domain. These two aspects are addressed below.

1.4.1 Evaluation Criteria

A variety of evaluation criteria have been used for CASA systems, which can be broadly divided into four categories: comparisons between segregated target and clean (premixing) target, changes in automatic speech/speaker recognition score, human listening tests, and correspondence with biological data. Each category is described below.

Comparison with Clean Target Signal. This approach assumes that the acoustic mixture to be separated has been generated by the addition of one or more interfering signals to a clean target signal. A ground-truth target signal is therefore available, which can be compared with the segregated target in a number of ways. For example, Cooke [28] assesses the similarity between symbolic time-frequency representations of the clean target (a speech signal) and the segregated target. Brown and Cooke [16] use binary T-F masks to resynthesize two signals, which correspond to the segregated target speech and the residual noise that it contains. These two waveforms are then compared in order to derive a normalized SNR in the range 0 to 1. Wang and Brown [133] use the same strategy, but quantify the performance using a conventional SNR metric that is expressed in decibels. Nakatani et al. [99] use the distortion between the short-term spectra of a segregated source and those of the original source to measure performance for speech separation.

Automatic Recognition Measure. One of the main motivations for speech separation is to improve ASR performance in the presence of acoustic interference. Hence, it is natural to evaluate a CASA system as the front end for an ASR device, so that the recognition score can be compared before and after the target speech is segregated (for example, see [136]). A similar approach can be applied to the evaluation of automatic speaker recognition systems. In the case of automatic music transcription systems, the evaluation typically involves a comparison between the notes in a musical score and the fundamental frequencies detected in the corresponding audio waveform (e.g., [48, 70]).

Human Listening. Human listeners can be used to evaluate CASA systems, by conducting formal tests that compare the intelligiblity of segregated speech and the unprocessed mixture of speech and noise (e.g., [123]). However, a potential pitfall needs to be considered here: we cannot "turn off" the ASA process of listeners when they are presented with an acoustic stimulus. As a result, we should not expect that a CASA system will yield an intelligibility gain until machine performance *exceeds* human performance. Hearing-impaired listeners are probably better suited to such an evaluation approach, as it is well known that such listeners have difficulty in recognizing speech in a noisy environment [95]. Certainly, if the intended application of the CASA system is to improve the hearing of impaired listeners, or that of normal listeners in very noisy environments, then this evaluation methodology is an appropriate choice. A different approach to CASA evaluation using human listeners has also been described by Ellis [39]. He asked listeners to score the resemblance between segregated sounds and the corresponding component sounds in an acoustic mixture.

Correspondence with Biological Data. If biological plausibility is an objective, then the evaluation criterion is usually how well the computer model of ASA accounts for known perceptual or neurobiological data. Models that employ this criterion include those of Wang [131], who simulated several qualitative ASA phenomena on the basis of a neural oscillator network (see [101] for a recent extension that produces quantitative results); McCabe and Denham [86] who proposed a neural network to model auditory streaming data; and Wrigley and Brown [138], who put forward a neural oscillator model to quantitatively simulate auditory attention data.

It has often been pointed out that there is no consensus on how to evaluate a CASA system [114, 40, 132]. This partly reflects the fact that idiosyncratic evaluation criteria are largely unavoidable in a maturing field. Also, the diverse range of CASA applications (see Section 1.2.3) necessitates different performance measures (for example, human intelligibility testing would be appropriate if the intended application were a hearing aid, but not if the application were ASR). These considerations notwithstanding, it is important to make the computational objective explicit and evaluate accordingly. For example, if the ideal binary mask is the goal, then one can measure the SNR using the signal resynthesized from the ideal mask [58]. Also, wherever possible, one should make an effort to use common evaluation tasks [40].

1.4.2 Corpora

A diverse range of corpora have been used for evaluating CASA systems. For voiced speech segregation, Cooke [28] compiled a list of 10 voiced utterances—

five sentences spoken by two male speakers—and 10 noise intrusions. The noise samples vary in their bandwidth and temporal structure, and include a recording from a busy teaching laboratory (which is similar to "cocktail party" noise), rock music, narrowband and broadband sounds, and other speech utterances. Combining each voiced utterance with each noise intrusion gives 100 mixtures, which have been commonly employed in CASA studies and facilitated quantitative comparison of different approaches (see Chapter 3). For general speech segregation, one can retain Cooke's noise intrusions while replacing the voiced utterances with more naturalistic speech material, such as those from the TIMIT corpus [44] or the TIDigits corpus [74].

Other corpora of noise intrusions include the NOISEX database [128], which contains various noise recordings such as factory noise and tank noise. Hu [56] collected 100 environmental sounds such as rain, wind, and crowd sounds, and used them for training and testing during the evaluation of an unvoiced speech separation system.

For evaluating ASR in noise, AURORA has become a standard series of corpora [107]. Presently there are three AURORA corpora of increasing complexity, in which the target speech is either digits or the 5000 word *Wall Street Journal* corpus [106]. The background noise (e.g., car noise) is either added digitally or recorded together with target speech, and a range of different SNR conditions are available. To facilitate comparison among front-end processing methods, standard back-end speech recognizers are also provided for reference.

Many tests (and corpora) have been devised to evaluate speech intelligibility in noise by both normal-hearing and hearing-impaired listeners. For example, the speech perception in noise test (SPIN) [67, 9] evaluates word recognition in context. Each SPIN list includes 50 sentences with 25 high-context sentences and 25 low-context sentences, and the listener's task is to repeat the last word of each sentence that is mixed with multitalker babble. Another example is the hearing in noise test (HINT) corpus [100]. The HINT corpus consists of 25 phonemically balanced lists, each containing 10 short sentences, mixed with speech-shaped noise. The sentences in HINT were adapted for use in American English from the BKB (Bamford–Kowal–Bench) corpus that contains short, semantically predictable sentences spoken by British speakers [5].

Relatively few corpora have been recorded in real acoustic environments with CASA evaluation in mind. A notable exception is the ShATR corpus [69], which was designed specifically for evaluating CASA systems. Recordings were made via multiple microphones, including close-talking microphones for each participant, an omnidirectional microphone, and a binaural dummy head. Related is the ICSI meeting corpus [61], which records multiparty meetings via both near-field and far-field microphones. In significant portions of the ICSI corpus, speech is mixed with other speech and nonspeech noise. Evaluation is facilitated by the availability of reference signals recorded by near-field microphones, but even these can contain some degree of interference from other speakers ("crosstalk") [139].

The above corpora are specifically targeted at speech processing. Corpora exist for other sounds, including bird calls and music. In the latter case, the RWC music database [49] is a good resource for CASA systems concerned with music processing, since it contains audio files and corresponding ground-truth data in the form of MIDI transcriptions and lyrics. Other useful corpora include collections of head-related transfer functions (HRTFs) recorded from dummy heads, which can be used to spatialize sound sources for evaluating binaural CASA systems. The website `www.casabook.org` provides a comprehensive list of these resources in addition to the aforementioned corpora.

Let us end this section by asking the question, What evaluation results will lead us to declare that the CASA problem is solved? The answer to this question obviously depends upon what is expected from CASA. If the purpose is an application domain, such as noise-robust ASR, then success can be more readily verified. Indeed, in certain restricted conditions, machine performance has already matched or surpassed that of human listeners. However, we favor a general criterion: we consider the problem solved *when a CASA system achieves human ASA performance in all listening situations.* It is the versatility of the human auditory system that, we believe, CASA should aim for.

1.5 OTHER SOUND SEPARATION APPROACHES

The sound separation problem has been investigated in audio signal processing for many years, and is usually motivated by specific engineering problems (e.g., robust ASR in the cockpit of an aircraft, where the statistics of the background noise are known). Two main approaches have been proposed: beamforming and independent component analysis. Each is explained and contrasted with CASA below. We also consider speech enhancement, which is related to source separation but has a narrower goal—to enhance target speech that is contaminated with noise.

Beamforming achieves sound separation by using the principle of spatial filtering ([127, 72, 12]; see also Chapter 6). Spatial filtering aims to boost the signal coming from a specific direction by appropriate configuration of a microphone array, and in doing so it attenuates interfering signals from other directions. The simplest is a delay-and-sum beamformer, which adds multiple microphone signals from the target direction in phase and uses phase differences to attenuate signals from other directions. The amount of noise attenuation increases as the number of microphones and the array length increase. Adaptive beamforming attempts to cancel noise sources via a weight adaptation or training process. In general, an adaptive beamformer with L microphones can remove only $L - 1$ different noise sources. To overcome this limitation, subband versions of adaptive beamforming have been developed, which allow cancellation of more noise sources whose spectra do not substantially overlap. With a properly configured array, spatial filtering can produce high-fidelity separation. Another advantage of spatial filtering is its ability to attenuate reverberation, because the target signal has a specific direction, whereas reverberant signals arrive from diffuse directions (see Chapter 7). The main limitation of beamforming is configuration stationarity [132]; it is difficult to separate a target that moves around or switches between different sound sources, both of which occur in "cocktail party" situations. Also, no separation is possible when multiple sounds come from directions that are the same or near to each other.

An approach related to beamforming is blind source separation using independent component analysis (ICA) [66, 4, 60], which combines adaptive filtering and machine learning techniques. Like beamforming, mixture signals are modeled in the standard form as a linear superposition of source signals. In other words, a mixing model of the form $x(t) = As(t)$ is assumed, where $s(t)$ is a vector of unknown source signals, A is a mixing matrix, and $x(t)$ is a vector of the mixed signals recorded by multiple microphones. In ICA, the main assumption is that sound sources are statistically independent. The separation problem is formulated as that of estimating a demixing matrix (i.e., the inverse of A). To make this formulation work requires a number of assumptions about the mixing process and the number of microphones [125]. ICA gives impressive separation results when the assumptions are satisfied, but its scope is limited as a result. For example, there must be at least as many microphones as the number of sources (although efforts have been made to relax this constraint [73, 137]). A more fundamental limitation is that the mixing matrix A needs to be stationary for a period of time in order to allow estimation of a large number of parameters. This assumption—akin to the configuration stationarity assumption in adaptive beamforming—is difficult to satisfy in situations in which speakers turn their heads or move around. Like spatial filtering, to be separable, source signals must originate from different spatial directions.

The goal of speech enhancement is to enhance target speech in the presence of background noise [78, 6]. Generally intended to apply to single-channel audio, this approach is based on statistical analysis of speech and noise, followed by estimation of clean speech from noisy speech. Many speech enhancement techniques have been proposed, including spectral subtraction and mean-square error estimation. The widely used method of spectral subtraction subtracts the power spectral density of the estimated background from that of the mixture [11]. The mean-square error estimator, proposed by Ephraim and Malah [41], models speech and noise spectra as statistically independent Gaussian random variables, and optimally estimates clean speech according to the minimum mean-square error criterion. To apply speech enhancement requires a good estimate of the interference, which is difficult to obtain unless the interference is stationary. To relax the stationarity assumption, algorithms typically involve detection of speech silence and subsequent noise estimation within speech gaps. Generally speaking, speech enhancement views a mixture as speech plus noise. This perspective is fundamentally narrower than the ASA perspective, which views a mixture as an auditory scene in which a variety of sounds can appear or disappear unpredictably. Speech enhancement cannot deal with such situations.

In visual processing, there is a subtle distinction between the terms "computer vision" and "computational vision." The former is more concerned with image processing and is more application-driven, whereas the latter is more concerned with modeling human vision as epitomized by Marr's research [85]. The distinction is analogous to that between other sound separation approaches and CASA.

1.6 A BRIEF HISTORY OF CASA (PRIOR TO 2000)

The following is not intended to be an exhaustive review of CASA development; rather, we focus on a number of key studies and emphasize their historical relationship and the main innovations introduced by each. Our account covers the period before the year 2000, as it takes time for a study to show its historical significance; also the subsequent chapters highlight many recent developments.

1.6.1 Monaural CASA Systems

The problem of monaural sound separation was first brought to a wide audience by Parsons [103], who describes a frequency-domain approach for the segregation of concurrent voices recorded by a single audio channel. Although Parsons makes reference to the "cocktail party effect," his approach is founded upon conventional signal-processing techniques rather than auditory modeling, and is principally motivated by the problem of crosstalk on communication channels. His system separates two talkers by selecting the harmonics belonging to each, and is therefore limited in its scope because each utterance must be continuously voiced. However, Parson's study established some directions for research that have proved influential, namely multipitch tracking, $F0$-based sound segregation, and the use of temporal continuity constraints. His work also represents an important conceptual advance, since its goal was to separate *both* voices from an acoustic mixture. It is therefore distinct from the preceding literature on speech enhancement, which aimed only to enhance a target voice (for a collection of early speech enhancement papers, see [78]).

A system for segregating the voices of two talkers described by Weintraub [136] may be regarded as the first monaural CASA study of any sophistication. In particular, it stands as the first approach that was grounded in a theory of auditory processing, rather than purely engineering concerns. Weintraub considers a harder problem than Parsons, in that each of the two speech signals can be voiced, unvoiced, or silent at any given time. To address this, Weintraub uses a state-based tracking procedure in which the number of voices present and their characteristics are represented by a Markov model. This approach enables sequential and simultaneous grouping to be represented within a common framework, and represents a significant advance over the simple continuity constraints used by Parsons. Voiced speech is separated on the basis of $F0$, but a joint time-frequency approach is employed—the correlogram—rather than a frequency-domain approach. Weintraub also develops a framework for evaluating his system. A waveform for each separated voice is resynthesized by refiltering the input waveform, weighted by spectral estimates from the Markov model. This waveform is then used as the input to an ASR system, allowing a comparison of speech recognition accuracy before and after processing by the system.

Weintraub's model has several limitations. For example, the state-based tracking requires that the two voices have a different average pitch range (a male and female talker were used), and therefore avoids the general problem of sequential grouping. Although unvoiced speech is considered, the model actually does little to segregate it. Also, the results of the evaluation are inconclusive; in some conditions, processing by the system actually causes a deterioration in ASR performance. Weintraub acknowledges that the link between source separation and ASR is a weakness of his system, but makes a number of suggestions as to how this could be improved. In particular, his suggestion that ASR systems should treat silent and masked intervals differently during decoding ([136], p. 145) comes very close to the notion of "missing data" speech recognition adopted by subsequent workers (see Chapter 9).

The study by Cooke [28] appeared shortly after the publication of Bregman's book [13], and is strongly influenced by his account. Additionally, Cooke was one of the first to note the relationship between Bregman's ASA theory and Marr's computational framework of vision [85], and to argue for a greater emphasis on the *representation* of the acoustic signal than his predecessors. More specifically, he proposes a time-frequency representation called synchrony strands, which is derived by applying local similarity and continuity constraints to the output of a cochlear model. Frequency, amplitude, and amplitude modulation rate are computed at each point along the strands, forming a rich representation that is amenable to search. This approach can potentially separate two voiced utterances with crossing $F0$s, a challenging problem for earlier systems. Cooke's system achieves separation of the two sources by grouping their harmonics in the time-frequency plane, so no explicit $F0$ tracking is necessary. Unlike previous approaches, Cooke's system does not require the number of sound sources to be specified a priori; rather, grouping continues until all the synchrony strands are accounted for, and the number of sources found is therefore an emergent property of the grouping process.

It should be noted that Cooke's system is not able to sequentially group acoustic events from the same source that are separated by silent intervals. It therefore does not address the problem of allocating a sequence of voiced and unvoiced speech sounds to the same source. Accordingly, Cooke evaluates his system using a corpus of acoustic mixtures in which the target sound is continuously voiced speech. His evaluation scheme is based on a comparison of the synchrony strand representations of two signals before they are mixed and after they have been mixed and grouped by his system. A criticism of this evaluation approach is that it is tightly bound to the synchrony strand representation; it is therefore difficult to compare the performance of Cooke's system with other CASA approaches that use different representations.

Mellinger [92] completed his study in the same year as Cooke, and also cites Bregman and Marr as theoretical influences. However, Mellinger's study differs in a number of respects from Cooke's, not least because his system is specialized for grouping the resolved harmonics of pitched musical instruments. Additionally, Mellinger's system does not use a discrete representation such as synchrony strands; rather, the output of a cochlear model is processed directly to form "feature maps" that encode acoustic onsets and frequency modulation. Harmonic components are grouped by considering the affinity between them, which is initially determined by their onset synchrony and subsequently by the coherence of their FM. Mellinger's system works in an online manner, so that signal representations are computed and grouped as time progresses. This is distinct from the "batch processing" of Cooke's system, in which the synchrony strand representation is computed first for the whole input, and then grouped. Online processing is clearly more suited to real-world applications.

Mellinger notes that his system "fails to work in some musically significant situations" (p. 183). This can be attributed to a number of factors. First, despite the important role of harmonic relations in music, his system has no way of grouping harmonics unless they share a common onset time and exhibit the same FM. Second, in his system the tracking of a harmonic is triggered by the detection of an acoustic onset. This approach is somewhat fragile; failure to detect an onset means that an entire acoustic event is missed by his tracking procedure.

The study of Brown [15] (see also [16]) combines a number of ideas from previous systems. Like Mellinger, his system employs maps of feature detectors that are justified on physiological grounds. In Brown's system, these maps extract onset, offset, periodicity, and frequency transition information from the output of a cochlear model. Periodicity information is derived from the correlogram, and is used for detecting resonances due to harmonics and formants. However, in Brown's system these intermediate representations are used to construct discrete time-frequency elements (or segments), which are similar in concept to Cooke's synchrony strands. Recall that in Cooke's system, specific grouping mechanisms are required to combine resolved and unresolved harmonics of the same voice. This step is avoided in Brown's study by deriving a pitch contour for each segment using "local" correlograms. This mechanism works in the same way for both resolved and unresolved harmonics, thus allowing them to be considered by a single search algorithm. Brown's system therefore represents an interesting fusion of ideas from Weintraub and Cooke's approaches; it is based on the correlogram, but performs grouping on the cochleagram using pitch contours that are computed for individual segments. Again, this means that there is no requirement for a priori knowledge of the number of sound sources that are present in the input.

Brown evaluates his system on the same corpus of voiced speech and noise intrusions used by Cooke, although the evaluation approach is quite different. It is assumed that the energy at each point in the time-frequency plane is allocated to a single source (the principle of "exclusive allocation" [13]) and hence the result of grouping is represented as a binary time-frequency mask. A signal for a target source is resynthesized by refiltering the input and weighting it by the time-frequency mask, in a similar approach to that described by Weintraub (see Section 1.3.6). However, Brown extends this approach by resynthesizing the original sources (before mixing) from the mask, so that the proportion of signal and noise energy in the separated target signal can be determined. This allows Brown to express the performance of his system in terms of an intuitive signal-to-noise ratio (SNR) metric.

Subsequently, innovations in CASA tended to focus on the system architecture. A novel contribution in this regard is the residue-driven architecture introduced by Nakatani et al. [99]. Critical of the "batch processing" approach of Cooke and Brown, these authors proposed an online scheme within a multi-agent paradigm. In their system, grouping processes are formulated as (largely independent) agents that interact to explain the auditory scene. Agents are of three types: event detectors, tracer generators, and tracers. Event detectors subtract the predicted input from the actual input, creating a residue. Tracer generators evaluate the residue and determine whether there is sufficient evidence to spawn a new tracer. Tracers group the features corresponding to a single stream, and also predict what the next input will be. The predicted input is then communicated to the event detector, thus completing the processing cycle.

Nakatani et al. describe agents for tracing static background noise and harmonics. For example, noisy speech might be accounted for by a noise tracer and a harmonic tracer. However, if a substantial harmonic residue is found during processing of the input, this suggests that another source has appeared and a new harmonic tracer is generated to track it. The authors evaluate the residue-driven system by comparing the $F0$ contours derived by harmonic tracers with ground-truth $F0$ contours, and by measuring the spectral distortion between the original and recovered sources. Additionally, they demonstrate the flexibility of the agent-based architecture by showing the variation in its performance when different types of agents are added and removed (e.g., its performance with and without noise tracers).

The use of a feedback loop in the residue-driven architecture was a development that subsequently motivated Ellis [39] to propose a prediction-driven approach to CASA. Ellis characterizes previous CASA systems as "data-driven," and argues that certain phenomena (such as the interpretation of masked signals in the auditory continuity effect) cannot be modeled without top-down information. He therefore proposed a system that is based on an internal world model, in which the auditory scene is explained in terms of a hierarchy of sound elements. The world model is constantly reconciled with the acoustic input, according to hypotheses of the sound sources present.

The prediction-driven architecture is implemented as a blackboard system, in which independent knowledge sources communicate via a shared database of hypotheses (the "blackboard"). In this respect, the prediction-driven approach to CASA may be regarded as a descendent of earlier blackboard-based signal understanding systems such as IPUS [75]. Ellis notes that the blackboard may have a conceptual advantage over the residue-driven approach of Nakatani et al. In the residue-driven approach, the detection, prediction and search for certain kinds of signals (such as harmonically related components) are tightly bound by relatively constrained interactions between specialized agents. Thus, it is not immediately clear how different properties of the same signal (such as its harmonic relations and spatial location) should be combined. In contrast, the blackboard architecture provides a relatively unconstrained framework in which various knowledge sources can influence alternative hypotheses about the input.

Ellis made a number of other contributions in this work. First, his CASA system uses a broader signal model than previous approaches, and includes representations of periodic elements derived from the correlogram ("wefts"), aperiodic elements ("noise clouds"), and transients. Second, his system allows the energy in a specific time-frequency region to be shared between sources, since harmonic sounds are identified from global pitch tracks first, rather than by finding elements directly in the time-frequency plane. Finally, Ellis stresses the importance of proper evaluation; in doing so, he develops an evaluation framework that is based upon a comparison of human and CASA performance on the same corpus, which is sufficiently general to be applied to other systems. On the other hand, the evaluation of his approach is rather tentative.

Another blackboard architecture was proposed by Godsmark and Brown [48, 47]. Their CASA system is designed for processing musical signals, in order to achieve the automatic transcription of polyphonic music. As in Ellis' system, top-down predictions (in this case concerning meter and familiar musical motifs) can influence organization at lower levels of the blackboard. In addition, their system employs a novel "organizational window" that slides over the input signal. Within this window, grouping principles interact in order to form multiple hypotheses about the auditory scene, which are then scored according to the acoustic evidence available. Once the window moves past a section of the input signal, its organization (in terms of the streams that are formed) is fixed. Their system also derives properties from groups of harmonics that are assumed to correspond to single musical notes. In particular, they compute a representation of timbre for each note, and use this to sequentially group notes that have originated from the same musical instrument.

1.6.2 Binaural CASA Systems

Human listeners have two ears, and it is known that binaural processing contributes to sound localization, auditory grouping, and robustness to reverberation. This has led a number of workers to develop binaural CASA systems, which take their inputs from two spatially separated microphones.

Lyon [83] was the first to demonstrate a binaural CASA system, which used spatial location information to separate and dereverberate two sounds originating from different directions. Time-varying gains were used to extract the desired source from each frequency channel, with gains set according to ITD estimates from a cross-correlogram. Subsequent binaural systems, such as the "cocktail party processor" described by Bodden [10], have largely followed the same architecture. Additional enhancements added by Bodden include adaptation to head-related transfer functions and a time-varying Wiener filter to extract the sources from the desired azimuth. He also employs a modified cross-correlogram that is sensitive to interaural intensity differences, as suggested by Lyon [83].

Two significant advances on Lyon's approach are described by Denbigh and Zhao [33] and Kollmeier and Koch [71], both of which involve the combination of spatial location with other cues. In the former study, the core of the system is an extension of Parsons' [103] approach for separating the voiced speech of two talkers according to harmonicity. In addition, Denbigh and Zhao use directional cues derived from two microphones to determine which pitch period should be allocated to which of the two talkers at each time frame. Kollmeier and Koch [71] also proposed the integration of monaural and binaural cues. More specifically, the first stage of their system identifies low-frequency envelope modulations that are typical of speech signals. The resulting AM spectrum is then weighted by a function derived from a localization system, so that speech energy that arises from a particular direction is passed, and other energy is suppressed.

There have also been attempts to incorporate binaural cues into more sophisticated CASA architectures. In particular, Nakatani et al. [98] have incorporated binaural cues into their residue-driven architecture. Their motivation for doing so was to address the poor performance of their system using harmonic tracking only, when the *F*0s of two sound sources are very close. Their binaural system initially uses harmonicity to separate a sound source, until a stable estimate of its location has been determined over a period of time using intensity and phase differences received at two microphones. Information about the source direction is then used to predict the next input, so that harmonic fragments are allocated to the correct location even if their *F*0s are close together and harmonic tracking fails.

1.6.3 Neural CASA Models

Little in Bregman's account [13] alludes to the neural mechanisms that underlie ASA. Similarly, most workers developing CASA systems have made clear that their systems are functional models rather than physiologically plausible ones. What, then, is the neural basis for ASA?

A partial answer to this question is suggested by the studies of Beauvois and Meddis [3], McCabe and Denham [86], and Grossberg [50], who describe neural models that exhibit the auditory streaming phenomenon. However, a more general account of the neural mechanisms underlying ASA is suggested by von der Malsburg and Schneider [129], who propose a neural network model of auditory source separation in which the allocation of features to streams is determined by neurons with an oscillatory response. Specifically, oscillators that represent features of the same stream are synchronized (phase locked with zero phase lag), and are desynchronized from oscillators that represent different streams.

A similar approach was subsequently adopted by Wang [130, 131], who models auditory stream segregation using a network of locally connected excitatory units with a global inhibitor. However, Wang addresses a fundamental problem apparent in von der Malsburg and Schneider's approach, namely that their oscillator network is dimensionless and fully connected; it therefore indiscriminately connects oscillators that are activated simultaneously by different acoustic sources. Instead, Wang hypothesizes the use of delay lines to construct a neural time axis, such that the auditory scene is represented within a time-frequency grid of oscillators.

These two studies established the principle of oscillatory correlation as a neural basis for ASA, but both required input in the form of symbolic patterns. Subsequently, Brown and Cooke developed two systems that paired a cochlear model with a network of chaotic oscillators, and processed sampled audio signals. The first modeled the grouping of acoustic components by harmonic relations and onset synchrony [18], and the second modeled a variety of two-tone streaming phenomena [19].

Following this, Wang and Brown [20] combined their approaches in a model of double vowel separation that took acoustic signals as input, and was based on the LEGION (locally excitatory, globally inhibitory) network of Terman and Wang [134]. This led to the development of a complete CASA system within the neural oscillator framework [133]. In this system, the two conceptual stages of ASA—segmentation and grouping—are represented in two layers of a LEGION network. Each layer is a two-dimensional time-frequency grid of oscillators. In the first layer, synchronized blocks of oscillators (segments) are formed, which correspond to connected regions of energy in the time-frequency plane (such as harmonics and formants). Different segments are desynchronized from one another. In the second (grouping) layer, connections are formed between segments if they are compatible with a fundamental frequency estimate obtained from a correlogram.

Wang and Brown's system is of interest because of its neurobiological basis. In terms of system architecture, one potential advantage of the neural oscillator approach is its parallel and distributed nature, which should make it more suitable for direct hardware implementation than other CASA architectures. However, complex oscillatory dynamics makes it difficult to incorporate more grouping principles.

1.7 CONCLUSIONS

Systems for CASA vary in their architecture, biological motivation, and dependence upon top-down and bottom-up grouping processes. All, however, have the same overriding goal—to computationally extract descriptions of individual sound sources from one or two recordings of an acoustic scene.

CASA is a relatively young discipline, and although much progress has been made in recent years, we are still far from achieving machine performance that is comparable with that of human listeners. The above review has identified a number of aspects of CASA systems (multipitch tracking, feature-based processing, binaural source localization and grouping, model-based segregation, and neural/perceptual modeling) which are key areas for further research. Robustness to reverberation is also an important goal, as is the application of CASA techniques to automatic speech recognition and music processing. The following chapters discuss these issues in more detail.

ACKNOWLEDGMENTS

Thanks to Guoning Hu and Soundararajan Srinivasan for their assistance with the preparation of this chapter, and to Albert Bregman and Stuart Wrigley for commenting on earlier drafts. Wang's research was supported in part by an AFOSR grant (FA9550-04-01-0117) and an AFRL grant (FA8750-04-1-0093). Brown's research was supported by EPSRC grant GR/R47400/01.

REFERENCES

1. J. I. Alcantara, B. C. J. Moore, V. Kuhnel, and S. Launer. Evaluation of the noise reduction system in a commercial digital hearing aid. *International Journal of Audiology,* 42:34–42, 2003.
2. P. Assmann and Q. Summerfield. The perception of speech under adverse acoustic conditions. In S. Greenberg, W. A. Ainsworth, A. N. Popper, and R. R. Fay, editors, *Speech Processing in the Auditory System,* volume 18 of *Springer Handbook of Auditory Research.* Springer-Verlag, Berlin, 2004.
3. M. W. Beauvois and R. Meddis. Computer simulation of auditory stream segregation in alternating tone sequences. *Journal of the Acoustical Society of America,* 99(4):2270–2280, 1996.
4. A. J. Bell and T. J. Sejnowski. An information-maximisation approach to blind separation and blind deconvolution. *Neural Computation,* 7:1129–1159, 1995.
5. J. Bench and J. Bamford. *Speech/Hearing Tests and the Spoken Language of Hearing-Impaired Children.* Academic Press, London, 1979.
6. J. Benesty, S. Makino, and J. Chen, editors. *Speech Enhancement.* Springer-Verlag, New York, 2005.
7. D. A. Berkley and J. B. Allen. Normal listening in typical rooms: The physical and psychophysical correlates of reverberation. In G. A. Studebaker and I. Hochberg, editors, *Acoustical Factors Affecting Hearing Aid Performance,* pp. 3–14. Allyn and Bacon, Needham Heights, MA, 1993.
8. C. Bey and S. McAdams. Schema-based processing in auditory scene analysis. *Perception and Psychophysics,* 64:844–854, 2002.
9. R. Bilger, J. M. Nuentzeq, W. M. Rabinowitz, and C. Rzeczkowski. Standardization of a test of speech perception in noise. *Journal of Speech and Hearing Research,* 27:32–48, 1984.
10. M. Bodden. Modelling human sound-source localization and the cocktail party effect. *Acta Acustica,* 1:43–55, 1993.
11. S. F. Boll. Suppression of acoustic noise in speech using spectral subtraction. *IEEE Transactions on Acoustics, Speech and Signal Processing,* 27:113–120, 1979.
12. M. S. Brandstein and D. B. Ward, editors. *Microphone Arrays: Signal Processing Techniques and Applications.* Springer-Verlag, New York, 2001.
13. A. S. Bregman. *Auditory Scene Analysis.* MIT Press, Cambridge, MA, 1990.
14. A. W. Bronkhorst and R. Plomp. Effect of multiple speechlike maskers on binaural speech recognition in normal and impaired hearing. *Journal of the Acoustical Society of America,* 92(6):3132–3139, 1992.
15. G. J. Brown. *Computational Auditory Scene Analysis: A Representational Approach.* PhD thesis, University of Sheffield, 1992.
16. G. J. Brown and M. P. Cooke. Computational auditory scene analysis. *Computer Speech and Language,* 8:297–336, 1994.
17. G. J. Brown and M. P. Cooke. Perceptual grouping of musical sounds: A computational model. *Journal of New Music Research,* 23(2):107–132, 1994.
18. G. J. Brown and M. P. Cooke. A neural oscillator model of primitive auditory grouping. In *Proceedings of the IEEE Workshop on Applications of Signal Processing to Audio and Acoustics,* pp. 35–38, Mohonk, NY, September 1995.

19. G. J. Brown and M. P. Cooke. Temporal synchronisation in a neural oscillator model of primitive auditory stream segregation. In D. F. Rosenthal and H. G. Okuno, editors, *Computational Auditory Scene Analysis,* pp. 87–101. Lawrence Erlbaum, Mahwah, NJ, 1998.

20. G. J. Brown and D. L. Wang. Modelling the perceptual segregation of concurrent vowels with a network of neural oscillators. *Neural Networks,* 10(9):1547–1558, 1997.

21. D. Brungart, P. S. Chang, B. D. Simpson, and D. L. Wang. Isolating the energetic component of speech-on-speech masking with an ideal binary time-frequency mask. Submitted, 2005.

22. D. S. Brungart. Information and energetic masking effects in the perception of two simultaneous talkers. *Journal of the Acoustical Society of America,* 109:1101–1109, 2001.

23. P. Chang. *Exploration of behavioural, physiological and computational approaches to auditory scene analysis.* Master's thesis, The Ohio State University, Department of Computer Science and Engineering. Available at http://www.cse.ohio-state.edu/pnl/theses.html, 2004.

24. E. C. Cherry. Some experiments on the recognition of speech, with one and with two ears. *Journal of the Acoustical Society of America,* 25(5):975–979, 1953.

25. E. C. Cherry. *On Human Communication.* MIT Press, Cambridge, MA, 1957.

26. H. S. Colburn. Theory of binaural interaction based on auditory-nerve data: II. Detection of tones in noise. *Journal of the Acoustical Society of America,* 61:525–533, 1987.

27. M. Cooke. Glimpsing speech. *Journal of Phonetics,* 31:579–584, 2003.

28. M. P. Cooke. *Modelling Auditory Processing and Organization.* PhD thesis, University of Sheffield, 1991 (also published by Cambridge University Press, Cambridge, UK, 1993).

29. M. P. Cooke, P. Green, L. Josifovski, and A. Vizinho. Robust automatic speech recognition with missing and unreliable acoustic data. *Speech Communication,* 34:267–285, 2001.

30. J. F. Culling, K. I. Hodder, and C. Y. Toh. Effects of reverberation on perceptual segregation of competing voices. *Journal of the Acoustical Society of America,* 114(5):2871–2876, 2003.

31. C. J. Darwin. Auditory grouping. *Trends in Cognitive Science,* 1:327–333, 1997.

32. E. de Boer. Synthetic whole-nerve action potentials for the cat. *Journal of the Acoustical Society of America,* 58:1030–1045, 1975.

33. P. N. Denbigh and J. Zhao. Pitch extraction and separation of overlapping speech. *Speech Communication,* 11(2–3):119–125, 1992.

34. L. Deng and C. D. Geisler. A composite auditory model for processing speech sounds. *Journal of the Acoustical Society of America,* 82(6):2001–2012, 1987.

35. H. Dillon. *Hearing Aids.* Thieme, New York, 2001.

36. W. J. Dowling. The perception of interleaved melodies. *Cognitive Psychology,* 5:322–337, 1973.

37. W. J. Dowling, K. M.-T. Lund, and S. Herrbold. Aiming attention in pitch and time in the perception of interleaved melodies. *Perception and Psychophysics,* 41:642–656, 1987.

38. R. Drullman and A. W. Bronkhorst. Multichannel speech intelligibility and speaker

recognition using monaural, binaural and 3D auditory presentation. *Journal of the Acoustical Society of America,* 107:2224–2235, 2000.

39. D. P. W. Ellis. *Prediction-Driven Computational Auditory Scene Analysis.* PhD thesis, MIT Department of Electrical Engineering and Computer Science, 1996.
40. D. P. W. Ellis. Evaluating speech separation systems. In P. Divenyi, editor, *Speech Separation by Humans and Machines.* Springer-Verlag, New York, 2005.
41. Y. Ephraim and D. Malah. Speech enhancement using a minimum mean-square error short-time spectral amplitude estimator. *IEEE Transactions on Acoustics, Speech and Signal Processing,* 32:1109–1121, 1984.
42. J. L. Flanagan. Models for approximating basilar membrane displacement. *Bell System Technical Journal,* 39:1163–1191, September 1960.
43. D. A. Forsyth and J. Ponce. *Computer Vision: A Modern Approach.* Prentice-Hall, Upper Saddle River, NJ, 2002.
44. J.Garofolo, L. Lamel, W. Fisher, J. Fiscus, D. Pallett, and N. Dahlgren. DARPA TIMIT acoustic-phonetic continuous speech corpus. Technical Report NISTIR 4930, National Institute of Standards and Technology, Gaithersburg, MD, 1993.
45. J. J. Gibson. *The Senses Considered as Perceptual Systems.* Greenwood Press, Westport, CT, 1966.
46. B. R. Glasberg and B. C. J. Moore. Derivation of auditory filter shapes from notched-noise data. *Hearing Research,* 47:103–138, 1990.
47. D. J. Godsmark. *A Computational Model of the Perceptual Organisation of Polyphonic Music.* PhD thesis, University of Sheffield, 2000.
48. D. J. Godsmark and G. J. Brown. A blackboard architecture for computational auditory scene analysis. *Speech Communication,* 27:351–366, 1999.
49. M. Goto. Development of the RWC music database. In *Proceedings of the 18th International Congress on Acoustics,* volume I, pp. 553–556, Kyoto, April 2004.
50. S. Grossberg. Pitch-based streaming in auditory perception. In N. Griffith and P. Todd, editors, *Musical Networks: Parallel Distributed Perception and Performance,* pp. 117–140. MIT Press, Cambridge, MA, 1999.
51. W. M. Hartmann. Pitch perception and the segregation and integration of auditory entities. In G. M. Edelman, W. E. Gall, and W. M. Cowan, editors, *Auditory Function: Neurobiological Bases of Hearing,* pp. 623–645. Wiley, New York, 1988.
52. W. M. Hartmann and D. Johnson. Stream segregation and peripheral channeling. *Music Perception,* 9:155–184, 1991.
53. H. Helmholtz. *On the Sensation of Tone.* Dover Publishers, New York, second English edition, 1863.
54. M. J. Hewitt and R. Meddis. An evaluation of eight computer models of mammalian inner hair-cell function. *Journal of the Acoustical Society of America,* 90(2):904–917, 1991.
55. J. Holdsworth, I. Nimmo-Smith, R. Patterson, and P. Rice. Implementing a gammatone filter bank. Technical report, MRC Applied Psychology Unit, Cambridge, February 1988.
56. G. Hu. *Monaural Speech Organization and Segregation.* PhD thesis, The Ohio State University, Biophysics Program, 2006.
57. G. Hu and D. L. Wang. Speech segregation based on pitch tracking and amplitude

modulation. In *Proceedings of the IEEE Workshop on Applications of Signal Processing to Audio and Acoustics,* pp. 79–82, Mohonk, NY, October 2001.

58. G. Hu and D. L. Wang. Monaural speech segregation based on pitch tracking and amplitude modulation. *IEEE Transactions on Neural Networks,* 15(5):1135–1150, 2004.
59. X. Huang, A. Acero, and H.-W. Hon. *Spoken Language Processing: A Guide to Theory, Algorithms, and System Development.* Prentice-Hall, Upper Saddle River, NJ, 2001.
60. A. Hyvărinen, J. Karbunen, and E. Oja. *Independent Component Analysis.* Wiley, New York, 2001.
61. A. Janin, D. Baron, J. Edwards, D. Ellis, D. Gelbart, N. Morgan, B. Peskin, T. Pfau, E. Shriberg, A. Stolcke, and C. Wooters. The ICSI meeting corpus. In *IEEE International Conference on Acoustics, Speech and Signal Processing,* volume 1, pp. 364–367, Hong Kong, 2003.
62. L. A. Jeffress. A place theory of sound localization. *Journal of Comparative and Physiological Psychology,* 41:35–39, 1948.
63. P. I. M. Johannesma. The pre-response stimulus ensemble of neurons in the cochlear nucleus. In *Proceedings of the Symposium on Hearing Theory,* pp. 58–69, IPO, Eindhoven, Netherlands, June 1972.
64. M. R. Jones and W. Yee. Attending to auditory events: The role of temporal organization. In S. McAdams and E. Bigand, editors, *Thinking in Sound,* pp. 69–112. Clarendon Press, Oxford, UK, 1993.
65. A. Jourjine, S. Rickard, and O. Yilmaz. Blind separation of disjoint orthogonal signals: Demixing N sources from 2 mixtures. In *Proceedings of the IEEE International Conference on Acoustics, Speech and Signal Processing,* pp. 2985–2988, Istanbul, June 2000.
66. C. Jutten and J. Herault. Blind separation of sources, part 1: An adaptive algorithm based on neuromimetic architecture. *Signal Processing,* 24:1–10, 1991.
67. D. N. Kalikow, K. N. Stevens, and L. L. Elliott. Development of a test of speech intelligibility in noise using sentence materials with controlled word predictability. *Journal of the Acoustical Society of America,* 61:1337–1351, 1977.
68. E. R. Kandel, J. H. Schwartz, and T. M. Jessell. *Principles of Neural Science,* 3rd ed,. Elsevier, New York, 1991.
69. B. L. Karlsen, G. J. Brown, M. P. Cooke, M. D. Crawford, P. D. Green, and S. J. Renals. Analysis of a multi-simultaneous-speaker corpus. In D. F. Rosenthal and H. G. Okuno, editors, *Computational Auditory Scene Analysis,* pp. 321–333. Lawrence Erlbaum, Mahwah, NJ, 1998.
70. A. P. Klapuri. *Signal Processing Methods for the Automatic Transcription of Music.* PhD thesis, Tampere University of Technology, Institute of Signal Processing, 2004.
71. B. Kollmeier and R. Koch. Speech enhancement based on physiological and psychoacoustical models of modulation perception and binaural interaction. *Journal of the Acoustical Society of America,* 95(3):1593–1602, 1994.
72. H. Krim and M. Viberg. Two decades of array signal processing research: The parametric approach. *IEEE Signal Processing Magazine,* 13:67–94, 1996.
73. T. W. Lee, M. S. Lewicki, M. Girolami, and T. J. Sejnowski. Blind source separation of more sources than mixtures using overcomplete representations. *IEEE Signal Processing Letters,* 6:87–90, 1999.

74. R. Leonard. A database for speaker-independent digit recognition. In *IEEE International Conference on Acoustics, Speech and Signal Processing,* pp. 328–331, San Diego, CA, March 1984.
75. V. R. Lesser, S. H. Nawab, and F. I. Klassner. IPUS: An architecture for the integrated processing and understanding of signals. *Artificial Intelligence,* 77(1):129–171, 1995.
76. H. Levitt. Noise reduction in hearing aids: A review. *Journal of Rehabilitation Research and Development,* 38(1):111–121, 2001.
77. J. C. R. Licklider. A duplex theory of pitch perception. *Experimentia,* 7:128–133, 1951.
78. J. S. Lim, editor. *Speech Enhancement.* Prentice-Hall, Englewood Cliffs, NJ, 1983.
79. W. Lindemann. Extension of a binaural cross-correlation model by contralateral inhibition. I. Simulation of lateralization for stationary signals. *Journal of the Acoustical Society of America,* 80(6):1608–1622, 1986.
80. W. Lindemann. Extension of a binaural cross-correlation model by contralateral inhibition. II. The law of the first wave front. *Journal of the Acoustical Society of America,* 80(6):1623–1630, 1986.
81. R. P. Lippmann. Speech recognition by machines and humans. *Speech Communication,* 22:1–16, 1997.
82. R. F. Lyon. A computational model of filtering, detection and compression in the cochlea. In *Proceedings of the IEEE International Conference on Acoustics, Speech and Signal Processing,* pp. 1282–1285, Paris, May 1982.
83. R. F. Lyon. A computational model of binaural localization and separation. In *Proceedings of the International Conference on Acoustics, Speech and Signal Processing,* pp. 1148–1151, 1983.
84. R. F. Lyon. Computational models of neural auditory processing. In *Proceedings of the IEEE International Conference on Acoustics, Speech and Signal Processing,* pp. 41–44, San Diego, CA, March 1984.
85. D. Marr. *Vision.* W. H. Freeman, San Francisco, CA, 1982.
86. S. L. McCabe and M. J. Denham. A model of auditory streaming. *Journal of the Acoustical Society of America,* 101(3):1611–1621, 1997.
87. R. Meddis. Simulation of mechanical to neural transduction in the auditory receptor. *Journal of the Acoustical Society of America,* 79(3):702–711, 1986.
88. R. Meddis. Simulation of auditory-neural transduction: Further studies. *Journal of the Acoustical Society of America,* 83(3):1056–1063, 1988.
89. R.Meddis and M. J. Hewitt. Virtual pitch and phase sensitivity of a computer model of the auditory periphery. I. Pitch identification. *Journal of the Acoustical Society of America,* 89(6):2866–2882, 1991.
90. R. Meddis and M. J. Hewitt. Virtual pitch and phase sensitivity of a computer model of the auditory periphery. II. Phase sensitivity. *Journal of the Acoustical Society of America,* 89(6):2883–2894, 1991.
91. R. Meddis, M. J. Hewitt, and T. M. Shackleton. Implementation details of a computational model of the inner hair-cell/auditory-nerve synapse. *Journal of the Acoustical Society of America,* 87(4):1813–1816, 1990.
92. D. Mellinger. *Event Formation and Separation in Musical Sound.* PhD thesis, Stanford University, 1991.
93. G. A. Miller. The masking of speech. *Psychological Bulletin,* 44:105–129, 1947.

94. G. A. Miller and J. C. R. Licklider. The intelligibility of interrupted speech. *Journal of the Acoustical Society of America,* 22(2):167–173, 1950.
95. B. C. J. Moore. *Cochlear Hearing Loss.* Whurr Publishers, London, 1998.
96. B. C. J. Moore. *An Introduction to the Psychology of Hearing,* 5th ed., Academic Press, London, 2003.
97. A. K. Nabelek and P. K. Robinson. Monaural and binaural speech perception in reverberation for listeners of various ages. *Journal of the Acoustical Society of America,* 71:1242–1248, 1982.
98. T. Nakatani, M. Goto, and H. G. Okuno. Localization by harmonic structure and its application to harmonic sound stream segregation. In *Proceedings of the IEEE International Conference on Acoustics, Speech and Signal Processing,* pp. 653–656, Atlanta, May 1996.
99. T. Nakatani, H. G. Okuno, and T. Kawabata. Residue-driven architecture for computational auditory scene analysis. In *Proceedings of the International Joint Conference on Artificial Intelligence (IJCAI),* pp. 165–172, Montreal, 1995.
100. M. Nilsson, S. Soli, and J. A. Sullivan. Development of the hearing in noise test for the measurement of speech reception thresholds in quiet and in noise. *Journal of the Acoustical Society of America,* 95:1085–1099, 1994.
101. M. Norris. *Assessment and Extension of Wang's Oscillatory Model of Auditory Stream Segregation.* PhD thesis, University of Queensland, School of Information Technology and Electrical Engineering, 2003.
102. S. E. Palmer. *Vision Science.* MIT Press, Cambridge, MA, 1999.
103. T. W. Parsons. Separation of speech from interfering speech by means of harmonic selection. *Journal of the Acoustical Society of America,* 60:911–918, 1976.
104. R. D. Patterson, M. H. Allerhand, and C. Giguère. Time-domain modeling of peripheral auditory processing: A modular architecture and a software platform. *Journal of the Acoustical Society of America,* 98(4):1890–1894, 1995.
105. R. D. Patterson, I. Nimmo-Smith, J. Holdsworth, and P. Rice. An efficient auditory filterbank based on the gammatone function. Technical report, MRC Applied Psychology Unit, Cambridge, 1987.
106. D. Paul and J. Baker. The design for the Wall Street Journal-based CSR corpus. In *Proceedings of the International Conference on Spoken Language Processing,* pp. 899–902, 1992.
107. D. Pearce and H.-G. Hirsch. The AURORA experimental framework for the performance evaluation of speech recognition systems under noisy conditions. In *Proceedings of the International Conference on Spoken Language Processing,* pp. iv, 29–32, 2000.
108. J. Pickles. *An Introduction to the Physiology of Hearing.* Academic Press, London, 1988.
109. R. Plomp. Binaural and monaural speech intelligibility of connected discourse in reverberation as a function of azimuth of a single competing sound source (speech or noise). *Acustica,* 34:200–211, 1976.
110. R. Plomp. Auditory handicap of hearing impairment and the limited benefit of hearing aids. *Journal of the Acoustical Society of America,* 63(2):533–549, 1978.
111. R. Plomp. A signal-to-noise ratio model for the speech reception threshold of the hearing impaired. *Journal of Speech and Hearing Research,* 29:146–154, 1986.
112. R. A. Rensink, J. K. O'Regan, and J. J. Clark. To see or not to see: The need for attention to perceive changes in scenes. *Psychological Science,* 8:368–373, 1997.

113. N. Roman, D. L. Wang, and G. J. Brown. Speech segregation based on sound localization. *Journal of the Acoustical Society of America,* 114(4):2236–2252, 2003.
114. D. F. Rosenthal and H. G. Okuno, editors. *Computational Auditory Scene Analysis.* Lawrence Erlbaum, Mahwah, NJ, 1998.
115. S. T. Roweis. One microphone source separation. In *Advances in Neural Information Processing Systems 13,* pp. 793–799. MIT Press, Cambridge, MA, 2000.
116. S. T. Roweis. Factorial models and refiltering for speech separation and denoising. In *Proceedings of Eurospeech,* pp. 1009–1012, Geneva, September 2003.
117. S. Seneff. A joint synchrony/mean-rate model of auditory speech processing. *Journal of Phonetics,* 16:55–76, 1988.
118. M. Slaney. Auditory toolbox (version 2). Technical Report, Interval Research Corporation, 1998.
119. M. Slaney and R. F. Lyon. A perceptual pitch detector. In *Proceedings of the IEEE International Conference on Acoustics, Speech and Signal Processing,* pp. 357–360, Albuquerque, NM, April 1990.
120. M. Slaney, D. Naar, and R. F. Lyon. Auditory model inversion for sound separation. In *IEEE International Conference on Acoustics, Speech and Signal Processing,* pp. 77–80, 1994.
121. S. Srinivasan and D. L. Wang. Modeling the perception of multitalker speech. In *Proceedings of Interspeech,* Lisbon, September 2005.
122. H. J. M. Steeneken. *On Measuring and Predicting Speech Intelligibility.* PhD thesis, University of Amsterdam, 1992.
123. R. J. Stubbs and Q. Summerfield. Evaluation of two voice-separation algorithms using normal-hearing and hearing-impaired listeners. *Journal of the Acoustical Society of America,* 84:1236–1249, 1988.
124. T. Tolonen and M. Karjalainen. A computationally efficient multi-pitch analysis model. *IEEE Transactions on Speech and Audio Processing,* 8(6):708–716, 2000.
125. A. J. W. van der Kouwe, D. L. Wang, and G. J. Brown. A comparison of auditory and blind separation techniques for speech segregation. *IEEE Transactions on Speech and Audio Processing,* 9:189–195, 2001.
126. L. P. A. S. van Noorden. *Temporal Coherence in the Perception of Tone Sequences.* PhD thesis, Eindhoven University of Technology, 1975.
127. B. D. van Veen and K. M. Buckley. Beamforming: A versatile approach to spatial filtering. *IEEE ASSP Magazine,* pp. 4–24, April 1988.
128. A. P. Varga, H. J. M. Steeneken, M. Tomlinson, and D. Jones. The NOISEX-92 study on the effect of additive noise on automatic speech recognition. Technical Report, Defence Research Agency Speech Research Unit, 1992.
129. C. von der Malsburg and W. Schneider. A neural cocktail-party processor. *Biological Cybernetics,* 54:29–40, 1986.
130. D. L. Wang. Auditory stream segregation based on oscillatory correlation. In *Proceedings of the IEEE International Workshop on Neural Networks for Signal Processing,* pp. 624–632, Ermioni, Greece, September 1994.
131. D. L. Wang. Primitive auditory segregation based on oscillatory correlation. *Cognitive Science,* 20:409–456, 1996.
132. D. L. Wang. On ideal binary mask as the computational goal of auditory scene analysis.

In P. Divenyi, editor, *Speech Separation by Humans and Machines,* pp. 181–197. Kluwer Academic, Norwell, MA, 2005.

133. D. L. Wang and G. J. Brown. Separation of speech from interfering sounds using oscillatory correlation. *IEEE Transactions on Neural Networks,* 10(3):684–697, 1999.
134. D. L. Wang and D. Terman. Locally excitatory globally inhibitory oscillator networks. *IEEE Transactions on Neural Networks,* 6(1):283–286, 1995.
135. M. Weintraub. The GRASP sound separation system. In *Proceedings of the IEEE International Conference on Acoustics, Speech and Signal Processing,* pp. 69–72, San Diego, CA, March 1984.
136. M. Weintraub. *A Theory and Computational Model of Auditory Monaural Sound Separation.* PhD thesis, Stanford University, August 1985.
137. S. Winter, H. Sawada, S. Araki, and S. Makino. Overcomplete BSS for convolutive mixtures based on hierarchical clustering. In *Proceedings of the Fifth International Conference on Independent Component Analysis,* pp. 652–660, Granada, September 2004.
138. S. N. Wrigley and G. J. Brown. A computational model of auditory selective attention. *IEEE Transactions on Neural Networks,* 15:1151–1163, 2004.
139. S. N. Wrigley, G. J. Brown, S. Renals, and V. Wan. Speech and crosstalk detection in multichannel audio. *IEEE Transactions on Speech and Audio Processing,* 13(1):84–91, 2005.
140. T. C. T. Yin and J. C. K. Chan. Interaural time sensitivity in medial superior olive of cat. *Journal of Neurophysiology,* 64(2):465–488, 1990.
141. W. A. Yost. The cocktail party problem: Forty years later. In R. H. Gilkey and T. R. Anderson, editors, *Binaural and Spatial Hearing in Real and Virtual Environments,* pp. 329–347. Lawrence Erlbaum, Mahwah, NJ, 1997.
142. X. Zhang, M. G. Heinz, I. C. Bruce, and L. H. Carney. A phenomenological model for the responses of auditory-nerve fibers: I. Nonlinear tuning with compression and suppression. *Journal of the Acoustical Society of America,* 109(2):648–670, 2001.

CHAPTER 2

Multiple *F*0 Estimation

ALAIN DE CHEVEIGNÉ

2.1 INTRODUCTION

This chapter is about the estimation of multiple fundamental frequencies (*F*0) from a waveform, such as the compound sound of several people speaking at the same time, or several musical instruments playing together. That information may be needed to transcribe a musical score, to extract intonation patterns for speech recognition, or as an ingredient for computational auditory scene analysis. The task of estimating the single *F*0 of an isolated voice has motivated a surprising amount of effort over the years [45]. Work on the harder task of estimating multiple *F*0s is now gaining momentum, fueled by progress in signal processing techniques on the one hand, and new applications such as interactive processing or indexing of music, multimedia, and speech on the other.

A multiple *F*0 estimation method is typically assembled from two elements: a single-voice *F*0 estimator, and a voice-segregation scheme. Here "voice" is used in a wide sense to designate the periodic signal produced by a source (human voice, instrument sound, etc.). Some space is accordingly devoted to the topic of single voice *F*0 estimation, but the reader should refer to the excellent treatise of Hess [45] for more details. Segregation techniques too are evoked, but the reader should follow pointers to other chapters of this book wherever possible.

A sound with a periodic waveform evokes a pitch that varies with *F*0, the inverse of the period [87]. The pitch may be salient and musical as long as the *F*0 is within about 30 Hz to 5 kHz [92, 105]. Sounds with the same period evoke the same pitch despite their diverse timbres, so pitch can be understood as an equivalence class. The auditory system is able to extract the period despite very different waveforms or spectra of sounds at the ears. Explanations of how this is done have been elaborated since antiquity [27]. Modern theories can be classified into two families: pattern matching and autocorrelation [27]. These theories are a source of inspiration for the development of *F*0 estimation methods that likewise can be organized according to a small number of basic principles, as we shall see in Section 2.3. Quite good solutions now exist for the task of single *F*0 estimation [45, 31].

Computational Auditory Scene Analysis. Edited by DeLiang Wang and Guy J. Brown

A musically inclined listener can often follow the melodic line of each instrument in a polyphonic ensemble. This implies that several pitches may be heard from a single compound waveform. Psychophysical data on this capability are fragmentary (e.g. [7, 8, 51]), so the limits of this capability, and the parameters that determine them, are not well known. This perceptual "proof of feasability" has nevertheless encouraged the search for algorithms for multiple $F0$ estimation. Multiple $F0$ estimation in essence involves two tasks: source separation and $F0$ estimation. If the compound signal representing the mixture were separated into streams, then it would be a simple matter to derive an $F0$ estimate from each stream using a single-voice estimation algorithm. On the other hand, if $F0$ estimates were known in advance, they could feed some of the separation algorithms described elsewhere in the book. This leads to a "chicken and egg" situation: estimation and segregation are each a prerequisite of the other, and a difficulty is to "bootstrap" this process.

There are other difficulties: the variety of signals and applications, the diversity of requirements and configurations that need evaluation, the existence of certain "degenerate" situations for which the problem is hard, and so on. Many polyphonic estimation schemes have been proposed, and beginners in this field may be bewilderd by the wide range and sophistication of methods. Is this complexity really necessary? In this chapter, I show how most methods sprout from a few simple ideas. Once those are understood, the jungle of methods should seem less wild. The rest of the chapter reviews the main approaches to multiple $F0$ estimation, trying wherever possible to extract the underlying insights and basic principles. A useful concept is that of the "signal model."

2.2 SIGNAL MODELS

By definition, a signal $x(t)$ is periodic iff there exists T such that

$$x(t) = x(t + T), \forall t. \tag{2.1}$$

If there exists one such T, there exist an infinity; the period is the smallest positive member of this set. Real signals differ from this description in various ways: they are of finite duration, their parameters may vary, there may be noise, and so on. In this sense, we speak of the periodic signal as a model that approximates signals found in the world. This model is parametrized by the period T (or its inverse $F0$), and by the shape of the waveform over a period-sized interval: $(x(t)\ t \in [0, T])$. It is useful in that it fits many sounds such as speech or musical sounds, because the parameter $F0 = 1/T$ is a good predictor of musical pitch or speech intonation, and because that same parameter is a useful ingredient in acoustic scene analysis algorithms (see Chapters 3 and 8).

An example of a periodic signal is the sinusoid $x(t) = A \cos(2\pi F0t + \phi)$. It is parametrized by the triplet $(F0, A, \phi)$, where A is the amplitude and ϕ the starting phase. Sinusoids are useful in the context of linear systems: the output of a linear system for sinusoidal input is another sinusoid of the same frequency but usually different amplitude and phase. Sinusoids (more precisely, complex exponentials) are eigenvectors of linear transforms. This property makes the sinusoid a very convenient model, as the effect of the linear system can be summarized by its effect on A and ϕ. The sum of sinusoids $x(t) = \Sigma_k A_k \cos(2\pi f_k t + \phi_k)$ is useful for the same reason, as the effect the linear system is simply described by its effect on A_k and ϕ_k for all k.

A special case of the sum-of-sines model is the harmonic complex for which all component frequencies are multiples of a common fundamental frequency $f_k = kF0$. It is parameterized by specifying $F0$ and (A_k, ϕ_k) for all k. The theorem of Fourier [35] states that this and the periodic signal model (Eq. 2.1) are equivalent, and fit exactly the same set of signals. Their parametrizations are related by the Fourier transform. $F0$ estimation methods are divided into time-domain and frequency-domain according to whether they adopt one or the other of these signal models. Figures 2.1**A–E** and 2.2**A–E** show examples that illustrate both models. Estimation involves finding the parameter T of the periodic model, or the parameter $F0$ of the harmonic model, that best fits the signal. Section 2.3 reviews a few simple ideas for doing so. Note that Fourier's theorem does *not* imply that there exists within the spectrum a component at $F0$ with nonzero amplitude. Confusion on this point has led to much effort being diverted to solving the "missing fundamental" problem.

The periodic (or equivalently harmonic) signal is the most basic model involved in $F0$ estimation, but other models may be of use. Examples are a periodic signal that varies slowly in amplitude or frequency, or a model of instrumental or voice production, or syntactic models of tone progression, and so on. They are useful for two reasons: (1) the extra parameters allow a better fit to the signal and thus ease the estimation of $F0$, and (2) other sources of knowledge may be brought in to constrain these parameters, again to get a more reliable estimate of $F0$. That knowledge is either "hard-wired" into the algorithm, or else learned from the data at run time. There is a continuum between methods that process only information from the signal within the analysis frame, and those that bring in context, source models, grammars, expectations, and so on.

2.3 SINGLE-VOICE *F*0 ESTIMATION

Before considering multiple voices, let us look at the simpler task of single-voice $F0$ estimation. Most polyphonic methods extend (or include) a single-voice method, and, therefore, schemes for that purpose are highly relevant. There are two basic approaches: spectral and temporal. In the former, a short-term Fourier transform is first applied to a frame of the waveform to obtain a spectrum, whereas in the latter the waveform is examined directly in the time domain. There are many variants of both approaches [45], but most flow from the same ideas. Note that most algorithms expect $F0$ to vary over time and attempt to produce a time series of estimates, $F0(t)$.

2.3.1 Spectral Approach

Figure 2.1**A** shows the short-term spectrum of a sinusoid. An obvious way to estimate its fundamental frequency is to measure the position of the spectral peak. However, this procedure fails for the spectrum in Fig. 2.1**B** that contains multiple peaks. A simple modification is to accept only the largest peak, but this algorithm fails for the spectrum in Fig. 2.1**C** for which the largest peak falls on a multiple of $F0$. A simple extension is to select the peak of lowest frequency, but this algorithm fails for the signal illustrated in Fig. 2.1**D** for which the lowest peak falls on a higher harmonic (the so-called "missing fundamental" waveform). Another cue, spacing

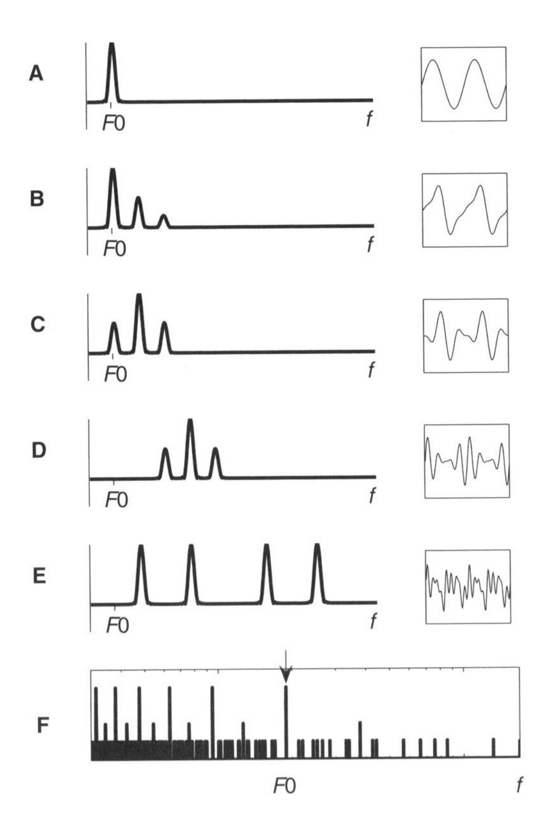

Figure 2.1 Spectra of simple signals that illustrate basic spectral $F0$ estimation schemes. Corresponding waveforms are shown as insets to the right. The spectrum peak determines the $F0$ of a pure tone (**A**) but a complex tone (**B**) has several such peaks. The largest peak determines the $F0$ of the waveform in **B**, but not **C**. The lowest-frequency peak determines the $F0$ of the waveform of **C** but not **D**. Interpartial spacing determines the $F0$ of **D** but not **E**. The Schroeder histogram (**F**) determines the $F0$ of the signal in **E** and of any periodic sound. The Schroeder histogram counts the subharmonics of every partial and accumulates them in a histogram. The cue to $F0$ is the rightmost of the series of maximum values of this histogram (arrow). Note that the abscissa of **F** is logarithmic.

between partials, indicates the correct *F*0 for this signal, but not for the signal illustrated in Fig. 2.1**E**.

A final strategy that works for this signal and all others is pattern matching. For each peak in the spectrum, divide its frequency by successive positive integers and distribute the resulting values among the bins of a histogram [Fig. 2.1**F**]. The largest counts are found in bins at frequencies that divide all partial frequencies. There is an infinite series of such bins, which all have the same count but vanishingly small abscissas. All are situated at subharmonics of the rightmost bin of the series, and the position of that bin thus indicates the fundamental. This idea was first applied to speech *F*0 estimation by Schroeder [104], but it has earlier roots in "pattern matching" models of pitch perception ([20]; see [21, 27] for reviews) that themselves evolved from the concept of "unconscious inference" of Helmholtz [123]. The idea has been proposed in many variants, such as the "spectral comb," "harmonic sieve," or "subharmonic summation" methods [78, 33, 44].

Most spectral *F*0 estimation methods now use pattern matching. Those that do not usually incorporate some form of preprocessing (nonlinearity and/or filtering) to generate or enhance cues such as interpartial spacing, or a fundamental component. For example, the method of [32] splits the signal over a bank of low-pass filters, selects the lowest-frequency channel with significant power, and measures the frequency of its output. Filtering reduces the signal to a sinusoid, so that the strategy of Fig. 2.1**A** can be applied to that output (see also [45]). Another recent example is the "TEMPO" algorithm of Kawahara et al. [61], which measures instantaneous frequency at the output of an array of bandpass filters. The channel that best responds to the fundamental is found on the basis of a "carrier-to-noise" measure. These algorithms are effective as long as the signal contains a sinusoidal component at *F*0. Such is often, but not always, the case. If that partial is absent, as in Fig. 2.1**D**, it may be reintroduced by nonlinear distortion (see, e.g., [101, 108]). Nonlinear distortion is not without problems, as one can find cases in which it instead suppresses the *F*0 component (for example, squaring a sinusoid would double its frequency and give an incorrect result).

Interpartial spacing was used, for example, in the methods of Lahat et al. [70], Chilton and Evans [17], or Kunieda et al. [68] that calculate the autocorrelation of the positive-frequency part of the spectrum. Any two spectral components spaced by *F*0 contribute to a peak at *F*0 in the spectrum autocorrelation. As argued by Klapuri [64], the spacing between adjacent components determines the rate of beating between them, and thus it can also be measured in the time domain (see next section). Algorithms based on interpartial spacing (or beats) fail if the spectrum is sparse; for example, if it consists of a single component at *F*0 (Fig. 2.1**A**) or of components at noncontiguous frequencies (Fig. 2.1**E**) but, again, one can use a nonlinearity to reintroduce power at harmonic frequencies within the gaps.

The strength of spectral methods is that they benefit from the theoretical power of Fourier analysis and from the efficiency of the fast fourier transform (FFT) to implement them. A weakness is their dependency on the shape and size of the analysis window. These remain as nuisance parameters of the estimation. These pros and cons are discussed in more detail below in the context of multiple *F*0 estimation. It may seem somewhat strange to go to the trouble of splitting the signal into partials, and then apply pattern matching to find the period that is, after all, obvious in the time-domain waveform. This reasoning motivates time-domain methods.

2.3.2 Temporal Approach

Figure 2.2**A** shows the waveform of a sinusoid. An obvious way to measure its period is to measure the interval between "landmarks" such as waveform peaks. This

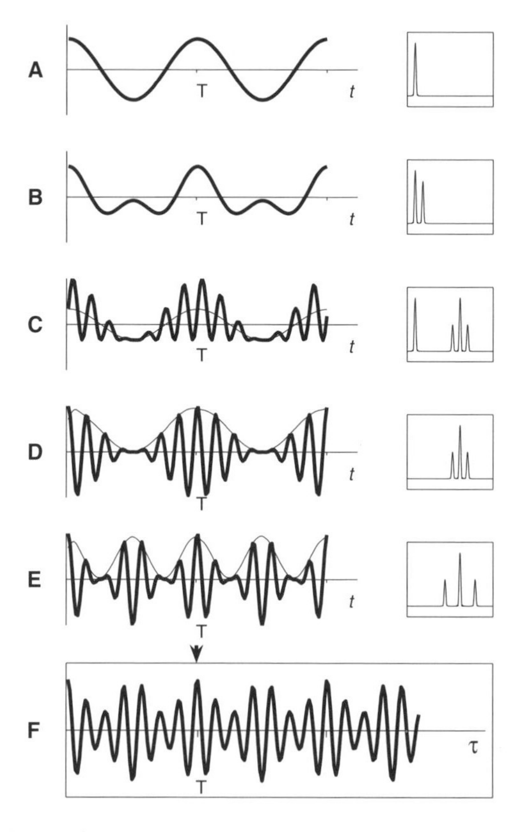

Figure 2.2 Waveforms of simple signals that illustrate time-domain *F*0 estimation schemes. Corresponding spectra are shown as insets to the right. The interval between waveform peaks indicates the period for the pure tone (**A**), but the complex tone (**B**) has several peaks per period. The largest peaks work for the complex tone (**B**), but would not work for the opposite waveform that has two "largest" peaks per period. Positive-going zero crossings work for **B** but not **C**. Low-pass filtering the signal in **C** would reduce it to its fundamental component (thin line) that has one peak or zero crossing per period, but the waveform in **D** has no fundamental component. The envelope may be obtained by full-wave rectification (or some other nonlinearity) followed by low-pass filtering (thin line). This works for **D**, but the envelope of **E** oscillates at twice its *F*0. The first major peak with nonzero lag τ (arrow) of the autocorrelation function (ACF) (f) can indicate the period of **E** or any other periodic waveform.

simple algorithm fails for the waveform in **B** that has several peaks per period. A modification is to take the largest peak, but this would fail for this same waveform if it were negated, as it would then have two "largest" peaks per period. Positive-going zero crossings would work for this waveform but fail for that of **C**, which has many crossings (and peaks) per period, as a consequence of a relatively large proportion of high-frequency power. An option is to apply low-pass filtering (thin line), but this strategy fails for the waveform of **D**, which lacks any low-frequency power. An option is to apply a nonlinearity; for example, full-wave rectification or squaring (thin line) and low-pass filtering to extract the envelope. However, this fails for the waveform of **E**, for which the envelope period is half the waveform period.

A final strategy works for this and all other periodic waveforms: self-similarity across time. Each waveform sample may be used, as it were, as a "landmark" to measure similarity for temporal spans of various sizes. For example, using the cross product between waveforms as a measure of similarity yields the familiar autocorrelation function (ACF), defined as

$$r_t(\tau) = (1/W) \sum_{j=t+1}^{t+W} x(j)x(j+\tau) \tag{2.2}$$

where τ is the "lag" (or delay), t is time at which the calculation is made, and W is the size of the window over which the product is integrated. The purpose of integration is to ensure that the measure is stable over time. Figure 2.2**F** shows the ACF of the waveform in Fig. 2.2**E**. The function has a series of global maxima at zero, at the period (arrow), and at all multiples of the period. The period is determined by scanning this pattern, starting at zero, and stopping at the first global maximum with nonzero abscissa. Autocorrelation was introduced for speech $F0$ estimation by Rabiner [93], but Licklider [73] had earlier suggested it to explain pitch perception, and the idea can be traced back even earlier [52] (see review in [27]).

Self-similarity methods such as the ACF can handle any periodic waveform. In contrast, strategies based on particular landmarks (peaks, etc.) must be associated with preprocessing to increase their salience or stability. For example, Dologlou and Carayannis [32] applied low-pass filtering to obtain a sinusoidal waveform with one peak per period, Howard [47] applied nonlinear filtering to "simplify" the waveform, and Howard [48] applied a neural network to learn a mapping between the voiced speech waveform and the glottal pulses that produced it. Earlier examples are reviewed by Hess [45]. The difficulty is to ensure that (a) at least one landmark occurs per period, (b) no more than one occurs per period, and (c) the landmark's position does not jump around within the period. These goals are impossible to guarantee in the general case; for any type of marker one can find examples such that an infinitesimal change in waveform produces a jump in marker position.

A detail must be mentioned at this point. We defined the ACF as in Eq. 2.2, but it is quite common to find a slightly different definition:

$$r_t'(\tau) = (1/W) \sum_{j=t+1}^{t+W-\tau} x(j)x(j+\tau) \tag{2.3}$$

in which W is replaced by $W - \tau$ as the upper limit of summation. This is often referred to as the "short-term ACF," whereas the definition of Eq. 2.2 has been diversely called "running ACF," "autocovariance," or "cross-correlation" [50]. The advantage of Eq. 2.3 is that it allows efficient implementation by the FFT. Its drawback is that for large τ the statistic is integrated over a small window, and thus is less stable over time. Figure 2.3 illustrates both definitions. The "short-term" ACF is plotted in **B** and the corresponding "running" ACF in **C**. Replacing $1/W$ by $1/(W - \tau)$ in Eq. 2.3 produces the so-called "unbiased" short-term ACF. In aspect it resembles the running ACF of Fig. 2.3**C**, but it is plagued by the same problem of insufficient temporal smoothing at large τ.

A useful variant of the ACF is the squared-difference function (SDF):

$$d_t(\tau) = (1/W) \sum_{j=t+1}^{t+W} [x(j) - x(j+\tau)]^2 \tag{2.4}$$

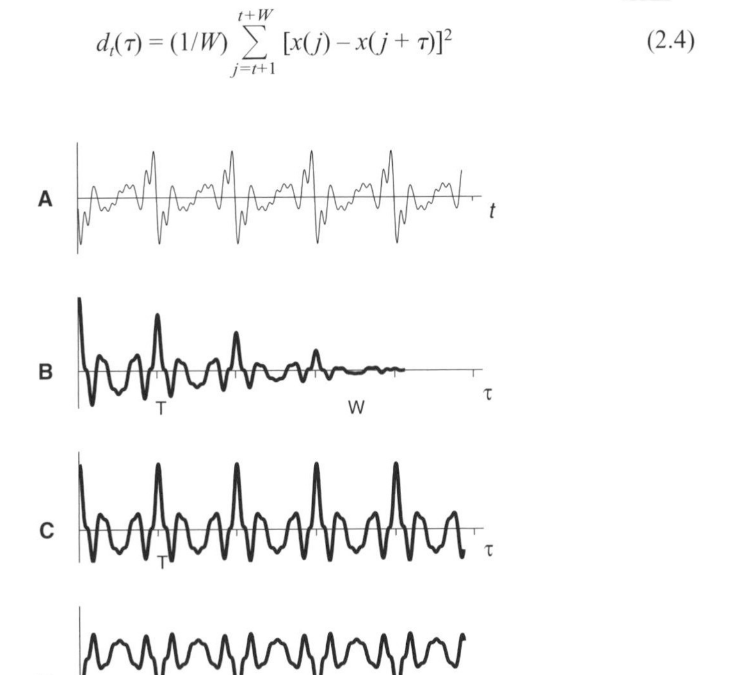

Figure 2.3 Illustration of the autocorrelation function. (**A**) Waveform of a periodic complex tone. (**B**) ACF calculated according to Eq. 2.3 ("short-term" ACF). Note that the function vanishes beyond $\tau = W$. (**C**) ACF calculated according to Eq. 2.2 ("running" ACF). (**D**) SDF. (**E**) Same ACF as in (c) but plotted as a function of an inverse log-lag scale [*log*$(1/\tau)$] [34]. Note the similarity of **E** with the Schroeder histogram plotted in Fig. 2.1**F**.

which is simply the squared Euclidean distance between a chunk of signal of size W and a similar chunk time shifted by τ. It is used, for example, by the cancellation model of [24], or the YIN method of [31]. Replacing Euclidean distance by city-block distance (sum of absolute values instead of squares) would yield the well-known AMDF, or average magnitude difference function [96]. ACF and SDF are related by the relation

$$d_t(\tau) = r_t(0) + r_{t+\tau}(0) - 2\, r_t(\tau) \tag{2.5}$$

The two first terms are local estimates of signal power, and to the degree that they are constant as a function of τ (i.e., if W is large enough), autocorrelation and squared difference function carry the same information. The cue to the period for the SDF is a dip rather than a peak, as illustrated in Fig. 2.3**D**. The nice thing about the SDF, as we shall see in Section 2.4.2, is that it can be generalized to estimate multiple periods. Note that the relation between ACF and SDF in Eq. 2.5 holds only if the ACF is calculated as in Eq. 2.2.

The strength of temporal methods is their conceptual simplicity, close to the mathematical definition of periodicity. There is, nevertheless, a deep link between spectral and temporal methods, and in particular between pattern matching and the ACF. To understand this link, recall that according to the Wiener–Khinchine theorem, the ACF is the inverse Fourier transform (IFT) of the power spectrum. As the waveform is real and its power spectrum symmetrical, the IFT is equivalent to cross correlation with a family of cosine functions. A cosine has regularly spaced peaks at integer multiples of its period, and can be understood as a particular form of harmonic template. Thus, the ACF can be seen as a form of pattern matching. This parallel is obvious if the ACF is plotted as a function of a log-lag scale as proposed by Ellis [34] (compare Fig. 2.3**E** with Fig. 2.1**F**).

Based on this reasoning, useful variants of the ACF are obtained by replacing the IFT by convolution with periodic templates that have sharper peaks than cosines (to increase their spectral selectivity), or peaks that decrease in amplitude (to discount the contribution of partials of higher frequency or rank) [65, 66]. A problem with the ACF is that the power spectrum puts strong emphasis on high-amplitude portions of the spectrum, and thus is sensitive to the presence of strong harmonics. This is alleviated by taking the logarithm before the cosine transform to obtain the well-known cepstrum [85]. Raising to the power 1/2 or 1/3 has a similar balancing effect [56], as reviewed recently by Klapuri [65]. These details are of limited theoretical importance but they have an impact on performance, particularly when the method is used within the context of multiple *F*0 estimation.

2.3.3 Spectrotemporal Approach

A variant of the temporal approach, inspired by auditory processing, involves splitting the signal over a filterbank. Each channel is treated as a waveform function of time, rather than as a sample along a slowly varying profile of spectral coefficients as in spectral methods. Each channel, dominated by a limited range of frequencies, is processed by time-domain methods as above, and the results are aggregated over channels. Typically, channel-wise ACFs may be added to obtain a summary autocorrelation function (SACF), as illustrated in Fig. 2.4. The idea was originally proposed in the pitch perception model of Licklider [73] and further developed by Meddis and Hewitt [79] and others [110, 74, 12]. It was applied to *F*0 estimation, for example, by [106, 22, 98].

There are several advantages of the spectrotemporal over the temporal approach. First, the weight of each channel may be adjusted to compensate for amplitude mismatches between spectral regions, which would otherwise be accentuated by the ACF [65]. Doing so is similar to the process of "spectral whitening" by inverse filtering that was applied in several early methods [45]. Second, channels dominated by noise, or by a competing source, can be discounted in the summary. We shall see how this can be put to use for multiple *F*0 estimation. A third advantage was pointed out by Klapuri [64, 65]. If higher-frequency channels have larger bandwidths (as is the case for models of cochlear filtering), then adjacent partials of high order in-

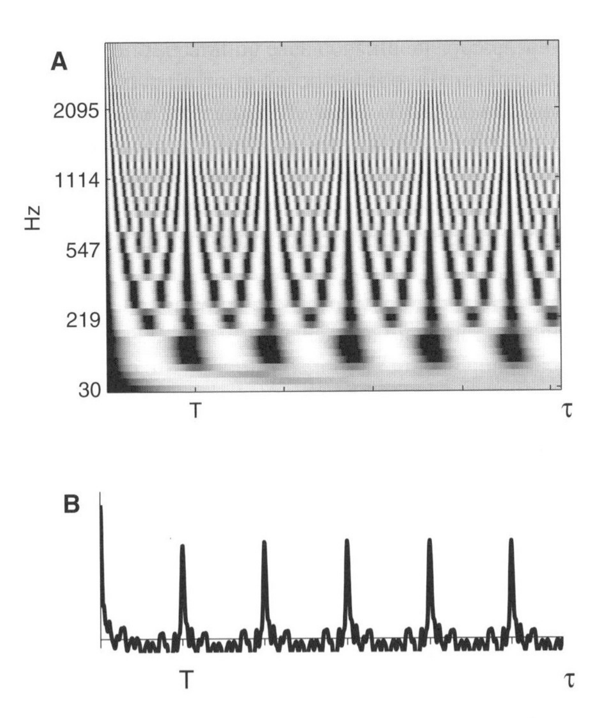

Figure 2.4 Spectrotemporal method of single-voice *F*0 estimation. (**A**) Array of ACFs calculated within channels of a filterbank. The filters are fourth-order gammatone filters with bandwidths based on psychophysical estimates of auditory selectivity [82] and center frequencies spaced equally in terms of bandwidth. Each channel is amplitude-normalized before the ACF calculation. (**B**) Summary ACF (SACF). These plots were calculated from the same waveform as in Fig. 2.3**A**. The difference from Fig. 2.3**C** is the result of amplitude normalization that emphasizes low-amplitude portions of the spectrum.

teract within those channels to create beats. Beat rate depends on interpartial spacing, and for high-order partials it may provide a cue to $F0$ that is more robust than the exact frequencies of the partials themselves, particularly if the spectrum is slightly inharmonic and/or $F0$ varies with time. Demodulation of higher-frequency channels (by a nonlinearity followed by low-pass filtering) allows cues from those beats to be incorporated into the SACF. Beats could actually be exploited without the filterbank, but what filtering buys in this context is to reduce the sensitivity of the beats to phase relations between partials that fall in different channels.

To summarize, many methods of single-voice $F0$ estimation have been proposed. Estimation can be understood in terms of fitting a model to the waveform. The most basic model is that of a periodic signal (Section 2.2), but more complex models may be used, for example instrument models that specify in detail the spectrotemporal shape of a "note," or dynamic models that constrain the variation of $F0$ over time. An estimation error occurs when the signal fits the model for an inappropriate set of parameters. The art of $F0$ estimation is to tweak the model (or the signal) to make such an erroneous fit less likely. This point of view is all the more useful in the case of multiple $F0$ estimation.

2.4 MULTIPLE-VOICE *F0* ESTIMATION

Several factors conspire to make multiple-voice $F0$ estimation more difficult than single-voice $F0$ estimation. Mutual overlap between voices weakens their pitch cues, and the cues must further compete with cues to other voices. There exist degenerate situations in which available information is ambiguous, as when the $F0$s are in simple ratios. Also, the diversity of situations to be considered (number and type of sources, relative amplitudes and timing, etc.) makes progress harder to evaluate than in the single $F0$ case.

The basic signal model is the sum of periodic signals. For example, in the case of two voices, the observable signal $z(t)$ is the sum of signals $x(t)$ and $y(t)$ of periods T and U:

$$z(t) = x(t) + y(t), \qquad x(t) = x(t + T), \qquad y(t) = y(t + U), \qquad \forall t \tag{2.6}$$

$F0$ estimation consists of finding parameters T and U that best fit the signal z. More complex models are discussed later on.

Three basic strategies have been used. In the first, a single-voice estimation algorithm is applied in the hope that it will find cues to several $F0$s. In the second strategy (iterative estimation), a single-voice algorithm is applied to estimate the $F0$ of *one* voice, and that information is then used to *suppress* that voice from the mixture so that the $F0$s of the other voices can be estimated. Suppression of those voices in turn may allow the first estimate to be refined, and so on. In a third strategy (joint estimation), all the voices are estimated at the same time.

As an example of the first strategy, the speech separation system of Weintraub [125] searched the ACF for cues to multiple periods. In the system of Stubbs and

Summerfield [115], the same was done for the cepstrum. It is rather challenging to make this strategy work. Looking at representations such as the Schroeder histogram of Fig. 2.1**F** or the ACF of Fig. 2.2**F**, it is obvious that they already contain multiple cues even for a single voice. Distinguishing these from cues to multiple voices is bound to be hard. Schemes have been proposed to attenuate spurious cues [56, 117, 34], but the conditions under which they are successful appear to be limited. We will concentrate instead on the two other strategies: iterative and joint estimation. As before, approaches can be classified as spectral, temporal, and spectrotemporal.

2.4.1 Spectral Approach

In a seminal paper, Parsons [89] calculated the short-term magnitude spectrum of mixed speech (sum of two talkers) over a 51.2 ms windows, and applied Schroeder's subharmonic histogram method, mentioned earlier, to determine the harmonic series that best matched the spectrum. A first *F*0 was derived, spectral peaks that matched its harmonic series were removed from the spectrum, and a second *F*0 was estimated from the remainder. The second voice could then be removed in turn to refine the estimate of the first. This process is illustrated in Fig. 2.5. The aim of Parsons was voice separation, but *F*0 extraction was a major subtask and his was one of the first multiple-*F*0 estimation systems. Many since Parsons have proposed to apply harmonic templates iteratively to dissect the short-term spectrum [103, 114, 59, 129, 38, 5, 19,

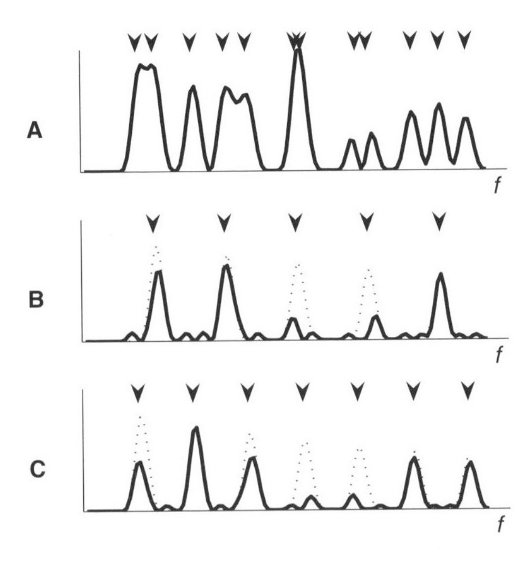

Figure 2.5 Spectral method of two-voice *F*0 estimation, based on Parsons [89]. (**A**) Spectrum of the sum of two concurrent voices. A first *F*0 estimate is derived from this spectrum and used to suppress one voice (voice A). (**B**) Thick line: result of suppressing voice A. The *F*0 of voice B can be estimated from this remainder, and used to suppress that voice in turn. (**C**) Thick line: result of suppressing voice B. The arrows indicate the harmonic series of each voice, and the dotted lines represent the spectra of the voices before mixing. Note that only part of the spectrum has been retrieved in each case.

55, 64, 100, 124, 121, 84]. These methods use the spectrum representation both as a source of cues to the $F0$ of a voice, and as a substrate from which it is possible to discount those cues so that the other $F0$s can be determined. In some methods, the estimation and suppression steps are performed in sequence; in others, they are performed jointly by fitting the compound spectrum to a model of overlapping spectra.

2.4.2 Temporal Approach

Supposing the period T of one voice has been determined, that voice can be suppressed by applying to the compound waveform a time-domain comb filter with impulse response $h_T(t) = \delta(t) - \delta(t - T)$. The impulse response and its power transfer function are illustrated in Fig. 2.6**A** and **B**. The transfer function has zeros at $1/T$

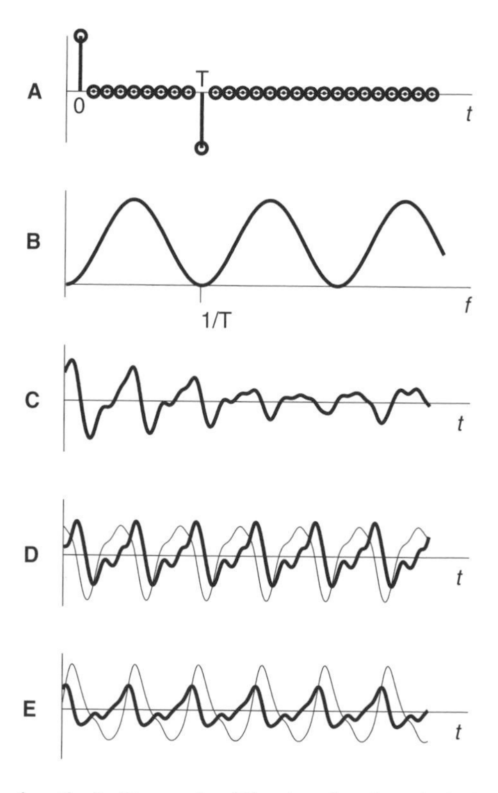

Figure 2.6 Temporal method of two-voice $F0$ estimation (iterative). (**A**): Impulse response of time-domain comb filter. (**B**) Power transfer function of the same filter. Zeros at multiples of $1/T$ cancel all harmonics of $F0 = 1/T$. (**C**) Sum of two complex tones with $F0$s one semitone apart (6%). A first $F0$ estimate is derived from this waveform and used to suppress one voice (voice A). (**D**) Thick line: result of suppressing voice A. The $F0$ of voice B can be estimated from this remainder and used to suppress voice B from the compound. (**E**) Thick line: result of suppressing voice B. The thin lines represent the complexes before mixing. Note that the filtered waveforms have the same period as voices A or B, respectively, but not the same shape.

and all its multiples, and these can suppress all the partials of a voice with $F0 = 1/T$. Tuning this filter to the period of voice A, that voice may be suppressed and the $F0$ of voice B estimated. Tuning the filter to the period of voice B, the estimate of voice A may be refined. This process is illustrated in Fig. 2.6**C**–**E**.

The idea was first proposed by Frazier et al. [36] for voice separation, and later used for multiple $F0$ estimation by de Cheveigné and others [23, 30, 56]. The period estimate may be obtained by any single-voice $F0$ estimation method; for example, by the ACF or SDF (Section 2.3.2). The latter option is of interest as the same operation (cancellation) serves in turn to measure cues to the $F0$ of a voice, and then to suppress them. Indeed, both steps may be performed jointly rather than in succession [23, 30, 29].

In the MMM method of [29], the period is found by forming the double difference function (DDF):

$$dd_t(\tau, \nu) = (1/W) \sum_{j=t+1}^{t+W} [x(j) - x(j+\tau) - x(j+\nu) + x(j+\tau+\nu)]^2 \qquad (2.7)$$

It is easy to see that this function is zero for $(\tau, \nu) = (jT, kU)$ for all integers (j, k), and, conversely, if periods (T, U) are unknown, they may be found by searching the (τ, ν) parameter space for the first minimum with nonzero coordinates. The function is illustrated in Fig. 2.7 for a mixture of two periodic sounds with periods that differ

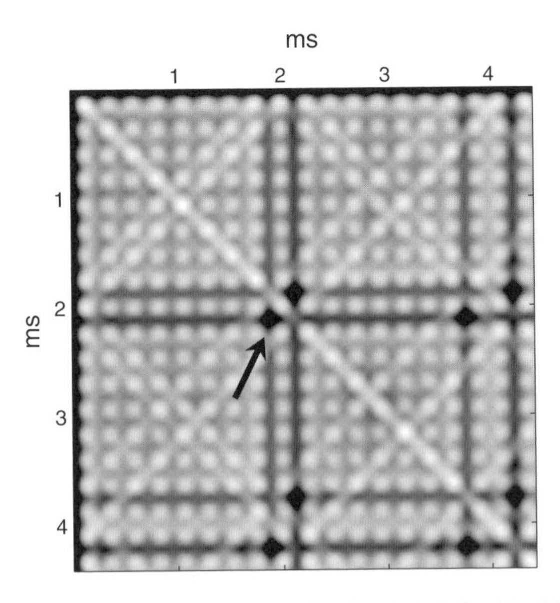

Figure 2.7 Temporal method of two-voice estimation (joint). Double difference function (DDF) in response to a mixture of two periodic sounds as a function of its two lag parameters, τ and ν. Darker means smaller. The coordinates of the minimum with smallest nonzero lag (arrow) indicate the periods T and U.

by two semitones (about 12%). Minima are visible at period multiples, as well as along the axes $\tau = 0$ and $\nu = 0$.

2.4.3 Spectrotemporal Approach

A third approach, intermediate between spectral and temporal, is to split the waveform over a bank of band-pass filters (Fig. 2.8). Meddis and Hewitt [80] extended their spectrotemporal model of single-pitch perception [79] to explain voiced speech segregation by using a cochlear filter bank to split acoustic information into channels belonging to either of two sources. ACFs calculated within each channel were initially summed across all channels to obtain a summary ACF (SACF) from

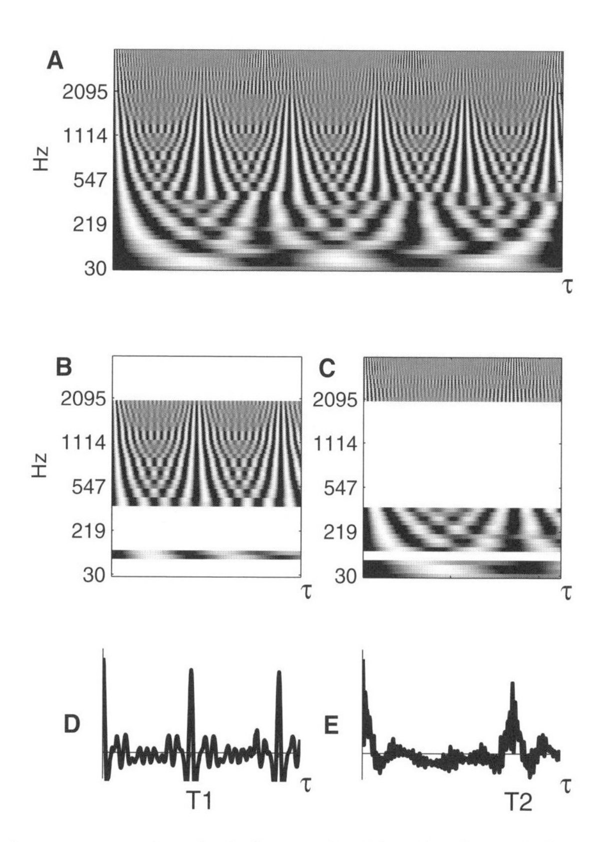

Figure 2.8 Spectrotemporal method of two-voice *F*0 estimation. (**A**) Illustration of a spectrotemporal two-voice estimation algorithm. (**A**) Array of ACFs at the output of a filterbank in response to the sum of two periodic signals (synthetic vowels "a" and "i"). (b) ACFs of channels dominated by one voice. (**C**) ACFs of channels dominated by the other voice. (**D**) SACF calculated from channels dominated by the first voice. (**E**) SACF calculated from channels dominated by the second voice. The *F*0s of both voices can be estimated from these SACFs.

which a dominant period was derived. Channels with peaks at that particular period were then assigned to the dominant voice, and the remaining channels used to estimate the identity of the second voice. Although not elaborated by the authors, a second period could also be estimated from those remaining channels. Channel selection had previously been proposed by Lyon [75] and Weintraub [125] for sound separation. The idea has since been used for multiple *F*0 estimation by Wu et al. [128, 126] and others [49, 76, 72].

How do spectral and spectrotemporal methods compare? Both split the signal into spectral "elements" (spectrum bins in one case, filter channels in the other) on the basis of their spectral properties. However, whereas spectral methods assign bins according to their position along the frequency axis, spectrotemporal methods assign channels according to the periodicity that dominates them. They thus differ in resolution requirements: spectral methods must resolve individual partials, and this requires a long analysis window, whereas spectrotemporal methods need merely to isolate spectral regions dominated by one or the other voice (Fig. 2.4**B**,**C**). Long analysis windows cannot be used if the signal is nonstationary; in that case, spectrotemporal methods may have the advantage.

How do temporal and spectrotemporal methods compare? Both estimate *F*0s based on temporal information. They differ in how the correlates of an unwanted voice are suppressed: channel selection for the former, and comb filtering for the latter. For signals that are perfectly periodic, comb filtering provides perfect rejection, whereas the degree of rejection of most filterbanks is limited by the slope of filter characteristics. Nevertheless, channel selection may be more effective in the presence of noise, or for slightly inharmonic sources for which harmonic cancellation is less effective. One might expect a combination of the two approaches (for example, time-domain cancellation at the output of filterbank channels) to be most effective, but it seems that this idea remains to be fully explored.

For slightly inharmonic sources such as strings, or in the event of slight *F*0 estimation errors or nonstationarity, it may be hard to segregate higher-order partials on the basis of their position relative to a *F*0-based harmonic series. This is particularly the case for high-frequency components, and so spectral approaches may have difficulty making use of higher-order partials. Temporal approaches based on comb filtering also run into problems in the same situation. However, the spectrotemporal approach allows the additional cue of interharmonic spacing. Spacing determines the beat rate between partials that interact within a channel, and that rate can be measured by applying a nonlinearity to the filter output followed by low-pass filtering to isolate the low-frequency beat components [65, 66, 128]. For this to work, the channel must contain partials of only one voice, and for that the filters must be narrow compared to features of the spectral envelope (e.g., formants) of each sound. The ability to extract this extra cue gives the spectrotemporal approach an edge over spectral and temporal methods.

Various criteria may be used to recognize the channels that belong to a voice. For example Wu and colleagues [127, 128, 126] use heuristic quality criteria to eliminate channels dominated by noise. Hu and Wang [49] include cross-channel correlation to group channels likely to belong to the same source. Klapuri [65, 66] discounts higher-frequency channels in which partials may be unresolved, and thus dominated by beats at the "chord root" frequency. The chord root is a common subharmonic of the voices present. If it is high enough to fall within the search range, it may be mistaken for the $F0$ of a primary voice.

2.5 ISSUES

This section deals with a number of "nitty-gritty" considerations that must be addressed for processing to be effective. Algorithms are sensitive to imperfections in the calculations, or to a mismatch between the signal model and the signal. It is important to distinguish between processing issues (for example, spectral resolution) and application-dependent issues (for example, imperfect periodicity or noise). For multiple $F0$ estimation, the "devil is in the details."

2.5.1 Spectral Resolution

The main issue with frequency-domain methods is spectral resolution. Supposing a temporal analysis window of duration D, short-term spectra are sampled in frequency with a resolution of $1/D$. This means that, according to Parseval's relation, signal power within the analysis window is partitioned among spectral coefficients. Spectral methods can use this partition to segregate voices and thus measure their $F0$s. More precisely, if partials of a voice fall on multiples of $1/D$, that voice can be removed so as to estimate the other voice's $F0$s. Such is the case only if that voice's $F0$ is an integer multiple of $1/D$, unfortunately an unlikely event. In general there is a mismatch between partials and the frequency grid. This may interfere with estimation of $F0$ of each voice and, more importantly, reduce the effectiveness of source suppression because each individual spectral coefficient contains power from several sources. Larger analysis windows allow finer spectral resolution, at the expense of temporal resolution and the ability to deal with time-varying signals. The need for power-of-two block sizes for FFT efficiency further restricts the choice of window size.

There are several ways to enhance spectral features. The short-term spectrum may be interpolated in the vicinity of spectral peaks (e.g., by fitting a smooth function such a parabola, or the Fourier transform of the analysis window, or a Gaussian, etc.) [59, 118]. In place of the Fourier transform, the waveform may be fitted to a sinusoidal model (e.g., [107, 11]) or a sum of damped sinusoids (e.g., [116]). The complex spectra of successive bins may be paired to obtain an instantaneous frequency estimate for each frequency bin. This is then used, rather than bin position, as a measure of frequency of the power within the bin. Instantaneous frequency has been used for single voice $F0$ estimation (e.g., [2, 3, 61, 4, 83]) and multiple-voice $F0$ estimation (e.g., [40, 9, 109]). Mapping power according to instantaneous frequency produces a spectrogram with sharper features than the Fourier spectrogram [16, 18, 61, 83, 43]. These techniques have been reviewed recently by Hainsworth [42] and Virtanen [118].

It is important to understand that these techniques improve the accuracy of cues to partials that are resolved, but do not address the problem of partials that are too close to have individual cues. Cues to partials that are close may undergo mutual distortion, or even merge into a single hybrid cue. To some extent, overlapping cues may be separated by modeling the superposition process. However, the effectiveness of this operation is limited by uncertainty as to phase relations between partials (see further on). In addition to these factors that relate purely to processing constraints, there are other factors related to stimulus imperfections, such as aperiodicity or noise, that contribute to make the compound spectrum difficult to partition among voices. Related to spectral resolution is the problem of temporal resolution of spectral analysis, as determined by the size, shape, and position of analysis windows. Kashino and colleagues [57] optimize the trade-off between these conflicting constraints with the use of "snapshots," windows starting from a discontinuity such as note onset, and extending as far as the signal is stable.

To summarize, performance of spectral approaches is limited by spectral resolution, itself determined by the short analysis window size required to follow changes in the signal. Many techniques exist to overcome these limits, but (a) they add to conceptual complexity and difficulty of implementation, and (b) they are not always as effective as needed.

2.5.2 Temporal Resolution

Limited sampling resolution. The accuracy of cues such as ACF peak position is limited by sampling resolution. Worse, suppression of a voice by comb filtering may be imperfect, thus impairing the estimation of the other voices. Resolution of ACF peaks may be improved by three-point parabolic interpolation, as the vicinity of an ACF peak is well approximated as a sum of cosines, each of which can be expanded as a Taylor series with terms of even order. Interpolation refines the value at the peak, which determines whether it wins over competing peaks, and its position, which determines the precise value of the period estimate. The same interpolation technique is applicable to the dip of the SDF (Eq. 2.4), and it may be extended to two-dimensional interpolation of the DDF pattern (Eq. 2.7) in the joint cancellation method: five samples (the minimum and its four immediate neighbors) constrain a paraboloid with no cross terms from which the global minimum may be interpolated [29]. Interpolation is also needed for voice suppression. A voice with a noninteger period can be suppressed by applying a time-domain comb filter with fractional delays, implemented either by an interpolating filter [69] or more simply, if less accurately, by linear interpolation.

Efficiency. Multiple $F0$ estimation is computationally expensive, and it is important to understand the factors that determine the cost. Estimation involves search within the space of possible periods. Supposing N expected periods, the size of the space varies as $O(K^N)$, where K is the number of points at which each period dimension is sampled. Joint estimation methods (e.g., [29]) search this space exhaustively. Iterative methods (e.g., [30]) search a subset of size $0(KNk)$, where k is the number of iterations. Search is indifferent to permutation of lags, so cost may be reduced by a large factor by ordering lags as $\tau_1 < \tau_2 < \ldots < \tau_N$. The asymptotic trends, however, remain the same. Each lag dimension is typically sampled uniformly at the same resolution as the waveform, so $K = f_s \tau_{\max}$, where f_s is the sampling rate and $\tau_{\max}$ the largest expected period. Nonuniform sampling such as logarithmic (Fig. 2.3) has also been proposed [34]. The appropriate degree of temporal integration also depends on $\tau_{\max}$. Specifically, the window of integration (W in Eq. 2.2, $W - \tau$ in Eq. 2.3) should be at least $\tau_{\max}$ in order to guarantee the stability over time of $F0$ estimates.

The short-term ACF, inverse Fourier transform of the short-term power spectrum, is best calculated by FFT. According to the previous reasoning, the window size W in Eq. 2.3 should be at least equal to $2\tau_{\max}$. The running ACF of Eq. 2.2 can likewise be calculated by FFT, as the inverse Fourier transform of the cross-spectrum between two windowed chunks of signal of size W and $W + \tau_{\max}$. The computational cost of an FFT of size N, $O(N \log N)$, is cheaper than the $O(N^2)$ cost of implementing Eqs. 2.2 or 2.3 directly. However, if it is necessary to repeat the calculation at a high frame rate, a recursive formula may be faster than the FFT. For example the formula $r_{t+1}(\tau) = r_t(\tau) - x(t)x(t + \tau) + x(t + W)x(t + W + \tau)$ updates the ACF at a frame rate equal to the waveform sampling rate. For exhaustive search, Eq. 2.7 needs to be evaluated repeatedly. The cost of doing so may be reduced by applying a computational formula such as $d_t(v, \nu) = d_t(v) + d_{t-\nu}(v) + d_t(\nu) - d_t(v + \nu) - d_{t-\nu}(\nu - v) + d_{t-\nu}(\nu)$ in which the DDF is expressed as a linear combination of DFs. Similar formulae are available involving ACFs [29]. This leads to computational savings if the necessary DFs (or ACFs) are precalculated.

Efficiency considerations are important in that computational costs may prohibit otherwise effective schemes.

2.5.3 Spectrotemporal Resolution

Spectrotemporal methods use an initial filterbank to split the waveform into channels, each of which is then processed in the time domain. Selectivity requirements are less stringent than for spectral methods. Rather than partials, it is sufficient to resolve spectral regions dominated by one or another voice. Increasing filter selectivity allows off-frequency components belonging to noise or other voices to be better attenuated. However, sharp skirts entail a long impulse response that may "smear" features over time, and thus limit the ability to track a time-varying voice. Also, if filters are narrow, more channels are required to cover the useful spectrum. The choice of filterbank is a trade-off between these conflicting requirements.

A common choice is a filterbank with characteristics similar to the human ear (see, e.g., [127, 128, 65]). Auditory filters are typically modeled as gammatone filters for which efficient implementations exist (see, e.g., [111, 18, 46, 91]). Bandwidths are usually set according to estimates of human "critical bandwidth" [130] or "equivalent rectangular bandwidths" (ERB) [82] that are roughly constant below 1 kHz (about 50–100 Hz) and proportional to frequency beyond 1 kHz (about 10%). There is no guarantee, however, that characteristics close to the human ear will ensure optimal multiple $F0$ estimation. Indeed, Karjalainen and Tolonen [56, 117] used only two bands covering the regions below and above 1 kHz, and Goto [38, 40] likewise used filtering to separate a low-frequency region (<262 Hz) from which a bass line was derived, from a high-frequency region (>262 Hz) from which a melody line was derived. No studies seem to have searched for optimal filtering characteristics, whether theoretically or empirically. A system could conceivably incorporate multiple filter characteristics so as to satisfy a wider range of constraints [28].

A weakness of the spectrotemporal approach is the cost of processing multiple channels in parallel. Efficient schemes exist to implement processing that is functionally similar in the frequency domain via standard FFT-based methods [62].

2.6 OTHER SOURCES OF INFORMATION

Up to this point, we have reviewed methods that exploit only one source of information: the signal within the analysis frame. This information is fragile and fragmentary. Other sources of information may contribute both to improve the accuracy of a voice's $F0$ estimate, and to better suppress that voice and estimate the others. This information is brought to bear via models of what to expect of the signal. It is important to realize that, if a model does not fit the signal being treated, this process may instead increase the risk of error.

2.6.1 Temporal and Spectral Continuity

A common assumption is that voices change slowly. Continuity over time of $F0$ estimates is exploited in postprocessing algorithms [45] such as median smoothing, dynamic programming, hidden Markov models (HMMs, see, e.g., [128]), or multiple "agents" [40]. The value for the current frame given by the "bottom-up" algorithm is tested for consistency with past (or future) values. Proximity of value may be complemented by a measure of quality to give more weight to reliable estimates. Processing may occur post hoc after the estimation stage, or else it may be integrated into the estimation algorithm itself (e.g., [119]). Estimation is improved directly, as a result of interpolating over errors and missing values, and also indirectly by (hopefully) increasing the likelihood that the voice is accurately suppressed so that other voice $F0$s can be estimated.

In addition to continuity of $F0$ tracks, the assumption that partial amplitudes vary smoothly can be used to track voices over instants when $F0$s cross or fall into a ratio for which the separation task is ambiguous. A different but related assumption is that all partial amplitudes vary according to the same function of time (to a fixed factor) [129]. Granted this assumption, amplitude variations that do not follow this function may be assigned to beats between closely spaced partials, and partial amplitudes can then be estimated from the minima and maxima of the beats [67, 121]. The assumption amounts to saying that the time-frequency envelope is the outer product of a spectral shape (common to all times) and a temporal shape (common to all frequencies). Spectrograms usually have more complex shapes, but techniques exist to decompose them into a sum of such simple envelopes [55, 13, 112]. The time course of amplitude itself can be modeled as a sum of smooth basis functions such as Gaussians or raised cosines [55, 19]. Cross-time dependencies can be modeled within the context of Bayesian models [124].

An assumption that has been used recently is spectral smoothness, that is, limited variation of partial amplitudes across the frequency axis [63, 129, 122, 5, 14, 71]. Many (but not all) musical instruments indeed have smooth spectral envelopes. Irregularity of the compound spectrum then signals the presence of multiple voices, and smoothness allows the contribution of a voice to shared partials to be discounted. For example if two voices are at an octave from each other, partials of even rank are the superposition of partials of both voices. Based on spectral smoothness, the contribution of the lower voice can be inferred from the amplitude of partials of odd rank, and subtracted to reveal the presence of the higher voice. The effectiveness of this strategy is nevertheless limited by uncertainty as to the relative phase of coinciding partials (see below). Spectral smoothness has also been used to reduce the likelihood of subharmonic errors [5, 63]. Beats between adjacent partials are strongest if the partials are of similar amplitude, and thus spectral smoothness enhances beat-related cues (see, e.g., [65]).

The effectiveness of the spectral smoothness assumption depends, of course, on its validity. If voices have irregular spectral envelopes, as in Fig. 2.1**E**, the assumption is likely instead to favor incorrect interpretations of the data. Some natural sources produce irregular spectra, such as the clarinet (for which even partials are weak) or the human voice (if harmonic spacing is wide relative to formant bandwidth), and, of course, there is no constraint at all on sounds produced electronically.

2.6.2 Instrument Models

Similar to continuity constraints but more sophisticated are instrument models. These can be of a general nature, such as the source-filter model that grounds inverse-filtering methods [45], or models of naturally amortized sounds as sums of exponentially decreasing sinusoids [116]. They can also be instrument-specific such as the piano models of [86, 97]. They may be predetermined (for example, from the physics of the instrument) or else learned from the data. Like spectral or temporal continuity, instrument models allow the contribution of one voice to be discounted from a compound spectrum. In some cases, a model of the time-domain waveform (or complex spectrum) of an instrumental note can be acquired [84, 9, 14] and used to perform exact subtraction. In general, however, exact phase information is lacking. In that case, it is usual to assume summation of power.

At this point, it is worth examining more closely the issue of phase-dependent summation and subtraction. Fig. 2.9**A** illustrates the magnitude spectral envelopes of two voices A and B with unknown phase, and Fig. 2.9**B** plots in gray the range occupied by the phase-dependent magnitude of their sum $A + B$. It illustrates an annoying fact: vector summation with unknown phase can produce values scattered

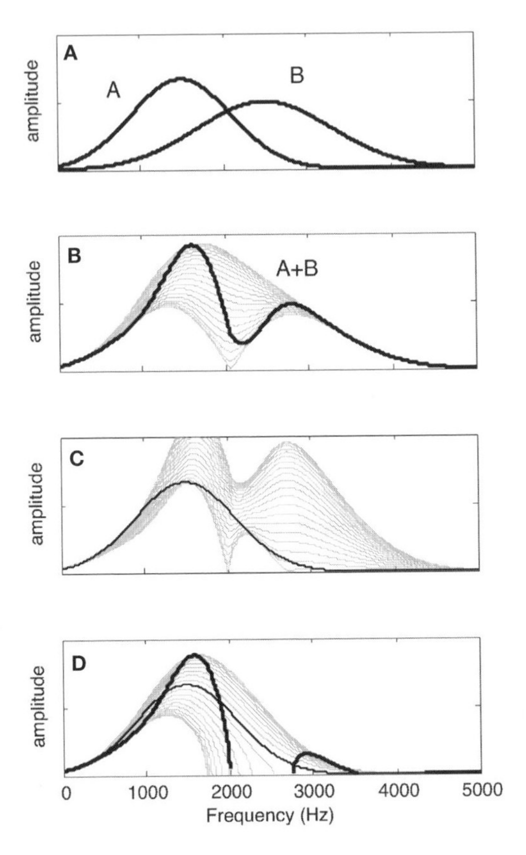

Figure 2.9 Phase dependency of spectral summation and subtraction. (**A**) Spectral envelopes of two sources A and B. (**B**) Gray: range of the possible phase-dependent amplitude of $A + B$. Thick line: a possible outcome. (**C**) Gray: range of values of A that might have produced the previous observation of $A + B$, given knowledge of B but no knowledge of the phase. Thin line: actual value of A. (**D**) Thick line: value of A that would have given rise to the previous observation of $A + B$ on the incorrect assumption of summation of power, and given knowledge of B. Gray: range of possible estimates of A based on the assumption of power summation, for all possible values of $A + B$. Thin line: actual value of A. The density of gray lines gives an idea of the probability densities. These can be used as a basis for inference according to Bayesian methods.

over a wide range (the spacing of the curves gives an indication of the probability density). Shown in black is the outcome for some arbitrary phase relation.

Suppose now that we wish to estimate A from this observation of $A + B$ and from prior knowledge of B (for example, from a source model). The range of magnitudes of A that could have produced the observation is plotted in gray in Fig. 2.9**C** (the correct value is in black). On the upside, this distribution is limited: We know for sure that A is *not* outside its range. On the downside, we do not know the exact value, and the range is rather wide. This illustrates a second annoying fact: knowledge of the magnitude spectra of the sum of two voices, and of one term, does not guarantee accurate estimation of the other term.

Lacking better knowledge, it is common to assume quadrature phase so that the power of the sum is the sum of the powers (this corresponds to its expected value, or average over trials with random phase). If we believe that such was the case here (i.e., we assume that the thick curve in Fig. 2.9**B** is the sum of powers of A and B), then A can be inferred exactly as illustrated by the thick curve in Fig. 2.9**D**. Part of this curve is missing, indicating that no value of A could have produced the observed value with this interval. The range of estimates of A across all phase-dependent values of $A + B$ is plotted in gray. The thin black line indicates the correct value of A, and as the density of gray plots suggests, it is not a likely outcome. This illustrates a third annoying fact: the assumption of power summation rarely leads to an accurate estimate. Nevertheless, values are constrained by the probability distributions (illustrated by the density of gray lines), and can be further constrained by other sources of information using Bayesian methods (see below) to lead to accurate estimates.

Thus, instrument models are less effective than we would hope for factoring a compound spectrum. To obtain useful results, there are at least three options: (1) use a model of the waveform or complex spectrum instead of the magnitude spectrum, (2) use a sparse representation (such as high-resolution spectrum) in which correlates of each voice occupy distinct dimensions so that vector summation does not occur, or (3) fall back on estimates of the distributions of values (as in Fig. 2.9) to inform probabilistic models such as those discussed in the next section.

2.6.3 Learning-Based Techniques

Many recent methods may be loosely described as based on "machine learning." Chapter 4 treats these techniques in greater detail; they are evoked here briefly with respect to $F0$ estimation. The idea is to learn the parameters of underlying models (for example, instrument models) from a database, or to structure the observable waveform or spectrum according to a generative model. Methods differ according to their required assumptions, the nature of the information derived from the database to constrain the model parameters, and the methodological framework. In some cases, the harmonic nature of voices is assumed beforehand, in others it is "learned" from the data, and in yet others it is not needed for separation (i.e., the method would work with nonharmonic voices). Once the voices have been separated, their $F0$s can be derived.

In deconvolutive sparse coding [120] the observed signal is assumed to be the sum of a relatively small number of recurring events such as musical notes. The waveform is modeled as a series of pulses (note onsets) convolved with time-frequency magnitude spectrograms (notes), and the method tries to derive both the pulse positions and the shape of the notes.

In nonnegative sparse coding [119, 120, 1, 112] the power or magnitude spectrum is modeled as a sum of spectra of individual events that occur relatively rarely ("sparsely"). For [1, 119] each event has an invariant spectral shape with time-varying amplitude (outer product of time and spectrum shapes; see above), but for [120] events are allowed to have an arbitrarily complex spectrotemporal shape (nevertheless stereotyped from event to event). Notes from the same voice are treated as distinct events, and successive events then need to be clustered into "voices." In the absence of phase information, it is reasonable to assume summation of power spectra (see above), but magnitude spectra are reported to work as well or better [120]. Actually, the choice of power versus magnitude is important only if both sources have significant energy within the same time-frequency pixels. It is moot for sources in which energy is sparsely distributed within the time-frequency plane (as is often the case for speech or environmental sounds [99]). In that case, summation of spectra can just as well be modeled by taking their maximum within each pixel, and each pixel thus "belongs" to a different source. Nonnegative decomposition has the advantage that components are guaranteed to be nonnegative everywhere, as is appropriate for magnitude or power spectra. This is in contrast to methods such as deconvolution, independent component analysis, or independent subspace analysis [13] that may well produce negative values.

Sagayama and colleagues [100] take a very different approach, by modeling the log-frequency spectrum as the result of convolution of a common harmonic spectrum by a number of "impulses" that each represent an individual $F0$. Multiple $F0$ estimates are obtained by deconvolution along the spectral axis. Saul and colleagues [102] perform a similar operation using nonnegative deconvolution in the spectral domain, and the log-lag ACF of [34] might allow a similar operation in the lag domain.

Bayesian approaches [58, 15, 59, 42, 113, 19, 94, 95, 124] treat the spectrum as the result of a number of "causes," and use a generative model to infer them from the observations [90]. Parameters of the generative model may be derived from previous knowledge, or they may be learned from isolated sounds [59, 109] or from repeated exposure to mixtures [90, 95]. An important distinction is between rule-based systems [77] and systems that undergo unsupervised learning [95, 99]. Hidden Markov models are used to model transitions between numbers of voices [125, 128], values of $F0$, or spectral shapes [5]. The models themselves may embody typical spectral shapes, simple transition probability rules (bigrams), or more complex grammars of music or language. Spectrum distributions are often modeled as Gaussians, although there is reason to believe that typical distributions may have rather different shapes (see, e.g., Fig. 2.9).

2.7 ESTIMATING THE NUMBER OF SOURCES

Up to this point, no concern was given to finding the number of voices present within a mixture. This is a difficult aspect of $F0$ estimation. Many studies ignore it and concentrate on the simpler task of producing some fixed number of estimates, regardless of the number of voices actually present. Human listeners, too, have difficulty counting voices. When asked to estimate the number of voices in mixtures of one to ten, subjects asymptoted at about three [60]. When asked to count voices in four-voice polyphony, musicians underestimated their number on half the trials [51].

Some signals are inherently ambiguous, and may be interpreted either as a single voice with low $F0$, or as the sum of several voices with higher, harmonically related $F0$s. An algorithm tuned to find as many voices as possible (or to favor the shortest possible periods) may "dismember" a voice into subsets of partials. Tuned to find as few as possible (or the longest possible periods), it may coalesce multiple voices. The voice count is accordingly over- or underestimated.

Iterative estimation methods typically apply a model at each iteration, and assign as much signal power to a voice as fits this model. Iteration continues on the remainder, and stops when the spectrum (or waveform) has been depleted of power. In the presence of noise, it may be hard to distinguish between residual noise and yet another source. It is necessary to set a threshold, and depending on its value, the algorithm is prone to either miss voices, or to find spurious voices. The aperiodicity of "real-world" sources has an effect similar to noise.

In the method of [29], cancellation filters are applied successively to remove each periodic voice. The algorithm stops when application of a new filter reduces power by less than a criterion ratio. Klapuri [62] evaluates the "global weight" of the $F0$ candidate derived from the residual after a voice has been suppressed, and stops the search if that weight falls below threshold. The system of Goto [40] uses "agents" that are created and terminated at each note onset and offset, whereas Martin [77] tracks hypotheses within a "blackboard" system [81, 90]. In nonnegative deconvolution [102, 109], the number of voices is given by the number of elements of the deconvolved matrix with amplitudes greater than some threshold. Wu and colleagues [127] use an HMM to model transitions between states of 0, 1, or 2 voices. Probabilities of state-to-state transitions, and value-to-value transitions within a state, were estimated from a speech database. The Bayesian system of Walmsley and colleagues [124] estimates the number of voices together with their $F0$s. Information criteria (BIC, AIC) offer a principled basis to decide whether the complexity of a model (such as a sum of periodic voices) is justified by the data [53, 54]. Some music transcription systems detect note onsets and offsets; this information is also of use in determining the number of notes.

2.8 EVALUATION

Evaluation is perhaps the weakest aspect of research on multiple $F0$ estimation. The many schemes suggested so far are difficult to rank on any common scale, for lack of widely accepted databases and metrics. Factors of difficulty are the diversity of target tasks (number of and nature of voices, application requirements, available knowledge, etc.), the lack of databases, and the large size of the parameter space to be tested. The situation is complicated by paradoxical effects; for example, the fact that random answers are more likely to be correct if the task is to estimate many voices (a priori a harder task). The authors of each algorithm tend to choose, as it were, a "niche" in which the algorithm has the best chances of behaving favorably.

Without universally established methodology, progress is difficult to measure. One can propose that:

- Evaluation should be differential, proceeding by comparison of different algorithms (hopefully, some of which are well known and competitive) on the same task and database. Widely available databases (e.g., [41]) and software implementations of algorithms are among the most useful contributions that a researcher can make to this field.
- Ground truth should be available and agreed upon. For estimation of single-voice $F0$, this may involve speech labeled with laryngograph signals, or music produced from known scores. For polyphonic $F0$ estimation, the situation is easier, as it is usually easy to assemble polyphonic test data from monophonic material that has been labeled by a well-known single-voice algorithm. The labels must nevertheless be made public to allow meaningful comparison.
- Evaluation should be designed to exercise each part (or knowledge source) individually. For example, to assess the effectiveness of an instrument model, it is necessary to include also a condition in which this model is disabled. In addition, it is instructive to know how far an algorithm can go on purely bottom-up data. Estimation of the number of sources should be evaluated separately from estimation of their values. Evaluation should include conditions with synthetic data that fit perfectly the underlying model, as well as "real-world" data with imperfections.
- Evaluation should strive to distinguish between performance limits due to algorithm limitations (e.g., insufficient spectral resolution) and those due to imperfections of the material (e.g., inharmonicity or imperfect repeatability of natural sounds). Noise, if present, must be fully described. For example, the mere specification of noise RMS is meaningless if the bandwidth of the noise is unknown.

Hopefully, as the field matures, the evaluation methodology will converge onto a set of widely accepted procedures that will allow progress to be evaluated.

2.9 APPLICATION SCENARIOS

As mentioned in the Introduction, multiple $F0$ estimation is useful for a wide range of applications. In automatic transcription, audio data are processed to obtain a score. In score following, the score preexists and the task is simply to align it temporally with incoming music. These two tasks differ widely in difficulty, and the same is true of each of them applied to different musical genres. Some applications involve a simpler task, such as extraction of the "predominant $F0$" [38, 41, 39], bass line [39], or melody [37, 88] from polyphonic audio. In interactive music applications, $F0$ estimation may need to be performed in real time.

A relatively new application is content-based indexing of audio databases. In the often-cited task of "query by humming," a piece of music is retrieved on the basis of a short extract hummed or whistled (more or less accurately) by the user. A straightforward approach is to label polyphonic material with a multiple *F*0 estimation algorithm, and compare the query to the multiple-track labels. However, a more effective alternative may be to match the query directly to the material, or to a "mid-level representation." Query by humming is just one example of a wide range of useful operations involving content-based indexing. An important issue is the scalability of such indexing representations [25, 10].

This book is about computational auditory scene analysis, for which *F*0 is an important ingredient. It can be used in two ways: to help extract a periodic source from a background of interfering sources (themselves not necessarily periodic), or else to help extract a source (itself not necessarily periodic) from a background of periodic sources. The first strategy is termed "harmonic enhancement," the second "harmonic cancellation." Both are a priori possible, but it has been argued that the second (cancellation) is more readily used by the auditory system [23] and also more effective in many situations [26]. It is not always necessary to extract the *F*0 of all sources: enhancement requires only the *F*0 (s) of the source(s) to be extracted, while cancellation requires only the *F*0 (s) of the interfering source(s). *F*0 may be combined with other cues such as cross-channel correlation in a multimicrophone setup, and possibly jointly estimated with those cues [6].

The diversity of applications illustrates the fact that the term "multiple *F*0 estimation" actually subsumes a wide variety of tasks. It underscores the difficulty of evaluation, and indeed, of defining the goal to be attained by future efforts. In principle, there is no hard limit on the number of *F*0s that can be extracted from a complex mixture, supposing their values are not in simple ratios for which the task is undetermined. In practice, noise and imperfect periodicity will always conspire to make the task difficult.

2.10 CONCLUSION

To summarize, a wide range of methods may be found in the literature. While many may appear extremely sophisticated, most boil down to a handful of very simple ideas. A major divide is between spectral approaches based on the FFT, and temporal methods based directly on the waveform. Intermediate between them are spectrotemporal approaches inspired from models of auditory processing. Another divide is between methods that rely purely on information derived from the signal within a particular analysis frame, and those that incorporate other sources of knowledge. Among them one may distinguish methods based on learning from those based on "hardwired" rules. Within each method one can distinguish two logical steps: *F*0 estimation of individual voices, and separation (or suppression) of voices so that the single-*F*0 estimation may be effective. These two steps may be performed iteratively, each voice being estimated and suppressed in turn, or jointly, all steps being performed at once.

The difficulty of the task is its diversity, as there is little unity between the situations and applications that call for estimation. For example, polyphonic transcription (production of a score from audio) and score following (alignment of a score to music) are of incommensurable difficulty, as are each of these tasks applied to different musical genres. A major obstacle is the lack of good evaluation methodology.

If I were to speculate on future progress in this field, I would suggest that it would come from (a) better evaluation, (b) better understanding of the basic issues involved in bottom-up estimation from the signal, and (c) principled incorporation of "high-level" sources of knowledge, most likely within a Bayesian framework.

ACKNOWLEDGMENTS

Malcolm Slaney's AuditoryToolbox [111] was used for simulations. Thanks are due to Hideki Kawahara and the editors for useful comments on initial versions of this chapter.

REFERENCES

1. S. A. Abdallah and M. D. Plumbley. Polyphonic music transcription by non-negative sparse coding of power spectra. In *International Symposium on Music Information Retrieval,* pp. 318–325, 2004.
2. T. Abe, T. Kobayashi, and S. Imai. Harmonics tracking and pitch extraction based on instantaneous frequency. In *IEEE International Conference on Acoustics, Speech, and Signal Processing,* pp. 756–759, 1995.
3. T. Abe, T. Kobayashi, and S. Imai. The IF spectrogram: A new spectral representation. In *International Conference on Simulation, Visualization and Auralization for Acoustic Research and Education,* pp. 423–440, Tokyo, 1997.
4. Y. Atake, T. Irino, H. Kawahara, J. Lu, S. Nakamura, and K. Shikano. Robust fundamental frequency estimation using instantaneous frequencies of harmonic components. In *International Conference on Spoken Language Processing,* volume II, pp. 907–910, 2000.
5. F. Bach and M. Jordan. Discriminative training of hidden markov models for multiple pitch tracking. In *International Conference on Acoustics, Speech, and Signal Processing,* volume V, pp. 489–492, 2005.
6. A. Baskind and A. de Cheveigné. Pitch-tracking of reverberant sounds, application to spatial description of sound scenes. In *Advanced Encryption Standard,* Banff Centre, Canada, 2003.
7. J. G. Beerends. *Pitches of Simultaneous Complex Tones.* PhD thesis, Technical University of Eindhoven, 1989.
8. J. G. Beerends and A. J. M. Houtsma. Pitch identification of simultaneous diotic and dichotic two-tone complexes. *Journal of the Acoustical Society of America,* 85: 813–819, 1989.
9. J. P. Bello Correa. *Towards the Automated Analysis of Simple Polyphonic Music: A Knowledge-Based Approach.* PhD thesis, University of London, 2003.
10. N. Bertin and A. de Cheveigné. Scalable metadata and quick retrieval of audio signals. In *International Symposium on Music Information Retrieval,* London, 2005.

11. P. Cano. Fundamental frequency estimation in the SMS analysis. In *COST G6 Conference on Digital Audio Effects,* pp. 99–102, 1998.
12. P. A. Cariani and B. Delgutte. Neural correlates of the pitch of complex tones. I. Pitch and pitch salience. *Journal of Neurophysiology,* 76:1698–1716, 1996.
13. M. A. Casey and A. Westner. Separation of mixed audio sources by independent subspace analysis. In *International Computer Music Conference,* pp. 154–161, 2000.
14. G. Cauwenberghs. Monaural separation of independent acoustical components. In *IEEE Symposium on Circuit and Systems (ISCAS),* 1999.
15. A. T. Cemgil. *Bayesian Music Transcription.* PhD thesis, Radboud Universiteit Nijmegen, 2004.
16. F. J. Charpentier. Pitch detection using the short-term phase spectrum. In *International Conference on Acoustics, Speech, and Signal Processing,* pp. 113–116, 1986.
17. E. Chilton and B. G. Evans. The spectral autocorrelation applied to the linear prediction residual of speech for robust pitch detection. In *IEEE International Conference on Acoustics, Speech, and Signal Processing,* pp. 358–361, 1988.
18. M. P. Cooke. *Modeling Auditory Processing and Organisation.* PhD thesis, University of Sheffield, Department of Computer Science, 1991.
19. M. Davy and S. Godsill. Bayesian harmonic models for musical signal analysis. In *Bayesian Statistics,* volume 7, pp. 105–124. Oxford University Press, Oxford, 2003.
20. E. de Boer. *On the "Residue" in Hearing.* PhD thesis, University of Amsterdam, 1956.
21. E. de Boer. On the "residue" and auditory pitch perception. In W. D. Keidel and W. D. Neff, editors, *Handbook of Sensory Physiology, volume V-3,* pp. 479–583. Springer-Verlag, Berlin, 1976.
22. A. de Cheveigné. Speech F0 extraction based on Licklider's pitch perception model. In *ICPhS,* volume 4, pp. 218–221, 1991.
23. A. de Cheveigné. Separation of concurrent harmonic sounds: Fundamental frequency estimation and a time-domain cancellation model of auditory processing. *Journal of the Acoustical Society of America,* 93:3271–3290, 1993.
24. A. de Cheveigné. Cancellation model of pitch perception. *Journal of the Acoustical Society of America,* 103:1261–1271, 1998.
25. A. de Cheveigné. Scalable metadata for search, sonification and display. In *International Conference on Auditory Display (ICAD 2002),* pp. 279–284, Kyoto, June 2002.
26. A. de Cheveigné. The cancellation principle in acoustic scene analysis. In P. Divenyi, editor, *Perspectives on Speech Separation,* pp. 243–257. Kluwer, New York, 2004.
27. A. de Cheveigné. Pitch perception models. In C. J. Plack, A. Oxenham, R. R. Fay, and A. N. Popper, editors, *Pitch—Neural Coding and Perception.* Springer-Verlag, New York, 2005.
28. A. de Cheveigné. Separable representations for cocktail party processing. In *Forum Acusticum (FA2005),* Budapest, 2005.
29. A. de Cheveigné and A. Baskind. F0 estimation of one or several voices. In *Eurospeech,* pp. 833–836, 2003.
30. A. de Cheveigné and H. Kawahara. Multiple period estimation and pitch perception model. *Speech Communication,* 27:175–185, 1999.
31. A. de Cheveigné and H. Kawahara. YIN, a fundamental frequency estimator for speech and music. *Journal of the Acoustical Society of America,* 111:1917–1930, 2002.

32. I Dologlou and G. Carayannis. Pitch detection based on zero-phase filtering. *Speech Communication,* 8:309–318, 1990.
33. H. Duifhuis, L. F. Willems, and R. J. Sluyter. Measurement of pitch in speech: An implementation of Goldstein's theory of pitch perception. *Journal of the Acoustical Society of America,* 71:1568–1580, 1982.
34. D. Ellis. *Prediction-Driven Computational Auditory Scene Analysis.* PhD thesis, MIT, 1996.
35. J. B. J. Fourier. *Traité Analytique de la Chaleur.* Didot, Paris, 1820.
36. R. H. Frazier, S. Samsam, L. D. Braida, and A. V. Oppenheim. Enhancement of speech by adaptive filtering. In *Proceedings of IEEE International Conference on Acoustics, Speech, and Signal Processing,* pp. 251–253, 1976.
37. E. Gomez, S. Streich, B. Ong, R. P. Paiva, S. Tappert, and J.-M. Batke. A quantitative comparison of different approaches for melody extraction from polyphonic audio recordings. Submitted.
38. M. Goto. A predominant-F0 estimation method for CD recordings: MAP estimation using EM algorithm for adaptive tone models. In *International Conference on Acoustics, Speech, and Signal Processing,* volume V, pp. 3365–3368, 2001.
39. M. Goto. A predominant-F0 estimation method for polyphonic musical audio signals. In *Independent Component Analysis and Blind Source Separation,* volume II, pp. 1085–1088, 2004.
40. M. Goto. A real-time music-scene-description system: predominant-F0 estimation for detecting melody and bass lines in real-world audio signals. *Speech Communication,* 43:311–329, 2004.
41. M. Goto, H. Hashiguchi, and S. Hayamizu. RWC music database: Popular, classic, and jazz music databases. In *International Symposium on Music Information Retrieval,* pp. 287–288, 2002.
42. S. Hainsworth. *Techniques for the Automated Analysis of Musical Audio.* PhD thesis, University of Cambridge, 2003.
43. S. W. Hainsworth and P. J. Wolfe. Time-frequency reassignment for musical analysis. In *International Computer Music Conference,* pp. 14–17, 2001.
44. D. J. Hermes. Measurement of pitch by subharmonic summation. *Journal of the Acoustical Society of America,* 83:257–264, 1988.
45. W. Hess. *Pitch Determination of Speech Signals.* Springer-Verlag, Berlin, 1983.
46. V. Hohmann. Frequency analysis and synthesis using a gammatone filterbank. *Acta Acustica United with Acustica,* 88:433–442, 2002.
47. D. M. Howard. Peak-picking fundamental period estimation for hearing prostheses. *Journal of the Acoustical Society of America,* 86:902–910, 1989.
48. I Howard. *Speech Fundamental Period Estimation Using Pattern Classification.* PhD thesis, London, 1991.
49. G. Hu and D. L. Wang. Monaural speech segregation based on pitch tracking and amplitude modulation. *IEEE Transactions on Neural Networks,* 15:1135–1150, 2004.
50. X. Huang, A. Acero, and H.-W. Hon. *Spoken Language Processing.* Prentice-Hall, Upper Saddle River, NJ, 2001.
51. D. Huron. Voice denumerability in polyphonic music of homogenous timbres. *Music Perception,* 6:361–382, 1989.

52. C. H. Hurst. A new theory of hearing. *Transactions of the Liverpool Biological Society,* 9:321–353 (and plate XX), 1895.
53. R. A. Irizarry. Local harmonic estimation in musical sound signals. *Journal of the American Statistical Association,* 96:357–367, 2001.
54. H. Kameoka, T. Nishimoto, and S. Sagayama. Separation of harmonic structures based on tied Gaussian mixture model and information criterion for concurrent sounds. In *International Conference on Acoustics, Speech, and Signal Processing,* volume IV, pp. 297–300, 2004.
55. H. Kameoka, T. Nishimoto, and S. Sagayama. Audio stream segregation of multi-pitch music signal based on time-space clustering using gaussian kernel 2-dimensional model. In *International Conference on Acoustics, Speech, and Signal Processing,* volume III, pp. 5–8, 2005.
56. M. Karjalainen and T. Tolonen. Multi-pitch and periodicity analysis model for sound separation and auditory scene analysis. In *International Conference on Acoustics, Speech, and Signal Processing,* pp. 929–932, 1999.
57. K. Kashino and S. J. Godsill. Bayesian estimation of simultaneous musical notes based on frequency domain modelling. In *International Conference on Acoustics, Speech, and Signal Processing,* volume IV, pp. 305–308, 2004.
58. K. Kashino, K. Nakadai, T. Kinoshita, and H. Tanaka. Organization of hierarchical perceptual sounds: Music scene analysis with autonomous processing modules and a quantitative information integration mechanism. In *International Joint Conference on Artificial Intelligence,* volume 1, pp. 158–164, 1995.
59. K. Kashino, K. Nakadai, T. Kinoshita, and H Tanaka. Application of the bayesian probability network to music scene analysis. In D. F. Rosenthal and H. G. Okuno, editors, *Computational Auditory Scene Analysis,* pp. 115–137. Lawrence Erlbaum, Mahwah, NJ, 1998.
60. M. Kashino and T. Hirahara. How many concurrent talkers can we hear out? In *Acoustical Society of Japan Autumn meeting* (in Japanese), pp. 467–468, 1995.
61. H. Kawahara, A. de Cheveigné, and R. D. Patterson. An instantaneous-frequency-based pitch extraction method for high quality speech transformation: Revised tempo in the straight-suite. In *International Conference on Spoken Language Processing,* 1998.
62. A. Klapuri. Multiple fundamental frequency estimation based on harmonicity and spectral smoothness. *IEEE Transactions on Speech and Audio Processing,* 11:804–816, 2003.
63. A. Klapuri. Multipitch estimation and sound separation by the spectral smoothness principle. In *International Conference on Acoustics, Speech, and Signal Processing,* 2001.
64. A. Klapuri. *Signal Processing Methods for the Automatic Transcription of Music.* PhD thesis, Tampere, 2002.
65. A. Klapuri. Auditory-model based methods for multiple fundamental frequency estimation. In A. Klapuri and M. Davy, editors, *Signal Processing Methods for Music Transcription,* Springer-Verlag, New York, 2005.
66. A. Klapuri. A perceptually-motivated multiple-F0 estimation method. In *IEEE Workshop on Applications of Signal Processing to Audio and Acoustics,* New Palz, NY, 2005.

67. A. Klapuri, T. Virtanen, and J.-M. Holm. Robust multipitch estimation for the analysis and manipulation of polyphonic musical signals. In *COST-G6 Conference on Digital Audio Effects,* Verona, Italy, 2000.

68. N. Kunieda, T. Shimamura, and J. Suzuki. Robust method of measurement of fundamental frequency by ACLOS—Autocorrelation of log spectrum. In *International Conference on Acoustics, Speech, and Signal Processing,* 1996.

69. T. I. Laakso, V. Välimäki, M. Karjalainen, and U. K. Laine. Splitting the unit delay tools for fractional delay filter design. *IEEE Signal Processing Magazine,* 13:30–60, 1996.

70. M. Lahat, R. J. Niederjohn, and D. A. Krubsack. A spectral autocorrelation method for measurement of the fundamental frequency of noise-corrupted speech. *IEEE Transactions on Acoustics, Speech and Signal Processing,* 35:741–750, 1987.

71. R. J. Leistikow, H. D. Thornburg, J. O. Smith III, and J. Berger. Bayesian identification of closely-spaced chords from single-fram stft peaks. In *Digital Audio Effects,* pp. 5–8, 2004.

72. Y. Li and D. Wang. Detecting pitch of singing voice in polyphonic audio. In *International Conference on Acoustics, Speech, and Signal Processing,* volume III, pp. 17–20, 2005.

73. J. C. R. Licklider. A duplex theory of pitch perception. *Experientia,* 7:128–134, 1951.

74. R. Lyon. Computational models of neural auditory processing. In *IEEE International Conference on Acoustics, Speech, and Signal Processing,* 1984.

75. R. F. Lyon. A computational model of binaural localization and separation. In W. Richards, editor, *Natural Computation,* pp. 319–327. MIT Press, Cambridge, MA, 1983–1988. Reprinted from *Proceedings of International Conference on Acoustics, Speech, and Signal Processing* 83:1148–1151.

76. K. D. Martin. Automatic transcription of simple polyphonic music: Robust front end processing. Technical Report 399, MIT Media Laboratory Perceptual Computing Section, 1996.

77. K. D. Martin. A blackboard system for automatic transcription of simple polyphonic music. Technical Report 385, MIT Media Laboratory Perceptual Computing, 1996.

78. P. Martin. Comparison of pitch detection by cepstrum and spectral comb analysis. In *IEEE International Conference on Acoustics, Speech, and Signal Processing,* pp. 180–183, 1982.

79. R. Meddis and M. J. Hewitt. Virtual pitch and phase sensitivity of a computer model of the auditory periphery. I: Pitch identification. *Journal of the Acoustical Society of America,* 89:2866–2882, 1991.

80. R. Meddis and M. J. Hewitt. Modeling the identification of concurrent vowels with different fundamental frequencies. *Journal of the Acoustical Society of America,* 91:233–245, 1992.

81. D. K. Mellinger. *Event Formation and Separation in Musical Sound.* PhD thesis, Stanford Center for Computer Research in Music and Acoustics, 1991.

82. B. C. J. Moore and B. R. Glasberg. Suggested formulae for calculating auditory-filter bandwidths and excitation patterns. *Journal of the Acoustical Society of America,* 74:750–753, 1983.

83. T. Nakatani and T. Irino. Robust and accurate fundamental frequency estimation based

on dominant harmonic components. *Journal of the Acoustical Society of America,* 116:3690–3700, 2004.

84. T. Nakatani, H. G. Okuno, and T. Kawabata. Residue-driven architecture for computational auditory scene analysis. In *International Joint Conference on Artificial Intelligence,* volume 1, pp. 165–172, 1995.
85. A. M. Noll. Cepstrum pitch determination. *Journal of the Acoustical Society of America,* 41:293–309, 1967.
86. L. I. Ortiz-Berenguer and Casajús-Quirós. Polyphonic transcription using piano modeling for spectral pattern recognition. In *Digital Audio Effects,* pp. 45–50, Hamburg, Germany, 2002.
87. A. Oxenham and C. J. Plack. *Pitch—Neural Coding and Perception.* Springer-Verlag, New York, 2005.
88. R. P. Paiva, T. Mendes, and A. Cardoso. On the detection of melody notes in polyphonic audio. In *International Symposium on Music Information Retrieval,* 2005.
89. T. W. Parsons. Separation of speech from interfering speech by means of harmonic selection. *Journal of the Acoustical Society of America,* 60:911–918, 1976.
90. M. D. Plumbley, S. A. Abdallah, J. P. Bello, M. E. Davies, G. Monti, and M. B. Sandler. Automatic music transcription and audio source separation. *Cybernetics and Systems,* 33:603–627, 2002.
91. D. Pressnitzer and D. Gnansia. Real-time auditory models. In *International Computer Music Conference,* Barcelona, Spain, 2005.
92. D. Pressnitzer, R. D. Patterson, and K. Krumbholz. The lower limit of melodic pitch. *Journal of the Acoustical Society of America,* 109:2074–2084, 2001.
93. L. R. Rabiner. On the use of autocorrelation analysis for pitch detection. *IEEE Transactions on Acoustics, Speech and Signal Processing,* 25:24–33, 1977.
94. C. Raphael. Automatic transcription of piano music. In *International Symposium on Music Information Retrieval,* pp. 15–19, 2002.
95. C. Raphael and J. Stoddard. Harmonic analysis with probabilistic graphical models. *Computer Music Journal,* 28:45–52, 2004.
96. M. J. Ross, H. L. Shaffer, A. Cohen, R. Freudberg, and H. J. Manley. Average magnitude difference function pitch extractor. *IEEE Transactions on Acoustics, Speech and Signal Processing,* 22:353–362, 1974.
97. L. Rossi, G. Girolami, and M. Leca. Identification of polyphonic piano signals. *Acustica,* 83:1077–1084, 1997.
98. J. Rouat, C. Y. Liu, and D. Morissette. A pitch determination and voiced/unvoiced decision algorithm for noisy speech. *Speech Communication,* 21:191–207, 1997.
99. S. Roweis. One-microphone source separation. In *Advances in Neural Information Processing Systems,* pp. 609–616. MIT Press, Cambridge MA, 2000.
100. S. Sagayama, K. Takahashi, H. Kameoka, and T. Nishimoto. Specmurt Analysis: A piano-roll-visualization of polyphonic music signal by deconvolution of log-frequency spectrum. In *SAPA* (ISCA tutorial and research workshop on statistical and perceptual audio processing), Jeju, Korea, 2004.
101. L. K. Saul, L. L. Lee, C. L. Isbel, and Y. LeCun. Real time voice processing with audiovisual feedback: Toward autonomous agents with perfect pitch. In S. Becker, S. Thrun, and K. Obermayer, editors, *Advances in Neural Information Processing Systems, volume 15,* pp. 1205–1212. MIT Press, Cambridge, MA, 2003.

102. L. K. Saul, F. Sha, and D. D. Lee. Statistical signal processing with nonnegativity constraints. In *Eighth European Conference on Speech Communication and Technology*, volume 2, pp. 1001–1004, 2003.
103. M. T. M. Scheffers. *Sifting Vowels.* PhD thesis, University of Gröningen, 1983.
104. M. R. Schroeder. Period histogram and product spectrum: New methods for fundamental-frequency measurement. *Journal of the Acoustical Society of America,* 43:829–834, 1968.
105. C. Semal and L. Demany. The upper limit of musical pitch. *Music Perception,* 8:165–176, 1990.
106. S. Seneff. Pitch and spectral estimation of speech based on auditory synchrony model. In *IEEE International Conference on Acoustics, Speech, and Signal Processing,* pp. 36.2.1–4, 1984.
107. X. Serra. Musical sound modeling with sinusoids plus noise. In C. Roads, S. Pope, A. Picialli, and G. De Poli, editors, *Musical Signal Processing.* Swets & Zeitlinger, 1997.
108. F. Sha, J. A. Burgoyne, and L. K. Saul. Multiband statistical learning for F0 estimation in speech. In *International Conference on Acoustics, Speech, and Signal Processing,* pp. 661–664, 2004.
109. F. Sha and L. K. Saul. Real-time pitch determination of one or more signals by nonnegative matrix factorization. In L. K. Saul, Y. Weiss, and L. Bottou, editors, *Advances in Neural Processing Systems, volume 17.* MIT Press, Cambridge, MA, 2005.
110. M. Slaney. A perceptual pitch detector. In *International Conference on Acoustics, Speech, and Signal Processing,* pp. 357–360, 1990.
111. M. Slaney. An efficient implementation of the Patterson-Holdsworth auditory filter bank. Technical Report 35, Apple Computer, 1993.
112. P. Smaragdis. Discovering auditory objects through non-negativity constraints. In *ISCA Workshop on Statistical and Perceptual Audio Processing,* Jeju, Korea, 2004.
113. A. D. Sterian. *Model-based Segmentation of Time–Frequency Images for Music Transcription.* PhD thesis, The University of Michigan, 1999.
114. R. J. Stubbs and Q. Summerfield. Evaluation of two voice-separation algorithms using normal-hearing and hearing-impaired listeners. *Journal of the Acoustical Society of America,* 84:1236–1249, 1988.
115. R. J. Stubbs and Q. Summerfield. Algorithms for separating the speech of interfering talkers: Evaluations with voiced sentences, and normal-hearing and hearing-impaired listeners. *Journal of the Acoustical Society of America,* 87:359–372, 1990.
116. H. D. Thornburg and R. J. Lestikow. An iterative filterbank approach for extracting sinusoidal parameters from quasi-harmonic sounds. In *IEEE Workshop on Applications of Signal Processing to Audio and Acoustics,* 2003.
117. T. Tolonen and M. Karjalainen. Computationally efficient multipitch analysis model. *IEEE Transactions on Speech and Audio Processing,* 8:708–716, 2000.
118. T. Virtanen. *Audio Signal Modeling with Sinusoids Plus Noise.* MS thesis, Tampere University of Technology, 2000.
119. T. Virtanen. Sound source separation using sparse coding with temporal continuity objective. In *International Computer Music Conference,* pp. 231–234, Singapore, 2003.
120. T. Virtanen. Separation of sound sources by convolutive sparse coding. In *ISCA Tutorial And Research Workshop on Statistical and Perceptual Audio Process—SAPA 2004,* Jeju, Korea, 2004.

121. T. Virtanen and A. Klapuri. Separation of harmonic sounds using multipitch analysis and iterative parameter estimation. In *IEEE Workshop on Applications of Signal Processing to Audio and Acoustics,* pp. 83–86, 2001.

122. T. Virtanen and A. Klapuri. Separation of harmonic sounds using linear models for the overtone series. In *IEEE International Conference on Acoustics, Speech, and Signal Processing,* volume 2, pp. 1757–1760, 2002.

123. H. von Helmholtz. *On the Sensations of Tone* (English translation A. J. Ellis, 1885, 1954). Dover, New York, 1877.

124. P. J. Walmsley, S. Godsill, and P. J. W. Rayner. Bayesian graphical models for polyphonic pitch tracking. In *Diderot Forum,* pp. 1–26, Vienna, 1999.

125. M. Weintraub. *A Theory and Computational Model of Auditory Monaural Sound Separation.* PhD thesis, Stanford, 1985.

126. M. Wu. *Pitch Tracking and Speech Enhancement in Noisy and Reverberant Environments.* PhD thesis, The Ohio State University, 2003.

127. M. Wu, D. Wang, and G. J. Brown. A multi-pitch tracking algorithm for noisy speech. In *IEEE International Conference on Acoustics, Speech, and Signal Processing,* volume I, pp. 369–372, 2002.

128. M. Wu, D. Wang, and G. J. Brown. A multipitch tracking algorithm for noisy speech. *IEEE Transactions on Speech and Audio Processing,* 11:229–241, 2003.

129. C. Yeh, A. Roebel, and X. Rodet. Multiple fundamental frequency estimation of polyphonic music signals. In *International Conference on Acoustics, Speech, and Signal Processing,* volume III, pp. 225–228, 2005.

130. E. Zwicker, G. Flottorp, and S. S. Stevens. Critical band width in loudness summation. *Journal of the Acoustical Society of America,* 29:548–557, 1957.

CHAPTER 3

Feature-Based Speech Segregation

DELIANG WANG

3.1 INTRODUCTION

As discussed in Chapter 1, there are two kinds of auditory scene analysis (ASA), namely, primitive ASA and schema-based ASA [7]. Primitive analysis relies on intrinsic sound attributes such as fundamental frequency and temporal continuity, and is regarded as an innate, bottom-up process. Schema-based analysis, on the other hand, relies on learned patterns (schemas) or trained models, and is regarded as a top-down process. This chapter covers CASA studies that aim at primitive organization, focusing on segregation of a target voice from acoustic interference using auditory features (see Chapter 4 for model-based organization).

Primitive ASA is grounded in the fact that there are properties of the auditory environment that are common to every listener everywhere, regardless of differences in language, culture, or ecology [7]. The auditory system has evolved to exploit such properties of environmental sounds, and developed a general capacity to segregate sound sources by utilizing their acoustic constraints.

Section 1.1 lists a number of acoustic features that contribute to auditory scene analysis. These include proximity in frequency and time, temporal continuity, periodicity (harmonicity), onset and offset, amplitude modulation (AM), frequency modulation (FM), and spatial location. Since the extraction of location cues involves binaural processing, we will not discuss the role of sound location in CASA in this chapter, which concentrates on monaural processing. Chapters 5 and 6 are devoted to sound localization and location-based segregation.

One often marvels at the human ability to segregate and understand speech in a "cocktail party" environment. An interesting question is: Can a listener perform speech segregation in a foreign environment? In other words, is linguistic knowledge required for human speech segregation? Although many experiments have been conducted to examine listeners' ability to perceive a nonnative spoken language in noise (see, e.g., [51]), as far as I know, no experiment has been published that addresses this very question. To get some idea of this, we mixed utterances in different languages and played the mixtures to listeners who did not know the lan-

Computational Auditory Scene Analysis. Edited by DeLiang Wang and Guy J. Brown

guage at all. Specifically, pairs of equally loud sentences in French, German, Hindi, Japanese, Mandarin, and Spanish were created. Informal tests suggest that listeners seem to be *able* to separate and follow a speaker like they do when listening to mixtures in their familiar language. Speaker characteristics such as pitch, time-frequency gaps present in natural speech, and distinct onsets and offsets appear to play a dominant role in our ability to segregate a target voice. For example, it is noticeably easier to separate utterances of opposite sexes, much as in their native language [10]. The point here is that language familiarity does not appear necessary for human speech segregation.

According to the ASA account (see Section 1.1.4), human sound organization takes place in two conceptual stages. The first stage, the so-called segmentation stage, decomposes the auditory scene into a collection of sensory elements (segments) and the second stage groups segments into streams. Furthermore, auditory scene analysis is broadly divided into two processes: a simultaneous process that integrates concurrent sounds and a sequential process that integrates sounds across time. Mimicking human ASA, a feature-based CASA system for speech segregation typically goes through the following stages, as shown in Fig. 3.1. In the first stage, an input mixture of speech and interference is processed by a model that corresponds to auditory peripheral processing, resulting in a peripheral representation like the cochleagram described in Section 1.3. The basic element of the peripheral representation is a time-frequency (T-F) unit corresponding to some measure of activity within a specific filter and time frame. Then auditory feature extraction takes place, leading to a number of feature representations. The next stage performs auditory segmentation, producing a collection of segments, each of which is a contiguous region in a two-dimensional T-F representation of the mixture. On the basis of extracted features and segments, simultaneous grouping is applied and it is followed by sequential grouping. By sequential grouping we mean integration of sounds from the same source that are separated in time. The

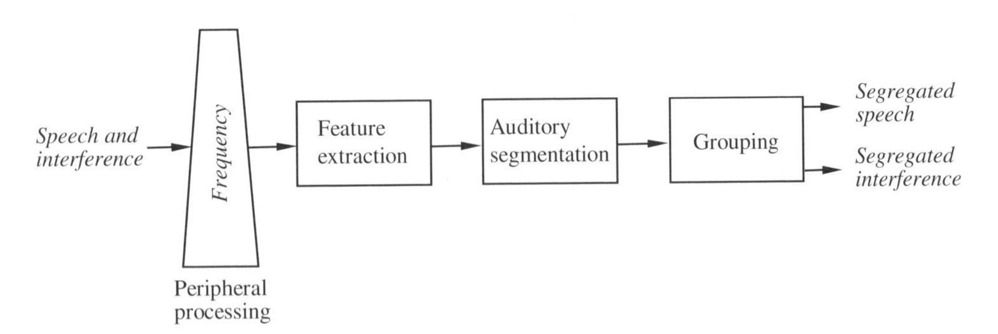

Figure 3.1 Schematic diagram of a typical feature-based CASA system. A mixture of speech and interference is processed in subsequent stages until target speech is finally segregated. The grouping stage includes simultaneous and sequential grouping. Note that the arrows indicate the main flows of processing, and backward flows are certainly possible.

result of grouping is a segregated target speech stream* and segregated interference; the latter may itself be speech utterances.

As will be seen later, a CASA system may not contain every processing stage in Fig. 3.1. In particular, researchers have been preoccupied with simultaneous organization of voiced speech; much less attention has been given to unvoiced speech segregation and sequential grouping. In the following, Section 3.2 discusses auditory feature extraction and Section 3.3 describes auditory segmentation. Sections 3.4 and 3.5 treat simultaneous grouping and sequential grouping, respectively. A concluding discussion is given in Section 3.6.

3.2 FEATURE EXTRACTION

In this section, we describe algorithms for extracting auditory features. In particular, the following features have been studied in the CASA literature: pitch or periodicity, sound onset and offset, amplitude modulation, and frequency modulation. Each is described below.

3.2.1 Pitch Detection

Pitch is the most studied feature in CASA, and pitch extraction has been treated extensively in Chapter 2. We mention it here for completeness. Note that pitch extraction generally occurs in individual time frames. As a result of pitch tracking, one or more pitch values may be extracted within a single time frame corresponding to distinct periodic sound sources. A common representation for both pitch detection and pitch-based grouping is the correlogram described in Chapters 1 and 2. The correlogram representation reveals the periodicity of a signal in individual frequency bands, and the correlogram response within a single T-F unit gives a periodicity pattern for a specific filter at a specific frame. Overall, the correlogram may be viewed as a three-dimensional volume with the three axes of time, frequency, and time lag.

3.2.2 Onset and Offset Detection

An onset of an acoustic event corresponds to a sudden intensity increase. Likewise, an offset corresponds to a sudden intensity decrease. A standard way to identify sudden intensity changes is to take the first-order derivative of intensity with respect to time and then find the peaks and valleys of the derivative. Because of intensity fluctuations within an event itself or caused by background noise, many peaks and valleys of the derivative do not correspond to actual onsets and offsets. To reduce such fluctuations, we can smooth the intensity over time, as is commonly done

*The term "stream" was originally used in psychophysical literature to refer a perceptual representation of an auditory event. It is used here to also refer to a formal representation of a sound source.

in edge detection for image analysis (see also [50]). Smoothing can be performed with a low-pass filter. Specifically, we can convolve the time-varying intensity $s(t)$ with a Gaussian function,

$$G(t, \sigma) = \frac{1}{\sqrt{2\pi}\sigma} \exp\left(-\frac{t^2}{2\sigma^2}\right) \tag{3.1}$$

where σ indicates the Gaussian width. Note that for any functions g and h, we have $(g * h)' = g * h'$, where "$*$" indicates convolution and the prime symbol indicates differentiation. Therefore, we obtain a smoothed and differentiated intensity $O(t)$ that equals $s(t) * G'(t, \sigma)$, where

$$G'(t, \sigma) = \frac{-t}{\sqrt{2\pi}\sigma^3} \exp\left(-\frac{t^2}{2\sigma^2}\right) \tag{3.2}$$

Therefore, onset and offset detection becomes a three-step procedure [28]:

1. Convolve the intensity $s(t)$ with G' to obtain $O(t)$.
2. Identify the peaks and the valleys of $O(t)$.
3. Mark those peaks that are above a certain threshold as onsets and those valleys that are below a certain threshold as offsets.

The thresholding in Step 3 is for the purpose of removing spurious peaks and valleys caused by noise. Note that the Gaussian width, σ, determines the extent of smoothing—a point to be discussed more in Section 3.3.3. The above procedure is similar to the standard Canny edge detector in image processing [11]. Figure 3.2 shows an example of onset and offset detection by the above procedure.

A number of studies have attempted to extract sound onsets and offsets. Mellinger [39] proposed an onset kernel consisting of a positive side preceded by a negative side. Brown and Cooke [8] computed onset and offset maps as part of their emphasis on auditory representations for CASA. Their onset and offset detection is computed on the response envelope (see Section 3.2.3 for envelope extraction) of a hair cell model through a combination of excitatory and inhibitory interaction akin to Mellinger's kernel. In a neural model of detecting amplitude transients, Fishbach et al. [17] proposed to combine delay kernels and a temporal receptive field to find sudden amplitude changes. The receptive field consists of an excitatory component with short delays and an inhibitory component with longer delays, mathematically equivalent to G'. Instead of convolving with G', their model uses G' as connection weights to a neuron model. Although the model of Fishbach et al. may result in something similar to the above three-step algorithm, the complex details such as neuronal refractory period and spike generation make it difficult to understand precisely what the model computes. Smith and Fraser [47] studied a similar neural model for onset detection. Their onset detection model employs the mechanism of synaptic depression, caused by dynamic changes in neurotransmitter levels. As a result, their model

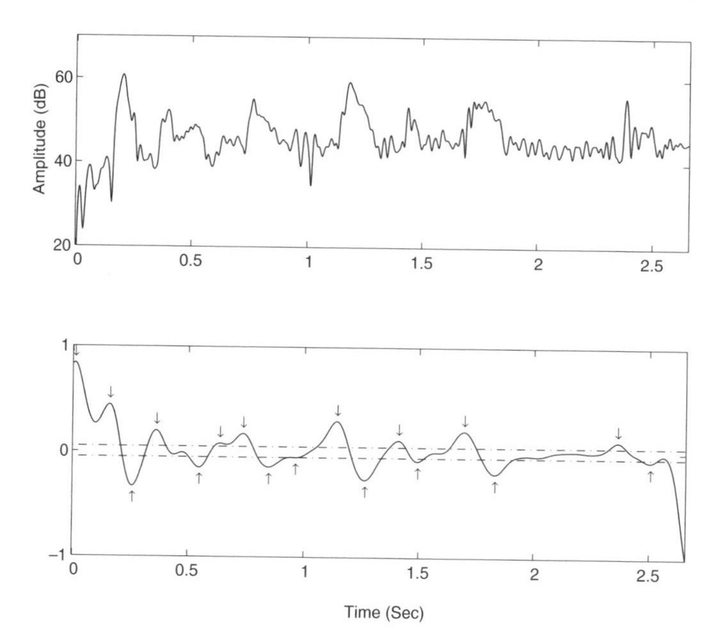

Figure 3.2 Onset and offset detection. The input is the response of a gammatone filter to a mixture of speech and crowd noise. The center frequency of the filter is 560 Hz. The upper panel shows the response intensity, and the lower panel shows the results of onset and offset detection where $\sigma = 16$ (corresponding to 40 ms). The threshold for onset detection is 0.05 and for offset detection it is –0.05, indicated by the dashed lines. Detected onsets are indicated by downward arrows and offsets by upward arrows.

responds to a sudden increase of sound energy but not sustained activity. The above three-step procedure for detecting onsets and offsets was proposed in the context of auditory segmentation (see Section 3.3) by Hu and Wang [28].

3.2.3 Amplitude Modulation Extraction

A steady tone has a constant amplitude or envelope. When two steady tones of close frequencies are input to an auditory filter whose passband includes the frequencies, the amplitude of the filter response becomes modulated as a result of the acoustic interaction between the two sinusoidal waves, and the amplitude modulation has a frequency equal to the difference between the two tone frequencies [25] (see also Figure 3.3).

The extraction of the AM (or the envelope) of a signal is a common operation in signal processing. For meaningful AM extraction, the amplitude of the signal is assumed to change at much slower rates than the carrier frequency. To introduce the AM concept, let us write a signal $x(t)$ with a finite number of components as a sum of cosine waves:

$$x(t) = \sum_{k=1}^{K} C_k \cos(\omega_k t + \phi_k) \tag{3.3}$$

where the amplitude C_k is a positive real number, and ω_k and ϕ_k are the angular frequency and the phase, respectively. The envelope of $x(t)$ is given by a function called the analytic signal, denoted as $\tilde{x}(t)$:

$$\tilde{x}(t) = \sum_{k=1}^{K} C_k\, e^{i(\omega_k t + \phi_k)} \tag{3.4}$$

In other words, the analytic signal is obtained by replacing each cosine component in the signal by a complex exponential. Note that the analytic signal is complex. According to the Euler formula,

$$\cos(\omega_k t + \phi_k) = \frac{e^{i(\omega_k t + \phi_k)} + e^{-i(\omega_k t + \phi_k)}}{2}$$

one can interpret the analytic signal as omitting the negative frequency terms from the real signal and then multiplying by two. The envelope of $x(t)$, $E(t)$, is then given by [24]

$$E(t) = |\tilde{x}(t)| \tag{3.5}$$

We now express an arbitrary real signal $x(t)$ as the inverse Fourier transform of its Fourier transform $X(\omega)$,

$$x(t) = \frac{1}{2\pi} \int_{-\infty}^{\infty} e^{i\omega t} X(\omega)\, d\omega \tag{3.6}$$

Its analytic signal $\tilde{x}(t)$ is given by removing the negative frequencies and multiplying by 2:

$$\tilde{x}(t) = \frac{1}{\pi} \int_{0}^{\infty} e^{i\omega t} X(\omega)\, d\omega \tag{3.7}$$

The above equation can be further written by introducing the theta function,

$$\tilde{x}(t) = \frac{1}{\pi} \int_{-\infty}^{\infty} e^{i\omega t} X(\omega)\theta(\omega)\, d\omega \tag{3.8}$$

where

$$\theta(\omega) = \begin{cases} 0 & \omega < 0 \\ 1/2 & \omega = 0 \\ 1 & \omega > 0 \end{cases}$$

The introduction of the theta function in Eq. 3.8 gives a natural link between the analytic signal and the Hilbert transform of the real signal $H[x(t)]$,

$$\tilde{x}(t) = x(t) + iH[x(t)] \tag{3.9}$$

where $H[x(t)] = x(t) * 1/\pi t$ [24]. Note that with the help of the Hilbert transform, the above expression of the analytic signal is entirely in the time domain.

Given Eq. 3.9, the envelope can be simply computed by taking the absolute value of the analytic signal. Computationally, however, finding the analytic signal directly from Eq. 3.9 leads to slow convergence and the result is often unreliable. With the efficiency of the fast Fourier transform (FFT), the common approach to compute the envelope is the following procedure:

1. Take the FFT of $x(t)$.
2. Set the negative frequency components to zero, and then take the inverse FFT.
3. Take the absolute value and multiply it by two.

This procedure results from Eq. 3.7 directly. Given the equivalent expressions of Eqs. 3.7 and 3.9 for the analytic signal, the procedure also gives a reliable way to compute the Hilbert transform. The above procedure to compute the envelope is frequently referred to as the Hilbert transform method, although no Hilbert transform is actually taken. Figure 3.3 illustrates envelope extraction using this method.

Another commonly used method to extract the envelope of a signal is to first perform half-wave rectification (setting negative signal values to zero), and then apply lowpass filtering of the rectified signal. This simple method amounts to an approxi-

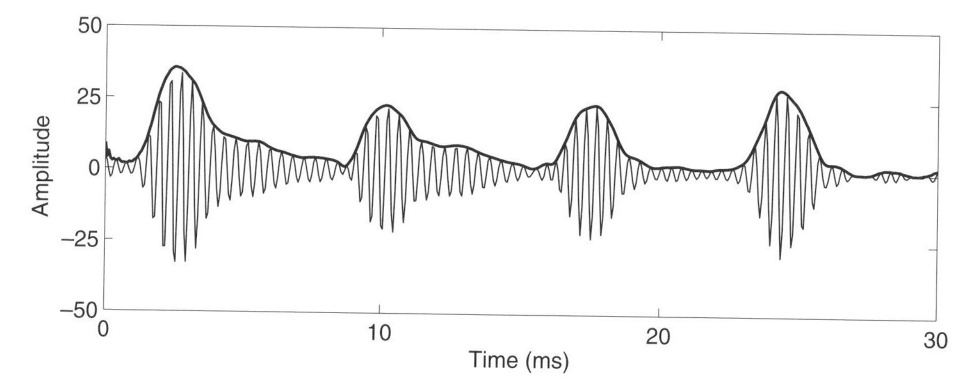

Figure 3.3 Envelope extraction via the Hilbert transform. The input is the response of a gammatone filter to clean speech. The center frequency of the filter is 2.6 kHz. The extracted envelope is shown as the thick curve.

mation of the Hilbert transform method, with the assurance that high-frequency components will be removed.

Taking the AM feature a step further, the amplitude modulation spectrum (AMS) has been advocated as a useful representation for speech separation [32, 5, 6, 3]. The introduction of the AMS is motivated by the biological evidence supporting a potentially separate modulation frequency dimension [35], and the observation that AM in the frequency range of pitch and syllabic variation provides long-term information for the representation of speech that is robust to background noise and room reverberation [23] (see also Section 7.4.1). The AMS is the spectral representation of the signal envelope. Typically, a signal (e.g., speech) is first decomposed by a filterbank or FFT. Then we extract the envelope of each filter response, which is a time signal. A subsequent Fourier transform gives the modulation spectrum of the filter channel. The resulting AMS is a two-dimensional representation in which the ordinate indicates the standard frequency dimension and the abscissa the modulation frequency dimension.

The idea of using AMS for sound separation has been suggested by Kollmeier and Koch [32], Berthommier and Meyer [5], and Atlas [3]. Voiced speech produces intense activity in the AMS near the $F0$ frequency and its multiples along the modulation dimension and near formant frequencies along the frequency dimension. It is interesting to contrast the AMS with the correlogram for representing periodicity information. In the former, the activity related to $F0$ is distributed across the modulation frequency axis, whereas in the latter the activity is centered on the same period. If the response to target speech is well separated from that to interference in the AMS, in principle one can separate voiced speech from background interference. However, because of the low resolution of the AMS and the fact that the pitch of the same speaker varies within a considerable range, it is unclear whether the AMS can offer any advantage over standard pitch-based segregation.

3.2.4 Frequency Modulation Detection

In a way similar to amplitude modulation, the carrier frequency can also be varied at FM rates much slower than the carrier frequency itself. Whereas FM analysis in signal processing generally deals with periodic or near-periodic modulation, FM as a CASA feature refers to a change in frequency of a sound component. Such frequency transitions can be detected in two ways. The first is the detection of a spatial contour on a two-dimensional cochleagram corresponding to an FM pattern and the second is the FM detection from the response of a bandpass filter. Each is described below.

The detection of a spatial contour on a cochleagram can be performed by convolution with a set of two-dimensional time-frequency kernels, which are assumed to correspond to T-F receptive fields in the auditory system. Specifically, we can choose a set of Laplacian-of-Gaussian filters commonly used in computational vision for edge detection. Let us start with a two-dimensional Gaussian function

$$G(t, f, \sigma_t, \sigma_f) = \frac{1}{2\pi\sigma_t\sigma_f} e^{-[(t^2/2\sigma_t^2)+(f^2/2\sigma_f^2)]} \tag{3.10}$$

where σ_t and σ_f are the Gaussian widths along the time (horizontal) and frequency (vertical) directions, respectively. We choose $\sigma_t > \sigma_f$ for an elongated horizontal orientation. To detect a frequency change, Riley [45] suggests that the Laplacian operation should be applied along the frequency dimension. Hence, the FM operator becomes

$$FM(t, f) = -\frac{\partial^2}{\partial f^2} G(t, f, \sigma_t, \sigma_f) \tag{3.11}$$

Such a kernel function shows a center-surround structure with a positive center lobe oriented along the time axis flanked by two negative lobes with the same orientation, as shown in Fig. 3.4. Other orientations can be obtained by rotating the coordinate system. Such kernel functions have been used by Mellinger [39] and Brown and Cooke [8] to compute frequency modulation maps from the cochleagram.

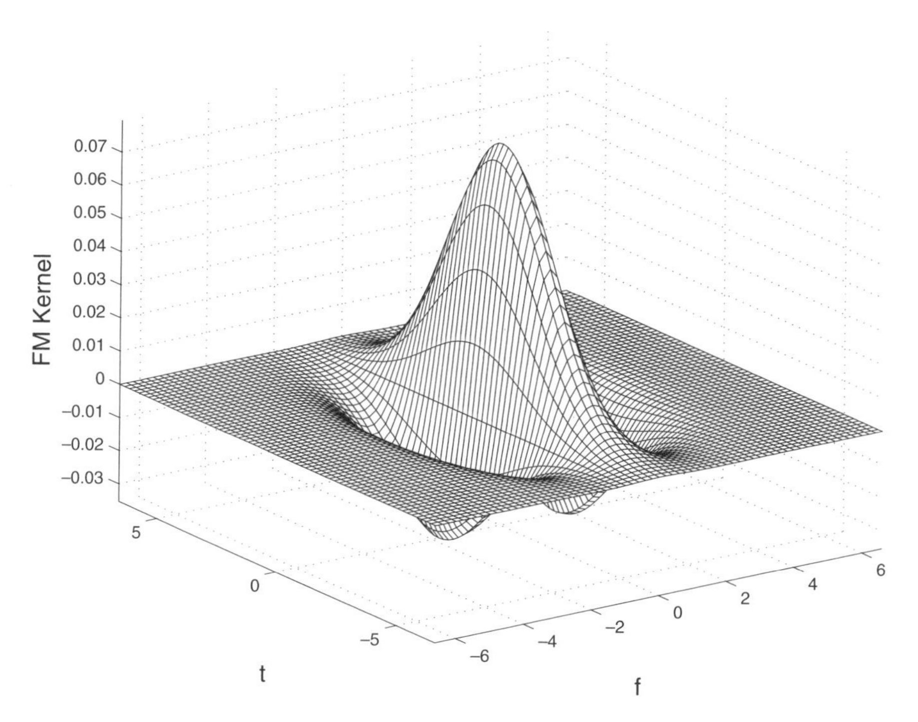

Figure 3.4 FM kernel. A two-dimensional kernel function defined in Eq. 3.11 is displayed with $\sigma_t = 2$ and $\sigma_f = 1$.

Within a single frequency channel, FM can be computed from the instantaneous frequency (IF) of the response. The IF of a real signal is defined as the time derivative of the instantaneous phase of its corresponding analytic signal. The IF computed in this way is measured in radians, and it needs to be divided by 2π to convert to Hz. Hence, given a channel response $x(t)$, which is a real signal, to compute its IF we first find $\tilde{x}(t)$, and then proceed to find the instantaneous phase of $\tilde{x}(t)$ and its first derivative. However, this direct calculation may lead to unstable IF varying in a range of positive and negative values. A technique introduced by Kumaresan and Rao [33] guarantees to yield positive IF. Their method decomposes an analytic signal into two analytic signals, one of which has a constant envelope and positive IF. As part of the computation, the envelope of the signal is approximated using an approach analogous to linear prediction, which is commonly used for spectral analysis. The criterion to be optimized here is a measure of flatness in the temporal envelope, rather than the spectral flatness measure used in linear prediction analysis in the time domain. As a result, their algorithm is referred to as linear prediction in the spectral domain. When this algorithm is applied to $x(t)$ (a bandpass signal) the resulting IF is not only positive but also smoothly varying. The variation in IF then indicates the frequency modulation within the channel.

3.3 AUDITORY SEGMENTATION

3.3.1 What Is the Goal of Auditory Segmentation?

As discussed in Section 3.1, the task of auditory segmentation is to integrate individual T-F units into segments, each of which forms a contiguous region on a two-dimensional cochleagram. Segments give an intermediate representation between T-F units and streams. The defining property of a segment is that *its underlying acoustic energy primarily originates from the same sound source.* As a result, a segment should not further segregate into parts but can group with other segments to form a stream. Since a T-F unit by itself satisfies the property of indivisibility as far as CASA is concerned, we need to place an additional constraint on segmentation that *adjacent segments either originate from different sound sources or at least have different properties.* So an ideal segment is the *largest* T-F region within which the intensity of one sound is higher than the combined intensity of all other sounds in each T-F unit. Two more qualifications are in order. First, since there are infinite acoustic events occurring simultaneously in the physical world at a given time, we have to limit the definition to an acoustic environment relative to a listener (see Section 1.2.2). In other words, only sounds audible to a listener should be considered; one can use the absolute threshold in a frequency band [40] to determine the audibility of a sound. Second, the same sound source may change its acoustic properties abruptly without a pause in between. For example, coarticulation in speaking may connect speech sounds with very different acoustic–phonetic properties. Hence, for auditory segmentation we should allow a boundary in time within the same stream if the boundary marks an abrupt change in sound properties. The type of target sound can further limit the output of auditory segmentation; for instance, for speech segregation we may only be interested in forming segments that correspond to speech.

Computationally, a segment provides a level of auditory description with which a number of sound attributes may be meaningfully computed, such as onset and offset, amplitude and frequency variation, and periodicity contour. For speech, a segment may correspond to a resolved harmonic, a formant, a region of fricative energy, and so on. Even for features that can be directly computed from individual T-F units, such as interaural time difference (see Chapter 5), feature extraction at the segment level is expected to be more robust because feature estimates at the unit level can be pooled or averaged within the same segment. Hence, the segment representation should facilitate auditory scene analysis. The emphasis on auditory segmentation is a unique aspect of the CASA approach to sound separation, and is consistent with Marr's notion of levels of representation for visual analysis [37].

The above definition of auditory segmentation is intrinsically related to the ideal binary mask as a computational goal of CASA (see Section 1.2.2); the ideal binary mask is simply a collection of ideal segments that belong to the target source. The relationship between auditory masking and the ideal binary mask also indicates that a segment masks the overlapping sounds from other sources.

Most of the CASA models have adopted an intermediate representation akin to that of segments, although detailed definitions of a segment may differ. The above definition implies exclusive allocation of sound energy within individual T-F units, which conforms to the exclusive allocation principle in ASA [7]. As pointed out by Bregman [7], there are exceptions to this principle in auditory perception. Also, there are CASA models that allow intermediate representations to have overlapping T-F regions. For example, Ellis uses periodic elements, transient elements, and noise clouds to represent the auditory scene [16]. The periodic elements describe a wideband periodic signal such as a section of voiced speech. The transient elements describe click-like sounds, and the noise clouds are broadband noises such as wind or unvoiced fricatives. For each type of representational element, he starts with a signal model and then uses it to extract such instances from the input mixture. The grouping part of his CASA model then seeks to interpret the auditory scene in terms of these immediate-level elements, which may overlap in time and frequency.

Other intermediate representations include Mellinger's model, which keeps track of individual frequency partials across time and frequency for musical sounds [39]. In a systematic study of auditory segmentation, Cooke [13] computes a collection of what he called synchrony strands from a model of the auditory periphery. To form a strand, his model first computes dominant frequency components in each frequency channel through IF extraction. Then neighboring channels within a time frame are linked together if they show correlated IF, resulting in a collection of spectral regions. Finally, spectral regions in neighboring time frames are linked together if they form a smooth trajectory.

In the following, we describe in detail two different methods of auditory segmentation. The first one is based on cross-channel correlation and temporal continuity and the second one is based on onset and offset analysis.

3.3.2 Segmentation Based on Cross-Channel Correlation and Temporal Continuity

This method builds on the correlogram or autocorrelation response within a T-F unit. Specifically, let u_{cm} denote a T-F unit of frequency channel c and time frame m, and let $A(c, m, \tau)$ denote its correlogram response, where τ is the autocorrelation lag. Note that in an auditory filterbank, the frequency bands of neighboring filters overlap (see Section 1.3). The extent of the overlap increases as the center frequency increases; in the high-frequency range the overlap becomes substantial. As a result, multiple neighboring channels respond to the same harmonic or formant in general. Since they respond to the same frequency component, their responses are highly correlated; in particular, their response frequencies are locked to the frequency of the acoustic component. Exploiting the fact that the phase of a correlogram response is normalized at the zero lag, the cross-channel correlation—the cross-correlation between neighboring correlogram responses—should indicate the extent to which the corresponding channels respond to the same sound component. To remove the impact of the DC component as well as that of the response amplitude that depends on the distance between the center frequency of the channel and the frequency of the component, $A(c, m, \tau)$ should be normalized (e.g., to have zero mean and unity variance). The normalized response is denoted by $\hat{A}(c, m, \tau)$. The cross-channel correlation between u_{cm} and $u_{c+1,m}$ is given by

$$C(c, m) = \frac{1}{L} \sum_{\tau=0}^{L-1} \hat{A}(c, m, \tau)\hat{A}(c+1, m, \tau) \tag{3.12}$$

where L denotes the maximum lag of the correlogram in sampling steps. Figure 3.5 illustrates cross-channel correlation within a time frame in response to a mixture of speech and trill telephone. Given $C(c, m)$, one can simply link two adjacent T-F units if their cross-channel correlation exceeds a threshold.

To expand segments along the time dimension we can simply link neighboring T-F units in time. In addition, to prevent T-F units with low responses from bridging different segments, an example of undersegmentation, we should consider only T-F units that produce some minimum activity in response to the input. This can be ensured by setting a threshold of response energy, which is given by the correlogram value at the zero lag, that is, $A(c, m, 0)$. The following algorithm summarizes the procedure of segmentation:

1. Select u_{cm} if $A(c, m, 0)$ exceeds a certain threshold.
2. Iteratively expand from a selected T-F unit to its selected neighbors in time, and its selected neighbors in frequency if the cross-channel correlation exceeds a certain threshold. This gives one segment.
3. Repeat Step 2 until all selected units have been considered.

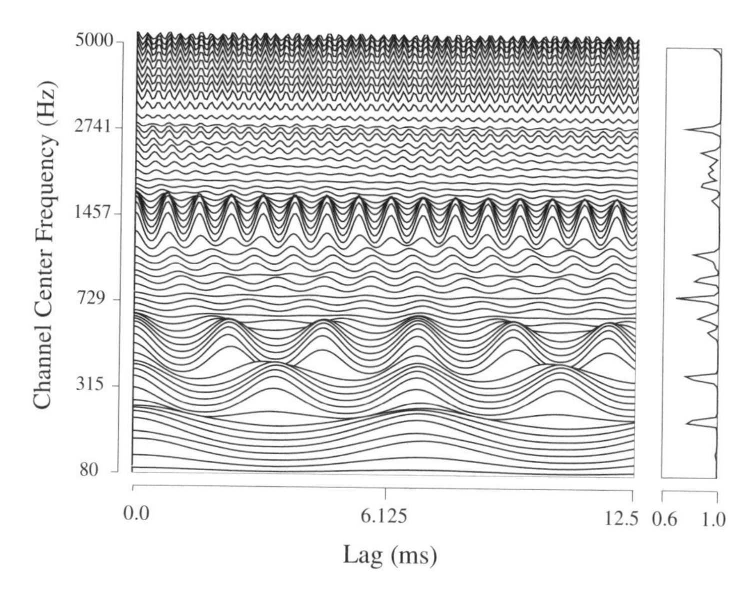

Figure 3.5 Correlogram and cross-channel correlation in response to a sound mixture of speech and telephone (from [52]). The correlogram is displayed within a 20 ms frame for 64 filter channels. The cross-channel correlation in the range of 0.6 and 1.0 is given in the right panel.

Brown and Cooke [8] studied a related method for auditory segmentation. In the frequency domain, their method computes a map with respect to cross-channel correlation and then forms so-called periodicity groups by applying a clustering technique to the cross-channel correlation map. In the time domain, their segmentation method tracks spectral peaks using a frequency transition map. Wang and Brown [52] subsequently proposed the use of the above segmentation algorithm, which is considerably simpler than that of Brown and Cooke. Hu and Wang [29] also used the above algorithm for segmentation, and further suggested that cross-channel correlation should be applied to the correlogram of response envelopes in the high-frequency range rather than original correlogram responses. This suggestion is based on the observation that amplitude modulation in high-frequency filter responses makes response envelope a more robust feature (see Section 3.2.3).

3.3.3 Segmentation Based on Onset and Offset Analysis

The cross-channel correlation method exploits the periodicity of the input sound, and hence is not an appropriate approach to segmenting aperiodic sounds such as unvoiced speech. A different approach to auditory segmentation is to view a segment as comprising an onset front, a steady region, and an offset front. Namely, it treats segmentation as identifying matching pairs of onset and offset fronts; an onset front and an offset front of a segment refer to the starting and the ending edges of the segment, respectively, which are contours along the frequency axis. From the computational standpoint, this perspective is intrinsically related to image segmentation, which identifies the bounding contours of visual objects. An advantage of such segmentation is its applicability to both voiced and unvoiced sounds. Recently, Hu and Wang [28, 30] proposed a segmentation method based on onset and offset analysis. The following description is based on their study, which consists of three stages: smoothing, onset/offset front matching, and multiscale integration. Each is described below.

Smoothing. We have discussed in Section 3.2.2 onset and offset detection within a single frequency channel, in which smoothing serves to reduce spurious onsets and offsets caused by response fluctuation and background noise. Smoothing in the frequency domain is also appropriate since an acoustic event tends to have onset and offset synchrony, and smoothing could help to reveal correlated onsets and offsets in neighboring frequency channels. A systematic way of performing smoothing at various scales is the scale space analysis via diffusion that is commonly used in image segmentation [44, 54]. Specifically, a one-dimensional diffusion of a variable v across a particular dimension x is given by the following partial differential equation:

$$\dot{v} = \frac{\partial}{\partial x}\left[D(v) \cdot \frac{\partial v}{\partial x}\right] \tag{3.13}$$

where $\dot{v}$ denotes the time derivative of v, and D denotes a control function. The above equation describes a diffusion process where the change of v is determined by its x gradient. For certain control functions, v will change in a way such that its x gradient gradually approaches a constant; in other words, v is gradually smoothed over x. The longer the diffusion time (called the scale parameter), the smoother is v. The set of the smoothed v values at different scales forms a scale space.

In a simple case in which $D = 1$, Eq. 3.13 becomes $\dot{v} = \partial^2 v / \partial x^2$. To understand how the equation performs smoothing, let us look at a local minimum point of v where by definition $\partial^2 v / \partial x^2 > 0$. So diffusion will increase the v value at this point. For the same reason, diffusion will decrease the v value at a local maximum point. In fact, in the case of $D = 1$, the diffusion process leads to Gaussian smoothing [54]. That is, $v(x, T) = v(x, 0) * G(0, \sqrt{2T})$ where $v(x, T)$ denotes the v value along x with the scale parameter T.

Because the time and frequency dimensions in a cochleagram have different physical meanings, diffusion should be applied to these two dimensions separately. One can apply diffusion first to the intensity within a frequency channel, and then to the frequency dimension. A critical issue for scale space analysis is how to choose scale parameters and control functions appropriately for different tasks. Too little smoothing does not help to remove insignificant intensity fluctuations, whereas too much smoothing blurs sharp intensity changes that signal onsets and offsets. Smoothing in the time dimension naturally corresponds to lowpass filtering and this connection helps to choose scale values for speech analysis.

Figure 3.6 illustrates the smoothing process in which the input is a mixture of speech and crowd noise, for three different scales. The top panel shows the initial

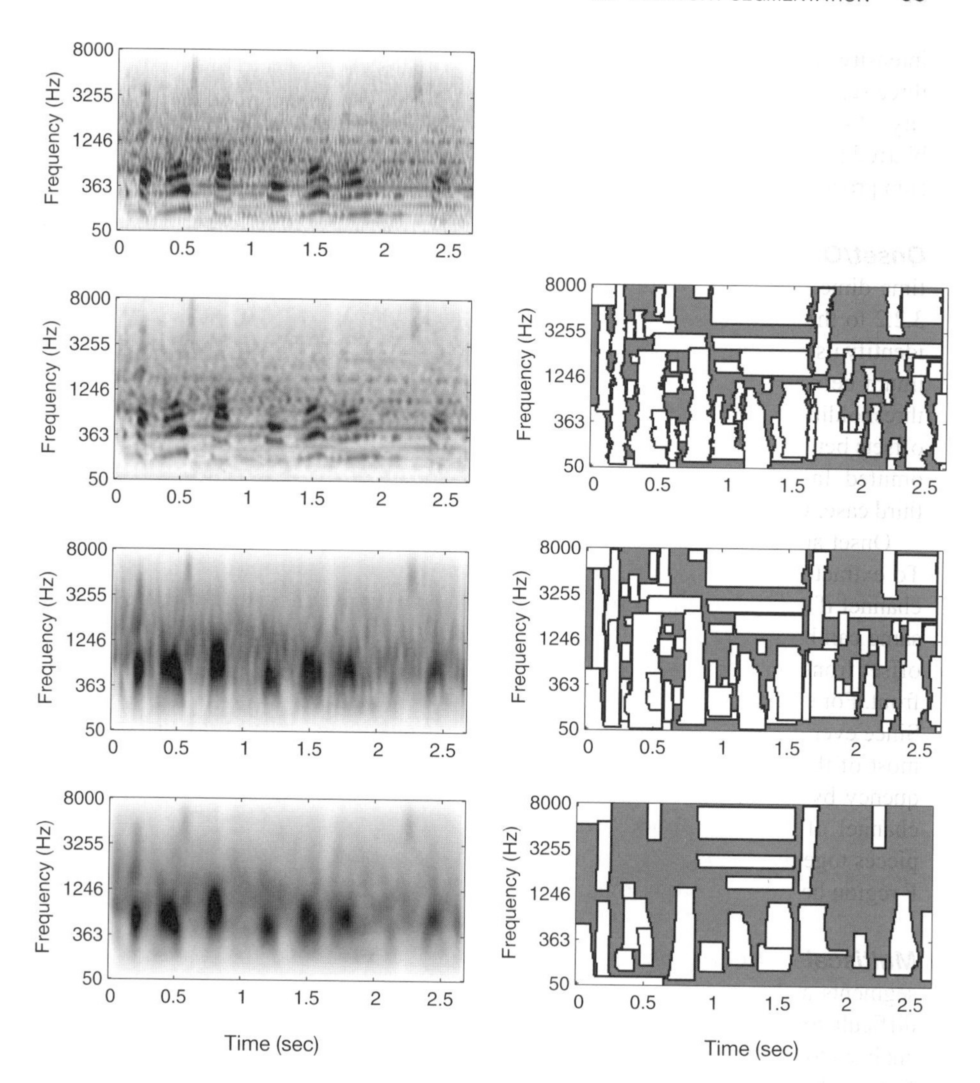

Figure 3.6 Smoothing and multiscale segmentation (from [30]). The input is a mixture of a female utterance and crowd noise. The top-left panel shows the initial intensity for the mixture. Frequency analysis is performed using a gammatone filterbank with 128 channels. The second-left panel shows the smoothed intensity at a fine scale, the third-left panel the smoothed intensity at a coarser scale, and the fourth-left panel the smoothed intensity at the coarsest scale. The first-right panel shows the segments generated from multiscale analysis at all of the above three scales, the second-right panel shows the segments generated from multiscale analysis at the two coarser scales, and the third-right panel shows the segments generated from single-scale analysis at the coarsest scale. On the right side, the background is indicated by gray.

intensity, and the lower three panels on the left side give the smoothed intensities at three scales from fine to coarse. In the figure, darker color indicates stronger intensity. As shown in the figure, local intensity variations become more and more blurred as scale increases but significant changes tend to be preserved in the diffusion process.

Onset/Offset Front Matching. At a given scale (T_f, T_t) along the frequency and time dimensions, we can apply onset and offset detection as described in Section 3.2.2 to the intensity of each frequency channel. For each detected onset, we then identify its matching offset as follows. For any signal there is precisely one local minimum between two consecutive local maxima (see Fig. 3.2). However, due to the thresholding operations in onset/offset detection, there can be zero, one, or multiple offsets between two consecutive onsets. In the first case, the second onset should be omitted. In the second case, the matching offset is the one between the onsets. In the third case, we should choose the offset that has the largest intensity decrease.

Onset and offset fronts correspond to potential bounding contours of segments. To extract an onset front, we merge an onset with its closest onset in an adjacent channel if their onset times are reasonably close in terms of onset synchrony*; the same can be applied to extract an offset front. The next step is to match onset and offset fronts for segmentation. To perform this matching, let us start with an onset front. For each onset along the front, we find its matching offset as discussed above. Since every offset must be on an offset front, we choose the offset front that crosses most of the matching offsets and then label the piece of the front enclosed in frequency by the onset front as the matching piece. This process repeats until every channel of the onset front is matched. Finally, we concatenate all the matching pieces together and they give the matching offset contour of the onset front. The T-F region between an onset front and its matching offset contour yields a segment.

Multiscale Integration. A large smoothing scale is good for extracting large segments and a small scale is good for extracting small segments. Therefore, it is difficult to obtain good segmentation at a single scale. This calls for multiscale analysis to reduce both undersegmentation and oversegmentation. One way to perform multiscale integration is to start segmentation at a large scale and then move to finer scales. The integration serves two purposes. First, it helps to locate more accurate onset and offset positions of an existing segment at a larger scale. Second, it includes new segments produced at a finer scale.

The right side of Figure 3.6 shows the segmentation results of multiscale analysis for the input on the top-left panel. The top-right panel gives the result by integrating the three scales shown on the left side. The middle-right panel gives the result by integrating the two larger scales corresponding to the lower two panels on

*Systematic phase shifts exist in responses of auditory filters across frequency. Such phase shifts should be considered when comparing onset times at different filter channels, although their effects are likely to be small here since the comparison is made between adjacent channels. Section 1.3.2 discusses how to compensate for phase shifts.

the left side, and the bottom-right panel gives the segmentation result at one large scale corresponding to the bottom-left panel. Although the bottom-right panel already captures a majority of speech segments, multiscale analysis produces both large and small segments.

3.4 SIMULTANEOUS GROUPING

Simultaneous grouping aims to group sound components or segments that overlap in time. The result of this stage of processing is a collection of streams, each of which is temporally continuous. We first discuss simultaneous grouping of voiced speech and then of unvoiced speech.

3.4.1 Voiced Speech Segregation

Voiced speech is characterized by the prominent feature of periodicity, which can be used to group the harmonics of a voiced utterance that are scattered in the frequency domain. Most of the CASA studies on simultaneous grouping have focused on voiced speech segregation.

Given the result of pitch tracking (Chapter 2), one can design a comb filter to extract from the mixture spectrum the frequency components at the multiples of the extracted $F0$. The name of the filter comes from the fact that its frequency response resembles a comb, with large values at the $F0$ and its integer multiples. Comb filtering is a commonly used technique in speech enhancement [36, 42]. An accurate pitch estimator is crucial for the performance of comb filtering as it is sensitive to pitch estimation errors. Also, the overlap between the speech spectrum and the interference spectrum presents a thorny issue for comb filtering. In the context of separating two voiced utterances, Parsons [43] developed several techniques to deal with overlapping spectral components. His method includes a stage that detects and resolves overlapping spectral peaks from two talkers. The overlap detection tests each spectral peak for its symmetry, its distance from adjacent peaks, and its phase pattern. Once a peak is determined to be an overlapping one, his method decomposes the peak into two component peaks by fitting the overlapping peak with an ideal peak of the same height and then fitting the difference with another ideal peak. If the two component peaks comprising an overlapping peak are too close to be resolvable, his method uses interpolation between adjacent harmonic peaks of the same speaker to allocate the overlapping peak to the two speakers.

Weintraub's CASA model [55] was designed to separate the simultaneous speech of a male speaker and a female speaker, whose distinct pitch ranges are used to perform sequential grouping (see Section 3.5). Once a pitch track is computed for a section of voiced speech, the model performs spectral amplitude estimation to separate a continuous stream. Specifically, within each frequency channel his model records a histogram of amplitude ratios between the two speakers from a training database of premixing utterances. Given a detected $F0$ at a time frame, the model first generates an initial estimate of the spectral amplitude using the detected $F0$ and trained histograms of correlogram responses at various $F0$ frequencies. After the initial estimate is determined, the model applies a complex algorithm that iteratively modifies amplitude estimates in order to satisfy the spectral continuity constraint within a time frame. This is done by locally searching for amplitude ratios within individual T-F units so as to maximally match spectral continuity and the $F0$ frequency. The iterative algorithm is not guaranteed to converge, so the algorithm must be stopped after a certain number of iterations.

A different approach was developed by Cooke [13]. After synchrony strands are formed from the input scene, Cooke's model performs simultaneous grouping in two stages. In the first stage, strands are organized into groups using the harmonicity principle and the AM principle separately. Specifically, his model starts with a seed component, which is the most dominant strand in the set of unorganized strands. Each seed then groups other strands that are harmonically related to the seed. A separate group is formed by grouping strands that have AM patterns similar to that of the seed. The first stage operates iteratively until all strands are organized into groups. In the second stage, two operations are performed. The first operation removes groups that are subsets of larger groups. In the second operation, a pitch contour is computed within each group of strands and then groups that have similar pitch contours are combined. Note that a pitch contour is a local property of a strand group, which should be distinguished from a pitch track of a sound source. The two-stage organization produces a number of continuous streams, each being a collection of strand groups. One of the segregated streams would correspond to a target source.

In the following, we describe two representative models for voiced speech segregation in some detail. The first is a model developed by Brown and Cooke [8] that emphasizes physiologically motivated representations. The second is a recent model by Hu and Wang [29] that exploits AM analysis to deal with unresolved harmonics and explicitly estimates the ideal binary mask.

Brown and Cooke Model. The model of Brown and Cooke [8] is a multistage CASA model for organization of voiced speech. A prominent characteristic of their model is that its components are justified in terms of physiological and psychoacoustic findings. After peripheral processing leading to the cochleagram, the model computes a number of auditory maps corresponding to FM, pitch, and onset/offset analysis as described in Section 3.2. The FM map together with cross-channel correlation provides the basis for segmentation. Specifically, each segment corresponds in time to a smooth track of spectral peaks on the FM map, and in frequency to a region of high cross-channel correlation (see Section 3.3.2).

With the auditory scene represented as a collection of T-F segments, simultaneous grouping in the Brown and Cooke model is based on pitch contour similarity as well as common onset and offset. The idea of grouping according to similarity of pitch contours was first proposed by Cooke [13]. However, the implementation in Brown and Cooke [8] is quite different from that in Cooke. Given a segment, the Brown and Cooke model first summates the correlogram responses of the segment within each time frame. Given the summated response at each frame, one could simply identify a segment pitch by finding the highest peak in the response much like correlogram-based pitch detection. However, this frame-by-frame detection does not take advantage of the temporal continuity of a pitch contour. Instead, their model applies a dynamic programming algorithm to identify a pitch contour over the duration of the segment, which implicitly enforces the continuity of pitch variation. Once a pitch contour is computed from a segment, we can quantify pitch contour similarity between two segments that overlap in time.

Let two pitch contours be denoted by $p_1(m)$ and $p_2(m)$, where m indicates time frame. Then the similarity between p_1 and p_2 is given by

$$\frac{1}{m_2 - m_1 + 1} \sum_{m=m_1}^{m_2} \exp\left\{-\frac{[p_1(m) - p_2(m)]^2}{2\sigma_p^2}\right\} \tag{3.14}$$

where m_1 and m_2 define the first and the last frames of the overlapping duration between p_1 and p_2, and σ_p is the Gaussian width that determines the amount of tolerance in the similarity measure. The similarity varies between 1 for identical pitch contours and 0 for very different ones.

Grouping according to common onset is realized as follows. First, the model checks whether the start of the segment triggers a corresponding onset on the onset map. If two segments both triggering an onset start at close times their similarity score is increased by a certain amount. The same operation is applied to offset. As a result, the similarity score between two segments is increased by a common onset and a common offset. The amount of increase is determined by two considerations. First, two segments that have both common onset and common offset will be grouped together regardless of their similarity in pitch contour. Second, although it enhances chances of grouping, a common onset or a common offset does not by itself provide sufficient evidence for grouping.

The above similarity measure applies to pairs of segments. For scene-level grouping, Brown and Cooke proposed an iterative algorithm that first selects the longest segment to start a new stream. Then each segment not in any stream is compared with every segment in the new stream. A segment is only added to the stream if its pairwise similarity score with every segment in the stream exceeds a threshold value. The iteration stops when the stream cannot recruit any more segments. The process of generating new streams repeats as long as there are segments remaining in the auditory scene.

Figure 3.7 illustrates the performance of the Brown and Cooke model on a mixture of voiced speech and a siren sound. The left side of the figure shows the segments and their pitch contours for the mixture. Two segregated streams with their sets of pitch contours are shown in the middle and the right side, corresponding to the segregated speech and the segregated siren, respectively. Systematic evaluation suggests that grouping by common onset and offset in their system contributes to only a small performance improvement.

Hu and Wang Model. While earlier CASA models perform well in the low-frequency range where harmonics are resolved, their performance is limited in the

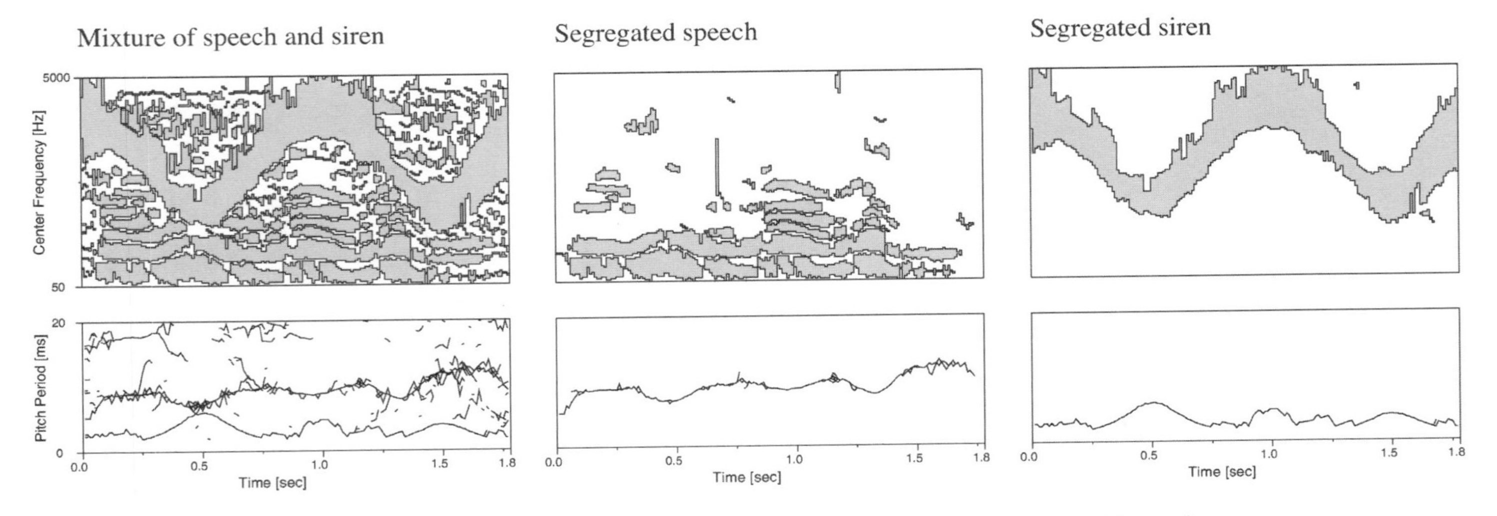

Figure 3.7 Voiced speech segregation by the Brown and Cooke model (Reprinted with permission by Elsevier from [8]). The top panels show segments and the bottom panels show their corresponding pitch contours. The left side shows a mixture of voiced speech and a synthetic siren, the middle a group of segments corresponding to segregated speech, and the right side a group of segments corresponding to a segregated siren.

high-frequency range where harmonics are unresolved. In particular, most of the target speech energy is lost as a result of a conservative strategy that includes a segment only when there is strong evidence to support grouping. Partly motivated by the psychophysical evidence that the human auditory system uses different mechanisms to deal with resolved and unresolved harmonics, Hu and Wang proposed a CASA model that groups resolved and unresolved harmonics differently [29]. Another novelty of the model is the use of a pitch track corresponding to target speech for simultaneous grouping.

The Hu and Wang model first segments the auditory scene using cross-channel correlation and temporal continuity as described in Section 3.3.2. It then performs initial grouping using the dominant pitch computed within each time frame of the scene. Initial grouping is done using the oscillatory correlation model of Wang and Brown [52] (see Chapter 10 for the detailed description of this model). From the algorithmic point of view, grouping in the Wang and Brown model takes place according to the similarity between the periodicity pattern (correlogram response) of a T-F unit and the dominant pitch of each frame (see Section 10.5 for more discussion). The result of the initial grouping is used to estimate a target pitch track.

Given the target pitch track, we label all T-F units within the duration of the track. Unit labeling is done using a periodicity criterion in the low-frequency range and an AM criterion in the high-frequency range. Specifically, for T-F unit u_{cm} the periodicity criterion labels it as a unit of target speech if the ratio $A[c, m, \tau_S(m)]/A[c, m, \tau_M(m)]$ is greater than a certain threshold, where τ_S denotes the lag period corresponding to the target pitch and τ_M indicates the lag that gives the maximum value of A over the plausible pitch range of speech (from 80 Hz to 500 Hz). Note that a high ratio suggests that the correlogram response is compatible with the target pitch period. A segment is then grouped to the target stream if most of its T-F units are labeled as belonging to the stream according to the periodicity criterion.

The motivation for using an AM criterion is the observation that, for a high-frequency filter responding to multiple harmonics of a periodic source, the filter response is amplitude modulated and the response envelope fluctuates at the $F0$ rate of the source [25] (see also Section 3.2.3). Figure 1.7 gives an illustration for a vowel stimulus (see also Fig. 3.3). To apply the AM criterion, the response of channel c is half-wave rectified and bandpass filtered to extract the $F0$ component corresponding to the target pitch. The filtered signal is normalized by its envelope to remove intensity fluctuations. We then model the normalized signal within frame m by the best-fitting sinusoid of the target pitch period. The fitting error of this sinusoidal model is given by

$$\frac{1}{M}\sum_{n=0}^{M-1}\left[\hat{r}(c,n)-\cos\left(\frac{2\pi n}{\tau_S(m)f_s}+\phi_{cm}\right)\right]^2 \tag{3.15}$$

where M denotes the frame length in sampling points. The normalized filter response is denoted by $\hat{r}$, whose samples within frame m are indexed by n. f_s indicates the sampling frequency. ϕ_{cm} denotes the optimal phase for the sinusoidal model and it can be analytically solved. u_{cm} is labeled by the AM criterion as belonging to target speech if the fitting error in Eq. 3.15 is less than a certain threshold. AM-based unit labeling is applied after periodicity-based labeling.

T-F units labeled by the AM criterion are organized into new segments. The Hu–Wang model includes a final grouping stage in which AM-labeled segments are grouped into the target stream. Furthermore, if a sizeable region within a target segment is unlabeled (which is only possible for the segments grouped by the periodicity criterion), it is removed from the target stream. Figure 3.8 shows the result of segregating a mixture of voiced speech and "cocktail-party" noise. The top-left panel shows the cochleagram of the mixture, and the top-right that of the target utterance. The segregation result is shown as the binary mask in the middle-left panel, where white indicates 1. The segregated target, shown in the lower-left panel, corresponds to the masked mixture using the binary mask. For comparison, the middle-right panel shows the ideal binary mask and the lower-right panel the ideally masked mixture.

3.4.2 Unvoiced Speech Segregation

Unvoiced speech lacks periodicity, which is the primary cue for voiced speech segregation. In addition, the sound energy of unvoiced speech is generally much weaker than that of voiced speech. As a result, segregation of unvoiced speech is particularly difficult. Before discussing recent attempts to solve this problem, let us first develop some understanding of unvoiced speech. Natural speech sounds consist of vowels and consonants. Vowels are all voiced, but consonants can be either voiced or unvoiced. There are three categories of unvoiced consonants: stops (e.g., /b/), fricatives (e.g., /f/), and affricates (e.g., /tʃ/). An affricate is a stop followed by a fricative [34], so stops and fricatives are the two main categories of unvoiced speech.

An extensive analysis by Dewey [15] finds that unvoiced consonants account for 21% of all phonemes used in written English. What about spoken English? For conversations over the telephone, a similar analysis by French, Carter, and Koenig [19] puts the use of unvoiced consonants at about 24% (see also [18], p. 92). In terms of read speech, we have performed a recent analysis [53] using the phonetically labeled data in the TIMIT corpus composed of a large number of read sentences representing major dialect regions in the United States [20]. Our analysis shows that unvoiced phonemes account for 23.1%. Another related question is the time proportion of unvoiced speech in speaking. Unfortunately, the reported data on telephone conversations do not provide durational information. Our analysis using the TIMIT corpus as well as conversational speech data with durations derived from a transcribed subset of the switchboard corpus [22] indicates that unvoiced speech accounts for about 26% in terms of time duration [53].

Because of the prominence of the periodicity cue, a sensible strategy for general speech segregation is to first separate voiced speech and then unvoiced speech. An advantage of this strategy is that segregated voiced speech may provide useful anchors for handling unvoiced sections. In the following, we discuss stops first, and

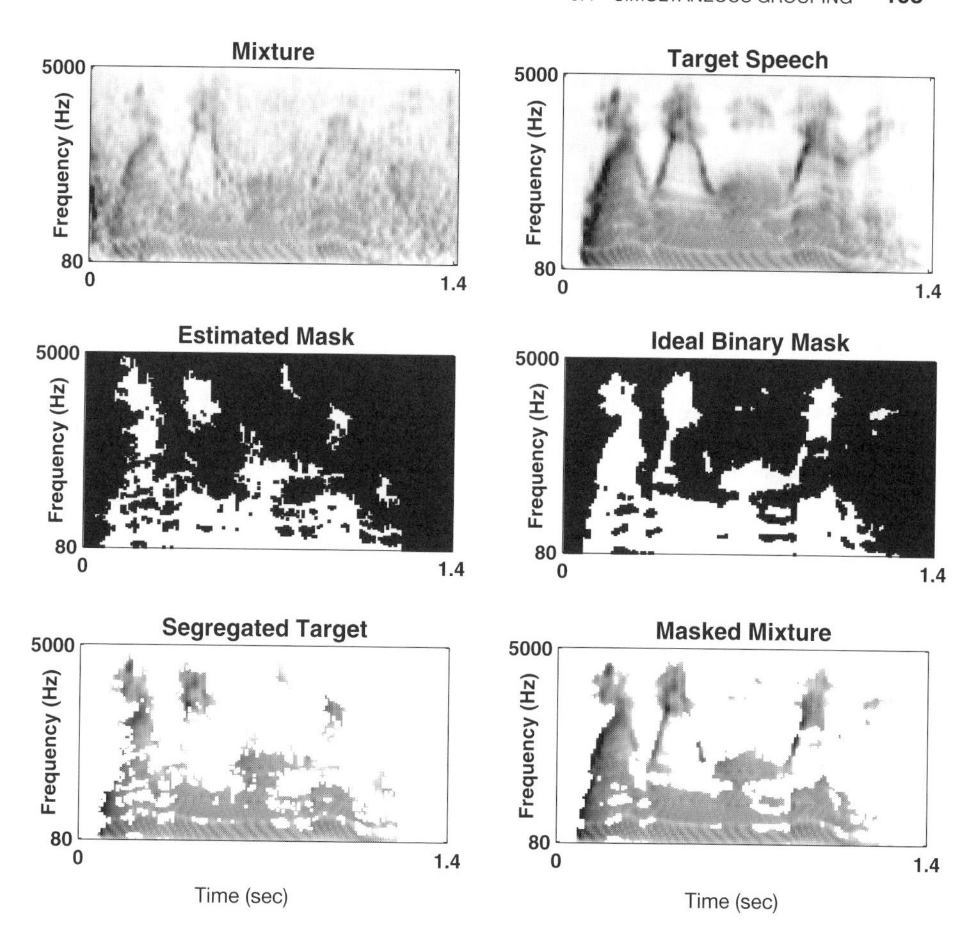

Figure 3.8 Voiced speech segregation by the Hu and Wang model. The top-left panel shows the cochleagram of a mixture of voiced speech and "cocktail party" noise. The top-right panel shows the cochleagram of clean speech. The middle-left panel shows the estimated binary mask generated by the model. The middle-right panel shows the ideal binary mask for the mixture. The bottom-left panel shows the masked mixture using the estimated mask, and the bottom-right panel shows the masked mixture using the ideal mask.

fricatives and affricates next. Our description is based on Hu's dissertation [26] (see also [27, 31]), which represents the first systematic study on unvoiced speech segregation.

Segregation of Stop Consonants. There are three pairs of stop consonants in English: /p/ and /b/, /t/ and /d/, and /k/ and /g/. The only difference between the two consonants of each pair is voicing: The first one is unvoiced whereas the second is voiced. Although half of the stops are considered voiced, their voicing is relatively weak [49] and they cannot be reliably segregated by a pitch-based analysis. Hence, all six stops are treated in stop consonant segregation. In terms of articulation, a stop consonant starts with a closure corresponding to the blockage of airflow in the vocal tract, followed by a burst corresponding to a sudden release of the airflow. As a result, onset is a major feature that distinguishes these consonants. Hu's system for stop segregation involves two steps, namely, stop detection and stop grouping.

In the first step, onset detection is performed in each frequency channel. Note that an onset event can be triggered by stops, other speech, or interference, and the challenge lies in identifying those triggered by stop consonants. A main characteristic of a stop onset is that it triggers a sudden response increase across a wide frequency range simultaneously. This suggests that we should consider only those onset events in which a synchronized onset is detected over a number of adjacent filter channels. Another characteristic of a stop burst is its duration. One can compute the duration of a stop burst in the following way. Observe that within a burst the spectral shape of the cochleagram response changes smoothly. Let $\mathbf{X}(n)$ be the vector of the cochleagram response across frequency at time step n. At a synchronized onset, the burst duration is the maximum interval around the onset time n_{on} where the cross-time correlation $\hat{\mathbf{X}}(n) \cdot \hat{\mathbf{X}}(n_{on})$ is greater than some value. Here $\hat{\mathbf{X}}$ indicates the normalized response with zero mean and unity variance. By analyzing the stops in a training set of speech signals, a synchronized onset is labeled a stop candidate if its corresponding burst duration is within a plausible range.

Some stop candidates originate from background noise and need to be removed. This part of stop detection is achieved by applying Bayesian classification to auditory–acoustic features. Stop bursts are characterized by the following features: spectrum, intensity, duration, formant transition, and voicing of the closure (see [1] for example). The formant transition from a stop to the following voiced phoneme is partly captured by the spectrum, and in any case it is very difficult to extract separately. Since voicing is weak and the burst duration is already considered in stop candidate identification, we consider spectrum and intensity for stop classification. Note that $\mathbf{X}$ encodes the spectrum feature. Figure 3.9 displays the average $\hat{\mathbf{X}}$ for each of the six stops from the training set of the TIMIT corpus. The intensity of a burst is obviously related to the overall intensity of an utterance. When the overall intensity of an utterance is normalized, the intensities of most stop bursts fall into a specific range. The intensity can be measured from $\mathbf{X}(n)$ and combined with the spectral shape of $\mathbf{X}(n)$ into a single feature for classification. Normalization can be performed on the basis of neighboring sections of segregated voiced speech which, according to the aforementioned strategy, is available prior to dealing with unvoiced speech. To perform classification, we train a classifier for each stop to discriminate a true stop from a false stop triggered by interference. Such training is done using the TIMIT corpus and an interference corpus that contains typical nonspeech sounds occurring in real environments such as traffic and crowd noise. The classification first finds the stop consonant that is most similar to a stop candidate using maximum likelihood, and then decides whether the candidate actually belongs to speech or interference by maximum a posteriori classification using the classifier of the consonant. This completes the stop detection step.

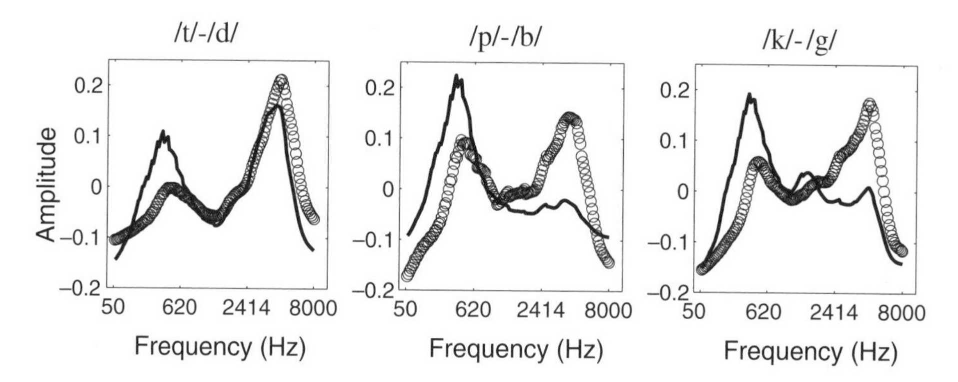

Figure 3.9 Average cochlear (spectral) responses to stop consonants. A gammatone filterbank of 150 channels, covering the frequency range from 50 Hz to 8 kHz, is used. The responses are normalized to have zero mean and unit variance. In each panel, circles indicate those of an unvoiced consonant and lines indicate those of the voiced counterpart.

For simultaneous grouping, the main ASA cue used is onset synchrony. Specifically, for each detected stop, the frequency channels that contain onsets synchronous with the onset of the stop burst are grouped together. The temporal boundary within each such channel is assumed to extend from the minimum point of the filter response immediately before the burst duration to the minimum point immediately after the burst. This pair of minima approximately marks the onset and the offset of the stop for the filter channel. The T-F units within this interval are hence labeled as belonging to the stop consonant.

Segregation of Fricatives and Affricates. In English, there are nine fricative consonants, consisting of /h/ and four unvoiced–voiced pairs: /f/ and /v/, /θ/ and /ð/, /s/ and /z/, and /ʃ/ and /ʒ/ [34]. A fricative is generated by a noise source that is forced through a narrow gap. The source signal of /h/ is the air turbulence deep within the vocal tract, and as a result this fricative sounds more like a noisy vowel. Depending on the context, the pronunciation of /h/ may be either voiced or unvoiced. There is one unvoiced–voiced pair of affricates, /tʃ/ and /dʒ/, and we treat the affricates the same way as fricatives for speech segregation. All fricatives are continuant [49], suggesting that their spectral envelope remains relatively steady for some duration. Because of this property, it is reasonable to apply segmentation before segregation. Hu's system for fricative/affricate segregation first performs auditory segmentation using onset/offset analysis (see Section 3.3.3). Once segments are produced, the system involves the following two steps. The first step removes those segments dominated by nonfricative, nonaffricate voiced energy. The second step applies Bayesian classification to determine whether each remaining segment belongs to a fricative or an affricate.

Both of the above steps engage feature-based classification. The two acoustic features used in classification are the spectrum, which includes both spectral shape and intensity, and duration. For segment removal, we first perform voiced speech segregation (see Section 3.4.1). For each time frame of the segregated target stream, spectrum-based classification labels whether the frame belongs to an affricate/affricate or any other phoneme. A Gaussian mixture model (GMM) is employed to model each phoneme at the frame level. For a T-F segment, it is removed if the acoustic energy corresponding to the frames labeled as nonfricative and nonaffricate is greater than half of the segment energy or greater than the average energy of phonetically labeled fricatives and affricates in the training corpus. Note that segment removal is a decision based on an average of frame-level decisions; although the frame-level GMM is a crude model for phonetic classification, it serves the purpose of segment removal.

For each retained segment s, it should belong to either a fricative/affricate or interference. So the second step applies binary Bayesian classification. The variable length of each segment creates a difficulty in segment modeling for classification. Clearly, individual frames within a segment are not statistically independent. Fortunately, incorporating only the dependency between two consecutive frames provides a good estimate for the likelihood $p(\mathbf{X_s}|H_s)$, where H_s is a hypothesis and $\mathbf{X_s}$ denotes the sequence of spectral vectors from frame m_1 to m_2 that comprises the segment s, that is, $\mathbf{X_s} = [X_s(m_1), X_s(m_1+1), \ldots, X_s(m_2)]$. In other words,

$$p(\mathbf{X_s}|H_s) \approx p(X_s(m_1)|H_s) \prod_{m=m_1}^{m_2-1} p(X_s(m+1)|X_s(m), H_s) \qquad (3.16)$$

One can explicitly include a duration variable in the conditional probabilities of Eq. (3.16). For each fricative and affricate as well as interference, the frame-level and pairwise likelihood functions are modeled as GMMs and trained using the TIMIT corpus and an interference corpus.

The result of segment classification is a collection of segments, each of which is labeled as belonging to either a fricative or an affricate. Figure 3.10 shows a segregation example in which the input is a mixture of a female utterance "That noise problem grows more annoying each day" and crowd noise that also contains some music. The utterance contains three fricatives: /ð/ in "That," /z/ in "noise," and /z/ in "grows," as well as an unvoiced affricate /tʃ/ as in "each." The middle-right panel of the figure shows the ideal binary mask, which is a collection of ideal segments belonging to the female utterance. From the ideal mask, it can be seen that the third fricative is severely corrupted by the crowd noise. As shown in the bottom-right panel, the model successfully separates the second fricative and the affricate. The first fricative, which is voiced, is partly separated by voiced speech segregation (the bottom-left panel).

3.5 SEQUENTIAL GROUPING

Given the results from simultaneous grouping, the task of sequential grouping is to link together the continuous streams originating from the same source that are separated in time. For speech segregation, this is the task of grouping successive speech

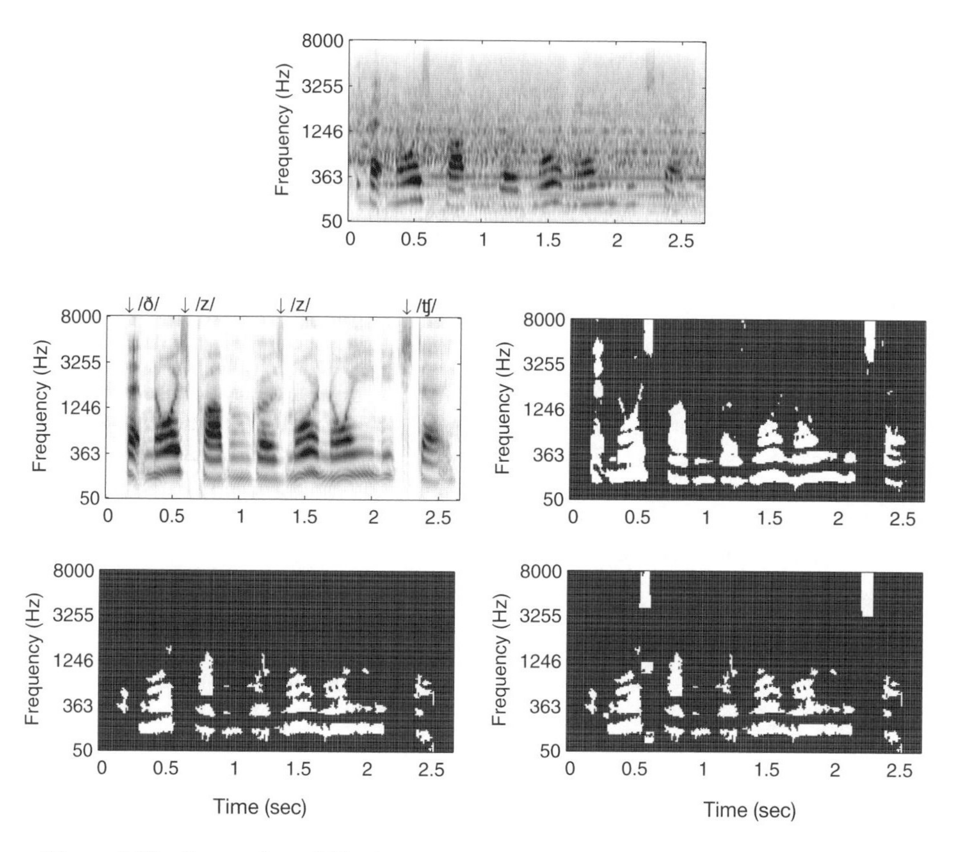

Figure 3.10 Segregation of fricatives and affricates (from [31]). The input mixture is the same as used in Fig. 3.6. The top panel shows the cochleagram of the input. The middle-left panel shows the cochleagram of the clean utterance: "That noise problem grows more annoying each day," where the locations of three fricatives and one affricate are indicated. The middle-right panel shows the ideal binary mask for the mixture. The bottom-left panel shows the binary mask obtained from voiced speech segregation, and the bottom-right panel the binary mask obtained from the Hu–Wang model for fricative/affricate segregation.

sections of the same speaker. Listeners rely on speaker characteristics, such as voice pitch and vocal tract information, for sequential organization. Compared to simultaneous grouping, sequential grouping is much less studied in CASA primarily for two reasons. First, simultaneous organization is considered an earlier stage. Second, sequential organization is likely more challenging because temporal continuity can no longer be utilized and features such as pitch and timbre tend to vary their values over time.

In the following, we will first discuss spectrum-based methods, and then pitch-based and model-based methods. Much discussion will be in the context of cochannel speech, in which utterances from two speakers are recorded on a signal communication channel. Namely, the auditory scene is a mixture of two independent utterances. Note that our discussion does not cover research aiming at restoring sounds that are masked by interference by taking advantage of temporal continuity; see [14, 38, 48] for such studies.

3.5.1 Spectrum-Based Sequential Grouping

Assuming that spectral features of the same speaker are more similar than those of different speakers, the key to spectrum-based sequential grouping is a similarity measure between two speech frames. Morgan et al. [41] described a method that uses the symmetrized divergence measure proposed by Carlson and Clements [12]. This measure employs LPC (linear predictive coding) coefficients and residual energy, and is computationally efficient. It can be shown that the widely used Itakura–Saito distance asymptotically approaches the symmetrized divergence measure.

In cochannel speech, sequential grouping amounts to determining which of the two speakers a current frame should be assigned to. In other words, the sequential grouping problem can be viewed as that of speaker assignment. Given the assignment of initial frames for each speaker in a mixture, we calculate the minimum divergence between the current frame and the preceding fifty voiced frames of each speaker, and the frame can then be assigned to the speaker with the smaller divergence. Though not described in Morgan et al., we can expect better performance by pooling frame-level decisions over all the frames of a continuous stream. This can be done by either averaging over all the frames within the stream or assigning the stream to the speaker who gives the minimum divergence of all the frames of the stream.

In the context of musical sound segregation, Brown and Cooke proposed a timbre-based approach for sequential grouping [9]. Timbre relates to the quality of a sound, but is notoriously hard to quantify. They suggested two dimensions to characterize timbre: brightness and onset asynchrony; the former is an average spectrum measure taken from the correlogram and the latter measures the shift of onset times among the frequency partials of a continuous stream. Two-dimensional vectors computed within individual groups are then subject to a cluster analysis, which yields the result of sequential grouping. A subsequent study by Godsmark and Brown [21] proposed a representation, called a "timbre track," to measure the dynamic changes of amplitude and spectral shape. Sequential grouping is then performed by comparing timbre tracks. The effectiveness of such timbre-based methods for sequential grouping of speech sounds is unknown, however.

3.5.2 Pitch-Based Sequential Grouping

In a classic study, Atal [2] demonstrated that dynamic pitch tracks can be used directly for speaker recognition with good performance. Obviously, pitch-based grouping only applies to continuous streams that are voiced. Pitch alone can help to separate speakers with very different pitch frequencies, such as speakers of different sexes (see, e.g., [55]), but this is clearly inadequate for general sequential organization since a speaker's pitch varies over a considerable range and the pitch ranges of different speakers can have substantial overlap. On the other hand, dynamic aspects of pitch may be more discriminative, which is at the core of a recent approach proposed by Shao and Wang [46] detailed below.

Shao and Wang [46] first apply a multipitch tracking algorithm [56] (see also Chapter 2) to detect continuous pitch tracks, which may overlap in time. Then overlapping portions of a track are eliminated, resulting in a set of nonoverlapping pitch tracks from a mixture. For sequential grouping between two consecutive pitch tracks, they consider the product of a pitch change between the tracks—the difference between the ending pitch of the first track and the beginning pitch of the second track—and the gap length between the tracks. The larger the product, the less likely that the tracks are produced by the same speaker. Hence the product is a pitch dynamic feature for two consecutive tracks. Furthermore, this feature shows a Gaussian-like distribution for pitch tracks of the same speaker. Hence, a training phase is employed to estimate the distribution parameters. For sequential grouping, a binary decision is made regarding whether the current track belongs to the same speaker as the preceding track on the basis of the likelihood of the pitch dynamic feature. This decision involves thresholding if only a single model is trained; the thresholding, however, can be eliminated by a discriminant analysis or by training a separate model for speaker change.

3.5.3 Model-Based Sequential Grouping

Model-based segregation is the topic of Chapter 4, where the trained models are used to organize the auditory scene. If the models whose outputs compose the scene are known in advance, sequential organization is implicitly given—T-F segments that are best interpreted by a given model form a complete stream (see, e.g., [4]). On the other hand, if the grouping decision cannot be made reliably within a short, continuous stream even with known speaker models, the issue of sequential grouping still arises. Here, the issue can be formulated as a joint problem of speaker identification (SID) and sequential organization [46].

Given a set of reference speaker models $\Lambda = \{\lambda_1, \lambda_2, \ldots, \lambda_K\}$, cochannel SID aims to identify two speaker models from Λ that maximize the posterior probability of the input observations. As discussed Section 3.5.2, we can assume that the input is a set of nonoverlapping streams, Ψ, each corresponding to a single pitch track. So cochannel SID becomes

$$\lambda_{\mathrm{I}}^*, \lambda_{\mathrm{II}}^* = \arg\max_{\lambda_{\mathrm{I}}, \lambda_{\mathrm{II}} \in \Lambda} p(\lambda_{\mathrm{I}}, \lambda_{\mathrm{II}} | \Psi)$$

The joint objective of sequential grouping and SID can then be formulated as finding a pair of speaker models, λ_{I}^* and λ_{II}^* along with a speaker assignment y^*, that jointly maximize the posterior probability,

$$\lambda_{\mathrm{I}}^*, \lambda_{\mathrm{II}}^*, y^* = \arg\max_{\lambda_{\mathrm{I}}, \lambda_{\mathrm{II}} \in \Lambda, y \in Y} p(\lambda_{\mathrm{I}}, \lambda_{\mathrm{II}}, y | \Psi) \tag{3.17}$$

where Y is the space of all possible sequential assignments. After making some reasonable assumptions, we can write the above equation as

$$\lambda_{\mathrm{I}}^{*}, \lambda_{\mathrm{II}}^{*}, y^{*} = \underset{\lambda_{\mathrm{I}}, \lambda_{\mathrm{II}} \in \Lambda, y \in Y}{\arg\max}\ p(\Psi | \lambda_{\mathrm{I}}, \lambda_{\mathrm{II}}, y) \tag{3.18}$$

so the computational objective is to find two speakers and a sequential assignment that yield the maximum likelihood.

For sentence-level utterances the number of continuous streams can often be greater than 20, so an exhaustive search is inefficient. Yang and Wang [46] found peaky distributions in the search space, suggesting that examining a relatively small portion of the space could be sufficient. They have subsequently devised an algorithm that keeps only the two best hypotheses when grouping a new stream. A comparison with exhaustive search demonstrates that this greedy algorithm achieves a level of performance close to that of the exhaustive search.

3.6 DISCUSSION

We have completed our description of the three main stages in feature-based speech segregation, namely, feature extraction, segmentation, and grouping (see Fig. 3.1). An inevitable question is: How well do these CASA systems work? The broad range of CASA tasks and the intrinsic system complexity make CASA evaluation a complex issue; see Section 1.4 for more discussion. Nonetheless, quantitative evaluation and comparison have been stressed in CASA development.

Since many studies have been conducted on simultaneous segregation of voiced speech, an assessment of progress for this task is possible. In what is regarded as the first systematic CASA study, Weintraub's model [55], designed to separate a mixture of a male and a female utterance, was evaluated on a digit recognition task. By comparing with recognizing the mixture directly, the segregation performed by his model leads to an improvement in the recognition accuracy for the male utterance (from 44.4% to 57%), but some degradation for the female utterance. A further evaluation in terms of spectral distance shows a similar pattern of results. Cooke [13] systematically evaluated his model on a database of 100 mixtures of voiced speech and a variety of noises (see Section 1.4.2). His evaluation metric is the percent match between the set of segments in the segregated stream and the set of segments in the clean target speech. Brown and Cooke [8] use a segregated stream in the form of a binary T-F mask to resynthesize a segregated target waveform from the clean speech and a residual noise waveform from the intrusion. They then calculate a normalized ratio between speech and noise. The evaluation performed by Wang and Brown [52] follows the same resynthesis strategy, but they measure the performance using a conventional SNR metric. A problem with such resynthesis strategy is that it does not punish the loss of target speech in the segregated stream, which has to be measured separately. Hu and Wang [29] introduced the use of the ideal binary mask as ground truth for quantitative CASA evaluation (see Section 1.4), and the resulting SNR shows good consistency with the performance in ASR performance and human speech intelligibility. In terms of comparison, the models of Brown and Cooke [8] and Wang and Brown [52] did not obtain significant improvements over Cooke's model. There are reasons to believe that these models perform better than Weintraub's model, although it is hard to give a rigorous statement since his model was evaluated in very different ways. A substantial improvement was made recently by Hu and Wang [29]. When evaluated on the same corpus, their model produces a 5.2 dB improvement over that of Wang and Brown, and a 6.4 dB improvement over a spectral subtraction method.

Hu's dissertation is the first comprehensive study on unvoiced speech segregation, and its performance has been systematically documented [26]. Regarding sequential grouping, a quantitative comparison by Shao and Wang [46] shows that the model-based approach performs substantially better than the pitch-based approach, which in turn performs slightly better than the spectrum-based approach. The inferior performance of spectrum- and pitch-based methods could reflect the limited discriminability of such features for sequential grouping.

Section 1.4 emphasizes the importance of systematic evaluation and comparison. Performance analysis is especially helpful for CASA systems, which typically involve multiple processing stages (see Fig. 3.1). Evaluation should be conducted on individual stages as well as on the system as a whole. Such an evaluation often reveals weak links in the system and raises precise questions for future research.

No matter how accurate feature-based analysis becomes, speech segregation will no doubt benefit from learned models of speech as well as the knowledge of the auditory scene in situations in which such models are available prior to scene analysis. This is the topic of the next chapter.

ACKNOWLEDGMENTS

Many thanks to Guoning Hu for his assistance in the preparation of this chapter. The research was supported in part by an AFOSR grant (FA9550-04-01-0117) and an AFRL grant (FA8750-04-1-0093).

REFERENCES

1. A. M. A. Ali and J. Van der Spiegel. Acoustic-phonetic features for the automatic classification of stop consonants. *IEEE Transactions on Speech and Audio Processing,* 9:833–841, 2001.
2. B. S. Atal, Automatic speaker recognition based on pitch contours. *Journal of the Acoustical Society of America,* 52:1687–1697, 1972.
3. L. Atlas. Modulation spectral filtering of speech. In *Proceedings of Eurospeech,* pp. 2577–2580, 2003.

4. J. Barker, M. Cooke, and D. Ellis. Decoding speech in the presence of other sources. *Speech Communication,* 45:5–25, 2005.
5. F. Berthommier and G. Meyer. Source separation by a functional model of amplitude demodulation. In *Proceedings of Eurospeech,* pp. 135–138, 1995.
6. F. Berthommier and G. Meyer. Improving of amplitude modulation maps for F0-dependent segregation of harmonic sounds. In *Proceedings of Eurospeech,* 1997.
7. A. S. Bregman. *Auditory Scene Analysis.* MIT Press, Cambridge, MA, 1990.
8. G. J. Brown and M. Cooke. Computational auditory scene analysis. *Computer Speech and Language,* 8:297–336, 1994.
9. G. J. Brown and M. Cooke. Perceptual grouping of musical sounds: A computational model. *Journal of New Music Research,* 23:107–132, 1994.
10. D. S. Brungart. Information and energetic masking effects in the perception of two simultaneous talkers. *Journal of the Acoustical Society of America,* 109:1101–1109, 2001.
11. J. Canny. A computational approach to edge detection. *IEEE Transactions on Pattern Analysis and Machine Intelligence,* 8:679–698, 1986.
12. B. A. Carlson and M. A. Clements. A computationally compact divergence measure for speech processing, *IEEE Transactions on Pattern Analysis and Machine Intelligence,* 13:1255–1260, 1991.
13. M. Cooke. *Modelling Auditory Processing and Organization.* Cambridge University Press, Cambridge, UK, 1993.
14. M. Cooke and G. J. Brown. Computational auditory scene analysis: Exploiting principles of perceived continuity. *Speech Commication,* 13:391–399, 1993.
15. G. Dewey. *Relative Frequency of English Speech Sounds.* Harvard University Press, Cambridge, MA, 1923.
16. D. P. W. Ellis. *Prediction-Driven Computational Auditory Scene Analysis.* PhD thesis, MIT Department of Electrical Engineering and Computer Science, 1996.
17. A. Fishbach, I. Nelken, and Y. Yeshurun. Auditory edge detection: A neural model for physiological and psychoacoustic responses to amplitude transients. *Journal of Neurophysiology,* 85:2303–2323, 2001.
18. H. Fletcher. *Speech and Hearing in Communication.* Van Nostrand, New York, 1953.
19. N. R. French, C. W. Carter, and W. Koenig. The words and sounds of telephone conversations. *Bell System Technical Journal,* 9:290–324, 1930.
20. J. Garofolo et al. DARPA TIMIT acoustic-phonetic continuous speech corpus. Technical Report NISTIR 4930, National Institute of Standards and Technology, 1993.
21. D. Godsmark and G. J. Brown. A blackboard architecture for computational auditory scene analysis. *Speech Communication,* 27:351–366, 1999.
22. S. Greenberg, J. Hollenback, and D. Ellis. Insights into spoken language gleaned from phonetic transcription of the switchboard corpus. In *Proceedings of International Conference on Spoken Language Processing,* 1996.
23. S. Greenberg and B. E. D. Kingsburg. The modulation spectrogram: In pursuit of an invariant representation of speech. In *Proceedings of IEEE International Conference on Acoustics, Speech, and Signal Processing,* pp. 1647–1650, 1997.
24. W. M. Hartmann. *Signals, Sound, and Sensation.* Springer-Verlag, New York, 1998.
25. H. Helmholtz. *On the Sensation of Tone,* 2nd English ed. Dover Publishers, New York, 1863.

26. G. Hu. *Monaural Speech Organization and Segregation.* PhD thesis, The Ohio State University Biophysics Program, 2006.
27. G. Hu and D. L. Wang. Separation of stop consonants. In *Proceedings of IEEE International Conference on Acoustics, Speech, and Signal Processing,* pp. II.749–752, 2003.
28. G. Hu and D. L. Wang. Auditory segmentation based on event detection. In *Proceedings of ISCA Tutorial and Research Workshop on Statistical and Perceptual Audio Processing,* 2004.
29. G. Hu and D. L. Wang. Monaural speech segregation based on pitch tracking and amplitude modulation. *IEEE Transactions on Neural Networks,* 15:1135–1150, 2004.
30. G. Hu and D. L. Wang. Auditory segmentation based on onset and offset analysis. *IEEE Transactions on Audio, Speech and Language Processing,* in press, 2006.
31. G. Hu and D. L. Wang. Separation of fricatives and affricates. In *Proceedings of IEEE International Conference on Acoustics, Speech, and Signal Processing,* pp. II.749–752, 2005.
32. B. Kollmeier and R. Koch. Speech enhancement based on physiological and psychoacoustical models of modulation perception and binaural interaction. *Journal of the Acoustical Society of America,* 95:1593–1602, 1994.
33. R. Kumaresan and A. Rao. Model-based approach to envelope and positive instantaneous frequency estimation of signals with speech applications. *Journal of the Acoustical Society of America,* 105:1912–1924, 1999.
34. P. Ladefoged. *Vowels and Consonants.* Blackwell, Oxford, UK, 2001.
35. G. Langner. Periodicity coding in the auditory system. *Hearing Research,* 60:115–142, 1992.
36. J. Lim, editor. *Speech enhancement.* Prentice-Hall, Englewood Cliffs, NJ, 1983.
37. D. Marr. *Vision.* Freeman, New York, 1982.
38. I. Masuda-Katsuse and H. Kawahara. Dynamic sound stream formation based on continuity of spectral change. *Speech Communication,* 27:235–259, 1999.
39. D. K. Mellinger. *Event Formation and Separation in Musical Sound.* PhD thesis, Stanford University Department of Computer Science, 1991.
40. B. C. J. Moore. *An Introduction to the Psychology of Hearing.* 5th ed. Academic Press, San Diego, CA, 2003.
41. D. P. Morgan, E. B. George, L. T. Lee, and S. M. Key. Cochannel speaker separation by harmonic enhancement and suppression, *IEEE Transactions on Speech and Audio Processing,* 5:407–424, 1997.
42. D. O'Shaughnessy. *Speech Communications: Human and Machine.* 2nd ed. IEEE Press, Piscataway, NJ, 2000.
43. T. W. Parsons. Separation of speech from interfering speech by means of harmonic selection. *Journal of the Acoustical Society of America,* 60(4):911–918, 1976.
44. P. Perona and J. Malik. Scale-space and edge detection using anisotropic diffusion. *IEEE Transactions on Pattern Analysis and Machine Intelligence,* 12:629–639, 1990.
45. M. D. Riley. *Speech Time-Frequency Representations.* Kluwer Academic, Boston, MA, 1989.
46. Y. Shao and D. L. Wang. Model-based sequential organization in cochannel speech. *IEEE Transactions on Audio, Speech, and Language Processing,* 14:289–298, 2006.

47. L. A. Smith and D. S. Fraser. Robust sound onset detection using leaky integrate-and-fire neurons with depressing synapses. *IEEE Transactions on Neural Networks,* 15:1125–1134, 2004.

48. S. Srinivasan and D. L. Wang. A schema-based model for phonemic restoration. *Speech Communication,* 45:63–87, 2005.

49. K. N. Stevens. *Acoustic Phonetics.* MIT Press, Cambridge, MA, 1998.

50. N. P. M. Todd and G. J. Brown. Visualization of rhythm, time, metre. *Artificial Intelligence Review,* 10:253–273, 1996.

51. S. J. van Wijngaarden, H. J. M. Steeneken, and T. Houtgast. Quantifying the intelligibility of speech in noise for non-native listeners. *Journal of the Acoustical Society of America,* 111:1906–1916, 2002.

52. D. L. Wang and G. J. Brown. Separation of speech from interfering sounds based on oscillatory correlation. *IEEE Transactions on Neural Networks,* 10:684–697, 1999.

53. D. L. Wang and G. Hu. Unvoiced speech segregation. *IEEE International Conference on Acoustics, Speech, and Signal Processing,* pp. V.953–956, 2006.

54. J. Weickert. A review of nonlinear diffusion filtering. In B. H. Romeny, L. Florack, J. Koenderink, and M. Viergever, editors, *Scale-Space Theory in Computer Vision,* pp. 3–28, Springer-Verlag, Berlin, 1997.

55. M. Weintraub. *A Theory and Computational Model of Auditory Monaural Sound Separation.* Ph.D. thesis, Stanford University Department of Electrical Engineering, 1985.

56. M. Wu, D. L. Wang, and G. J. Brown. A multipitch tracking algorithm for noisy speech. *IEEE Transactions on Speech and Audio Processing,* 11:229–241, 2003.

CHAPTER 4

Model-Based Scene Analysis

DANIEL P. W. ELLIS

4.1 INTRODUCTION

When multiple sound sources are mixed together into a single channel (or a small number of channels) it is in general impossible to recover the exact waveforms that were mixed; indeed, without some kind of constraints on the form of the component signals, it is impossible to separate them at all. These constraints could take several forms. For instance, given a particular family of processing algorithms (such as linear filtering, or selection of individual time-frequency cells in a spectrogram), constraints could be defined in terms of the relationships between the set of resulting output signals, such as statistical independence [3, 41], or clustering of a variety of properties that indicate distinct sources [1, 45]. These approaches are concerned with the relationships between the properties of the complete set of output signals, rather than the specific properties of any individual output; in general, the individual sources could take any form.

Another way to express the constraints is to specify the form that the individual sources can take, regardless of the rest of the signal. These restrictions may be viewed as "prior models" for the sources, and source separation then becomes the problem of finding a set of signals that combine together to give the observed mixture signal at the same time as conforming in some optimal sense to the prior models. This is the approach to be examined in this chapter.

4.2 SOURCE SEPARATION AS INFERENCE

It is helpful to start with a specific description. We will consider this search for well-formed signals from a probabilistic point of view. Assume that an observed mixture signal $x(t)$ is composed as the sum of a set of source signals $\{s_i(t)\}$ (where the braces denote a set, and each s_i is an individual source) via

$$x(t) = \sum_i s_i(t) + \nu(t) \tag{4.1}$$

Computational Auditory Scene Analysis. Edited by DeLiang Wang and Guy J. Brown where $\nu(t)$ is a noise signal (comprising sensor and/or background noise). If, for example, we assume ν is Gaussian and white with variance σ^2, then the probability density function (pdf) for x becomes a normal distribution whose mean is the sum of the sources,

$$p[x(t)|\{s_i\}] = \mathcal{N}\left[x(t); \sum_i s_i(t), \sigma^2\right] \tag{4.2}$$

that is, $x(t)$ is Gaussian distributed with mean $\Sigma_i s_i(t)$ and variance σ^2. [In our notation, $x(t)$ refers to the scalar random value at a particular time, whereas omitting the time dependency to leave x refers to the complete function for all time. Thus, the observation at a particular time, $x(t)$, may in general depend on the source signals at other times, which are included in s_i; for this particular example, however, $x(t)$ depends only on the values of the sources at that same instant, $\{s_i(t)\}$.] Different noise characteristics would give us different pdfs, but the principle is the same. Now, we can apply Bayes rule to get

$$p(\{s_i\}|x) = \frac{p(x|\{s_i\})p(\{s_i\})}{p(x)} \tag{4.3}$$

which introduces prior probability terms for both the observation $p(x)$ and for the set of component sources $p(\{s_i\})$. If we assume that the sources are more or less independent of one another, we can factor $p(\{s_i\}) = \Pi_i p(s_i)$, that is, the product of independent prior distributions for each source.

Equation 4.3 is a precise statement of our source separation problem: we have an observed signal x and we want to infer what we can about the individual sources s_i it contains. If the posterior distribution over the set $\{s_i\}$ is broad and equivocal, we do not have enough information to recover the sources from the mixture, but if there are compact regions in which particular values for $\{s_i\}$ have a large posterior probability, we can make useful inferences that the original source signals are most likely as predicted by those regions. If pressed to recover estimates of actual source signals (rather than simply expressing our uncertainty about them), we can search for the most likely sources given the observations,

$$\{\hat{s}_i\} = \underset{\{s_i\}}{\operatorname{argmax}}\ p(\{s_i\}|x) = \underset{\{s_i\}}{\operatorname{argmax}}\ p(x|\{s_i\})\prod_i p(s_i) \tag{4.4}$$

where $p(x)$ has disappeared since it does not influence the search for the maximum.

In theory, then, source separation could be performed by searching over all possible combinations of source signals to find the most likely set; in practice, this is an exponential search space over continuous signals, so direct search is impractical. Equation 4.4 can, however, be seen as a very general expression of the source separation problem, decomposed into two pieces: $p(x|\{s_i\})$, that is, to what extent the proposed components $\{s_i\}$ are consistent with the observations, and $p(s_i)$, the a priori likelihood of each candidate source signal. $p(x|\{s_i\})$ is relatively simple to compute given a complete set of source components (Eq. 4.2), so the power in solving this problem, if it can be solved, needs to come from $p(s_i)$, the likelihood of a particular signal s_i under our source model. To emphasize the role of this model by giving it a symbol M, we can write this probability as $p(s_i|M)$.

$p(s_i|M)$ embodies the constraints to the effect that s_i cannot be just any signal (if it could, the source separation problem would generally be insoluble), but instead is limited to take on some particular forms. Hence, even though mixing the sources into x has irretrievably lost dimensionality (e.g., from the $N \times T$ samples of N sources at T discrete time steps, down to just T samples of x), there may still be a unique, or tightly bounded, set of sources $\{s_i\}$ consistent both with the observed x and the prior model constraints $p(s_i|M)$.

As an example, consider the situation in which observations consist of multidimensional vectors—think, perhaps, of the short-time Fourier transform (STFT) magnitude columns of a spectrogram—and each source is constrained to consist of a sequence of frames drawn from a dictionary; in Roweis [36] these dictionaries are represented as vector-quantizer codebooks trained to capture the range of short-time spectral characteristic of a particular voice. Then separating a mixture of two sources is a finite search at each time step over all entries from the two codebooks to find the pair of vectors that combine to match the observation most closely, subject to learned and/or assumed uncertainty and noise.

Figure 4.1 illustrates this case. In an eight-dimensional feature vector, the signal emitted by source A can be described with just five prototype slices (which may occur in any order), and source B with three more. In the illustration, each prototype slice consists of a vector that is flat except for one dimension that is larger than the rest. Moreover, the dominant dimensions are taken from different subsets for the

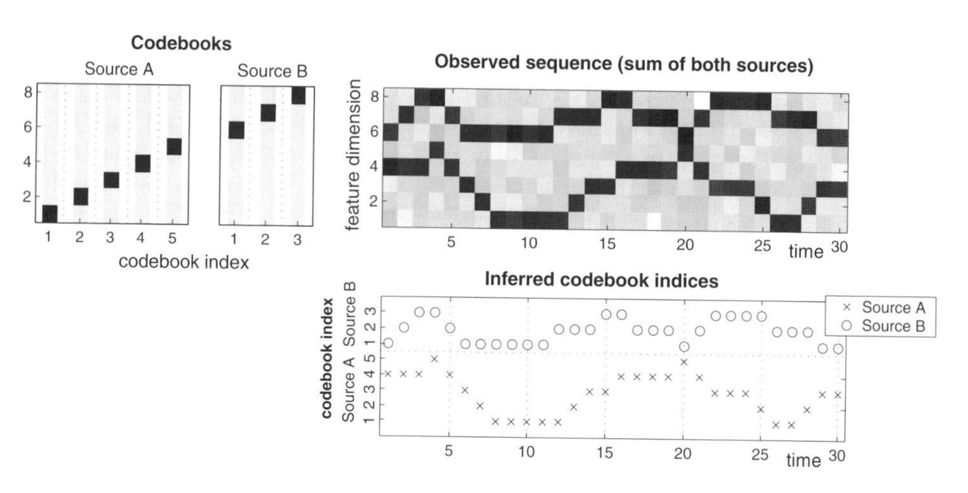

Figure 4.1 Example of a mixture composed of two sources defined by codebooks as shown on the left. The codebook states are distinct between the two sources, and the individual source states are readily inferred from the observed mixture.

two sources. Thus, in the noisy observations of a mixture (sum) of signals from both sources, it is trivial to identify which state from each source provides the best fit to each time frame; the choices are unambiguous, both between and within sources. In terms of Eq. 4.4, the likelihood of a "fit" of a possible isolated source signal s_i under the source model is simply a product of the likelihoods of each feature vector (since the model includes no sequential constraints). The behavior of each source s_i is defined by the state model denoted as M_i, and which we can introduce explicitly into the equations by making it a conditioning term, that is, $p(s_i) \equiv p(s_i|M_i)$. For the particular case of these memoryless codebooks, we have

$$\begin{aligned} p(s_i|M_i) &= \prod_t p[s_i(t)|M_i] \\ &= \prod_t \sum_{q_i(t)} p[s_i(t)|q_i(t)]\, p[q_i(t)|M_i] \end{aligned} \tag{4.5}$$

where $q_i(t)$ is the (discrete-valued) codebook state index for model M_i at time t; specifying $q_i(t)$ captures everything we need to know about the model M_i to calculate $p[s_i(t)]$, as indicated by the disappearance of M_i from the conditioning term for $s_i(t)$ once $q_i(t)$ is present. $p[q_i(t)|M_i]$ is a prior weight that can bias the model toward particular codewords, but we can assume it is essentially uniform for allowed codewords. We could express the codebook relationship as

$$p[s_i(t)|q_i(t)] = \mathcal{N}[s_i(t); C_i(q_i(t)), \Sigma_i] \tag{4.6}$$

where C_i is the codebook lookup table of prototype slices, and covariance Σ_i allows each codeword to cover a small volume of signal space rather than a discrete point; Σ_i could also be taken from a lookup table indexed by $q_i(t)$, or could be shared across the whole codebook as suggested in Eq. 4.6. As the elements of Σ_i reduce to zero, s_i becomes deterministically related to the codebook index, $p[s_i(t)|q_i t)] = \delta[s_i(t)\ C_i(q_i(t))]$ (the Dirac delta function), or simply $s_i(t) = C_i[q_i(t)]$.

If we assume the codewords are arranged to cover essentially nonoverlapping regions of signal space, then for any possible value of $s_i(t)$ there will be just one codebook index that dominates the summation in Eq. 4.5, so the summation can be well approximated by the single term for that index:

$$\begin{aligned} p(s_i|M_i) &= \prod_t \sum_{q_i(t)} p[s_i(t)|q_i t)]\, p[q_i(t)|M_i] \\ &\approx \prod_t \max_{q_i(t)} p[s_i(t)|q_i t)]\, p[q_i(t)|M_i] \\ &= \prod_t p[s_i(t)|q_i^*(t)]\, p[q_i^*(t)|M_i] \end{aligned} \tag{4.7}$$

where the best state index at each time

$$q_i^*(t) = \operatorname*{argmax}_{q_i(t)} p[s_i(t)|q_i(t)]\, p[q_i(t)|M_i] \tag{4.8}$$

or, assuming a spherical covariance for the codewords and uniform priors,

$$q_i^*(t) = \operatorname*{argmin}_{q_i(t)} \|s_i(t) - C_i[q_i(t)]\| \tag{4.9}$$

Thus, given the individual source signals s_i it is straightforward to find the state indices q_i [note that just as s_i indicates the source signal $s_i(t)$ across all time, we similarly use q_i to indicate the entire state sequence $q_i(t)$ for that source through all time]; however, in the source separation problem we do not of course observe the individual sources, only their combination x.

In general, a specific parameterization of a source model provides an adequate description of a source signal; thus, if we know the appropriate state sequence q_i for source i, we may know as much detail as we care about for the corresponding signal s_i. In this case, we can reexpress our original source separation problem as the problem of finding the model parameters, in this case, the best-fitting set of per-source codebook state sequences $\{\hat{q}_i\}$ instead of the best-fitting set of source signals $\{\hat{s}_i\}$. Now Eq. 4.4 becomes

$$\{\hat{q}_i\} = \operatorname*{argmax}_{\{q_i\}} p(\{q_i\}x, \{M_i\}) = \operatorname*{argmax}_{\{q_i\}} p(x|\{q_i\})\prod_i p(q_i|M_i) \tag{4.10}$$

which due to the "memoryless" source models can be decomposed into the product of per-time likelihoods:

$$p(x|\{q_i\})\prod_i p(q_i|M_i) = \prod_t \left(p[x(t)|\{q_i(t)\}]\prod_i p[q_i(t)|M_i]\right) \tag{4.11}$$

This can be maximized by maximizing independently at each value of t, that is, solving for the sets of state indices $\{q_i(t)\}$ separately for every t. The "generative" equation for observation given the unknown parameters (the analog of Eq. 4.2) becomes

$$p[x(t)|\{q_i\}] = \mathcal{N}[x(t); \sum_i C_i[q_i(t)], \sigma^2\mathbf{I} + \sum_i \Sigma_i] \tag{4.12}$$

Here, σ^2 is the variance due to the additive observation noise from Eq. 4.1, which has been extended to be a multidimensional observation by multiplying by identity matrix $\mathbf{I}$. This is added to the Σ_i terms, which provide the additional variance contributed by each of the individual sources in the mixture. Now, it is clear how to express mathematically the intuitively obvious source separation problem of Fig. 4.1. We can simply search across all possible combinations of the five codebook vectors of source A and the three vectors of source B to find which of those fifteen combinations maximizes Eq. 4.11, or equivalently Eq. 4.12 if the probabilities $p[q_i(t)|M_i]$ are uniform over allowable states $q_i(t)$. We do this at every time step, and gather the codebook indices for each source into the sequences plotted in the lower right of Fig. 4.1. We can use these index sequences to look up actual feature vector prototypes from the two codebooks to recover "ideal" source signal sequences (i.e., the expected values, with no observation or model noise).

4.3 HIDDEN MARKOV MODELS

The derivation above was simplified by the assumption of "memoryless" source models, so that we could separate the source signals one step at a time; the range of problems where this is sufficient to resolve sources is very limited and generally not very interesting (for instance, if the above example is interpreted as describing spectrogram slices, signal separation could be achieved with a simple fixed low-pass/highpass filter pair). However, the same approach can be applied when instantaneous observations are ambiguous—for instance when both sources are generated by the same model—provided that the models also provide adequate sequential constraints, and the observation avoids pathological ambiguities.

The most simple form of temporal constraint is a transition matrix, where the concept of the codebook index is broadened slightly to become a "state," and the transition matrix specifies which states can occur after which other states. A realization of a signal from a model now consists of assigning a specific state to each time frame, following the transition matrix constraints. Figure 4.2 shows a new example in which both sources are using the same model (so the feature dimensions containing information about the two states inevitably overlap), but this model includes the constraint that states must move left to right, that is, a particular state in one time frame can only be followed in the next time frame by the same state, or its immediate neighbor to the right (wrapping around from the rightmost state back to the beginning). This is illustrated by the state diagram below the codebook vectors, and equivalently with the transition matrix, where a shaded cell indicates an allowed transition.

Although still simple, this example illustrates several important points. Looking at the codebook (now referred to as the per-state expectations or means) along with the state diagram, we see that the source model again predicts a sequence of vectors, each with a single feature dimension larger than the rest. What is new is that the transition matrix constrains the sequence of states such that at any time, the next time frame can only either repeat the vector of the current frame (i.e., remain within the same state via the self-loop arc), or advance one step to the right in the state diagram to select the next vector along. This results in a pattern of the boosted dimension first moving upward (states 1 through 5) then moving back down again (states 5 through 8), then repeating; the self-loops imply that it can take an unknown number of time frames to move on to the next state, but the "direction of motion" of the prominent dimension will not change, except when in states 5 and 1.

The interesting consequence of this is that distinct states may have identical "emission" characteristics (state means). Thus, states 4 and 6 both have their state mean maximum in dimension 4, and by looking at one frame alone it is impossible to distinguish between them. If, however, you had a sequence of vectors from such a source, and you looked forward or backward from a particular point in time until you saw a vector whose maximum was not in dimension 4, you could immediately distinguish between the two possible states: state 4 is always preceded by a maximum in dimension 3 and followed (after any self-loop steps) by a maximum in dimension 5; state 6 has the reverse pattern. This is the sense in which the state generalizes the

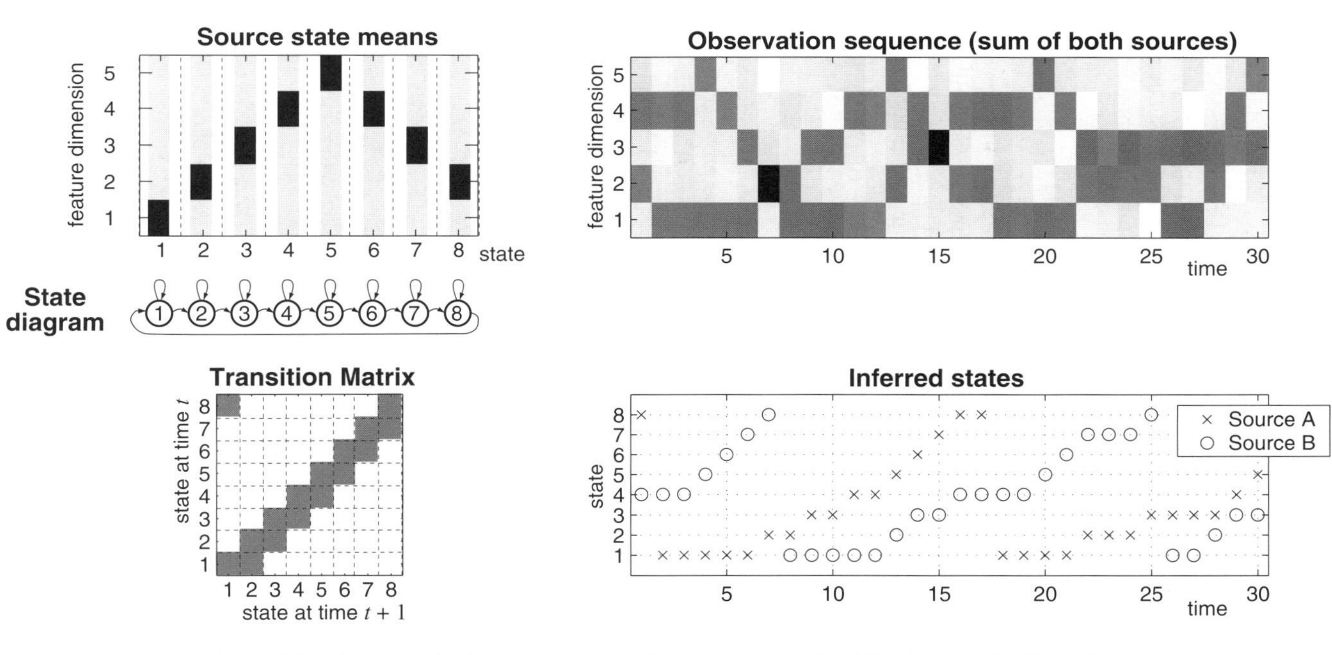

Figure 4.2 Example of mixture composed of two sources defined by the same hidden Markov model. Sequence constraints (indicated by the equivalent state diagram and transition matrix) mean that, for this observation sequence, each source can be uniquely recovered in spite of overlap between the state outputs.

codebook index; not only does it define the prototype output, but it also embodies information about what will happen in the future (and what has happened in the past).

This "memory" encoded in the state is known as a first-order Markov property, since the behavior of the model depends on the current state but on nothing earlier in the history of the signal once the current state is known. Thus, this combination of a set of probabilistic per-state emission distributions (the codebook state means and associated variances) and a matrix of state-to-state transition probabilities is known as the hidden Markov model (HMM) [29], since the state sequence is "hidden" by the uncertainty in the emission distributions. It turns out, however, that arbitrarily complex sequential structure can be encoded in an HMM, simply by expanding the range of values that the state can take on (the "state space") to capture more and more detail about the prior history.

Looking at the observed mixture of two sources (top right of Fig. 4.2) and knowing that each source is constrained to move the prominent dimension all the way to the top then all the way to the bottom, one dimension at a time, it becomes clear that the state trajectory of the two sources is largely unambiguous (although because of the symmetry occurring when both sources share a model, there is always a permutation ambiguity between the source labels). The bottom-right pane shows both inferred-state sequences, which are now plotted on the same axis since they come from the same model.

Note, however, that the inference is not completely unambiguous. At time step 22, source A must be in state 2 (since it was in state 1 at the previous time, but the observation is no longer consistent with state 1), and source B must be in state 7 (coming out of state 6). Similarly, by time 25, source B must be state 8 (ready to move into state 1 at step 26), so source A must be in state 3 to account for the combined observation. Since both dimensions 2 and 3 are large for all observations in between, the state transitions of the two models must happen at the same time, but we cannot tell whether it occurs at time 23, 24, or 25 (the inference procedure has arbitrarily chosen 25). This is, however, a narrowly bounded region of uncertainty, and soon disappears.

A more serious problem occurs if both models go into the same state at the same time. This would result in an observation similar to time step 15, with a single dimension even larger than the usual prominent dimensions. But unlike time step 15, which is explained by two different states that happen to have the same observation pattern (i.e., a collision in feature space), a collision in state space makes it impossible to find the correct correspondence between the discrete paths leading into and out of the collision, both future trajectories form valid continuations from the point of state collision, that is, valid continuations of either preceding path. The best hope for an unambiguous analysis is that the two sources never arrive at the same state at the same time; clearly, this hope becomes more and more likely as the total number of states in the model becomes larger and larger, so the prior probability of any state becomes lower, and the probability of both models being in the same state at the same time drops even faster.

For the formal analysis of this situation, we note that the component source signals are still adequately described by the state index sequences (although those sequences are now subject to the constraints of the transition matrix). Our source inference problem can thus still be reduced to solving for the state sequences for each model, that is, Eq. 4.10, repeated here:

$$\{\hat{q}_i\} = \underset{\{q_i\}}{\operatorname{argmax}}\ p(\{q_i|x, \{M_i\}) = \underset{\{q_i\}}{\operatorname{argmax}}\ p(x|\{q_i\})\prod_i p(q_i|M_i) \tag{4.13}$$

Now, however, we cannot simply decompose this into a set of independent time slices since the transition matrix introduces dependence between time slices. We can, however, consider the time observations to be independent given the state indices:

$$p(x|\{q_i\})\prod_i p(q_i|M_i) = \left(\prod_t p[x(t)|\{q_i(t)\}]\right)\prod_i p(q_i|M_i) \tag{4.14}$$

where $p(q_i|M_i)$ is the likelihood of an entire state sequence $\{q_i(1), q_i(2), \ldots q_i(T)\}$ under the model. We can use the chain rule to decompose this into the product of a set of conditional probabilities,

$$\begin{aligned} p(q_i|M_i) &= p[q_i(1), q_i(2), \ldots q_i(T)|M_i] \\ &= p[q_i(1)|M_i]p[q_i(2)|q_i(1), M_i] \ldots p[q_i(T)|q_i(T-1), q_i(T-2), \ldots q_i(1), M_i] \\ &= p[q_i(1)|M_i)p(q_i(2)|q_i(1), M_i] \ldots p[q_i(T)|q_i(T-1), M_i] \\ &= p[q_i(1)|M_i]\prod_{t=2}^{T} p[q_i(t)|q_i(t-1), M_i] = \prod_{t=1}^{T} p[q_i(t)|q_i(t-1), M_i] \end{aligned} \tag{4.15}$$

We hide the special case for $p[q_i(1)|M_i]$ in the final step by defining it as equal to the otherwise undefined $p[q_i(1)|q_i(0), M_i]$. The simplification of the longer conditions comes from the Markov property; everything about the possible behavior at time $t+1$ is captured in the state at time t, so once that state appears in a conditioning expression, no earlier states need appear, or, mathematically:

$$p[q_i(t+1)|q_i(t), q_i(t-1), \ldots q_i(1), M_i] = p[q_i(t+1)|q_i(t), M_i] \tag{4.16}$$

Substituting this into Eq. 4.14, gives:

$$p(x|\{q_i\})\prod_i p(q_i|M_i) = \prod_t \left(p[x(t)|\{q_i(t)\}]\prod_i p[q_i(t)|q_i(t-1), M_i]\right) \tag{4.17}$$

which is the same as Eq. 4.11 from the earlier example, except for the dependence of each state index on its predecessor, $q_i(t-1)$. To find the set of state sequences that maximize this, we could almost maximize for each t independently (as above) were it not for the reference to $q_i(t-1)$. However, we can solve for the maximizing sequences using the efficient dynamic programming approach known as the Viterbi algorithm. Although it is generally described for single Markov chains, we can easily convert our problem into this form by considering the set of state indices at a particular time, $\{q_i(t)\}$ (a single state index for each source), to define the state of a super model whose state space is composed of the outer product of all the component model state spaces (once again, as for the earlier example). The most likely path through this composed state space will simultaneously solve for the complementary best paths through all the individual models. Thus, we are solving for

$$\{\hat{q}_i\} = \underset{\{q_i\}}{\operatorname{argmax}} \prod_i p[x(t)|\{q_i(t)\}]p[\{q_i(t)\}|\{q_i(t-1)\}, \{M_i\}] \tag{4.18}$$

The dynamic programming trick of the Viterbi algorithm is to solve for the most-likely state sequence $\{\hat{q}_i\}$ recursively, by defining the best paths to all states at time $t + 1$ in terms of the best paths to all states at time t, repeating until the end of the observation sequence is reached, then back-tracing the sequence of states that led to the most likely state in the final step. The point to notice is that at any given time step prior to the end of a sequence, it is not possible to know which of the states will be the one on the final most-likely path, but by tracing the best paths to every one of them, we are sure to include the single one we actually want. Then, when we do get to the end of the sequence, we can quickly trace back from the state with the greatest likelihood at the final step to find all the preceding states on that best path. Thus, if $\hat{p}[\{q_i(t)\}|x]$ is the likelihood of the state sequence reaching some particular state set $\{q_i(t)\}$ at time t that best explains x (up to that time), we can define

$$\hat{p}[\{q_i(t)\}|x] = p[x(t)|\{q_i(t)\}] \max_{\{q_i(t-1)\}} \hat{p}[\{q_i(t-1)\}|x]p[\{q_i(t)\}|\{q_i(t-1)\}] \tag{4.19}$$

that is, the likelihood of the most likely state path to any particular state set at time t is the local likelihood of the current observation given this state set, $p[x(t)|\{q_i(t)\}]$ multiplied by the best likelihood to a state set at the preceding time step $\hat{p}[\{q_i(t-1)\}|x]$, scaled by the transition probability between those state sets $p[\{q_i(t)\}|\{q_i(t-1)\}]$, where all state sets $\{q_i(t-1)\}$ at the preceding time step are searched to find the largest possible product. The best preceding state set is saved for every state at every time step to allow reconstruction of the best path once the most likely state set at the end of the sequence has been found.

Thus, instead of trying to search over all possible state sequences for the best one (of order N^T calculations, for N states over T time steps), the Markov property permits the Viterbi algorithm to keep track of only one path per model state, then incrementally work through the time steps until the end is reached (order $T \times N^2$ calculations).

In practice, to solve the example in Fig. 4.2, we again construct a new state space of $8 \times 8 = 64$ source state means, corresponding to the expected outputs for every combination of the eight states of each of two models, and a $64 \times 64 = 4096$ element transition matrix, indicating all the allowable transitions between each of the 64 super states; since the transition matrices of the individual model states are so sparse in this example—only two out of eight allowable successors to any state—the super-state transition matrix is correspondingly sparse with only four out of 64 allowable successors from any state. (The Matlab code to reproduce this example is available at http: //www.casabook.org/.) Having reconstructed our problem as a single HMM, we can then use a standard implementation of the Viterbi algorithm to solve for the best state sequence.

In doing so, we notice one pitfall of this approach. The combined state space can grow quite large, particularly for more than two sources, as its size is the product of the sizes of the state spaces of all the individual source models. Moreover, the transition matrix for this composed model is the square of this larger number of states, which, of course, gets larger more quickly. In the absence of special measures to take advantage of whatever sparsity may exist in this matrix, simply representing the transition matrix may be the limiting factor in direct implementations of this kind of system.

It should be noted that this idea of inferring multiple Markov chains in parallel is precisely what was proposed by Varga and Moore [42] as HMM decomposition and by Gales and Young [12] as parallel model combination, both in the specific context of speech recognition, where the HMM models for speech already exist. The idea has had limited practical impact not only because of the tractability problems arising from the product state space, but also because observations of mixtures of signals can interfere with feature normalization schemes. In particular, it is normal in speech models to ignore the absolute energy of the signal (e.g., the zeroth cepstral coefficient), and focus instead on the relative distribution of energy in frequency and time. Even the spectral variation may be modeled only relative to some long-term average via feature normalizations such as cepstral mean subtraction (CMS, where every cepstral dimension is made zero mean within each utterance prior to recognition [34]). However, to employ the kind of factorial HMM processing we have described above, it is necessary to model explicitly the absolute levels of the sources in order to predict their combined effect, that is, the spectrum of two sources combined depends critically on their energy relative to each other, something generally not available from speech recognition models. And a normalization like CMS will in general require different offsets for each of the combined source signals, meaning that CMS-based source models cannot be directly combined, since they are modeling different domains. Although these feature normalization tricks are secondary to the main pattern recognition approach of speech recognition, they are critical to its effectiveness, since without them the distributions to be modeled become far less compact and regular, requiring many more parameters and much larger training sets for equivalent performance.

4.4 ASPECTS OF MODEL-BASED SYSTEMS

For the purposes of illustration, the preceding section focused on the particular instance of hidden Markov models for sources. However, as mentioned in Section 4.1, any source separation must employ some kind of constraints, and constraints on the form of individual sources (instead of the relationships between sources) can be expressed and discussed as source models. Thus, the model-based perspective is very general and can be made to fit very many source separation systems. Each system is differentiated by its approach to handling a number of key questions, namely: which kinds of constraints are to be expressed (including the domain in which the model is defined), and how to specify or learn those constraints (specifying the structure of actual sources); how to fit the model to the signal (i.e., how to search the model space); and how to use the inferred model parameters to generate an output such as a resynthesized isolated source signal. We now consider each of these questions in turn.

4.4.1 Constraints: Types and Representations

The essence of model-based separation is that some prior knowledge about the expected form of the sources has to be extracted. These are the constraints that make an otherwise underconstrained model solvable, and we consider the model to be whatever it is that captures these constraints. There are two major classes of model constraints that fall under this description. The first, as used in the examples of Sections 4.2 and 4.3, may be termed "memorization"; in some domain, with some degree of accuracy, the model strives to memorize all possible realizations of the signal class, and then the fit of a proposed signal to a model, $p(s_i|M_i)$, is evaluated by seeing if s_i matches one of the memorized instances, or falls into the set of signals that was inferred as their generalization. We will call these models "explicit," since the representation, such as the state means in the earlier examples, are generally directly expressed in the domain of the signal or some close relative.

The alternative to this is "implicit" models, where the subset of allowable signals is defined indirectly through some function evaluated on the signal. This is the case for many well-known computational auditory scene analysis (CASA) systems (such as Brown and Cooke [6]) which enforce an attribute such as periodicity of their outputs not by matching signals against a dictionary of known periodic signals, but by defining an algorithm (like harmonic grouping) that ensures that the only possible outputs of the system will exhibit this attribute. Such models have been discussed in Chapter 3.

We now consider in more detail the different kinds of constraints, and how they are defined or learned, for each of these two classes in turn.

Explicit Signal Models. The kind of signal model that is most easily understood is one in which, like our initial examples, instances or prototypes of the signal realizations are directly stored. An extreme example of such an explicit signal model would be if the waveform of the source signal were known and fixed, and only its time of occurrence in the mixture were unknown. In this case, a matched filter would be sufficient to estimate these times. The state-based model examples of Sections 4.2 and 4.3 are slightly more general, in that the signal model is the concatenation of a sequence of predefined vectors (or distributions within the space of feature vectors), and the source signal can be composed of these "dictionary elements" in an unspecified (but constrained) order.

Given enough states along with the transition probabilities, Markov models of this kind can in theory describe any kind of signal, in the limit by memorizing every possible individual sequence with its own set of states, although this is infeasible for interesting sound sources. Typically, the compromise is to use a dictionary that attempts to represent all possible signals with a minimum average approximation error, which will decrease as the size of the dictionary increases. As an example, the models in Roweis [35] use 8000 states (implying 64 million transition arcs), but the kinds of signals that can be constructed by sampling from the space described by the model (i.e., making random transitions and emissions according to the model distributions), though identifiably approximations of speech, are very far from fully capturing the source constraints. Nix et al. [26] go so far as to use 100,000 codebook entries; this large codebook forces them to use a stochastic search (particle filtering) to identify high-likelihood matches.

Magnitude–spectral values over windows long enough to span multiple pitch cycles (i.e., narrowband spectrogram slices, with windows in the range 20–30 ms for speech) are the most popular domain for these kinds of models because in such a representation a waveform with a static, pseudoperiodic "sound" (say, a sung vowel) will have a more-or-less constant representation; it is also broadly true that two waveforms that differ only in phase are perceived as the same, at least as long as the phase difference between them does not change too rapidly with frequency. The disadvantage of magnitude spectra is that they do not combine linearly (however, see the discussion of max-approximation in Section 4.4.2). Linearity can be recaptured by working directly in the waveform domain (e.g., with codebooks of short fixed-length segments of waveforms). The disadvantage is that a periodic signal will not result in a sequence of identical states, because in general the fixed-length windows will have different alignments with the cycle in successive frames and, hence, significantly different waveforms. An approach to this is to construct a codebook consisting of sets of codewords, where each set consists of a single waveshape realized at every possible temporal alignment relative to the analysis frames [4].

One way of thinking about these models is that they are defining a subspace containing the signals associated with the modeled source, within the space of all possible sounds. Learning a dictionary of specific feature vectors is covering this subspace with a set of discrete blobs (which permits us to enumerate allowable transitions between them), but it might also be described other ways, for instance by basis vectors. Jang and Lee [19] use independent component analysis (ICA) to learn statistically efficient basis functions for the waveforms of different sources, then perform separation by finding the waveforms that project onto these basis sets so as to jointly maximize the likelihood of the models constituted by the ICA-defined subspaces; their system achieves modest signal-to-noise ratio (SNR) improvements of 3–6 dB even without modeling any kind of temporal structure; each 8 ms frame is processed independently. Approaches of this kind could be based on many novel ideas in subspace learning, also known as nonlinear dimensionality reduction, such as those of Weinberger et al. [43].

Another approach to handling the impractically large sets of memorized states that a dictionary needs to represent if it is to provide an accurate, detailed signal model is to look for a factorization, so that a large number of effective memorized states are actually represented as the combinations of a smaller number of state factors. One way to do this is to break a large spectral vector into a number of separate subbands, spanning different frequency ranges [31]. Although this allows N prototypes for each of M subbands to represent N^M complete vectors, many of these will not correspond to observed (or likely) source spectra (unless the source's subbands are independent, which would be very unusual). This can be ameliorated to some extent through a coupled HMM, where a given state depends not only on its immediate predecessor in the same subband (a conventional HMM) but also on the states in the two flanking channels, allowing local influence to propagate all the way across the spectrum. A variational approximation can be used to decouple the bands, in the context of a particular set of states, and thus the same efficient Baum–Welch training [29] conventionally used for speech recognition HMMs can be used for each chain independently, repeated iteratively until all states converge.

A different way to factorize the prototype spectra is to model them as a combination of coarse and fine spectral structure. In speech, coarse structure arises primarily from the resonances of the vocal tract (the "formants"), and the fine structure comes from the excitation of the vocal folds (for "voiced" speech, which has an identifiable pitch) or from noise generated at constrictions in the vocal tract (for unvoiced speech such as fricatives). Factoring spectrogram frames into these two components might offer substantial gains in efficiency if it can be done in such a way that the excitation and resonance state choices are close to being independent—which seems plausible given their separate physical origins, but remains to be further investigated [20, 22].

Large, detailed, explicit signal models cannot be constructed by hand, of course. They are possible only because of the availability of effective machine-learning algorithms for generalizing ensembles of training examples into parametric models such as Gaussian mixtures (for independent values) and hidden Markov models (for sequences). Although finding the optimal solution to representing a large collection of feature vectors with a limited number of prototypes in order to minimize distortion is NP-hard [13], simple, greedy algorithms can be very successful. For instance, the standard algorithm for building a vector quantization codebook, where the objective is to define a finite set of vectors that can be used to represent a larger ensemble with the smallest average squared error, is due to Linde, Buzo, and Gray [16]. For an N entry codebook, first choose N vectors at random from the ensemble as the initial codebook. Next, quantize the entire training ensemble by assigning them to the closest codeword. Then rewrite each codeword as the mean of all the vectors that were assigned to it; repeat until the average distortion of the revised codebook ceases to improve.

Implicit Signal Models. Traditionally, CASA has referred to systems that attempt to duplicate the abilities of listeners to organize mixtures into separately perceived sources by using computational analogs of the organizational cues described by psychologists such as Bregman [5] or Darwin and Carlyon [8]. These systems, discussed in detail in Chapter 2, may be considered as falling into our model-based perspective since they extract signals by finding components within the mixture that match a restricted set of possible signals, the $p(s_i)$ of Eq. 4.4. For example, many CASA systems make heavy use of the "harmonicity cue," the inference that a set of simultaneous but distinct sinusoids, or subbands exhibiting periodic amplitude modulation, belong to a single common source if their frequencies or modulations are consistent with a common fundamental period [6, 17, 44]. One popular way to evaluate this is by autocorrelation; energy in any frequency band that arises from a common fundamental period will exhibit a large peak at that period in its autocorrelation. Evaluating the autocorrelation in multiple subbands and looking for a large peak at a common period is one way to calculate a model fit that could easily take the form of $p(s_i|M)$. This fits into the framework of Eq. 4.4 in that candidate sources lacking the required harmonic structure are not considered, that is, they are assigned a prior probability of zero. (Further enhancements to include CASA cues such as common onset or spatial location can similarly be expressed as priors over possible solutions.) Such a model could, of course, equivalently be "memorized"—represented in explicit form—by training a vector quantizer or Gaussian mixture model on a large number of the signals that scored highly on that metric. A sufficiently large model of this kind could be an arbitrarily accurate representation of the model, but it seems to be an inefficient and inelegant way to capture this simple property.

In an effort to broaden the scope of CASA beyond harmonic (pitched) signals, the system of Ellis [9] uses a hand-designed set of "atomic elements," parametric sound fragments able to provide pitched or unpitched energy with constrained forms (such as separable time and frequency envelopes, or constant energy decay slopes) that were based on observations of real-world sources and insights from human perception. The ambiguity of decomposing mixtures into these elements led to a complex blackboard system tracking and extending multiple alternative source-set hypotheses, pruned on the basis of their "quality" of explanation, which can be interpreted as incorporating both the match of signal to explanation $p(x|\{s_i\})$ and the match of the explanatory sources to the underlying signal models $p(s_i|M_i)$.

Considering these implicit models as mechanisms for calculating the probability of a signal fitting the model can be helpful. For instance, although much of the structure of these models is encoded by hand, there may be ways to tune a few parameters, such as weights for mapping near misses to intermediate probabilities, that can be set to maximize the match to a collection of training examples, just as explicit models are based more or less entirely on training examples.

The "deformable spectrogram" model is an interesting middle ground between directly encoded rules and parametric models inferred from data [30, 32]. Based on the observation that narrowband spectrograms of speech and other sources have a high degree of continuity along time, the model attempts to describe each small patch of a spectrogram as a transformation of its immediately preceding time slice, where the transformations are predominantly translations up or down in frequency. This results in a distribution of transformation indices across time and frequency, which is smoothed via local dependencies whose form is learned from examples to maximize the likelihood of training data under the model. This part of the model is similar to conventional CASA in that a particular signal property, in this case the continuity of time-frequency energy across small frequency shifts, is captured implicitly in an algorithm that can describe only signals that have this property. But when local transformations are unable to provide a good match for a particular time frame, as might happen when the properties of a sound change abruptly as a new source appears, the model has a separate "matching" mode, in which a new spectral slice is drawn from a dictionary, very much like the basic state-based models. The difference is that these memorized states are used only occasionally when tracking the deformations fails. Figure 4.3 illustrates the reconstruction of a segment of speech starting from a few sampled spectra, with the intervening frames inferred by propagating these spectra using the inferred transformation maps. This recourse to states is also key to the use of this model for signal separation, since, when applied in subbands, it marks potential boundaries between different states, very much like the elemental time-frequency regions used in earlier CASA systems. Source "separation" consists of attempting to group together the regions that appear to come from a common source; for instance, by similarities in pitch, or consistency with another model better able to capture structure at longer timescales such as the pronunciation and language models of a speech recognizer.

4.4.2 Fitting Models

As we have noted already, the problem with Eq. 4.4 is that it requires searching for the maximum over an exponentially sized space of all possible combinations of all possible source signals $\{s_i\}$, which is completely impractical for direct implementation. By assigning zero probability to many candidate source waveforms, source models can at least restrict the range of sets $\{s_i\}$ that needs to be considered, but exhaustive search is rarely practical. In our examples, we limited the temporal dependence of our source signals to at most one step into the past. This meant that we could use dynamic programming to effectively search over all possible sequences by considering only all possible sequential pairs of states at each time step, a complexity of $(N^M)^2 \times T$ for N states per source and M sources over T time steps, instead of $(N^M)^T$ for the full search over all sequences. But even with this considerable help, merely searching across all combinations of states from two or more models can quickly become unwieldy for models with useful amounts of detail. For instance, Roweis [35] had 4000 states in each model, so a full combined state space would involve 16,000,000 codewords, approaching the practical limit for current computers, and considering all possible adjacent pairs of these combined states would involve $(4000^2)^2 = 2.56 \times 10^{14}$ likelihood calculations per time step, amounting to several weeks of calculation using today's computers, even ignoring the storage issues. Clearly, some shortcuts are required, both for representing the joint state space, and for calculating the state sequence hypotheses (i.e., "pruning" or abandoning unpromising hypotheses early in their development).

One way to improve the efficiency of the large joint state space is to avoid precalculating all N^M state distributions, but create them on the fly from the individual models. With aggressive hypothesis pruning, it may be that many state combinations are never even considered and thus their expected templates need never be calculated. Since the expected observation for a combination of sources in some

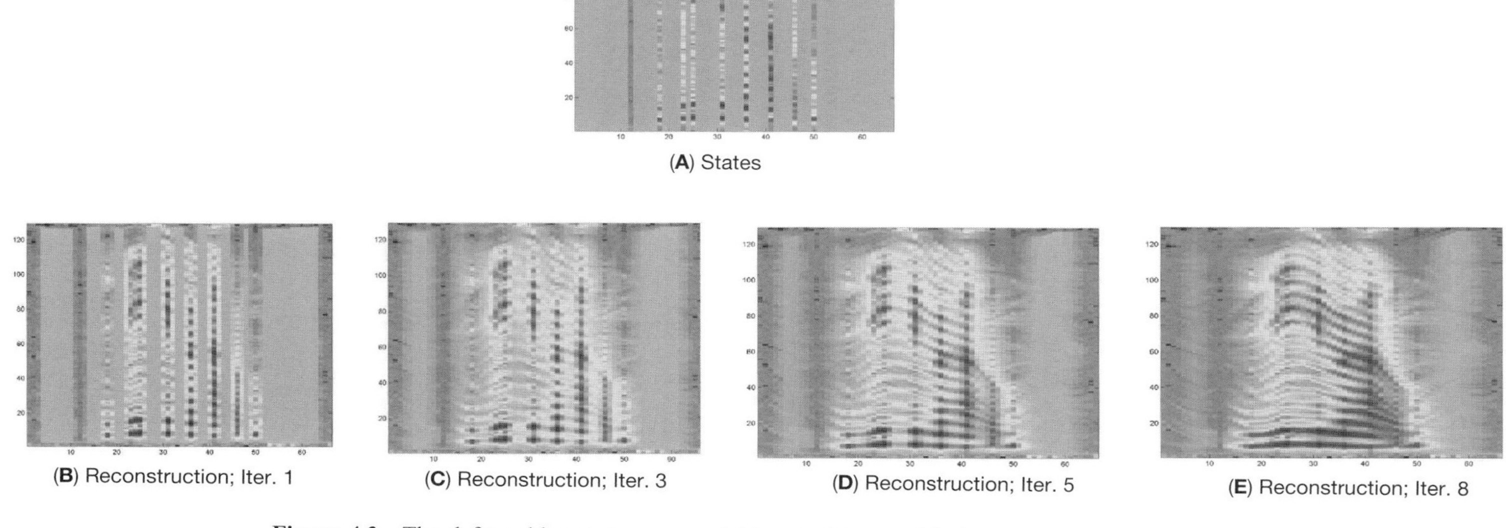

Figure 4.3 The deformable spectrogram matching–tracking model. A spectrogram is primarily modeled as a field of local transformations (not shown) describing how each part slice is related to its immediate predecessors via transformation, but the system also has a dictionary of seed states that can be inserted at points when local transformation fails to describe the observations. These panes show how a signal is reconstructed from this description, starting with the discrete states (**A**), through successive iterations of propagating those slices out into adjacent frames according to the transformation fields, through belief propagation (from Reyes-Gómez [30]).

particular state is simply the superposition of the observations from each individual source (e.g., sum in the amplitude domain of Eq. 4.1), we can exploit properties of natural sounds to represent the observation templates implicitly by reference to the per-source states, and to prune away large chunks of the candidate state space without careful search. The approximation discussed in the next section can make this particularly efficient.

The Max-Approximation. When matching spectral magnitudes (i.e., spectrogram slices), it is commonly assumed that one or other signal dominates in each cell, that is, $|A + B| \approx \max(|A|, |B|)$ at each point on the time-frequency distribution [36]. This is justified by the wide local variation in energy across even small local regions of source spectrograms, so that even if two sources have similar energy at the coarse scale, at a fine scale they will have comparable energy in only a small proportion of cells. In other cells, the magnitude ratio between the two sources is so great that the smaller can be safely ignored, and the magnitude of the combination is approximated as the magnitude of the larger component. Simple experiments show that for randomly selected mixtures of two voices with equal energy (recorded in quiet; for example, from the TIMIT database [14]), for STFT window lengths anywhere between 2 and 100 ms, the difference in dB magnitude between individual time-frequency cells is approximately normal distributed with a standard deviation of about 20 dB (bottom-left panel of Fig. 4.4). Adding two cells with unknown phase alignment but a difference in magnitude of 6.5 dB results in a cell with an expected magnitude equal to the maximum of the two, but a standard deviation of 3 dB; this deviation naturally grows smaller as the difference in magnitude gets larger. Thus, only 25% of cells in a spectrogram of the sum of two clean speech signals (i.e., the proportion of normal distribution lying within $6.5 \div 20$ deviations from the mean) will have a magnitude distributed with a standard deviation greater than 3 dB away from the maximum of the two mixed components. Empirically, mixtures of two TIMIT signals have log-spectral magnitudes that differ by more than 3 dB from the maximum of the two components in fewer than 20% of cells, and the average dB difference between the max over the sources and the mixture spectrogram is very close to zero, although it is a heavy-tailed distribution, and significantly skewed, as shown in the bottom-right panel of Fig. 4.4. Note that the skew between the mixture spectrogram and the maximum of individual sources is negative, meaning that the mix log spectrogram has a wider range of values when it is smaller than the sources, because two equal-magnitude sources in theory completely cancel to create a mixture cell with an unbounded negative value on a logarithmic scale. The most by which the log magnitude of the sum of two sources can exceed the larger of their magnitudes is 6 dB.

This max-approximation can be used to accelerate the search for well-matching pairs of states [35, 36]. On the assumption that the magnitude spectrum of an observed mixture will have energy that is (at least) the element-wise maximum of the spectra of the two components, we can sort each model's codebook by the total magnitude in each codeword in excess of the combined observation we are trying to match, related to the "bounded integration" that can be used to match spectral models in the presence of masking, as described in Chapter 9. Assuming that the max-approximation holds, codewords with elements that are larger than the corresponding mixture observation dimension cannot be a part of the mixture, or else the magnitudes in those observation elements would be larger. Thus, all code words from both models containing dimensions that significantly exceed the observation can be eliminated, and only combinations of the remaining states need be considered. Further, when searching for states to pair with a given state to complete an explanation, dimensions in which the first state is significantly below the observation can be used to further restrict the set of states from the second model to be considered, since the responsibility for explaining these so-far unaccounted for dimensions will fall fully on the second state. Once the codebooks have been so pruned, the actual best match can be found as the closest match under the max-approximation; it would even be possible to use a more sophisticated function to combine each candidate pair of states into an expected distribution for the joint distributions, since only a few of these state pairs need be considered.

Of course, the dynamic programming search relies not only on the match between the state's emission distribution and the actual observation ($p[x(t)|\{q_i(t)\}]$ from Eq. 4.19), but also the transition costs from all the preceding states being considered ($p[\{q_i(t)\}|\{q_i(t-1)\}]$). In theory, a wide variation in transition costs could completely reorder the likelihood of different states to be very different from an ordering based on emission match alone, although in practice many observation (mis)matches constitute such small probabilities that they will never be boosted back into viability through large transition likelihoods. In any case, estimating and storing the N^2 values of the full transition matrix for each model may be difficult. One alternative is to ignore transition probabilities altogether, but learn codebooks for overlapping windows of features (e.g., pairs of adjacent frames), and let the obligation to match the overlapping frame shared between two states introduce constraints on which states can follow others [36].

When the source magnitudes are comparable in particular dimensions (e.g., closer than 6.5 dB), the max assumption involves a fairly significant departure from reality, but it is possible to make a more accurate inference of the true underlying per-source magnitudes. Kristjansson et al. [21] use the "Algonquin" algorithm (iterative linearization of a more complex relationship) to infer source spectra that achieve about 6 dB reduction of interferer energy for separating single channel mixtures of male and female voices based on 512 component Gaussian mixtures and 128 dimension spectrogram frames, for a multiple-speaker, limited-vocabulary corpus (TIDIGITS [24]).

In cases in which hypothesis probabilities can be evaluated but there is no efficient algorithm for searching systematically through an intractable hypothesis space, there are also stochastic search methods that move around locally in the hypothesis space, searching for the best local optimum that can be found. Clearly, such approaches are very dependent on their initialization, and on the complexity of the hypothesis likelihood function. Nix et al. [26] use sequential Monte Carlo methods (also known as "particle filtering") to develop a fixed-size set of many thou-

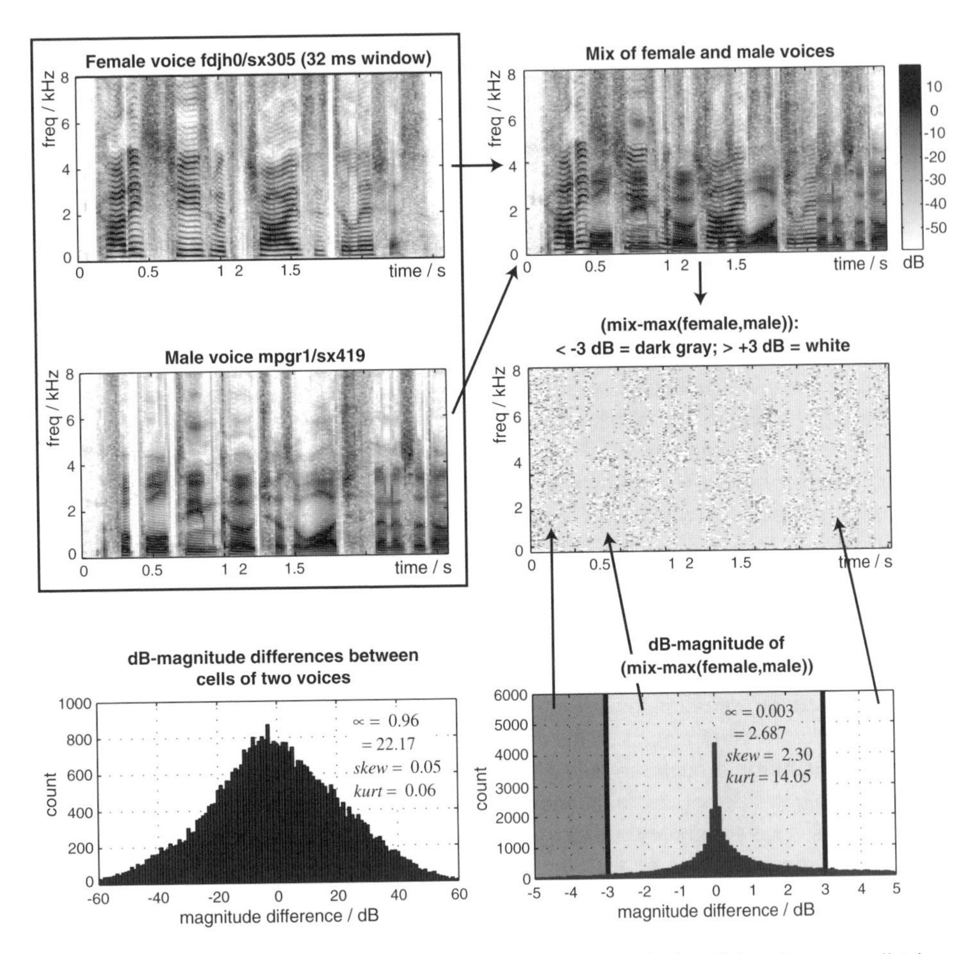

Figure 4.4 Illustration of the relationship between the magnitudes of time-frequency distributions of two voices. Top left: Spectrograms of female and male voices. Top right: Spectrogram of the sum of the two waveforms. Middle right: Time-frequency cells in which the log magnitude of the sum is more than 3 dB different from the maximum of the log magnitudes of the individual sources (18% of cells in this example). Bottom left: Histogram showing the distribution of log magnitude differences between corresponding cells in the two source spectrograms. Bottom right: Histogram of the differences between the spectrogram of the mixed signal and the maximum of the two component spectrograms; the region between the solid lines corresponds to the neutral-colored cells in the middle-right panel.

sands of individual samples in the hypothesis space. At each step, the least likely hypothesis "particles" are eliminated, and the most likely ones are duplicated, to evolve along slightly different random paths within those higher-likelihood areas. Although the algorithm can be computationally expensive (e.g., thousands of times slower than real time), it offers a relatively straightforward approach to searching intractable spaces.

These codebook models draw techniques and inspiration from the large hidden-Markov models used in contemporary large vocabulary speech recognizers, which frequently contain thousands of states and sometimes even hundreds of thousands of Gaussian components. Speech recognizers, however, have the advantage that, since just one source is present, the absolute level of the signal can be normalized, or simply ignored in modeling (for instance, by excluding the zeroth cepstral coefficient from models, or by normalizing feature dimensions along time within an utterance, which can also remove any static spectral coloration). When two or more sources are combined, however, their relative amplitude (and fixed spectral shaping) has a central impact on the resulting waveform, so individual source models must include these factors, either directly within the state set, or as separate factors that must somehow be inferred and tracked for each source during inference.

Search in Implicit Models. When we cast traditional CASA models such as Brown and Cooke [6] or Hu and Wang [17] into the model search of Eq. 4.4, it draws attention to the search over alternative candidate source signals, which is generally described as grouping in those models, and constructed in such a way as to minimize the number of alternatives that are explicitly considered. In both those systems, a source is defined by a pitch track extracted from the mixture, then all time-frequency cells "consistent" with that pitch track are assigned to the source mask. Because these models do not attempt to estimate the probabilities of their mask assignments, it would be difficult to compare and choose between alternative hypotheses.

Other models go beyond this. In the "speech fragment decoder" of Barker et al. [2], separating speech from interference is cast as finding the best subset to cover the target speech from among a collection of relatively large time-frequency regions (that might arise from a CASA-like initial segmentation, although most of their results come from a cruder breakup of regions poking out of a static noise-floor estimate). They do this by incorporating the search for the best "segregation hypothesis" (set of regions assigned to the target speech) into the existing speech recognizer search across phones and words for the best-fitting model. Since the outcome of this search is the pair of utterance and segregation mask that affords the greatest agreement between observed mixture and the recognizer's speech models (exploiting the missing-data matching of Cooke et al. [7]), the speech models are directly guiding the search by rejecting any segregation masks that imply target energy that is inconsistent with the utterance hypotheses being considered. This is a very neat way to combine the heuristic segmentation principles of CASA systems with the highly developed and powerful models of speech signals contained within current speech recognizers. However, these models actually do not attempt to enforce or exploit consistency in the character of the target voice because they are speaker independent, that is, deliberately trained on a range of speakers to accept any "typical speech" segments equally well. These issues are developed further in Chapter 9.

By contrast, Shao and Wang [38] use models drawn from contemporary work in speaker identification (instead of speech recognition) to organize smaller fragments. Given a set of segments extracted from a mixture of two voices by looking for partial pitch tracks, they identify both speakers by finding the two models that together best account for all the segments (from an ensemble of previously trained single-speaker models), and can then assign individual segments to each speaker based on likelihood, thereby segregating these partial representations of the sources. The speaker models, which are typically large Gaussian mixture distributions for the instantaneous spectra produced by each speaker, do not provide or encode any temporal sequence constraints or structure, which, while presumably making them weaker, also makes them much simpler to apply.

4.4.3 Generating Output

By approaching a problem as blind signal separation, the implication is that the desired outputs are the actual original waveforms comprising the mixture. But when a problem is described as scene analysis, it is no longer so clear what the desired outputs should be. If we are attempting to duplicate human perceptual organization, there is no strong reason to believe that the brain ever constructs a representation close to the raw waveform of the individually perceived sources. Perhaps a much more abstract description of distinct sources in terms of high-level attributes is sufficient. Such alternatives become increasingly clear in the context of model-based scene analysis, since models also represent waveforms in terms of a smaller set of more abstract parameters such as state paths, pitch tracks, or word-string hypotheses.

Of course, there are situations in which extracting waveforms is relatively easy, predominantly when multiple channels are available (e.g., from multiple microphones). In this case, there may exist combinations of time-invariant filters through which the channels may be combined to largely or completely cancel out interfering sources, although this generally relies on there being only a small number of spatially compact sources of interfering noise. Although this scenario, considered in more detail in Chapter 6, is amenable to a range of more general source separation algorithms such as beamforming and independent component analysis (ICA) [3, 18], there are opportunities for models to help still further. Seltzer et al. [37] obtain a target for gradient-descent optimization of the coefficients of their filter-and-sum microphone array by looking at the means of the HMM states from the best path of a conventional speech recognizer applied to the array output. As the speech signal is enhanced through incrementally refined filter coefficients, the recognizer alignment is reestimated to achieve iterative improvement of both speech recognizer result and filter coefficients. The system enhances the speech against nonspeech interference, and also ensures that the filter-and-sum operation results in a speech signal that is spectrally balanced to resemble the recognizer's training set. Reyes-Gómez et al. [33] use a similar approach to separate two simultaneous speakers, using a variational approximation to solve the factorial HMM problem of identifying the separate state paths for the two speakers. This approach is highly relevant to the model-based analysis presented in this chapter, and readers are directed to Ghahramani and Jordan [15].

It is, however, the single-channel case that this chapter is mostly concerned with. In that case, only trivial mixtures (those with nonoverlapping spectral support) can be separated via linear time-invariant filtering. If, however, we allow the filters to vary with time, we can still extract multiple, distinct waveforms by assigning the same spectral region to different sources at different times. This is the essence of time-frequency masking, where the magnitudes of the cells of an invertible time-frequency representation (i.e., a short-time Fourier transform in which the phase information is retained) are scaled by element-wise multiplication with a time-frequency mask. Masks are often binary, meaning that some cells are entirely deleted (reduced to zero magnitude), and some are left unchanged (multiplied by 1). Time-frequency masking has been used in the majority of single-channel signal separation systems, including Brown and Cooke [6], Roweis [35], and Hu and Wang [17], and can give startlingly good results, particularly when the mask is chosen "optimally" to pass only cells in which the local energy of target exceeds interference. From Wiener filtering, we know that optimizing the SNR (i.e., minimizing the total energy of any distortion relative to the original signal) is achieved by scaling the mixture of signal plus noise by $1/(1 + P_N/P_S)$, where P_N and P_S are the noise and signal power, respectively. Applying this to each time-frequency cell individually leads to a "soft mask" that optimizes the overall SNR by maximizing the contribution of each cell. Quantizing this optimal mask to the closest binary level (1 or 0) gives the best binary mask, which deletes all cells in which $1/(1 + P_N/P_S) < 1/2$, that is, $P_N > P_S$, or a local SNR below 0 dB. This is in line with our expectations, since each included cell increases the noise denominator by the amount of interference included in the cell at the same time as reducing it by the target energy that is saved from deletion; total noise is minimized by including only cells where the net noise contribution is negative i.e. target exceeds interference. Time-frequency masking can be viewed as the combination of a set of time-varying filters, one for each frequency channel, whose frequency response is fixed but whose gain switches dynamically between zero and one as successive time frames in that frequency band are passed or excluded. The limitation of time-frequency masking is that it cannot do anything substantial to separate target and interference energy that fall into a single cell. Even if "soft masks" are used (scaling cells by a real value in an effort to reconstruct the magnitude of one source even when a cell is significantly influenced by both), scaling only the magnitude will not give a particularly high-quality reconstruction of the target source, which would rely on recovering the source, not mixture, phase as well.

As we have explained, CASA systems usually operate by identifying larger time-frequency regions and assigning them to one of several inferred sources. Thus, the core representation of these sources is a binary time-frequency map that can be used directly for time-frequency masking. For explicit models, such as those that calculate the sequence of codewords that are inferred to best approximate the source, there are several options for resynthesis. Usually, the model states do not include phase information, so they cannot be directly resynthesized to a waveform. (Magnitude-only reconstruction procedures have been proposed [25], but they rely on consistency between overlapping windows, which is likely to be seriously disrupted in the quantized magnitude sequences derived from state-based models [28, sec. 7.5].)

One source for the missing phase information needed for conventional STFT overlap–add inversion is the original mixture; taking the state-based magnitude inferences and copying the phase from the time-frequency analysis of the original mixture will give some kind of reconstruction in which interference is reduced. However, cells in the original mixture that are not dominated by the target will have inappropriate phase, and the energy reconstructed from those cells is more likely to be perceived as interference. It would be better to eliminate those cells altogether, as in time-frequency masking. Roweis [35] converts his state-based inferences for the two component sources into a binary mask simply by comparing the codewords from the two models, and including in the mask only those cells in which the inferred target magnitude is larger than the inferred interference. After time-frequency masking, the magnitude of the resynthesis in these cells may not precisely match the inferred magnitude vectors (since both phase and magnitude are being taken from the mixture), but because they arise from the same underlying signal, the result is likely to have less distortion: any magnitude quantization in the codebook now has no direct impact. All the codebooks have to do is correctly model the relative magnitudes of target and interference in each cell; the exact levels are less important.

This approach, of using possibly crude estimates of multiple components to choose a time-frequency mask, could be used with any system that estimates the spectral magnitude of both source and interference, but more complex, parametric models have at least the possibility of recovering the signal even in time-frequency regions dominated by interference, provided that the model carries enough information for resynthesis. Thus, the "weft" elements of Ellis [10] modeled periodic energy with a single pitch and a smooth time-frequency envelope that could be interpolated across occlusions; this is sufficient parametric information to resynthesize a signal that, though different from the original in the fine detail of its phase structure, can carry much the same perceptual impression.

In many cases, however, it may be that the abstract, parametric internal representation is a useful and appropriate output from the scene analysis. Thus, although the system of Ellis [9] analyzed mixtures and broke them down into elements that could be individually resynthesized and recombined into waveform approximations, the higher-level description of the mixture in terms of these parametric elements would be more useful for many applications, including classifying the contents of the mixture, or searching for particular events within the audio content. Similarly, the "speech fragment decoder" of Barker et al. [2] performs an analysis of a source mixture, but its primary output is the word string from the speech recognizer it contains. Outputs other than waveforms have major implications for the evaluation of scene analysis systems since there is no longer any possibility of using an SNR-like measure of success; however, given that these systems are often working at a level abstracted away from the raw waveform, SNR is frequently an unsatisfying and misleading metric, and the effort to devise measures better matched to the task and processing would be well spent [11].

4.5 DISCUSSION

So far we have presented the basic ideas behind model-based scene analysis, and examined the common issues that arise in constructing systems that follow this approach: representing the constraints, searching the hypothesis space, and constructing suitable outputs. In this section, we discuss some further aspects of model-based separation and its relation to other blind signal separation approaches such as ICA and sparse decomposition.

4.5.1 Unknown Interference

So far, our analysis has been symmetric in all the sources that make up a mixture. Although the goal of our processing may be to find out about a particular target source, the procedure involved estimating *all* the sources $\{s_i\}$ in order to check their consistency with the mixed observations x. This is often seen as a weakness. For instance, in speech enhancement applications, although we are very interested in our target signal, and can often expect to have the detailed description of it implied by the model $p(s_i|M_i)$, we do not really care at all about the interference signal, and we do not know (or want to assume) very much about it. Requiring a detailed, generative model for this interference is either very limiting, if our enhancement will now only work when the interference matches our noise model, or very burdensome, if we somehow attempt to build a detailed model of a wide range of possible interference.

This raises a point that we have skirted over thus far, namely, how the match to the observation constrains the complete set of source signal hypotheses. Equation 4.4 searches over the product $p(x|\{s_i\})\Pi_i p(s_i|M_i)$, and we have focused mainly on how this is limited by the model match likelihoods $p(s_i|M_i)$, but, of course, it will also reject any hypotheses where $p(x|\{s_i\})$ is small, that is, the set of proposed sources would not combine to give something consistent with the observations. So another natural limitation on the search over $\{s_i\}$ is that once all but one of the sources has a hypothesis, the last remaining source is strongly constrained by the need to match the observation. Returning to the simple linear combination model of Eqs. 4.1 and 4.2, this could be a tighter constraint that the source model, that is, the full distribution over the Nth source given the observation and the hypotheses for the other sources becomes

$$p(s_N|x, \{s_1 \ldots s_{N-1}\}) \propto p(x_i|\{s_i\})p(s_N|M_N)$$
$$= \mathcal{N}\left(x; \sum_i s_i, \sigma^2\right) p(s_N|M_N)$$
$$= \mathcal{N}\left(s_N; x - \sum_{i \setminus N} s_i, \sigma^2\right) p(s_N|M_N) \quad (4.20)$$

Either the first term (the normal distribution arising from the model of the domain combination physics) or the second term (our familiar source model) can provide tight constraints, and one approach to finding a s_N is simply to take the mode of the normal distribution, $\hat{s}_N = x - \Sigma_{i \setminus N} \hat{s}_i$ (i.e. $x - \hat{s}_0$ in the simplest case of a single target with hypothesized value $\hat{s}_0$ and a single interfering source) and multiply its likelihood by $p(\hat{s}_N|M_N)$; if the model distribution is approximately constant over the scale of variation represented by σ^2, this single point will suffice to evaluate the total likelihood of any set of sources involving $\{s_1 \ldots s_{N-1}\}$.

The point to remember when considering the problem of unknown interference is that although we have focused on source models that provide tight constraints on the source components, the analysis still applies for much looser models of $p(s_i|M_i)$ able to accept a broad range of signals. In particular, if the hypothesized mix contains only one such "loose" model (for instance in the mix of a single well-characterized voice and single "garbage model" of everything else), then the source inference search is still straightforward when the $p(x|\{s_i\})$ constraint is applied.

The more specific a background noise model can be, the more discrimination it adds to the source separation problem. A simple, stationary noise model that posits a large variance in every dimension may give a similar likelihood to background noise with or without an added foreground spectrum, since the changes due to the foreground result in a spectrum that is still well within the spread covered by the noise model. A more sophisticated noise model, however, might be able to capture local structure such as slow variation in noise power; a running estimate of the current noise level would then allow a much tighter model of the expected noise at each frame, and better discrimination between target states embedded in that noise. Thus, even though the specific spectral form of interference may be unknown in advance, there may be other constraints or assumptions that can still be captured in a kind of model, to allow separation. By the same token, target-extraction systems that do not have an explicit source model for the interference may often be equivalent (or approximations) to a model-based analysis with a particular, implicit model for noise.

4.5.2 Ambiguity and Adaptation

When a signal is being analyzed as the combination of a number of sources without any prior distinction between the properties of each source (such as a mixture of otherwise unspecified voices), there is, of course, ambiguity over which source will appear under each of the system's source indices. As long as correspondence remains fixed throughout the signal, this is unlikely to cause a problem. However, if models really are identical, there may be cases in which it is possible to lose track of the sources and permute the source-to-output arrangement midstream; for instance, if two speakers both happen to fall silent at the same time, without any difference in context to constrain which of the continuations belongs with which history. This is the problem of state-space collision as mentioned in Section 4.3. In the case of two speakers, the problem could be resolved by noting that within the speaker-independent space covered by the generic models, each particular source was falling into a particular subspace, determined by the particular characteristics of each speaker. By updating the source models to incorporate that observation, the symmetry could be broken and the ambiguity resolved. This process bears strong similarities to model adaptation in speech recognition, where a generic model is altered (for instance, through an affine transformation of its codeword centers) to achieve a better fit to the signal so far observed [23]. In model-based separation, where separation power derives from the specificity of match between model and source, there is great benefit to be had from online tuning of certain model parameters such as gain, spectral balance, spatial location, and so on, that may vary independently of the core source signal and are thus amenable to being described via separate factors.

4.5.3 Relations to Other Separation Approaches

CASA is frequently contrasted with independent component analysis (ICA) [3, 18] as a very different approach to separating sources that is based on complex and heuristic perceptual models rather than simple, intrinsic properties. At the same time, Smaragdis [40] has argued that the psychoacoustic principles of auditory scene analysis are more compactly explained simply as aspects of more fundamental properties such as the independence of sources. Unlike ICA, the model-based approach we have described assumes the independence of the source signals (by multiplying together their individual probabilities $p[s_i|M_i]$), whereas ICA would calculate and attempt to maximize some measure of this independence. The approaches can, however, be complementary; model parameters may provide an alternative domain in which to pursue independence [19], and prior source models can help with problems such as correctly permuting the independent sources found in separate frequency bands in spectral-domain convolutive ICA [39].

Another closely related approach is sparse decomposition [27, 46], in which a signal or mixture is expressed as the sum of a set of scaled basis functions drawn from an overcomplete dictionary, that is, a dictionary with more entries than there are degrees of freedom in the domain. Because the dictionary is overcomplete, there are many combinations of basis functions that will match the observation with equal accuracy, so additional objectives must be specified. By finding the solution with the minimum total absolute magnitude of basis scaling coefficients, the solution is driven to be sparse, with the majority of values close to zero, that is, the signal has been modeled by combining the smallest number of dictionary elements. When the dictionary adequately covers the sources, and the sources have distinct bases, the source separation is of very high quality. This is in a sense a generalization of the large memorization models discussed in this chapter, since, for linear combination, if the basis sets consist of the codewords and the signal is a match to one of them, then sparse decomposition will find the appropriate codeword and the decomposition will be a single nonzero coefficient. In addition, because the coefficients are scalars, sparse decomposition automatically encompasses all scaled versions of its basis functions, as well as linear combinations, that is, the entire subspace spanned by its bases. At the same time, specification of the overcomplete dictionaries is a critical part of the problem with no single solution, and scaled and/or combined instances of the codewords may not be appropriate to the problem. Searching for the sparse solution is also a major computational challenge.

Given that CASA is motivated by the goal of reproducing the source organization achieved by listeners, we may ask if model-based scene analysis bears any relation to perceptual organization. In his seminal account of sound organization in listeners, Bregman [5] draws a sharp distinction between primitive and schema-based auditory scene analysis. Primitive processes include the fusion of simultaneous harmonics and energy sharing a common onset. Because these organizing principles reflect the intrinsic physics of sound-producing mechanisms in the world, Bregman argues that these abilities come ready-made in the auditory system, since, based on the engineering received wisdom of his day, he sees it as needlessly inefficient to require each individual to learn (acquire) these regularities from the environment when they could be preprogrammed by evolution. Schema-based organization, by contrast, is explicitly based on an individual's specific, recent and/or repeated experiences, and is needed to explain preferential abilities to segregate, for instance, voices in a particular language, as well as the way that certain segregation abilities can improve with practice or effort, not the kind of properties one would expect from low-level, preprogrammed neural hardware.

Although the model-based approach may seem like a natural fit to Bregman's schema, some caution is advised. First, the kinds of perceptual phenomena that schema are used to explain, such as learning of specific large-scale patterns like a particular melody or a new language, are not particularly addressed by the techniques described above, which are still mainly focused at low-level moment-to-moment signal separation. Second, we have argued that model-learning approaches can actually do away with the need for separate, hard-wired primitive mechanisms, since these regular patterns could be learned relatively easily from the environment. In the 15 years since Bregman's book was published, there has been a definite shift in favor of engineering systems that learn from data rather than coming preprogrammed with explicit rules. Although some level of environment-adapted hardware specialization is obviously needed, it is at least plausible to argue that evolution could have left the majority of real-world regularities to be acquired by the individual (rather than hard-coding them), finding this to be the most efficient balance between functionality and efficiency of genetic encoding. Although there is strong evidence for a distinction in listeners between how harmonics are fused into tones, versus higher-level phenomena such as how phonemes are fused into words, it is interesting to note that these could both be the result of making inferences based on patterns learned from the environment, where there might be a spectrum of timescales and prior confidences associated with different classes of learned patterns, yet each could constitute an instance of a single basic memorization process.

4.6 CONCLUSIONS

Source models are an effective and satisfying way to express the additional constraints needed to achieve single-channel source separation or scene analysis. Many approaches can be cast into the hypothesis search framework, and the probabilistic perspective can be a powerful domain for integrating different information and constraints. Major problems remain, however: simplistic modeling schemes, such as attempting to memorize every possible short time frame that a source can emit, are inadequate for the rich variety seen in real-world sources such as human speech, and any rich model presents an enormous search space that cannot really be covered except with special, domain-specific tricks to prune hypotheses under consideration. Both these areas (better models and more efficient search) present major areas for future research in model-based scene analysis, and in particular can benefit from innovations in signal description and analysis described elsewhere in this volume.

REFERENCES

1. F. R. Bach and M. I. Jordan. Blind one-microphone speech separation: A spectral learning approach. In *Advances in Neural Information Processing Systems (NIPS),* volume 17. MIT Press, Cambridge MA, 2004.
2. J. Barker, M. P. Cooke, and D. P. W. Ellis. Decoding speech in the presence of other sources. *Speech Communication,* 42:5–25, 2005.
3. A. J. Bell and T. J. Sejnowski. An information-maximization approach to blind separation and blind deconvolution. *Neural Computation,* 7(6):1129–1159, 1995.
4. T. Blumensath and M. Davies. Unsupervised learning of sparse and shift-invariant decompositions of polyphonic music. In *Proceedings of the IEEE International Conference on Acoustics, Speech, and Signal Processing,* volume V, pp. 497–500, Montreal, 2004.
5. A. S. Bregman. *Auditory Scene Analysis.* Bradford Books, MIT Press, Cambridge, MA, 1990.
6. G. J. Brown and M. P. Cooke. Computational auditory scene analysis. *Computer Speech and Language,* 8:297–336, 1994.
7. M. P. Cooke, P. D. Green, L. B. Josifovski, and A. Vizinho. Robust automatic speech recognition with missing and unreliable acoustic data. *Speech Communication,* 34(3): 267–285, 2001.

8. C. J. Darwin and R. P. Carlyon. Auditory grouping. In B. C. J. Moore, editor, *The Handbook of Perception and Cognition,* volume 6, *Hearing,* pp. 387–424. Academic Press, New York, 1995.
9. D. P. W. Ellis. *Prediction-Driven Computational Auditory Scene Analysis.* PhD thesis, Department of Electrical Engineering and Computer Science, MIT, 1996.
10. D. P. W. Ellis. The weft: A representation for periodic sounds. In *Proceedings of the IEEE International Conference on Acoustics, Speech, and Signal Processing,* volume II, pp. 1307–1310, Munich, 1997.
11. D. P. W. Ellis. Evaluating speech separation systems. In P. Divenyi, editor, *Speech Separation by Humans and Machines,* chapter 20, pp. 295–304. Kluwer, Norwell, MA, 2004.
12. M. J. F. Gales and S. J. Young. HMM recognition in noise using parallel model combination. In *Proceedings of Eurospeech-93,* volume 2, pp. 837–840, 1993.
13. M. R. Garey, D. S. Johnson, and H. S.Witsenhausen. The complexity of the generalized Lloyd-max problem. *IEEE Transactions on Information Theory,* 28(2):255–256, 1982.
14. J. Garofolo. Getting started with the DARPA TIMIT CDROM: An acoustic phonetic continuous speech database, 1993. Available at http://www.ldc.upenn.edu/Catalog/LDC93S1.html.
15. Z. Ghahramani and M. I. Jordan. Factorial hidden Markov models. *Machine Learning,* 29:245–273, 1997.
16. R. M. Gray. Vector quantization. *IEEE ASSP Magazine,* 1(2):4–29, 1984.
17. G. Hu and D. L. Wang. Monaural speech segregation based on pitch tracking and amplitude modulation. *IEEE Transactions on Neural Networks,* 15(5):1135–1150, 2004.
18. A. Hyvärinen. Survey on independent component analysis. *Neural Computing Surveys,* 2:94–128, 1999.
19. G.-J. Jang and T.-W. Lee. A probabilistic approach to single channel blind signal separation. In *Advances in Neural Information Processing Systems (NIPS),* pp. 1173–1180, 2002.
20. T. Kristjansson. Personal communication, 2003.
21. T. Kristjansson, H. Attias, and J. Hershey. Single microphone source separation using high resolution signal reconstruction. In *Proceedings of the IEEE International Conference on Acoustics, Speech, and Signal Processing,* volume II, pp. 817–820, Montreal, 2004.
22. T. Kristjansson and J. Hershey. High resolution signal reconstruction. In *Proceedings of the IEEE Workshop on Automatic Speech Recognition and Understanding ASRU,* 2003.
23. C. J. Leggetter and P. C. Woodland. Maximum likelihood linear regression for speaker adaptation of continuous density hidden Markov models. *Computer Speech and Language,* 9:171–186, 1995.
24. R. G. Leonard. A database for speaker independent digit recognition. In *Proceedings of the IEEE International Conference on Acoustics, Speech, and Signal Processing,* volume III, pp. 42–45, 1984.
25. S. H. Nawab, T. F. Quatieri, and J. S. Lim. Signal reconstruction from short-time Fourier transform magnitude. *IEEE Transactions on Acoustics, Speech, and Signal Processing,* 31(4):986–998, 1983.
26. J. Nix, M. Kleinschmidt, and V. Hohmann. Computational auditory scene analysis by

using statistics of high-dimensional speech dynamics and sound source direction. In *Proceedings of Eurospeech,* pp. 1441–1444, Geneva, 2003.

27. B. A. Pearlmutter and A. M. Zador. Monaural source separation using spectral cues. In *Proceedings of Fifth International Conference on Independent Component Analysis ICA-2004,* 2004.
28. T. F. Quatieri. *Discrete-Time Speech Signal Processing: Principles and Practice.* Prentice-Hall, Upper Saddle River, NJ, 2002.
29. L. R. Rabiner. A tutorial on hidden Markov models and selected applications in speech recognition. *Proceedings of the IEEE,* 77(2):257–286, 1989.
30. M. J. Reyes-Gómez. *Statistical Graphical Models for Scene Analysis, Source Separation and Other Audio Applications.* PhD thesis, Department of Electrical Engineering, Columbia University, New York, 2005.
31. M. J. Reyes-Gómez, D. P. W. Ellis, and N. Jojic. Multiband audio modeling for single channel acoustic source separation. In *Proceedings of the IEEE International Conference on Acoustics, Speech, and Signal Processing,* volume V, pp. 641–644, Montreal, 2004.
32. M. J. Reyes-Gómez, N. Jojic, and D. P. W. Ellis. Deformable spectrograms. In *Proceedings of AI and Statistics,* Barbados, pp. 285–292, 2005.
33. M. Reyes-Gómez, B. Raj, and D. P. W. Ellis. Multi-channel source separation by beamforming trained with factorial hmms. In *Proceedings of the IEEE Workshop on Applications of Signal Processing to Audio and Acoustics,* pp. 13–16, Mohonk NY, October 2003.
34. A. E. Rosenberg, C.-H. Lee, and F. K. Soong. Cepstral channel normalization techniques for HMM-based speaker verification. In *International Conference on Speech and Language Proceedings,* pp. 1835–1838, Yokohama, September 1994.
35. S. Roweis. One-microphone source separation. In *Advances in Neural Information Processing Systems 11,* pp. 609–616. MIT Press, Cambridge, MA, 2001.
36. S. Roweis. Factorial models and refiltering for speech separation and denoising. In *Proceedings of Eurospeech,* Geneva, 2003.
37. M. Seltzer, B. Raj, and R. M Stern. Speech recognizer-based microphone array processing for robust hands-free speech recognition. In *Proceedings of the IEEE International Conference on Acoustics, Speech, and Signal Processing,* volume I, pp. 897–900, Orlando, 2002.
38. Y. Shao and D. L. Wang. Model-based sequential organization in cochannel speech. *IEEE Transactions on Audio, Speech and Language Processing,* 14:289–298, 2006.
39. P. Smaragdis. Blind separation of convolved mixtures in the frequency domain. In *Proceedings of the International Workshop on Independent and Artificial. Neural Networks,* Tenerife, 1998.
40. P. Smaragdis. Exploiting redundancy to construct listening systems. In P. Divenyi, editor, *Speech Separation by Humans and Machines.* Kluwer, Norwell, MA, 2004.
41. P. Smaragdis. Non-negative matrix factor deconvolution: Extraction of multiple sound sources from monophonic inputs. In *Proceedings of the International Congress on Independent Component Analysis and Blind Signal Separation ICA-2004,* Granada, Spain, September 2004.
42. A. Varga and R. Moore. Hidden Markov model decomposition of speech and noise. In *Proceedings of the IEEE International Conference on Acoustics, Speech, and Signal Processing,* pp. 845–848, 1990.

43. K. Q. Weinberger, B. D. Packer, and L. K. Saul. Nonlinear dimensionality reduction by semidefinite programming and kernel matrix factorization. In *Proceedings of the Tenth International Workshop on Artificial Intelligence and Statistics,* pp. 381–388, Barbados, January 2005.

44. M. Weintraub. *A Theory and Computational Model of Auditory Monoaural Sound Separation.* PhD thesis, Department of Electrical Engineering, Stanford University, 1985.

45. O. Yilmaz and S. Rickard. Blind separation of speech mixtures via time–frequency masking. *IEEE Transactions on Signal Processing,* 52(7):1830–1847, July 2004.

46. M. Zibulevsky and B. A. Pearlmutter. Blind source separation by sparse decomposition in a signal dictionary. *Neural Computation,* 13(4):863–882, April 2001.

CHAPTER 5

Binaural Sound Localization

RICHARD M. STERN, GUY J. BROWN, and DELIANG WANG

5.1 INTRODUCTION

We listen to speech (as well as to other sounds) with two ears, and it is quite remarkable how well we can separate and selectively attend to individual sound sources in a cluttered acoustical environment. In fact, the familiar term "cocktail party processing" was coined in an early study of how the binaural system enables us to selectively attend to individual conversations when many are present, as in, of course, a cocktail party [23]. This phenomenon illustrates the important contribution that binaural hearing makes to auditory scene analysis, by enabling us to localize and separate sound sources. In addition, the binaural system plays a major role in improving speech intelligibility in noisy and reverberant environments.

The primary goal of this chapter is to provide an understanding of the basic mechanisms underlying binaural localization of sound, along with an appreciation of how binaural processing by the auditory system enhances the intelligibility of speech in noisy acoustical environments, and in the presence of competing talkers. Like so many aspects of sensory processing, the binaural system offers an existence proof of the possibility of extraordinary performance in sound localization and signal separation, but it does not yet provide a very complete picture of how this level of performance can be achieved with the contemporary tools of computational auditory scene analysis (CASA).

We first summarize in Section 5.2 the major physical factors that underly binaural perception, and we briefly summarize some of the classical physiological findings that describe mechanisms that could support some aspects of binaural processing. In Section 5.3 we briefly review some of the basic psychoacoustical results that have motivated the development of most of the popular models of binaural interaction. These studies have typically utilized very simple signals such as pure tones or broadband noise, rather than more interesting and ecologically relevant signals such as speech or music. We extend this discussion to results involving the spatial perception of multiple sound sources in Section 5.4. In Section 5.5 we introduce and discuss two of the most popular types of models of binaural interaction, cross-corre-

Computational Auditory Scene Analysis. Edited by DeLiang Wang and Guy J. Brown

lation-based models and the equalization–cancellation (EC) model. Section 5.6 then discusses how the cross-correlation model may be utilized within a CASA system, by providing a means for localizing multiple (and possibly moving) sound sources. Binaural processing is a key component of practical auditory scene analysis, and aspects of binaural processing will be discussed in greater detail later in Chapters 6 and 7.

5.2 PHYSICAL AND PHYSIOLOGICAL MECHANISMS UNDERLYING AUDITORY LOCALIZATION

5.2.1 Physical Cues

A number of factors affect the spatial aspects of how a sound is perceived. Lord Rayleigh's "duplex theory" [86] was the first comprehensive analysis of the physics of binaural perception, and his theory remains basically valid to this day, with some extensions. As Rayleigh noted, two physical cues dominate the perceived location of an incoming sound source, as illustrated in Fig. 5.1. Unless a sound source is located directly in front of or behind the head, sound arrives slightly earlier in time at the ear that is physically closer to the source, and with somewhat greater intensity. This *interaural time difference* (ITD) is produced because it takes longer for the sound to arrive at the ear that is farther from the source. The *interaural intensity difference* (IID) is produced because the "shadowing" effect of the head prevents some of the incoming sound energy from reaching the ear that is turned away from the direction of the source. The ITD and IID cues operate in complementary ranges

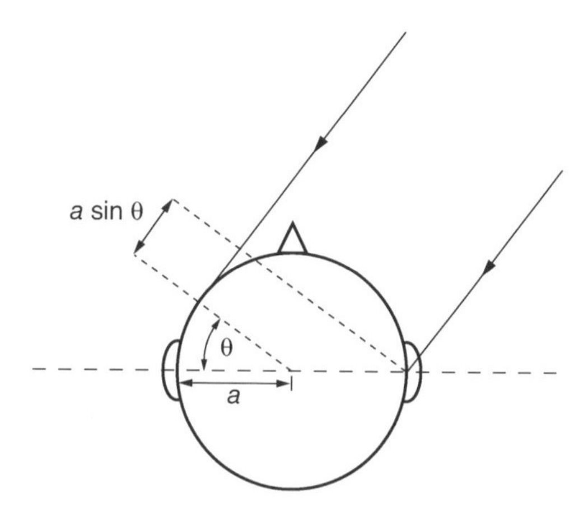

Figure 5.1 Interaural differences of time and intensity impinging on an ideal spherical head from a distant source. An interaural time delay (ITD) is produced because it takes longer for the signal to reach the more distant ear. An interaural intensity difference (IID) is produced because the head blocks some of the energy that would have reached the far ear, especially at higher frequencies.

of frequencies, at least for simple sources in a free-field (such as a location outdoors or in an anechoic chamber). Specifically, IIDs are most pronounced at frequencies above approximately 1.5 kHz because it is at those frequencies that the head is large compared to the wavelength of the incoming sound, producing substantial reflection (rather than total diffraction) of the incoming sound wave. Interaural timing cues, on the other hand, exist at all frequencies, but for periodic sounds they can be decoded unambiguously only for frequencies for which the maximum physically possible ITD is less than half the period of the waveform at that frequency. Since the maximum possible ITD is about 660 μs for a human head of typical size, ITDs are generally useful only for stimulus components below about 1.5 kHz. Note that for pure tones, the term *interaural phase delay* (IPD) is often used, since the ITD corresponds to a phase difference.

If the head had a completely spherical and uniform surface, as in Fig. 5.1, the ITD produced by a sound source that arrives from an azimuth of θ radians can be approximately described (e.g., [64, 65, 106]) using diffraction theory by the equation

$$\tau = (a/c)\, 2 \sin \theta \tag{5.1a}$$

for frequencies below approximately 500 Hz, and by the equation

$$\tau = (a/c)(\theta + \sin \theta) \tag{5.1b}$$

for frequencies above approximately 2 kHz. In the equations above, a represents the radius of the head (approximately 87.5 mm) and c represents the speed of sound.

The actual values of IIDs are not as well predicted by wave diffraction theory, but they can be measured using probe microphones in the ear and other techniques. They have been found empirically to depend on the angle of arrival of the sound source, frequency, and distance from the sound source (at least when the source is extremely close to the ear). IIDs produced by distant sound sources can become as large as 25 dB in magnitude at high frequencies, and the IID can become greater still when a sound source is very close to one of the two ears.

The fissures of the outer ears (or *pinnae*) impose further spectral coloration on the signals that arrive at the eardrums. This information is especially useful in a number of aspects of the localization of natural sounds occurring in a free-field, including localization in the vertical plane, and the resolution of front–back ambiguities in sound sources. Although measurement of and speculation about the spectral coloration imposed by the pinnae have taken place for decades (e.g., [2, 51, 65, 81, 102, 105]), the "modern era" of activity in this area began with the systematic and carefully controlled measurements of Wightman and Kistler and others (e.g., [82, 124, 125, 126]), who combined careful instrumentation with comprehensive psychoacoustical testing. Following procedures developed by Mehrgardt and Mellert [81] and others, Wightman and Kistler and others used probe microphones in the ear to measure and describe the transfer function from sound source to eardrum in anechoic environments. This transfer function is commonly referred to as the *head-*

related transfer function (HRTF), and its time-domain analog is the *head-related impulse response* (HRIR). Among other attributes, measured HRTFs show systematic variations of frequency response above 4 kHz as a function of azimuth and elevation. There are also substantial differences in frequency response from subject to subject. HRTF measurements and their inverse Fourier transforms have been key components in a number of laboratory and commercial systems that attempt to simulate natural three-dimensional acoustical environments using signals presented through headphones (e.g., [7, 43]).

As Wightman and Kistler note [126], the ITD as measured at the eardrum for broadband stimuli is approximately constant over frequency and it depends on azimuth and elevation in approximately the same way from subject to subject. Nevertheless, a number of different azimuths and elevations will produce the same ITD, so the ITD does not unambiguously signal source position. IIDs measured at the eardrum exhibit much more subject-to-subject variability, and for a given subject the IID is a much more complicated function of frequency, even for a given source position. Wightman and Kistler suggest that because of this, IID information in individual frequency bands is likely to be more useful than overall IID.

5.2.2 Physiological Estimation of ITD and IID

There have been a number of physiological studies that have described cells that are likely to be useful in extracting the basic cues used in auditory spatial perception, as have been described in a number of comprehensive reviews (e.g., [27, 29, 87, 66, 128]). Any consideration of neurophysiological mechanisms that potentially mediate binaural processing must begin with a brief discussion of the effects of processing of sound by the auditory periphery. As was first demonstrated by von Békésy [122] and by many others since, the mechanical action of the cochlea produces a frequency-to-place transformation. Each of the tens of thousands of fibers of the auditory nerve for each ear responds to mechanical stimulation along only a small region of the cochlea, so the neural response of each fiber is highly frequency specific, and the stimulus frequency to which each fiber is most sensitive is referred to as the *characteristic frequency* (CF) for that fiber. This frequency-specific "tonotopic" representation of sound in each channel of parallel processing is preserved at virtually all levels of auditory processing. In addition, the response of a fiber with a low CF is "synchronized" to the detailed time structure of a low-frequency sound, in that neural spikes are far more likely to occur during the negative portion of the pressure waveform than during the positive portion.

The auditory-nerve response to an incoming sound is frequently modeled by a bank of linear bandpass filters (that represent the frequency selectivity of cochlear processing), followed by a series of nonlinear operations at the outputs of each filter that include half-wave rectification, nonlinear compression and saturation, and "lateral" suppression of the outputs of adjacent frequency channels (that represent the subsequent transduction to a neural response). A number of computational models incorporating varying degrees of physiological detail or abstraction have been developed that describe these processes (e.g., [75, 80, 89, 133]; see also Section

1.3.2). One reason for the multiplicity of models is that it is presently unclear which aspects of nonlinear auditory processing are the most crucial for the development of improved features for robust automatic speech recognition, or for the separation of simultaneously presented signals for CASA. Any physiological mechanism that extracts the ITD or IID of a sound (as well as any other type of information about it) must operate on the ensemble of narrowband signals that emerge from the parallel channels of the auditory processing mechanism, rather than on the original sounds that are presented to the ear.

The estimation of ITD is probably the most critical aspect of binaural processing. As will be discussed below in Section 5.5.2, many models of binaural processing are based on the cross correlation of the signals to the two ears after processing by the auditory periphery, or based on other functions that are closely related to cross correlation. The physiological plausibility of this type of model is supported by the existence of cells first reported by Rose et al. [95] in the inferior colliculus in the brain stem. Such cells appear to be maximally sensitive to signals presented with a specific ITD, independent of frequency. This delay is referred to as a *characteristic delay* (CD). Cells exhibiting similar response have been reported by many others in other parts of the brain stem, including the medial superior olive and the dorsal nucleus of the lateral lemniscus.

Several series of measurements have been performed that characterize the distribution of the CDs of ITD-sensitive cells in the inferior colliculus (e.g., [129, 66]), and the medial geniculate body (e.g., [109]). The results of most of these studies indicate that ITD-sensitive cells tend to exhibit CDs that lie in a broad range of ITDs, with the density of CDs decreasing as the absolute value of the ITD increases. Although most of the CDs appear to occur within the maximum ITD that is physically possible for a point source in a free-field for a particular animal at a given frequency, there is also a substantial number of ITD-sensitive cells with CDs that fall outside this "physically plausible" range. In a recent series of studies, McAlpine and his collaborators have argued that most ITD-sensitive units exhibit characteristic delays that occur in a narrow range that is close to approximately one-eighth of the period of a cell's characteristic frequency (e.g., [78]), at least for some animals.

The anatomical origin of the characteristic delays has been the source of some speculation. Although many physiologists believe that the delays are of neural origin, caused either by slowed conduction delays or by synaptic delays (e.g., [22, 132]), other researchers (e.g., [101, 104]) have suggested that the characteristic delays could also be obtained if higher processing centers compare timing information derived from auditory-nerve fibers with different CFs. In general, the predictions of binaural models are unaffected by whether the internal delays are assumed to be caused by neural or mechanical phenomena.

A number of researchers have also reported cells that appear to respond to IIDs at several levels of the brain stem (e.g., [11, 19]). Since the rate of neural response to a sound increases with increasing intensity (at least over a limited range of intensities), IIDs could be detected by a unit that has an excitatory input from one ear and an inhibitory input from the other.

5.3 SPATIAL PERCEPTION OF SINGLE SOURCES

5.3.1 Sensitivity to Differences in Interaural Time and Intensity

Humans are remarkably sensitive to small differences in interaural time and intensity. For low-frequency pure tones, for example, the *just-noticeable difference* (JND) for ITDs is on the order of 10 μs, and the corresponding JND for IIDs is on the order of 1 dB (e.g., [34, 53]). The JND for ITD depends on the ITD, IID, and frequency with which a signal is presented. The binaural system is completely insensitive to ITD for narrowband stimuli above about 1.5 kHz, although it does respond to low-frequency envelopes of high-frequency stimuli, as will be noted below. JNDs for IID are a small number of decibels over a broad range of frequencies. Sensitivity to small differences in interaural correlation of broadband noise sources is also quite acute, as a decrease in interaural correlation from 1 to 0.96 is readily discernable (e.g., [40, 45, 90]).

5.3.2 Lateralization of Single Sources

Until fairly recently, most studies of binaural hearing have involved the presentation of single sources through headphones, such as clicks (e.g., [32, 49]), bandpass noise (e.g., [120]), pure tones (e.g., [34, 96, 130]), and amplitude-modulated tones (e.g., [3, 52]). Such experiments measure the *lateralization* of the source (i.e., its apparent lateral position within the head) and are therefore distinct from *localization* experiments (in which the task is to judge the apparent direction and distance of the source outside the head).

Unsurprisingly, the perceived lateral position of a narrowband binaural signal is a periodic (but not sinusoidal) function of the ITD with a period equal to the reciprocal of the center frequency. It was found that the perceived lateralization of a narrowband signal such as a pure tone was affected by its ITD only at frequencies below approximately 1.5 kHz, consistent with the comments on the utility of temporal cues at higher frequencies stated above. Nevertheless, the perceived laterality of broader-band signals, such as amplitude-modulated tones and bandpass-filtered clicks, can be affected by the ITD with which they are presented, even if all components are above 1.5 kHz, provided that the stimuli produce low-frequency envelopes (e.g., [4, 5, 52, 79]).

IIDs, in contrast, generally affect the lateral position of binaural signals of all frequencies. Under normal circumstances, the perceived laterality of stimuli presented through headphones is a monotonic function of IID, although the exact form of the function relating lateral position to IID depends upon the nature of the stimulus, including its frequency content, as well as the ITD with which the signal is presented. Under normal circumstances, IIDs of magnitude greater than approximately 20 dB are perceived close to one side of the head, although discrimination of small changes in IID based on lateral position cues can be made at IIDs of these and larger magnitudes (e.g., [34, 53]).

Many studies have been concerned with the ways in which information related to the ITD and IID of simple and complex sources interact with each other. If a binaural signal is presented with an ITD of less than approximately 200 μs and an IID of 5 dB in magnitude, its perceived lateral position can approximately be described by a linear combination of the two cues. Under such conditions, the relative salience of ITD and IID was frequently characterized by the time–intensity trading ratio, which can range from approximately 20 to 200 μs/dB, depending on the type of stimulus, its loudness, the magnitude of the ITDs and IIDs presented, and other factors (cf. [39]). Although it had been suggested by Jeffress [59] and others that this time–intensity conversion might be a consequence of the observed decrease in physiological latency in response to signals of greater intensity, lateralization studies involving ITDs and IIDs of greater magnitude (e.g., [34, 96]) indicate that response latency alone cannot account for the form of the data. Most contemporary models of binaural interaction (e.g., [8, 70, 110]) assume a more central form of time–intensity interaction.

5.3.3 Localization of Single Sources

As noted above, Wightman and Kistler developed a systematic and practical methodology for measuring the HRTFs that describe the transformation of sounds in the free-field to the ears. They used the measured HRTFs both to analyze the physical attributes of the sound pressure impinging on the eardrums, and to synthesize "virtual stimuli" that could be used to present through headphones a simulation of a particular free-field stimulus that was reasonably accurate (at least for the listener used to develop the HRTFs) [124, 125, 126]. These procedures have been adopted by many other researchers.

Wightman and Kistler and others have noted that listeners are able to describe the azimuth and elevation of free-field stimuli consistently and accurately. Localization judgments obtained using "virtual" headphone simulations of the free-field stimuli are generally consistent with the corresponding judgments for the actual free-field signals, although the effect of elevation change is less pronounced and a greater number of front-to-back confusions in location is observed [125]. On the basis of various manipulations of the virtual stimuli, they also conclude that under normal circumstances the localization of free-field stimuli is dominated by ITD information, especially at the lower frequencies, that ITD information must be consistent over frequency for it to play a role in sound localization, and that IID information appears to play a role in diminishing the ambiguities that give rise to front–back confusions of position [126]. Although the interaural cues for lateralization are relatively robust across subjects, fewer front-to-back and other confusions are experienced in the localization of simulated free-field stimuli if the HRTFs used for the virtual sound synthesis are based on a subject's own pinnae [123, 84]. Useful physical measurements of ITD, IID, and other stimulus attributes can also be obtained using an anatomically realistic manikin such as the popular Knowles Electronics Manikin for Acoustic Research (KEMAR) [18]. Examples of HRTFs (and the corresponding HRIRs) recorded from a KEMAR manikin are shown in Fig. 5.2. Experimental measurements indicate that localization judgments are more accurate when the virtual source is spatialized using HRTFs obtained by direct measurement

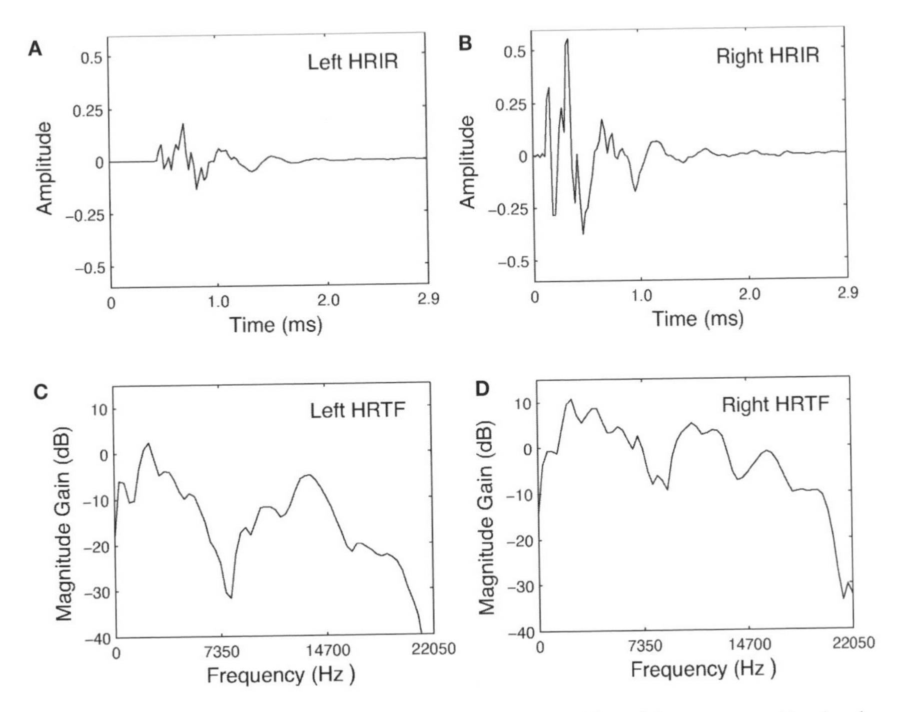

Figure 5.2 Head-related impulse responses (HRIRs) (**A, B**) and the corresponding head-related transfer functions (HRTFs) (**C, D**) recorded at the left and right ears of a KEMAR manikin, in response to a source placed at an azimuth of 40° to the right of the head with 0° elevation. Note that the stimulus is more intense in the right ear, and arrives first in the right ear before reaching the left ear. Plotted from data recorded by Gardner and Martin [47], with permission of the authors.

in human ears, rather than from an artificial head such as the KEMAR. However, there is a strong learning effect, in which listeners adapt to unfamiliar HRTFs over the course of several experimental sessions [83].

5.3.4 The Precedence Effect

A further complication associated with judging the location of a sound source in a natural acoustic environment, such as an enclosed room, is that sound is reflected from various surfaces before reaching the ears of the listener. However, despite the fact that reflections arise from many directions, listeners are able to determine the location of the direct sound quite accurately. Apparently, directional cues that are due to the direct sound (the "first wavefront") are given a higher perceptual weighting than those due to reflected sound. The term *precedence effect* is used to describe this phenomenon (see [72] for a review). Further discussion of the precedence effect can be found in Chapter 7, which deals with the effect of reverberation on human and machine hearing.

5.4 SPATIAL PERCEPTION OF MULTIPLE SOURCES

5.4.1 Localization of Multiple Sources

Human listeners can localize a single sound source rather accurately. One accuracy measure for azimuth localization is the minimum audible angle (MAA), which refers to the smallest detectable change in angular position. For sinusoidal signals presented on the horizontal plane, spatial resolution is highest for sounds coming from the median plane (directly in front of the listener) with about 1° MAA, and it deteriorates markedly when stimuli are moved to the side; for example, the MAA is about 7° for sounds originating at 75° to the side [8, 85]. In terms of making an absolute judgment of the spatial location of a sound, the average error is about 5° for broadband stimuli presented on the median plane, and increases up to 20° for sounds from the side [56, 67].

In the context of this book, a very relevant issue is the ability of human listeners to localize a sound in multisource scenarios. An early study was conducted by Jacobsen [58] who observed that, for pure tones presented with a white-noise masker, the MAA is comparable to that when no masker is presented so long as the signal-to-noise ratio (SNR) is relatively high (10 to 20 dB). In a comprehensive investigation, Good and Gilkey [48] examined the accuracy of localization judgments by systematically varying SNR levels and sound azimuths. In their study, the signal is a broadband click train that may originate from any of a large number of spatial locations, and a broadband masker is always located on the median plane. As expected, the accuracy of localization decreases almost monotonically when the SNR is lowered. However, Good and Gilkey found that azimuth localization judgments are not strongly influenced by the interference and remain accurate even at low SNRs. This holds until the SNR is in the negative range, near the detection threshold for the target, beyond which the target will be inaudible due to masking by the interference. The same study also reported that the effect of the masker on localization accuracy is a little stronger for judgments of elevation, and is strongest in the front–back dimension. Using multisource presentation of spoken words, letters and digits, Yost et al. [131] found that utterances that can correctly be identified tend to be correctly localized.

Similar conclusions have been drawn in a number of subsequent studies. For example, Lorenzi et al. [74] observed that localization accuracy is unaffected by the presence of a white-noise masker until the SNR is reduced to the 0–6 dB range. They also reported that the effect of noise is stronger when it is presented at the side than from the median plane. Hawley et al. [50] studied the localization of speech utterances in multisource configurations. Their results show that localization performance is very good even when three competing sentences, each at the same level as the target sentence, are presented at various azimuths. Moreover, the presence of the interfering utterances has little adverse impact on target localization accuracy so long as the target is clearly audible, consistent with the earlier observation by Yost et al. [131]. On the other hand, Drullman and Bronkhorst [35] reported rather poor performance in speech localization in the presence of one to four competing talkers, which may have been caused by the complexity of the task that required the subjects to first detect the target talker and then determine the talker location. They further observed that localization accuracy decreases when the number of interfering talkers increases from two to four.

Langendijk et al. [67] studied listeners' performance in localizing a train of noise bursts presented together with one or two complex tones. Their findings confirm that for positive SNR levels, azimuth localization performance is not much affected by the interference. In addition, they found that the impact of maskers on target localization is greater when azimuth separation between multiple sources is reduced. Using noise bursts as stimuli for both target and masker, Braasch [12] showed that localization performance is enhanced by introducing a difference in onset times between the target and the masker when the SNR is 0 dB.

In summary, the body of psychoacoustical evidence on sound localization in multisource situations suggests that localization and identification of a sound source are closely related. Furthermore, auditory cues that promote auditory scene analysis also appear to enhance localization performance.

5.4.2 Binaural Signal Detection

Although the role that the binaural system plays in sound localization is well known, binaural processing also plays a vital role in detecting sounds in complex acoustical environments. In the following, we review some classical studies on binaural signal detection, and then discuss the ability of binaural mechanisms to enhance the intelligibility of speech in noise.

Classical Binaural Detection. Most of the classical psychoacoustical studies of binaural detection have been performed with simple stimuli such as pure tones, clicks, or broadband noise. Generally, it has been found that the use of binaural processing significantly improves performance for signal detection tasks if the overall ITD, IID, or interaural correlation changes as a target is added to a masker. For example, if a low-frequency tone and a broadband masker are presented monaurally, the threshold SNR that is obtained will generally depend on various stimulus parameters such as the target duration and frequency, and the masker bandwidth. Diotic presentation of the same target and masker (i.e., with identical signals presented to the two ears) will produce virtually the same detection threshold. On the other hand, if either the target or masker are presented with a nonzero ITD, IID, or IPD, the target will become much easier to hear. For example, using 500 Hz tones as targets and broadband maskers, presentation of the stimuli with the masker interaurally in phase and the target interaurally out of phase (the "N_0S_π" configuration) produces a detection threshold that is about 12 to 15 dB lower than the detection threshold observed when the same target and masker are presented monaurally or diotically (the "N_0S_0" configuration) [54]. This "release" from masking is referred to as binaural "unmasking" or the binaural masking level difference (MLD or BMLD).

Detection performance improves for stimuli that produce an MLD because the change in net ITD or IID that occurs when the target is added to the masker is detectable by the binaural processing system at lower SNRs than those needed for monaural detection. The MLD is one of the most robust and extensively studied of all binaural phenomena, particularly for tonal targets and noise maskers. Durlach and Colburn's comprehensive review [39] describes how the MLD depends on stimulus parameters such as SNR, target frequency and duration, masker frequency and bandwidth, and the ITD, IID, and IPD of the target masker. These dependencies are for the most part well described by modern theories of binaural hearing (e.g., [15, 25, 26]).

Binaural Detection of Speech Signals. Although most of the MLD literature was concerned with the detection of tonal targets in noise maskers, the importance of interaural differences for improving speech intelligibility in a noisy environment has been known since at least the 1950s, when Hirsch, Kock, and Koenig first demonstrated that binaural processing can improve the intelligibility of speech in a noisy environment [55, 60, 61]. As reviewed by Zurek [134], subsequent studies (e.g., [20, 21, 33, 77]) identified two reasons why the use of two ears can improve the intelligibility of speech in noise. First, there is a head-shadow advantage that can be obtained by turning one ear toward the direction of the target sound, at the same time increasing the likelihood that masking sources will be attenuated by the shadowing effect of the head if they originate from directions that are away from that ear. The second potential advantage is a *binaural-interaction advantage* that results from the fact that the ITDs and IIDs of the target and masker are different when the sources originate from different azimuths. Many studies confirm that speech intelligibility improves when the spatial separation of a target speech source and competing maskers increases, both for reasons of head-shadowing and binaural interaction.

Levitt and Rabiner [68] developed a simple model that predicts the binaural-interaction advantage that will be obtained when speech is presented with an ITD or IPD in the presence of noise. They predicted speech intelligibility by first considering the ITD or IPD and SNR of the target and masker at each frequency of the speech sound, and assuming that the binaural-interaction advantage for each frequency component could be predicted by the MLD for a pure tone of the same frequency in noise, using the same interaural parameters for target and masker. The effects of the MLD are then combined across frequency using standard articulation index theory (e.g., [42, 44, 63]).

Zurek [134] quantified the relative effects of the head-shadow advantage and the binaural advantage for speech in the presence of a single masking source. He predicted the head-shadow of speech from the SNR at the "better" ear for a particular stimulus configuration, and the binaural-interaction advantage from the MLD expected from the stimulus components at a particular frequency, again combining this information across frequency by using articulation index theory. Zurek found the predictions of this model to be generally consistent with contemporary data that described the average dependence of intelligibility on source direction and listening mode, although it did not model variations in data from subject to subject. Hawley et al. [50] extended Zurek's approach in a series of experiments that employed multiple streams of competing speech maskers. The predictions of both models indicate that many phenomena can be accounted for simply by considering the monaural characteristics of the signals that arrive at the "better" ear, but that processing based on binaural interaction plays a significant role for certain configurations of targets and competing maskers.

5.5 MODELS OF BINAURAL PERCEPTION

In this section, we review some of the classical and more recent models of binaural interaction that have been applied to simple and more complex binaural phenomena. We begin with a discussion of two seminal theories of binaural interaction, the *coincidence-based model* of Jeffress [59] and the *equalization–cancellation model* (EC model) of Durlach [37, 38]. Most current binaural models trace their lineage to one (and in some cases both) of these theories. We continue with a discussion of modern realizations of the Jeffress model that typically include a model for auditory-nerve activity, a mechanism that extracts a frequency-dependent measurement of the interaural cross correlation of the signals, along with some additional processing to incorporate the effects of IIDs and to develop the perceptual representation that is needed to perform the particular psychoacoustical task at hand. In addition to this discussion, the reader is referred to a number of excellent reviews of models of binaural interaction that have appeared over the years [13, 27, 28, 29, 115, 116, 119].

5.5.1 Classical Models of Binaural Hearing

The Jeffress Hypothesis. Most modern computational models of binaural perception are based on Jeffress's description of a neural "place" mechanism that would enable the extraction of interaural timing information [59]. Jeffress postulated a mechanism that consisted of a number of central neural units that recorded coincidences in neural firings from two peripheral auditory-nerve fibers, one from each ear, with the same CF. He further postulated that the neural signal coming from one of the two fibers is delayed by a small amount that is fixed for a given fiber pair, as in the block diagram of Fig. 5.3. Because of the synchrony in the response of low-frequency fibers to low-frequency stimuli, a given binaural coincidence-counting unit at a particular frequency will produce maximal output when the external stimulus ITD at that frequency is exactly compensated for by the internal delay of the fiber pair. Hence, the external ITD of a simple stimulus could be inferred by determining the internal delay that has the greatest response over a range of frequencies. Although the delay mechanism was conceptualized by Jeffress and others in the form of the ladder-type delay shown in Fig. 5.3, such a structure is only one of several possible realizations. These coincidence-counting units can be thought of as mathematical abstractions of the ITD-sensitive units that were first described by Rose et al. decades later, as discussed in Section 5.2.2 above. The important characteristic-delay parameter of the ITD-sensitive units is represented by the difference in total delay incurred by the neural signals from the left and right

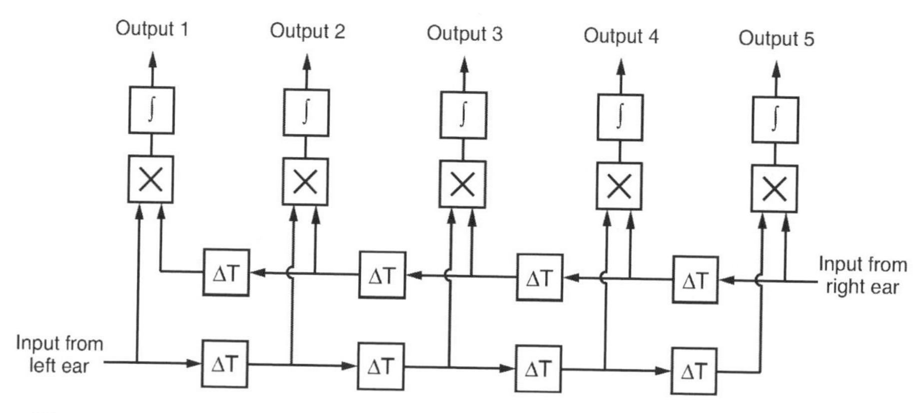

Figure 5.3 Schematic representation of the Jeffress place mechanism. Boxes containing crosses are correlators (multipliers) that record coincidences of neural activity from the two ears after the internal delays (ΔT).

ears that are input to a particular coincidence-counting unit. This parameter will be referred to as the net internal delay τ for that particular unit. As will be discussed below, the short-term average of a set of such coincidence outputs at a particular CF plotted as a function of their internal delay τ is an approximation to the short-term cross-correlation functions of the neural signals arriving at the coincidence detectors. Licklider [69] proposed that such a mechanism could also be used to achieve an autocorrelation of neural signals for use in models of pitch perception (see Section 2.3.2).

We note that the interaural coincidence operation cannot by itself account for effects related to the IID of the stimulus. Jeffress proposed the *latency hypothesis,* which is based on the common observation that the neural response to more intense sounds tends to be initiated more rapidly than the response to less intense sounds. Jeffress postulated that this effect enabled IIDs to be converted to ITDs at the level of the peripheral auditory response to a sound. It is now widely believed that the interaction between the effects of IIDs and ITDs is mediated at a more central level.

The Equalization–Cancellation Model. The equalization–cancellation model (or EC model) has also been extremely influential for decades, no doubt because of its conceptual simplicity and because of its ability to describe a number of interesting binaural phenomena. The model was first suggested by Kock [60] and was subsequently developed extensively by Durlach (e.g., [37, 38]). While the EC model was primarily developed to describe and predict the binaural masking-level differences described in Section 5.4.2, it has been applied with some success to other phenomena as well. In the classic application to binaural detection, the EC model assumes that the auditory system transforms the signals arriving at the two ears so that the masker components are "equalized," or made equal to one another to the extent possible. Detection of the target is achieved by "cancelling," or subtracting the signals to the two ears after the equalization operation. Considering the two classical binaural signal configurations described in Section 5.4.2 for MLD experiments, it is clear that signals in the N_0S_0 configuration will not be detectable by binaural processing because if the equalization and cancellation operations are accurate and successful, the target will be cancelled along with the masker. Binaural detection stimuli presented in the N_0S_π configuration should be easily detected, however, because the target is reinforced as the masker is cancelled. Quantitative predictions for the EC model are obtained by specifying limits to the operations used to achieve the cancellation process, as well as sources of internal noise. The performance and limitations of the EC model are discussed in detail in the excellent early review of binaural models by Colburn and Durlach [28].

5.5.2 Cross-Correlation-Based Models of Binaural Interaction

Stimulus-Based Cross-Correlation Models. Although both Jeffress and Licklider had introduced the concept of correlation in their models of binaural interaction and processing of complex signals, the work of Sayers and Cherry (e.g., [24, 96, 97, 98, 99]) represented the first comprehensive attempt to relate the fusion and lateralization of binaural stimuli to their interaural cross correlation. Sayers and Cherry considered the short-time cross-correlation function (or the "running" cross-correlation function) of the stimuli:

$$R(t, \tau) = \int_{-\infty}^{t} x_L(\alpha) x_R(\alpha - \tau) w(t - \alpha) p(\tau)\, d\alpha \tag{5.2}$$

where $x_L(t)$ and $x_R(t)$ are the signals to the left and right ears. The function $w(t)$ represents the temporal weighting of the short-time cross-correlation operation, and is exponential in form in most of Sayers and Cherry's calculations. The function $p(\tau)$ is typically a double-sided decaying exponential that serves to emphasize the contributions of internal delays τ that were small in magnitude. We refer to this type of emphasis as "centrality." The reader should note that Sayers and Cherry's function does not take into account any signal processing by the peripheral auditory system. As is true for all cross-correlation-based models, an additional mechanism is needed to account for the effects of IID.

Sayers and Cherry added a constant proportional to the intensity of the left-ear signal to values of the internal delay τ that were less than zero and a (generally different) constant proportional to the intensity of the right-ear signal to values of τ that were greater than zero. A judgment mechanism then extracted subjective lateral position using the statistic

$$\hat{P} = \frac{I_L - I_R}{I_L + I_R} \tag{5.3}$$

where I_L and I_R are the integrals of the intensity-weighted, short-time cross-correlation function over negative and positive values of τ, respectively. Sayers and Cherry considered the lateral position of a variety of stimuli including speech, pure tones of various frequencies, and click trains, and found that they were predicted at least qualitatively by the above lateralization function or variants of it. Furthermore, their data indicated that the latency hypothesis could not adequately describe all of the complexities of the dependence of perceived laterality on the IID of the stimulus.

Models Incorporating the Auditory-Nerve Response to the Stimuli. Although the models of Sayers and Cherry predicted binaural phenomena from the cross-correlation of the auditory stimulus itself, more recent theories of binaural interaction have been based on the interaural correlation of the *neural response* to the stimulus, rather than to the auditory stimulus itself. Early physiologically based models focused more on the development of closed-form mathematical analytical functions to describe and predict the data under consideration. Nevertheless, the trend over the last several decades has been to use computational models to calculate both the auditory-nerve response to sounds and the subsequent processing of the signal needed to obtain both the representation of the interaural timing information that is associated with the signal, as well as subjective variables such as lateral position that are developed from that representation. The use of computational simulation (rather than analytical description) has the advantage that models can make use of more complex and accurate characterization of the auditory-nerve response to the stimuli as well as the advantage that far more complex stimuli can be considered. On the other hand, the ever-increasing complexity of the computational models can make it more difficult to understand exactly which aspect of a model is most important in describing the data to which it is applied.

We first consider in some detail the influential quantification of Jeffress's hypothesis by Colburn [25, 26]. We then review several extensions to the Jeffress–Colburn model in the 1970s and 1980s by Stern and Trahiotis [110, 111, 113, 114, 115, 117, 118], by Blauert and his colleagues Lindemann and Gaik [6, 9, 46, 70, 71], and finally by Breebaart and colleagues [14, 15, 16].

Colburn's Quantification of the Jeffress Hypothesis. Colburn's model of binaural interaction based on auditory-nerve activity consisted of two components: a model of the auditory-nerve response to sound, and a "binaural displayer" that can be used to compare the auditory-nerve response to the signals of the two ears. The model of auditory-nerve activity used in the original Colburn model was simple in form to facilitate the development of analytic expressions for the time-varying means and variances of the neural spikes produced by the putative auditory-nerve fibers. Based on an earlier formulation of Siebert [107], Colburn's peripheral model consisted of a bandpass filter (to depict the frequency selectivity of individual fibers), an automatic gain control (which limits the average rate of response to stimuli), a low-pass filter (which serves to limit phase-locking to stimulus fine structure at higher frequencies), and an exponential rectifier (which roughly characterizes peripheral nonlinearities). The result of this series of operations is a function $r(t)$ that describes the putative instantaneous rate of firing of the auditory-nerve fiber in question. The responses themselves were modeled as nonhomogeneous Poisson processes. Similar functional models have been used by others (e.g., [9, 46, 70,

103]). In recent years, models of the peripheral auditory response to sound have become more computationally oriented and physiologically accurate (e.g., [80, 133]).

The heart of Colburn's binaural model [25, 26] is an ensemble of units describing the interaction of neural activity from the left and right ears generated by auditory-nerve fibers with the same CF, with input from one side delayed by an amount that is fixed for each fiber pair, as in Jeffress's model depicted in Fig. 5.3. Following Jeffress's hypothesis, each coincidence-counting mechanism emits a pulse if it receives two incoming pulses (after the internal delay) within a sufficiently short time of each other. If the duration of the coincidence window is sufficiently brief, the Poisson assumption enables us to compute statistics for the coincidence counts as a function of running time, CF, and internal delay. For example, it can be shown [25] that the average number of coincidences observed at time t from all fiber pairs with CF f and internal delay τ, $E[L(t, \tau, f)]$, is approximately

$$E[L(t, \tau, f)] = \int_{-\infty}^{t} r_L(\alpha)\, r_R(\alpha - \tau) w_c(t - \alpha) p(\tau, f)\, d\alpha \tag{5.4}$$

where $r_L(t)$ and $r_R(t)$ are now the functions that describe the instantaneous rates of the Poisson processes that describe the activity of the two auditory-nerve fibers that are the inputs to the coincidence-counting unit. Here, $L(t, \tau, f)$ is a binaural decision variable and $E[\cdot]$ denotes the expectation. The function $w_c(t)$ represents the temporal weighting function as before, and $p(\tau, f)$ now represents the relative number of fiber pairs with a particular internal delay τ and CF f. Comparing Eq. 5.4 to Eq. 5.2, it can readily be seen that the relative number of coincidence counts of the Jeffress–Colburn model, considered as a function of the internal-delay τ at a particular CF f, is an estimate of the short-time interaural cross-correlation of the *auditory-nerve responses to the stimuli* at each CF.

Colburn and Durlach [28] have noted that the cross-correlation mechanism shown in Fig. 5.3 can also be regarded as a generalization of the EC model of Durlach [37]. As described above, the EC model yields predictions concerning binaural detection thresholds by applying a combination of ITD and IID that produces the best equalization of the masker components of the stimuli presented to each of the two ears. Cancellation of the masker is then achieved by subtracting one of the resulting signals from the other. Predictions provided by the EC model are generally dominated by the effects of the ITD-equalization component rather than the IID-equalization component. Because the interaural delays of the fiber pairs of the Jeffress–Colburn model perform the same function as the ITD-equalizing operation of the EC model, most predictions of MLDs for the two models are similar. The Jeffress–Colburn model can explain most (but not all) binaural detection phenomena by the decrease in correlation of the stimuli (and the corresponding decrease in response by the coincidence-counting units) that is observed at the ITD of the masker and the frequency of the target when the target is added to the masker in a configuration that produces binaural unmasking.

Figure 5.4 is an example of the representation of simple stimuli by a contemporary implementation of the Jeffress–Colburn model. Responses are shown to a

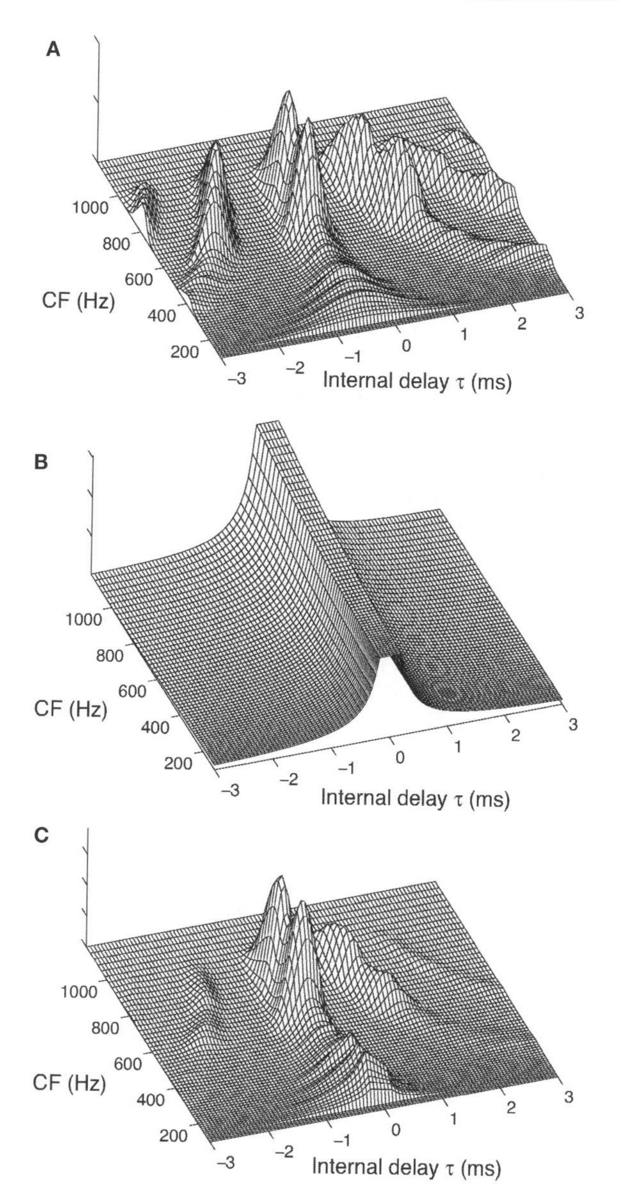

Figure 5.4 Representation of bandpass noise by the Jeffress–Colburn model. Panel **A** shows the rate of response of individual Jeffress–Colburn coincidence detectors to bandpass noise with nominal corner frequencies of 100 and 1000 Hz, presented with an ITD of –0.5 ms. Shown are the relative rate of coincidences of a fiber pair, reported as a joint function of CF and internal delay. Panel **B** shows the putative distribution of the fiber pairs, again in terms of CF and internal delay [113]. Panel **C** shows the product of the two functions depicted in the upper two panels, which represents the total number of coincidence counts recorded in response to the signal.

binaural band-pass noise with nominal corner frequencies of 100 and 1000 Hz, presented with an ITD such that the left ear is leading by 0.5 ms. The upper panel shows the relative rate of coincidences that would be produced by the coincidence-counting units as a function of their internal delay τ and CF f. The calculations in this figure were implemented by passing the incoming signals through a bank of gammatone filters as realized in Slaney's auditory toolbox [108], followed by a simple half-wave, square-law rectifier that squares positive input values and sets negative input values to zero. This model lacks many of the known complexities of the auditory-nerve response to sound. We note that maximum response is noted at internal delays equal to –0.5 ms over a broad range of frequencies, and that secondary maxima are observed at internal delays that are separated by integer multiples of the reciprocal of the CF for a particular fiber pair. Although these secondary ridges provide some ambiguity, it is clear that for natural stimuli the ITD can be inferred from the location of the "straight" ridge that is consistent over frequency [114, 117]. The central panel of the figure depicts the function $p(\tau)$, which specifies the distribution of the coincidence-counting units as a function of internal delay and CF. The specific function shown here was developed by Stern and Shear [113] and postulates a greater number of units with internal delays that are smaller in magnitude. The lower panel shows the total response of all the fiber pairs as a function of τ and f which reflects both the average response per fiber pair and the distribution per fiber pair; it is the product of the functions shown in the upper two panels.

5.5.3 Some Extensions to Cross-Correlation-Based Binaural Models

Extensions by Stern and Trahiotis. Stern and his colleagues (e.g., [110, 111, 114, 113, 117]) extended Colburn's model to describe the subjective lateral position of simple stimuli, and evaluated the extent to which the position cue could be used to account for performance in the largest possible set of psychoacoustical results in subjective lateralization, interaural discrimination, and binaural detection. Stern's extensions to Colburn's coincidence-counting mechanism include explicit assumptions concerning time–intensity interaction, and a mechanism for extracting subjective lateral position. These extensions of the Jeffress–Colburn model are referred to as the "position-variable model" by Stern and his colleagues. In addition, a second coincidence-based mechanism was proposed that serves to emphasize the impact of ITDs that are consistent over a range of frequencies [114].

One aspect of the model that received particular attention was the form of the function $p(\tau, f)$ that specifies the distribution of internal delays at each CF. The distribution function shown in the central panel of Fig. 5.4 was developed based on careful consideration of several key lateralization and detection results [113]. This function specifies a greater number of coincidence-counting units with internal interaural delays of smaller magnitude, which has been confirmed by physiological measurements (e.g., [66]). Nevertheless, a substantial fraction of the coincidence counters are assumed to have internal delays that are much greater in magnitude than the largest delays that are physically attainable with free-field stimuli. The existence of these very long internal delays is in accord with psychoacoustical as well as physiological data (see Section 5.2.2).

To account for the effects of IID, Stern and Colburn proposed that the representation of timing information shown in Fig. 5.4 be multiplied by a pulse-shaped weighting function with a location along the internal-delay axis that varied according to the IID of the stimulus. They further proposed that the subjective lateral position of a simple binaural stimulus could be predicted by the "center of mass" along the internal-delay axis of the combined function that reflects both ITD and IID. More recently, Stern and Trahiotis [114] incorporated an additional modification to the model called "straightness weighting" that is designed to emphasize the modes of the function that appear at the same internal delay over a range of CFs. This second-level mechanism, which emphasizes ITDs that are consistent over a range of frequencies, has the additional advantage of sharpening the ridges of the cross-correlation patterns in the Jeffress–Colburn model along the internal-delay axis. It should be noted that Stern and his colleagues considered only the lateral position of simple stimuli such as pure tones, clicks, or broadband noise, and never carefully addressed the problem of the separate lateralization of individual sources when they are presented simultaneously.

Extensions by Blauert and His Colleagues. Blauert and his colleagues have made important contributions to correlation-based models of binaural hearing over an extended period of time. Their efforts have been primarily directed toward understanding how the binaural system processes more complex sounds in real rooms and have tended to be computationally oriented. This approach is complementary to that of Colburn and his colleagues, who (at least in the early years) focused on explaining "classical" psychoacoustical phenomena using stimuli presented through earphones. More recently, Blauert's group worked to apply knowledge gleaned from fundamental research in binaural hearing toward the development of a "cocktail party processor" that can identify, separate, and enhance individual sources of sound in the presence of other interfering sounds (e.g., [10]).

In a relatively early study, Blauert and Cobben [9] combined the running cross correlator of Sayers and Cherry [97] with a model of the auditory periphery suggested by Duifhuis [36] that was similar to the model proposed by Siebert and adopted by Colburn. They subsequently developed a series of mechanisms that explicitly introduced the effects of stimulus IIDs into the modeling process. One of the most interesting and best known of these mechanisms was proposed by Lindemann [70], which may be regarded as an extension and elaboration of an earlier hypothesis of Blauert [6]. Lindemann extended the original Jeffress coincidence-counter model in two ways (see Fig. 5.5). First, he added a mechanism that inhibits outputs of the coincidence counters when there is activity produced by coincidence counters at adjacent internal delays. Second, he introduced monaural-processing mechanisms at the "edges" of the display of coincidence-counter outputs that become active when the intensity of the signal to one of the two ears is extremely small.

One of the properties of the Lindemann model is that the interaction of the inhibition mechanism and the monaural processing mechanism causes the locations of

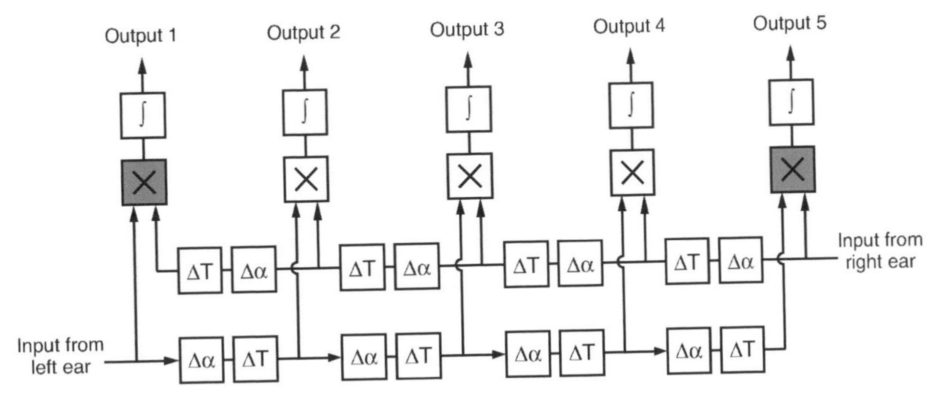

Figure 5.5 Schematic diagram of Lindemann's model. ΔT denotes a time delay and $\Delta\alpha$ denotes an attenuator. Boxes containing crosses are correlators (multipliers). At the two ends of the delay lines, the shaded boxes indicate correlators that are modified to function as monaural detectors. Redrawn with permission from Fig. 3(a) of [70]. Copyright 1986, Acoustical Society of America.

peaks of the coincidence-counter outputs along the internal-delay axis to shift with changes in IID. In other words, this model produces a time–intensity trading mechanism at the level of the coincidence-counter outputs. Although the net effect of IIDs on the patterns of coincidence-counter outputs in the Lindemann model is not unlike the effect of the intensity-weighting function in the model of Stern and Colburn [110], the time–intensity interaction of the Lindemann model arises more naturally from the fundamental assumptions of the model rather than as the result of the imposition of an arbitrary weighting function. In addition to the time–intensity trading properties, Lindemann also demonstrated that the contralateral-inhibition mechanism could also describe several interesting phenomena related to the precedence effect [71]. Finally, the inhibitory mechanisms of Lindemann's model produce a "sharpening" of the peaks of the coincidence-counter outputs along the internal-delay axis, similar to that achieved by the second-level coincidence layer of Stern's model that was designed to emphasize ITDs that are consistent over frequency.

Gaik [46] extended the Lindemann mechanism further by adding a second weighting to the coincidence-counter outputs that reinforces naturally occurring combinations of ITD and IID. This has the effect of causing physically plausible stimuli to produce coincidence outputs with a single prominent peak that is compact along the internal-delay axis and that is consistent over frequency. Conversely, very unnatural combinations of ITD and IID (which tend to give rise to multiple spatial images) produce response patterns with more than one prominent peak along the internal-delay axis. The Blauert–Lindemann–Gaik model has been used as the basis for several computational models for systems that localize and separate simultaneously presented sound sources (e.g., [10]).

Models that Incorporate Interaural Signal Cancellation. Although the coincidence-counting mechanism proposed by Jeffress and quantified by Colburn has dominated models of binaural interaction, the EC model developed by Durlach has retained its appeal for many years, and figures strongly in the conceptualization of data by Culling and Summerfield [30] and others (see Section 5.7). As noted above in Section 5.5.2, the Jeffress–Colburn model is loosely based on neural cells that are excited by inputs from both ears (called EE cells). Breebaart has recently proposed an elaboration of the Jeffress–Colburn model that includes an abstraction of cells that are excited by input from one ear but inhibited by input from the other ear (EI cells) [14, 15, 16]. The incorporation of EI units into binaural models provides an explicit mechanism that can in principle estimate the IID of a signal in a given frequency band, just as the EE-based mechanism of the Jeffress–Colburn model is typically used to provide an estimate of the ITD of a signal. In addition, the EI mechanism can provide the interaural cancellation in the EC model.

Figure 5.6 is a block diagram that summarizes the processing of Breebaart's model [14] that enables the simultaneous estimation of ITD and IID at a single frequency. In this diagram, units labeled $\Delta\alpha$ insert small attenuations to the signal path, just as the units labeled ΔT insert small time delays, as in the earlier models of Colburn, Blauert, and their colleagues. With this configuration, both the ITD and the IID of a signal component can be inferred by identifying which EI unit exhibits the minimal response. Specifically, the IID of the signal determines which row of EI units includes the minimal response, and the ITD of the signal determines which columns include the minimal response. Note that the response patterns will tend to repeat periodically along the horizontal axis, because the effective input to the network is narrowband after peripheral auditory filtering.

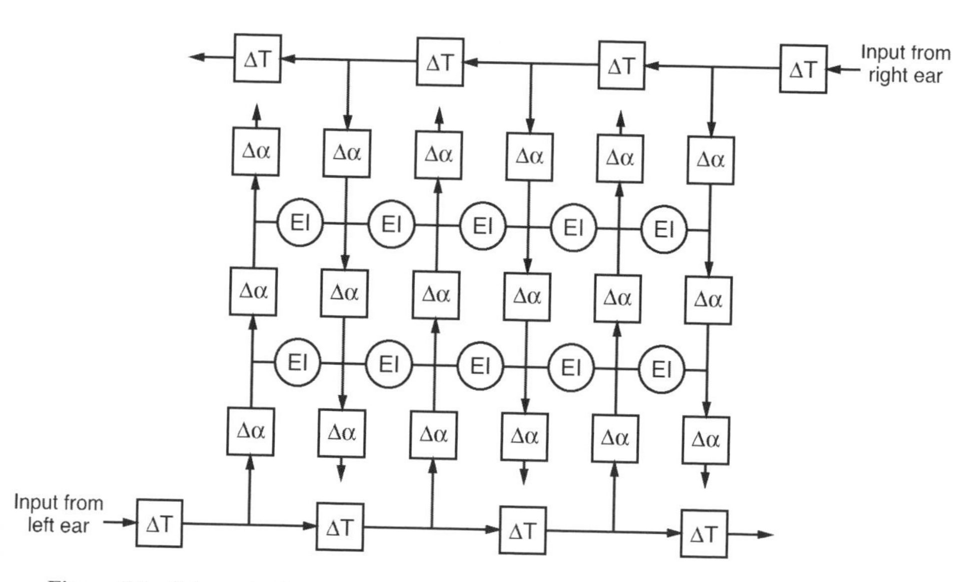

Figure 5.6 Schematic diagram of Breebaart's model. ΔT denotes a time delay and $\Delta\alpha$ denotes an attenuator. The circles denote excitation–inhibition (EI) elements. Redrawn with permission from Fig. 3 of [14]. Copyright 2001, Acoustical Society of America.

Figure 5.7 is an example of a response of such a network to a pure tone of 500 Hz presented with zero ITD and IID in the absence of internal noise. Note that the network exhibits a minimum of response at 0 μs ITD and 0 dB IID, with the minima repeating periodically along the internal delay axis with a period of 2 ms (the period of the 500 Hz tone). Since the pattern shifts horizontally with a change of ITD and vertically with a change of IID, this type of processing maintains an independent representation of ITD and IID for subsequent processing. In contrast, other models (e.g., [9, 46, 70, 110]) combine the effects of ITD and IID at or near the first stage of binaural interaction.

As Breebaart [14] notes, the predictions of this model will be very similar to the predictions of other models based solely on EE processing for many stimuli. Nevertheless, Breebaart also argues that predictions for EI-based models will differ from those of EE-based models in some respects. For example, Breebaart [14] argues that the dependence of binaural detection thresholds on target and masker duration is better described by EI-based processing. Furthermore, he argues that the parallel and independent availability of estimates of ITD and IID (as discussed in the previous paragraph) is likely to be useful for describing particular detection and discrimination results, and that EI-based models are better able to describe the dependence of binaural detection experiments on overall stimulus level.

5.6 MULTISOURCE SOUND LOCALIZATION

We now consider the ways in which the models discussed above have been applied to describe signals containing multiple sound sources. In particular, we focus on the application of such models within CASA systems, where they often form the basis for localizing and segregating speech that has been contaminated by interfering sounds.

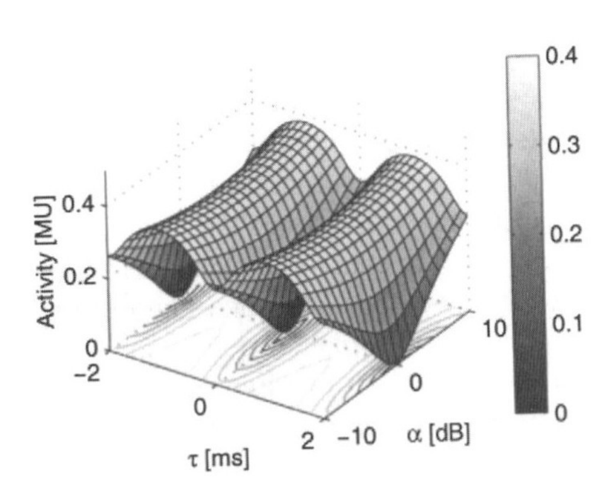

Figure 5.7 Activity of EI units in Breebaart's model, as a function of the characteristic IID (α) and ITD (τ) of each unit. The input was a 500 Hz tone presented diotically (i.e., with zero IID and ITD) and there is no internal noise. Redrawn with permission from Fig. 5 of [14]. Copyright 2001, Acoustical Society of America.

Although the following discussion of multisource localization is based on the type of binaural processing described in the previous section, it should be noted that the problem has been extensively studied in array signal processing (for reviews, see [121, 62]), where the main concern is direction-of-arrival estimation. For example, the classic MUSIC (MUltiple SIgnal Classification) algorithm [100] performs a principal component analysis on the array covariance matrix in order to separate a signal subspace and a noise subspace, which are used in a matrix operation that results in peak responses in the source directions. Although accurate source localization can be obtained with a large sensor array, the utility of such techniques is limited when the array is limited to only two sensors, as in the case of the auditory system. The MUSIC algorithm, for instance, can localize only one sound source with two microphones. General multisource localization must also consider room reverberation, which introduces multiple reflections of every sound source and complicates the localization problem considerably. The problem of reverberation will be treated extensively in Chapter 7, and hence we do not deal with room reverberation in this section.

As has been described previously, the human binaural system appears to make use of several different types of information to localize and separate signals, including ITDs, IIDs, and the ITDs of the low-frequency envelopes of high-frequency stimulus components. Nevertheless, of these cues, ITD information typically dominates localization (at least at low frequencies) and has been the basis for all of the models described above. As a consequence, the primary stage of binaural interaction in multisource localization algorithms is a representation of interaural timing information based on either an EE-type mechanism as in the Jeffress–Colburn model or an EI-type mechanism as in the Breebaart model. This representation can be subsequently modified to reflect the impact of IIDs and possibly other types of information. Figure 5.8 illustrates how the Jeffress–Colburn mechanism can be used to localize two signals according to ITD. Panels **A** and **B** of the figure show the magnitude spectra in decibels of the vowels /AH/ and /IH/ spoken by a male and a female speaker, respectively. Panel **C** shows the relative response of the binaural coincidence-counting units when these two vowels are presented simultaneously with ITDs of 0 and –0.5 ms, respectively. The 700 Hz first formant of the vowel /AH/ is clearly visible at the 0 ms internal delay, and the 300 Hz first formant of the vowel /IH/ is seen at the delay of –0.5 ms.

5.6.1 Estimating Source Azimuth from Interaural Cross-Correlation

The first computational system for joint localization and source separation was proposed by Lyon in 1983 [76]. His system begins with a cochleagram of a sound mixture, which is essentially a representation that incorporates cochlear filtering and mechanical-to-neural transduction of the auditory nerve (see Section 1.3). The system subsequently computes the cross-correlation between the left and the right cochleagram responses, resulting in a representation similar to that shown in Panel **C** of Fig. 5.8. Lyon termed this the *cross-correlogram*. He suggests summing the cross-correlation responses over all frequency channels, leading to a summary

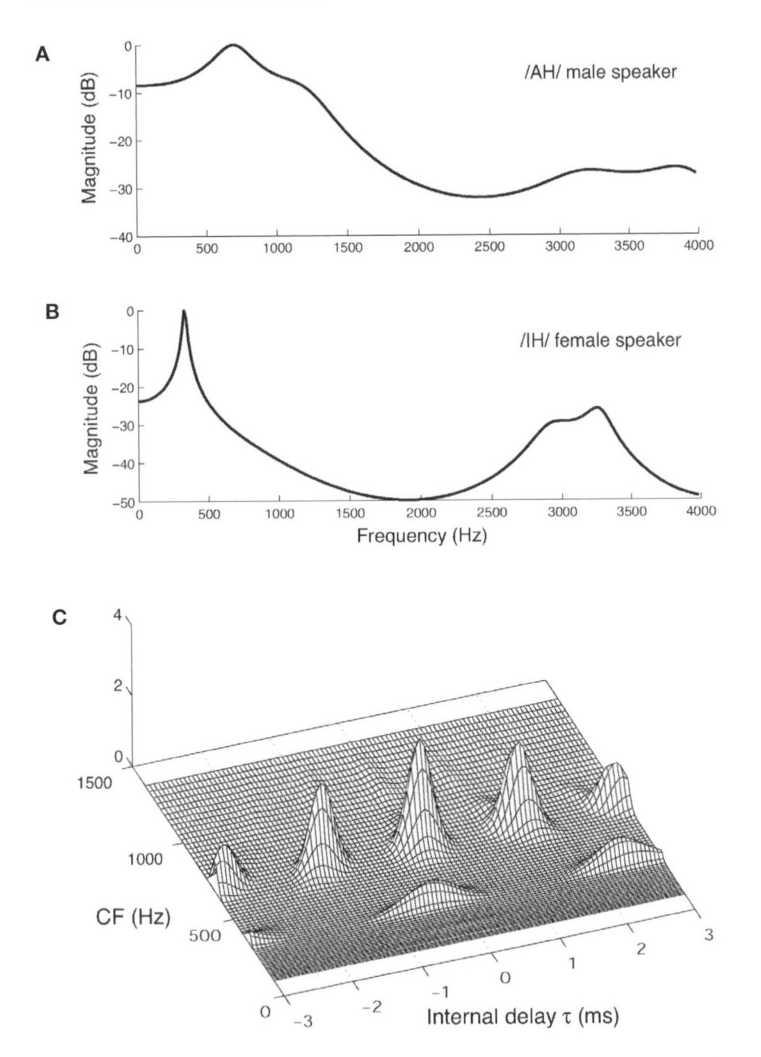

Figure 5.8 Upper and middle panels: Spectrum of the vowels /AH/ (**A**) and /IH/ (**B**) as recorded by a male and female speaker, respectively. (**C**) Response of an implementation of the Jeffress–Colburn model to the simultaneous presentation of the /AH/ presented with a 0 ms ITD and the /IH/ presented with a –0.5 ms ITD.

cross-correlogram in which prominent peaks indicate the ITDs of distinct sound sources. The idea of performing peak detection in the summary cross-correlogram for multisource localization has since been adopted in many subsequent systems, although Lyon's original study did not evaluate the idea in a systematic way.

A more systematic study of multisource localization was conducted by Bodden [10], again in the context of location-based source separation. The binaural processor used in Bodden's system follows the extension to Jeffress's model by Blauert and his colleagues as detailed in the previous section; in particular, it incorporates contralateral inhibition and adapts to HRTFs. As a result, his model for localization is based not only on ITD but also on IID. His system analyzes the acoustic input using a filterbank with 24 channels, intended to simulate the critical bands in human hearing. A mixture of broadband signals such as speech may have large spectral overlap. In other words, different sources of this kind are usually not well separated in frequency, as shown in Fig. 5.8. As a result, binaural responses to different sources interact in a complex way in the cross-correlogram, so that peaks in the cross-correlogram no longer reliably indicate ITDs of individual sounds. Hence, spectral overlap between multiple sound sources creates a major difficulty in localization.

To deal with the problem of spectral overlap, the Bodden model incorporates a number of computational stages. First, the cross-correlation function within each frequency band is converted from an internal-delay axis to an azimuth axis. This conversion is performed in a supervised training stage using white noise presented at various arrival angles between –90° to 90° in the frontal horizontal plane. The mapping between peak positions on the cross-correlation axis to the corresponding azimuths is first established within each frequency band, and linear interpolation is used to complete the conversion. Bodden has observed some frequency dependency in the conversion, which is consistent with the observed frequency dependence of physical measurements of ITD by Shaw and others, as summarized in Eq. 5.1. To perform localization, the model sums converted cross-correlation patterns across different frequency channels. Rather than simply adding them together as is done in Lyon's model, the Bodden system introduces another supervised training stage in order to determine the relative importance of different bands; this is done in a similar way to the conversion to the azimuth axis. This second training stage provides a weighting coefficient for each critical band, and the weighted sum across frequency is performed on the binaural responses in different frequency channels. To further enhance the reliability of multisource localization, his system also performs a running average across time within a short window (100 ms). These steps together result in a summary binaural pattern indexed by azimuth and running time.

A peak in the resulting summary pattern at a particular time is interpreted as a candidate for the azimuth of an acoustic event at that time. Bodden's model decides whether an azimuth candidate corresponds to a new event by tracking the time course of the amplitude of the summary binaural pattern; a new event should have an accompanying amplitude increase. Results on two-source localization were reported, and when the two sources are well separated in azimuth, the model gives good results. It should be noted that the sound sources used in Bodden's experiments were combined digitally from recordings of the individual sources in isolation (Bodden, personal communication); it is more difficult to obtain comparable improvements using sound sources recorded in a natural environment, especially if reverberation is a factor. This concern also applies to a number of subsequent studies (e.g., [94]).

5.6.2 Methods for Resolving Azimuth Ambiguity

When the azimuths of sound sources are not far apart, the Bodden model has difficulty in resolving them separately because, as discussed earlier, different sources may interact within individual frequency channels to produce misleading peaks in the binaural pattern. For example, two sounds may interact to produce a broad cross-correlogram peak that indicates a ghost azimuth in the middle of the two true azimuths. Several methods have been proposed to sharpen the cross-correlation modes along the internal-delay axis and resolve ambiguities in source azimuth, including the previously mentioned second level of coincidences across frequency proposed by Stern and Trahiotis [114, 116] and the contralateral-inhibition mechanism proposed by Lindemann [70, 71].

One approach that has been effective in computational systems for source separation is the computation of a "skeleton" cross-correlogram [88, 94], which is motivated by the observation that peaks in a cross-correlogram function are too broadly tuned to resolve small differences in azimuth (see Fig. 5.8). The basic idea is to replace the peaks in a cross-correlogram response by a Gaussian function with a narrower width. Specifically, each local peak in the cross-correlogram is reduced to an impulse of the same height. The resulting impulse train is then convolved with a Gaussian, whose width is inversely proportional to the center frequency of the corresponding filter channel. The resulting summary skeleton cross-correlogram is more sharply peaked and, therefore, represents multiple azimuths more distinctly.

Figure 5.9 illustrates the utility of the skeleton cross-correlogram for a mixture of a male speech utterance presented at 0° and a female utterance presented at 20°. Fig. 5.9**A** shows the cross-correlogram for a time window of 20 ms in response to the stimulus, where a 64 channel gammatone filterbank with filter center frequency ranging from 80 Hz to 5 kHz is used to perform peripheral analysis. The cross-correlogram is shown for the range [–1 ms, 1 ms]. For frequency bands in which energy from one source is dominant, the binaural response indicates the ITD of the true azimuth. On the other hand, for channels in which the energy from each source is about the same, peaks in the binaural response deviate from the true ITDs. In this case, the summary cross-correlogram shown in the lower panel of Fig. 5.9**A** does not provide an adequate basis for resolving the two azimuths. Figure 5.9**B** shows the corresponding skeleton cross-correlogram with sharper peaks (note that a conversion from the delay axis to the azimuth axis has been performed in this figure). Integrating over frequency yields a summary response that clearly indicates the two underlying azimuths, as shown in the lower panel of Fig. 5.9**B**. A skeleton cross-correlogram could be produced by applying a lateral inhibition mechanism to a conventional cross-correlogram; lateral inhibition should be contrasted with contralateral inhibition used in Lindemann's model [70].

Motivated by the psychoacoustical evidence that human performance in localizing multiple sources is enhanced when the sources have different onset times, Braasch [12] introduced a model, called interaural cross-correlation difference, for localizing a target source in the presence of background noise. When the target and the background noise are uncorrelated, the cross-correlogram of the mixture is the sum of the cross-correlogram of the target and that of the noise. If the cross-correlo-

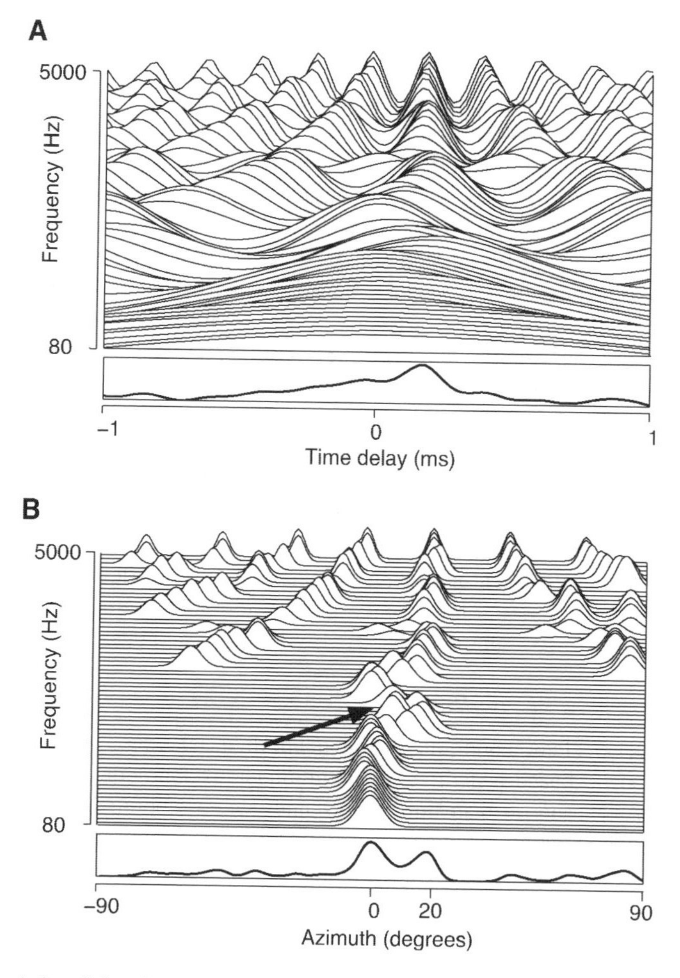

Figure 5.9 Azimuth localization for a mixture of two utterances: a male utterance presented at 0° and a female utterance at 20°. (**A**) Cross-correlogram for a time frame 400 ms after stimulus onset. The lower panel shows the summary cross-correlogram. (**B**) Skeleton cross-correlogram for the same time frame. The lower panel shows a summary plot along the azimuth axis. The arrow points to channels that contain roughly equal energy from the target and the interference. Reprinted with permission from Fig. 3 of [94]. Copyright 2003, Acoustical Society of America.

gram of the background is known, it can then be used to subtract from the mixture cross-correlogram, producing the cross-correlogram of the target alone. As a result, target-localization performance is better than that which is obtained using the traditional cross-correlogram. The basic idea behind Braasch's difference model is similar to that of spectral subtraction commonly used in speech enhancement (see Section 1.5). Consequently, we can expect that Braasch's model will work reasonably well for signals that are in the presence of stationary background noise, but not so well when the interference is nonstationary in nature, as in the case of a competing talker.

It is well known that the energy distribution of a speech utterance varies a great deal with respect to time and frequency. Hence, for a mixture of speech sources, or any other sources with similar characteristics such as musical sounds, a single source tends to dominate the responses within individual time-frequency units, each corresponding to a specific frequency channel and a specific time frame. This property has been well established in earlier parts of this book (see Chapter 1 and Chapter 4). Figure 5.8 illustrates such a situation for a mixture of two concurrent vowels. Because the two vowels show energy peaks in different spectral regions, they are well separated in the binaural pattern shown in Panel **C** of the figure. Capitalizing on this observation, Faller and Merimaa [41] recently proposed a method to automatically select the time-frequency (T-F) units dominated by a single source. The ITD and IID cues provided by such units resemble those elicited by a single source. Faller and Merimaa assume that only reliable binaural cues derived from such units are fed to the later stages of the auditory pathway.

Faller and Merimaa [41] propose an interaural coherence (IC) measure to select reliable T-F units. The IC is estimated from a normalized cross-correlation function, which may be defined for a specific frequency channel at time t as

$$\hat{R}(t, \tau) = \frac{\int_{-\infty}^{t} x_L(\alpha) x_R(\alpha - \tau) w(t - \alpha)\, d\alpha}{\sqrt{\int_{-\infty}^{t} x_L^2(\alpha) w(t - \alpha)\, d\alpha} \sqrt{\int_{-\infty}^{t} x_R^2(\alpha - t) w(t - \alpha)\, d\alpha}} \tag{5.5}$$

The notations in the above equation are the same as those in Eq. 5.5. $\hat{R}(t, \tau)$ is evaluated in the range [−1 ms, 1 ms]. The IC at time t is then given by

$$IC(t) = \max_{\tau} \hat{R}(t, \tau) \tag{5.6}$$

This measure of interaural coherence, which in many ways may be regarded as an amplification and implementation of an earlier approach by Allen et al. [1], is effective for the following reasons. When a single source dominates a frequency band at a particular time, the left and the right signals will be similar except for an ITD, which leads to a high IC value. On the other hand, if multiple sources have significant energy within the same time-frequency unit, the left and the right signals will be incoherent, leading to a low IC value. In the Faller and Merimaa model, binaural cues from T-F units with high IC values are retained, and the cues from other T-F units are discarded. Their evaluation results show that the selection of binaural cues based on interaural coherence yields sharper peaks in a joint ITD–IID feature space, which implies more robust localization of multiple sources. Faller and Merimaa have also evaluated their method on localizing reverberant sounds, and more discussion on this will be given in Chapter 7.

A major cause of azimuth ambiguity is the occurrence of multiple peaks in the high-frequency range of the cross-correlogram in response to a broadband source (e.g., speech), as shown in Fig. 5.9**A**. This is because at high frequencies the wavelength is shorter than the physical distance between the two ears. When multiple sources are active, their interaction may lead to a summary cross-correlogram in which the peaks no longer reflect the true ITDs, as discussed by Stern et al. [114, 117]. Liu et al. [73] divide the peaks into those on the primary trace that coincide with the true ITDs and those on the secondary traces that do not indicate true ITDs. Furthermore, they observe that the primary and secondary traces form a characteristic pattern on the cross-correlogram and devise a template-matching method that integrates peaks across frequency on the basis of the characteristic correlation pattern. They call their spectral integration method the "stencil" filter, and have reported good results for localizing up to four speech sources. A more detailed description of the stencil filter will be given in Chapter 6, in the context of location-based sound separation.

5.6.3 Localization of Moving Sources

Sound localization in real-world environments must consider the movement of sound sources. For example, a speaker often turns his or her head and walks while talking. Source movement introduces yet another dimension of complexity; in particular, it limits the window length for temporal integration, which has proven to be very useful for accurate localization. Few studies have addressed this important problem. Bodden [10] made an attempt to test his model on localizing two moving sources that cross each other in their azimuthal paths. However, the two sources in his evaluation are active alternatively, with little temporal overlap, presumably because of the inability of the model to deal with simultaneously active sources.

Roman and Wang [93] recently presented a binaural method for tracking multiple moving sources. They extend a hidden Markov model (HMM) for multipitch tracking [127] to the domain of multisource localization and tracking. Specifically, they extract ITD and IID cues and integrate these cues across different frequency channels to produce a likelihood function in the target space. The HMM is subsequently employed to form continuous azimuth tracks and detect the number of active sources across time. Their model combines instantaneous binaural information and an explicit dynamic model that simulates the motion trajectory of a sound source. They have reported good results for tracking three active sources whose azimuth paths may cross each other. A further evaluation shows a favorable comparison with a Kalman-filter-based approach [92].

5.7 GENERAL DISCUSSION

The ability to localize sound sources is a crucial aspect of auditory perception; it creates a sense of space for the listener, and provides information about the spatial location of objects in the environment, which can be critical for survival. The psychoacoustical studies reviewed here reveal that human binaural hearing can accurately localize sounds, particularly in the horizontal median plane, and that this capacity remains largely intact in the presence of other interfering sounds. Reproducing this ability in computational systems is a significant challenge, and an important one for the development of CASA systems.

Binaural sound localization is largely based on comparisons between the signals arriving at the two ears. Several cues contribute in this respect, including ITD, IID, as well as the ITD of the low-frequency envelopes of high-frequency components of the signals. Cues related to frequency-dependent filtering by the pinnae also play a role, particularly for localization in the vertical plane and for resolving front–back ambiguities. Although ITD appears to play a dominant role in the determination of the sound azimuth, it is clear that all of these cues interact in complex ways. At present, these interactions are not well understood; similarly, there is much that we do not know about how binaural cues are employed to process multiple sound sources.

The Jeffress coincidence mechanism [59], postulated almost 60 years ago, has dominated computational modeling of sound localization. However, the Jeffress scheme and its variants (see Section 5.5.2) were principally developed to explain the localization of simple stimuli such as a single sound source in an anechoic environment. Although these models have been successful in modeling the binaural perception of isolated sounds, there have been relatively few attempts to quantitatively evaluate them using multiple sound sources. Nonetheless, the key representation in Jeffress' scheme—referred to by some as the cross-correlogram—has been used as the basis for a number of multisource localization and sound-separation systems. Within such systems, integration of cross-correlation responses across frequency and time is necessary for good localization performance. Computational efforts in multisource localization have mostly concentrated on how to resolve azimuth ambiguity introduced by the presence of multiple sound sources.

Spatial location is an important cue for auditory scene analysis (see Section 1.1 and Bregman [17]), and many computational studies have addressed the problem of location-based grouping, the topic of the next chapter. In such studies, sound localization is often considered to be a prerequisite step. This, plus the fact that spectrotemporal integration is usually needed for accurate localization, raises an interesting conceptual issue of whether sound separation depends on localization or sound localization depends on separation. Experimental evidence discussed in Section 5.4.1 suggests that a target sound in background noise that is clearly audible is also easily localizable. This indicates a close link between sound localization and separation.

On the one hand, the robust MLD effect due to spatial separation (see Section 5.4.2) and other binaural advantages in sound separation would strongly imply the contribution of location-based grouping to auditory scene analysis. On the other hand, it is well known that binaural cues such as ITD and IID are susceptible to intrusion by background noise and room reverberation (see Chapter 7). Such intrusions make reliable estimates of binaural cues very difficult, if not impossible, in local time-frequency regions.

Indeed, the notion that the human auditory system groups sources according to ITD is far from universally accepted. For example, the results of several experiments using earphones indicate that listeners are unable to achieve separate identification of simultaneously presented vowel-like bandpass-noise sounds solely on the basis of their ITDs (e.g., [30, 31, 57]). Culling and Summerfield (and others) believe that these results indicate that the human auditory system does not make use of ITD information in achieving an initial grouping of signal components according to sound source. They suggest that signal components are extracted independently for each frequency using an interaural cancellation mechanism similar to that used in the EC model. In contrast, Stern et al. have found that identification according to ITD is easily accomplished if similar noise bands are modulated in amplitude or frequency in a fashion that is similar to the natural modulations of speech signals [91, 112]. One plausible interpretation of these results is that the human auditory system first groups signal components according to common modulation in amplitude and/or frequency (as well as other information such as the harmonicity of periodic signal components), and that ITD information is then used to segregate the grouped components, perhaps abetted by IID information to provide additional help in interpreting ambiguous inputs. Certainly, sound localization and sound separation are likely to interact in ways that we do not yet fully understand.

Although the question of how human listeners utilize ITDs to segregate sounds remains an interesting one, it is not necessary for a CASA system to mimic every aspect of auditory processing. As a practical matter, signal segregation is generally more easily achieved computationally on the basis of ITD information than it is based on common amplitude or frequency modulation, at least for anechoic signals. Hence, the use of ITD information as the basis for signal separation remains appealing for computational systems.

Future research in multisource localization will need to address the issue of room reverberation, which few computational studies have tackled directly (see Chapter 7). More consideration also needs to be given to moving sound sources, which frequently occur in natural acoustic scenes. Background interference, reverberation, and motion conspire to create a tremendous problem for computational sound localization systems that operate in real-world environments, and they will present a challenge for many years to come.

ACKNOWLEDGMENTS

Preparation of this chapter was supported by the National Science Foundation (Grant IIS-0420866) for Richard Stern. Guy Brown was supported by EPSRC grant GR/R47400/01, and DeLiang Wang was supported by an AFOSR grant (FA9550-04-01-0117) and an AFRL grant (FA8750-04-1-0093). The first author is especially grateful for many discussions over many years with Steve Colburn and Tino Trahiotis, which have been very instrumental in shaping his attitudes about many theoretical and experimental aspects of binaural hearing. Substantial portions of the discussion about binaural models in this chapter are based on previous chapters written with Trahiotis [115, 116].

REFERENCES

1. J. B. Allen, D. A. Berkley, and J. Blauert. Multimicrophone signal-processing technique to remove room reverberation from speech signals. *Journal of the Acoustical Society of America,* 62(4):912–915, 1977.
2. D. Batteau. The role of the pinna in human localization. *Proceedings of the Royal Society of London B, Biological Sciences,* 168:158–180, 1967.
3. L. R. Bernstein and C. Trahiotis. Lateralization of low-frequency complex waveforms: The use of envelope-based temporal disparities. *Journal of the Acoustical Society of America,* 77:1868–1880, 1985.
4. L. R. Bernstein and C. Trahiotis. Detection of interaural delay in high-frequency sinusoidally amplitude-modulated tones, two-tone complexes, and bands of noise. *Journal of the Acoustical Society of America,* 95:3561–3567, 1994.
5. L. R. Bernstein and C. Trahiotis. Enhancing sensitivity to interaural delays at high frequencies by using "transposed stimuli." *Journal of the Acoustical Society of America,* 112:1026–1036, 2002.
6. J. Blauert. Modeling of interaural time and intensity difference discrimination. In G. van den Brink and F. Bilsen, editors, *Psychophysical, Physiological, and Behavioural Studies in Hearing,* pp. 412–424. Delft University Press, Delft, 1980.
7. J. Blauert. Introduction to binaural technology. In R. H. Gilkey and T. R. Anderson, editors, *Binaural and Spatial Hearing in Real and Virtual Environments,* chapter 28, pp. 593–610. Lawrence Erlbaum, Mahwah, NJ, 1997.
8. J. Blauert. *Spatial Hearing,* revised ed. MIT Press, Cambridge, MA, 1997.
9. J. Blauert and W. Cobben. Some considerations of binaural cross-correlation analysis. *Acustica,* 39:96–103, 1978.
10. M. Bodden. Modelling human sound-source localization and the cocktail party effect. *Acta Acustica,* 1:43–55, 1993.
11. J. C. Boudreau and C. Tsuchitani. Binaural interaction in the cat superior olive S segment. *Journal of Neurophysiology,* 31:442–454, 1968.
12. J. Braasch. Localization in the presence of a distractor and reverberation in the frontal horizontal plane: II. Model algorithms. *Acta Acustica united with Acustica,* 88: 956–969, 2002.
13. J. Braasch. Modelling of binaural hearing. In J. Blauert, editor, *Communication Acoustics,* chapter 4, pp. 75–108. Springer-Verlag, Berlin, 2005.
14. J. Breebaart, S. van de Par, and A. Kohlrausch. Binaural processing model based on contralateral inhibition. I. Model structure. *Journal of the Acoustical Society of America,* 110:1074–1088, 2001.
15. J. Breebaart, S. van de Par, and A. Kohlrausch. Binaural processing model based on contralateral inhibition. II. Dependence on spectral parameters. *Journal of the Acoustical Society of America,* 110:1089–1103, 2001.
16. J. Breebaart, S. van de Par, and A. Kohlrausch. Binaural processing model based on

contralateral inhibition. III. Dependence on temporal parameters. *Journal of the Acoustical Society of America,* 110:1125–1117, 2001.

17. A. S. Bregman. *Auditory Scene Analysis.* MIT Press, Cambridge, MA, 1990.
18. M. D. Burkhard and R. M. Sachs. Anthropometric manikin for acoustic research. *Journal of the Acoustical Society of America,* 58(1):214–222, 1974.
19. D. Caird and R. Klinke. Processing of binaural stimuli by cat superior olivary complex neurons. *Experimental Brain Research,* 52:385–399, 1983.
20. R. Carhart. Monaural and binaural discrimination against competing sentences. *International Audiology,* 1:5–10, 1965.
21. R. Carhart, T. W. Tillman, and K. R. Johnson. Release from masking for speech through interaural time delay. *Journal of the Acoustical Society of America,* 42:124–138, 1967.
22. C. E. Carr and M. Konishi. Axonal delay lines for time measurement in the owl's brainstem. *Proceedings of the National Academy of Science USA,* 85:8311–8315, 1988.
23. C. Cherry. Some experiments on the recognition of speech, with one and two ears. *Journal of the Acoustical Society of America,* 26:554–559, 1953.
24. E. C. Cherry. Two ears—but one world. In W. A. Rosenblith, editor, *Sensory Communication,* pp. 99–117. MIT Press, Cambridge, MA, 1961.
25. H. S. Colburn. Theory of binaural interaction based on auditory-nerve data. I. General strategy and preliminary results on interaural discrimination. *Journal of the Acoustical Society of America,* 54:1458–1470, 1973.
26. H. S. Colburn. Theory of binaural interaction based on auditory-nerve data. II. Detection of tones in noise. *Journal of the Acoustical Society of America,* 61:525–533, 1977.
27. H. S. Colburn. Computational models of binaural processing. In H. Hawkins and T. McMullen, editors, *Auditory Computation,* Springer Handbook of Auditory Research, chapter 9, pp. 332–400. Springer-Verlag, New York, 1996.
28. H. S. Colburn and N. I. Durlach. Models of binaural interaction. In E. C. Carterette and M. P. Friedmann, editors, *Hearing,* volume IV of *Handbook of Perception,* chapter 11, pp. 467–518. Academic Press, New York, 1978.
29. H. S. Colburn and A. Kulkarni. Models of sound localization. In R. Fay and T. Popper, editors, *Sound Source Localization,* Springer Handbook of Auditory Research, chapter 8, pp. 272–316. Springer-Verlag, New York, 2005.
30. J. F. Culling and Q. Summerfield. Perceptual separation of concurrent speech sounds: Absence of across-frequency grouping by common interaural delay. *Journal of the Acoustical Society of America,* 98(2):785–797, 1995.
31. C. J. Darwin and R. W. Hukin. Perceptual segregation of a harmonic from a vowel by interaural time difference and frequency proximity. *Journal of the Acoustical Society of America,* 102(4):2316–2324, 1997.
32. E. E. David, N. Guttman, and W. A. van Bergeijk. On the mechanism of binaural fusion. *Journal of the Acoustical Society of America,* 30:801–802, 1958.
33. D.D.Dirks and R. H. Wilson. The effect of spatially-separated sound sources. *Journal of Speech and Hearing Research,* 12:5–38, 1969.
34. R. H. Domnitz and H. S. Colburn. Lateral position and interaural discrimination. *Journal of the Acoustical Society of America,* 61:1586–1598, 1977.

35. R. Drullman and A. W. Bronkhorst. Multichannel speech intelligibility and speaker recognition using monaural, binaural and 3D auditory presentation. *Journal of the Acoustical Society of America,* 107:2224–2235, 2000.
36. H. Duifhuis. Consequences of peripheral frequency selectivity for nonsimultaneous masking. *Journal of the Acoustical Society of America,* 54:1471–1488, 1973.
37. N. I. Durlach. Equalization and cancellation theory of binaural masking level differences. *Journal of the Acoustical Society of America,* 35(8):1206–1218, 1963.
38. N. I. Durlach. Binaural signal detection: Equalization and cancellation theory. In J. V. Tobias, editor, *Foundations of Modern Auditory Theory,* volume 2, pp. 369–462. Academic Press, New York, 1972.
39. N. I. Durlach and H. S. Colburn. Binaural phenomena. In E. C. Carterette and M. P. Friedman, editors, *Hearing,* volume IV of *Handbook of Perception,* chapter 10, pp. 365–466. Academic Press, New York, 1978.
40. N. I. Durlach, K. J.Gabriel, H. S.Colburn, and C. Trahiotis. Interaural correlation discrimination: II. Relation to binaural unmasking. *Journal of the Acoustical Society of America,* 79:1548–1557, 1986.
41. C. Faller and J. Merimaa. Sound localization in complex listening situations: Selection of binaural cues based on interaural coherence. *Journal of the Acoustical Society of America,* 116(5):3075–3089, 2004.
42. H. Fletcher and R. H. Galt. The perception of speech and its relation to telephony. *Journal of the Acoustical Society of America,* 22:89–151, 1950.
43. S. H. Foster, E. M. Wenzel, and R. M. Taylor. Real time synthesis of complex acoustic environments. In *Proceedings of the IEEE Workshop on Applications of Signal Processing to Audio and Acoustics,* New Paltz, NY, 1991.
44. N. R. French and J. C. Steinberg. Factors governing the intelligibility of speech sounds. *Journal of the Acoustical Society of America,* 19:90–119, 1947.
45. K. J. Gabriel and H. S. Colburn. Interaural correlation discrimination: I. Bandwidth and level dependence. *Journal of the Acoustical Society of America,* 69:1394–1401, 1981.
46. W. Gaik. Combined evaluation of interaural time and intensity differences: Psychoacoustic results and computer modeling. *Journal of the Acoustical Society of America,* 94:98–110, 1993.
47. B. Gardner and K. Martin. HRTF measurements of a KEMAR dummy-head microphone. Technical Report 280, MIT Media Lab Perceptual Computing Group, 1994. Available at http://sound.media.mit.edu/KEMAR.html.
48. M. D. Good and R. H. Gilkey. Sound localization in noise: The effect of signal-to-noise ratio. *Journal of the Acoustical Society of America,* 99:1108–1117, 1996.
49. G. G. Harris. Binaural interactions of impulsive stimuli and pure tones. *Journal of the Acoustical Society of America,* 32:685–692, 1960.
50. M. L. Hawley, R. Y. Litovsky, and H. S. Colburn. Speech intelligibility and localization in a multi-source environment. *Journal of the Acoustical Society of America,* 105:3436–3448, 1999.
51. J. Hebrank and D.Wright. Spectral cues used in the localization of sound sources in the median plane. *Journal of the Acoustical Society of America,* 56:1829–1834, 1974.
52. G. B. Henning. Detectability of interaural delay in high-frequency complex waveforms. *Journal of the Acoustical Society of America,* 55:84–90, 1974.

53. R. M. Hershkowitz and N. I. Durlach. Interaural time and amplitude bands for a 500-Hz tone. *Journal of the Acoustical Society of America,* 46:1464–1467, 1969.

54. I. J. Hirsch. The influence of interaural phase on interaural summation and inhibition. *Journal of the Acoustical Society of America,* 20:150–159, 1948.

55. I. J. Hirsch. Relation between localization and intelligibility. *Journal of the Acoustical Society of America,* 22:196–200, 1950.

56. P. M. Hofman and A. J. van Opstal. Spectro-temporal factors in two-dimensional human sound localization. *Journal of the Acoustical Society of America,* 103:2634–2648, 1998.

57. R. W. Hukin and C. J. Darwin. Effects of contralateral presentation and of interaural time differences in segregating a harmonic from a vowel. *Journal of the Acoustical Society of America,* 98:1380–1387, 1995.

58. T. Jacobsen. Localization in noise. Technical Report 10, Technical University of Denmark Acoustics Laboratory, 1976.

59. L. A. Jeffress. A place theory of sound localization. *Journal of Comparative Physiology and Psychology,* 41:35–39, 1948.

60. W. E. Kock. Binaural localization and masking. *Journal of the Acoustical Society of America,* 22:801–804, 1950.

61. W. Koenig. Subjective effects in binaural hearing. *Journal of the Acoustical Society of America,* 22:61–62(L), 1950.

62. H. Krim and M. Viberg. Two decades of array signal processing research: The parametric approach. *IEEE Signal Processing Magazine,* 13:67–94, 1996.

63. K. D. Kryter. Methods for the calculation and use of the articulation index. *Journal of the Acoustical Society of America,* 34:1689–1697, 1962.

64. G. F. Kuhn. Model for the interaural time differences in the azimuthal plane. *Journal of the Acoustical Society of America,* 62:157–167, 1977.

65. G. F. Kuhn. Physical acoustics and measurements pertaining to directional hearing. In W. A. Yost and G. Gourevitch, editors, *Directional Hearing.* Springer-Verlag, New York, 1987.

66. S. Kuwada, R. Batra, and D. C. Fitzpatrick. Neural processing of binaural temporal cues. In R. H. Gilkey and T. R. Anderson, editors, *Binaural and Spatial Hearing in Real and Virtual Environments,* chapter 20, pp. 399–425. Lawrence Erlbaum, Mahwah, NJ, 1997.

67. E. H. A. Langendijk, D. J. Kistler, and F. L. Wightman. Sound localization in the presence of one or two distractors. *Journal of the Acoustical Society of America,* 109:2123–2134, 2001.

68. H. Levitt and L. R. Rabiner. Predicting binaural gain in intelligibility and release from masking for speech. *Journal of the Acoustical Society of America,* 42:820–829, 1967.

69. J. C. R. Licklider. Three auditory theories. In S. Koch, editor, *Psychology: A Study of a Science,* pp. 41–144. McGraw-Hill, New York, 1959.

70. W. Lindemann. Extension of a binaural cross-correlation model by contralateral inhibition. I. Simulation of lateralization for stationary signals. *Journal of the Acoustical Society of America,* 80:1608–1622, 1986.

71. W. Lindemann. Extension of a binaural cross-correlation model by contralateral inhibition. II. The law of the first wavefront. *Journal of the Acoustical Society of America,* 80:1623–1630, 1986.

72. R. Y. Litovsky, S. H. Colburn, W. A. Yost, and S. J. Guzman. The precedence effect. *Journal of the Acoustical Society of America,* 106:1633–1654, 1999.

73. C. Liu, B. C. Wheeler, W. D. O'Brien, R. C. Bilger, C. R. Lansing, and A. S. Feng. Localization of multiple sound sources with two microphones. *Journal of the Acoustical Society of America,* 108(4):1888–1905, 2000.

74. C. Lorenzi, S. Gatehouse, and C. Lever. Sound localization in noise in normal-hearing listeners. *Journal of the Acoustical Society of America,* 105:1810–1820, 1999.

75. R. F. Lyon. A computational model of filtering, detection and compression in the cochlea. In *Proceedings of the IEEE International Conference on Acoustics, Speech and Signal Processing,* pp. 1282–1285, Paris, May 1982.

76. R. F. Lyon. A computational model of binaural localization and separation. In *Proceedings of the IEEE International Conference on Acoustics, Speech and Signal Processing,* pp. 1148–1151, 1983.

77. N. W. MacKeith and R. R. A. Coles. Binaural advantages in hearing of speech. *Journal of Laryngology and Otolaryngology,* 85:213–232, 1985.

78. D. McAlpine, D. Jiang, and A. R. Palmer. Interaural delay sensitivity and the classification of low best-frequency binaural responses in the inferior colliculus of the guinea pig. *Hearing Research,* 97:136–152, 1996.

79. D. McFadden and E. G. Pasanen. Lateralization of high frequencies based on interaural time differences. *Journal of the Acoustical Society of America,* 59:634–639, 1976.

80. R.Meddis and M. J. Hewitt. Virtual pitch and phase sensitivity of a computer model of the auditory periphery. I. Pitch identification. *Journal of the Acoustical Society of America,* 89(6):2866–2882, 1991.

81. S. Mehrgardt and V. Mellert. Transformation charactersitics of the external human ear. *Journal of the Acoustical Society of America,* 61:1567–1576, 1977.

82. J. C. Middlebrooks, J. C. Makous, and D. M. Green. Directional sensitivity of sound-pressure levels in the human ear canal. *Journal of the Acoustical Society of America,* 86:89–108, 1989.

83. P. Minnaar, S. K. Olesen, F. Christensen, and H. Møller. Localization with binaural recordings from artificial and human heads. *Journal of the Audio Engineering Society,* 49(5):323–336, May 2001.

84. H. Møller, M. F. Sørensen, C. B. Jensen, and D. Hammershøi. Binaural technique: Do we need individual recordings? *Journal of the Audio Engineering Society,* 44:451–469, June 1996.

85. B. C. J. Moore. *An Introduction to the Psychology of Hearing,* 5th ed. Academic Press, London, 2003.

86. J. W. Strutt (Third Baron of Rayleigh). On our perception of sound direction. *Philosophical Magazine,* 13:214–232, 1907.

87. A. R. Palmer. Neural signal processing. In B. C. J. Moore, editor, *Hearing, Handbook of Perception and Cognition,* chapter 3, pp. 75–121. Academic Press, New York, 1995.

88. K. J. Palomäki, G. J. Brown, and D. L. Wang. A binaural processor for missing data speech recognition in the presence of noise and small-room reverberation. *Speech Communication,* 43(4):361–378, 2004.

89. R. D. Patterson, M. H. Allerhand, and C. Giguere. Time-domain modeling of peripher-

al auditory processing: A modular architecture and a software platform. *Journal of the Acoustical Society of America,* 98:1980–1984, 1995.

90. I. Pollack and W. J. Trittipoe. Binaural listening and interaural cross correlation. *Journal of the Acoustical Society of America,* 31:1250–1252, 1959.
91. A. M. Ripepi. *Lateralization and Identification of Simultaneously-Presented Whispered Vowels and Speech Sounds.* Master's thesis, Carnegie Mellon University, Pittsburgh, PA, May 1999.
92. N. Roman. *Auditory-Based Algorithms for Sound Segregation in Multisource and Reverberant Environments.* PhD thesis, Ohio State University Department of Computer Science and Engineering, 2005. Available at http://www.cse.ohio-state.edu/pnl/theses.html.
93. N. Roman and D. L. Wang. Binaural tracking of multiple moving sources. In *Proceedings of the IEEE International Conference on Acoustics, Speech and Signal Processing,* volume V, pp. 149–152, 2003.
94. N. Roman, D. L. Wang, and G. J. Brown. Speech segregation based on sound localization. *Journal of the Acoustical Society of America,* 114(4):2236–2252, 2003.
95. J. E. Rose, N. B. Gross, C. D. Geisler, and J. E. Hind. Some neural mechanisms in the inferior colliculus of the cat which may be relevant to localization of a sound source. *Journal of Neurophysiology,* 29:288–314, 1966.
96. B. McA. Sayers. Acoustic-image lateralization judgments with binaural tones. *Journal of the Acoustical Society of America,* 36:923–926, 1964.
97. B. McA. Sayers and E. C. Cherry. Mechanism of binaural fusion in the hearing of speech. *Journal of the Acoustical Society of America,* 36:923–926, 1957.
98. B. McA. Sayers and P. A. Lynn. Interaural amplitude effects in binaural hearing. *Journal of the Acoustical Society of America,* 44:973–978, 1968.
99. B. McA. Sayers and F. E. Toole. Acoustic-image judgments with binaural transients. *Journal of the Acoustical Society of America,* 36:1199–1205, 1964.
100. R. Schmidt. Multiple emitter location and signal parameter estimation. *IEEE Transactions on Antennas and Propagation,* 34:276–280, 1986.
101. M. R. Schroeder. New viewpoints in binaural interactions. In E. F. Evans and J. P. Wilson, editors, *Psychophysics and Physiology of Hearing,* pp. 455–467. Academic Press, London, 1977.
102. C. L. Searle, L. D. Braida, D. R. Cuddy, and M. F. Davis. Binaural pinna disparity: Another auditory localization cue. *Journal of the Acoustical Society of America,* 57:448–455, 1975.
103. S. Seneff. A joint synchrony/mean-rate model of auditory speech processing. *Journal of Phonetics,* 15:55–76, 1988.
104. S. A. Shamma, N. Shen, and P. Gopalaswamy. Binaural processing without neural delays. *Journal of the Acoustical Society of America,* 86:987–1006, 1989.
105. E. A. G. Shaw. Transformation of sound pressure level from the free field to the ear drum in the horizontal plane. *Journal of the Acoustical Society of America,* 39:465–470, 1974.
106. E. A. G. Shaw. Acoustical features of the human external ear. In R. H. Gilkey and T. R. Anderson, editors, *Binaural and Spatial Hearing in Real and Virtual Environments,* chapter 2, pp. 25–47. Lawrence Erlbaum, Mahwah, NJ, 1997.

107. W. M. Siebert. Frequency discrimination in the auditory system: Place or periodicity mechanisms. *Proceedings of the IEEE,* 58:723–730, 1970.

108. M. Slaney. *Auditory Toolbox (V.2),* 1998. Available at http://www.slaney.org/malcolm/pubs.html.

109. T. R. Stanford, S. Kuwada, and R. Batra. A comparison of the interaural time sensitivity of neurons in the inferior colliculus and thalamus of the unanesthetized rabbit. *Journal of Neuroscience,* 12:3200–3216, 1992.

110. R. M. Stern and H. S. Colburn. Theory of binaural interaction based on auditory-nerve data. IV. A model for subjective lateral position. *Journal of the Acoustical Society of America,* 64:127–140, 1978.

111. R. M. Stern and H. S. Colburn. Lateral-position based models of interaural discrimination. *Journal of the Acoustical Society of America,* 77:753–755, 1985.

112. R. M. Stern, A. M. Ripepi, and C. Trahiotis. Fluctuations in amplitude and frequency fluctuations in amplitude and frequency enable interaural delays to foster the identification of speech-like stimuli. In P. Divenyi, editor, *Dynamics of Speech Production and Perception.* IOS Press, Amsterdam, 2005.

113. R. M. Stern and G. D. Shear. Lateralization and detection of low-frequency binaural stimuli: Effects of distribution of internal delay. *Journal of the Acoustical Society of America,* 100:2278–2288, 1996.

114. R. M. Stern and C. Trahiotis. The role of consistency of interaural timing over frequency in binaural lateralization. In Y. Cazals, K. Horner, and L. Demany, editors, *Auditory Physiology and Perception,* pp. 547–554. Pergamon Press, Oxford, 1992.

115. R. M. Stern and C. Trahiotis. Models of binaural interaction. In B. C. J.Moore, editor, *Hearing, Handbook of Perception and Cognition,* chapter 10, pp. 347–386. Academic Press, New York, 1995.

116. R. M. Stern and C. Trahiotis. Models of binaural perception. In R. Gilkey and T. R. Anderson, editors, *Binaural and Spatial Hearing in Real and Virtual Environments,* chapter 24, pp. 499–531. Lawrence Erlbaum, Mahwah, NJ, 1996.

117. R. M. Stern, A. S. Zeiberg, and C. Trahiotis. Lateralization of complex binaural stimuli: A weighted image model. *Journal of the Acoustical Society of America,* 84:156–165, 1988.

118. R. M. Stern, T. Zeppenfeld, and G. D. Shear. Lateralization of rectangularly-modulated noise: Explanations for counterintuitive reversals. *Journal of the Acoustical Society of America,* 90:1901–1907, 1991.

119. C. Trahiotis, L. R. Bernstein, R. M. Stern, and T. N. Buell. Interaural correlation as the basis of a working model of binaural processing: An introduction. In R. Fay and T. Popper, editors, *Sound Source Localization,* Springer Handbook of Auditory Research, chapter 7, pp. 238–271. Springer-Verlag, Heidelberg, 2005.

120. C. Trahiotis and R. M. Stern. Lateralization of bands of noise: Effects of bandwidth and differences of interaural time and intensity. *Journal of the Acoustical Society of America,* 86:1285–1293, 1989.

121. B. D. van Veen and K. M. Buckley. Beamforming: A versatile approach to spatial filtering. *IEEE ASSP Magazine,* pp. 4–24, April 1988.

122. G. von Békésy. *Experiments in Hearing.* McGraw-Hill, New York; reprinted by the Acoustical Society of America, 1989, 1960.

123. E. M. Wenzel, M. Arruda, D. J. Kistler, and F. L. Wightman. Localization using nonin-

dividualized head-related transfer functions. *Journal of the Acoustical Society of America,* 94:111–123, 1993.

124. F. L. Wightman and D. J. Kistler. Headphone simulation of free-field listening. I: Stimulus synthesis. *Journal of the Acoustical Society of America,* 85:858–867, 1989.
125. F. L. Wightman and D. J. Kistler. Headphone simulation of free-field listening. II: Psychophysical validation. *Journal of the Acoustical Society of America,* 87:868–878, 1989.
126. F. L. Wightman and D. J. Kistler. Factors affecting the relative salience of sound localization cues. In R. H. Gilkey and T. R. Anderson, editors, *Binaural and Spatial Hearing in Real and Virtual Environments,* pp. 1–23. Lawrence Erlbaum Associates, Mahwah, NJ, 1997.
127. M. Wu, D. L. Wang, and G. J. Brown. A multipitch tracking algorithm for noisy speech. *IEEE Transactions on Speech and Audio Processing,* 11(3):229–241, 2003.
128. T. C. T. Yin, P. X. Joris, P. H. Smith, and J. C. K. Chan. Neuronal processing for coding interaural time disparities. In R. H. Gilkey and T. R. Anderson, editors, *Binaural and Spatial Hearing in Real and Virtual Environments,* chapter 21, pp. 427–445. Lawrence Erlbaum, Mahwah, NJ, 1997.
129. T. C. T. Yin and S. Kuwada. Neuronal mechanisms of binaural interaction. In G. M. Edelman, W. E. Gall, and W. M. Cowan, editors, *Dynamic Aspects of Neocortical Function,* page 263. Wiley, New York, 1984.
130. W. A. Yost. Lateral position of sinusoids presented with intensitive and temporal differences. *Journal of the Acoustical Society of America,* 70:397–409, 1981.
131. W. A. Yost, R. H. Dye, and S. Sheft. A simulated "cocktail party" with up to three sources. *Perception and Psychophysics,* 58:1026–1036, 1996.
132. S. R. Young and E. W. Rubel. Frequency-specific projections of individual neurons in chick brainstem auditory nuclei. *Journal of Neuroscience,* 3:1373–1378, 1983.
133. X. Zhang, M. G. Heinz, I. C. Bruce, and L. H. Carney. A phenomenological model for the response of auditory-nerve fibers: I. Nonlinear tuning with compression and suppresion. *Journal of the Acoustical Society of America,* 109:648–670, 2001.
134. P. M. Zurek. Binaural advantages and directional effects in speech intelligibility. In G. A. Studebaker and I. Hochberg, editors, *Acoustical Factors Affecting Hearing Aid Performance.* Allyn and Bacon, Boston, 1993.

CHAPTER 6

Localization-Based Grouping

ALBERT S. FENG and DOUGLAS L. JONES

6.1 INTRODUCTION

Real-world auditory scenes are generally complex and comprise multiple auditory sources, each producing a distinct sound pattern. Sounds from neighboring sources often overlap in time and spectrum, making it difficult to isolate, localize, and characterize the individual sources. Because all sounds in the environment converge onto our two eardrums, the brain must perform signal sorting, to determine "what" is out there and "where" it is. The acoustic components associated with one source must be grouped, and the resulting stream of sound must be segregated from other streams associated with neighboring sources [8]. In spite of the challenges to listening, we and other living organisms (e.g., frogs, birds, bats) have been shown to have the ability to communicate effectively in complex auditory scenes.

Two basic listening strategies may be used to extract signals embedded in competing sounds. One involves grouping on the basis of the source's acoustic characteristics (e.g., pitch or harmonic structure), and the other involves grouping by the physical location of the source. This chapter will describe approaches that involve localization-based grouping, and will focus on two-sensor biological as well as engineering-based schemes.

The principle of localization-based grouping in biological systems is also the primary engineering approach for extraction of signals in acoustically cluttered environments. The engineering approach generally involves using an array of spatially distributed microphones to perform beamforming, and a larger array size is used to achieve higher spatial selectivity [20]. To cancel out noise effectively, the array must contain $N + 1$ microphones if there are N interfering sources in the ambiance [30]. In contrast, humans can localize as many as six concurrent sources [9] and can effectively cancel out noise from multiple sound sources using just two receivers [5, 12]. Binaural processing has advantages over monaural processing in terms of sound localization, echo suppression, and noise cancellation a la the "cocktail party" effect [12, 23, 26], and, thus, a two-receiver system will be the focus of this

Computational Auditory Scene Analysis. Edited by DeLiang Wang and Guy J. Brown

chapter. Below we will first describe the beamforming techniques for signal extraction in order to illuminate the differences and similarities in the engineering- and biologically based schemes.

6.2 CLASSICAL BEAMFORMING TECHNIQUES

Signal processing approaches to localization generally begin with beamforming, the process of using multiple sensors arrayed in a known geometric pattern to determine the direction of sources in the far field (i.e., at a distance considerably larger than the array aperture, or span), or even the exact location of sources in the near field. Often, the sensors are equally spaced along a line or in a plane, creating a linear, equally spaced array, which leads to simplifications in the mathematical computations that support faster or more sophisticated processing techniques. A two-element array is necessarily a linear, equally spaced array.

6.2.1 Fixed Beamforming Techniques

The simplest beamforming algorithms sum the individual signals from each array element (e.g., microphone) with sets of fixed weights and time delays to constructively enhance signals from specific "look directions." For example, simply adding all the signals from a linear, equally spaced array constructively sums all signals from a direction perpendicular to the array (i.e., "broadside"). Signals from other directions arrive at different times at the different sensors, so their peaks and valleys do not coincide and they are not enhanced as much by the summing operation. For certain directions and frequencies, they may cancel out completely; this is known as a "null" in the array response. Signals from any desired direction can be enhanced by time delaying the signals on each sensor to time align them such that they sum constructively. Sources are located by beamforming in all directions and looking for peaks in the energy of the output signals as a function of direction. In the very important special case in which the signals are sinusoids or occupy only a narrow range of frequencies, the time delay corresponds simply to a phase shift and can often be implemented much more simply via multiplication by a phase-shifted, complex-valued weight, hence the prevalence of narrowband beamforming techniques. The beam pattern of an array is the relative amplitude and phase of the processed array output to a signal arriving from various directions.

Fixed beamformers are simple to implement; robust to errors in the positions, gains, or phases of the sensor elements; and can even be optimal for equal interference or noise from all directions. Particularly for small arrays, however, they have significant limitations. The fixed beam pattern means that interference from certain directions is not completely cancelled; these secondary regions in the beam pattern from which interferers are passed with lesser gain are known as sidelobes. As a rule of thumb, the main-lobe span, or the directional resolution of the array as a function of the full range of unambiguous directions (e.g., 180° in a half-wavelength, linearly spaced array), is proportional to $1/M$, where M is the number of sensor elements;

for binaural or other small arrays, this is clearly very poor. Hence, more sophisticated methods have been developed to improve the resolution and interference rejection of arrays.

6.2.2 Adaptive Beamforming Techniques

Adaptive beamforming is the most commonly used enhancement. Almost all adaptive beamformers are based on the minimum variance principle first introduced by Capon [10], and are variously known as minimum-variance distortionless-response (MVDR) or linearly constrained minimum-variance (LCMV) adaptive beamformers. The key ideas are to constrain the combining weights to preserve any signal from the "target" (or "look") direction with no distortion (no change in gain or phase, hence "linearly constrained" or "distortionless response") while otherwise minimizing the average energy of the output (hence "minimum variance"). For narrowband signals, Capon derived the optimal combining filter weights, which are

$$\mathbf{w}_{\mathbf{opt}} = \mathbf{R}^{-1}\mathbf{e}/\mathbf{e}^{\mathbf{H}}\mathbf{R}^{-1}\mathbf{e}$$

where $\mathbf{w}_{\mathbf{opt}}$ is the vector of optimal combining weights, $\mathbf{R}^{-1}$ is the inverse of the cross-correlation matrix between the signals at the different sensors, $\mathbf{e}$ is the "steering vector" indicating the relative phases and amplitudes of a source from the target or look direction, and $\mathbf{e}^{\mathbf{H}}$ denotes the complex-conjugate transpose, or Hilbert transpose, of the steering vector. The distortionless response (target-preserving) constraint is $\mathbf{e}^{\mathbf{H}}\mathbf{w}_{\mathbf{opt}} = 1$.

Capon's adaptive beamformer has two primary drawbacks: (1) it applies only to narrowband signals, precluding important wideband signals such as human speech, and (2) computation of the correlation matrix inverse is both computationally expensive and numerically sensitive to errors. The first problem can be managed by splitting the signal into a number of narrow frequency bands by use of a filterbank or an FFT-based, short-time Fourier transform and processing each separately [62] using Capon's approach. For example, the wideband MVDR beamformer introduced by Frost [21] overcomes both of these problems. Frost applies an adaptive digital filter, rather than a single complex weight, to the input from each sensor element; this allows a different, optimal response at each frequency. Frost also introduces an iterative, adaptive update of the filter weights along the lines of the least mean squares (LMS) adaptive filter algorithm [27, 64, 65].

Frost's method, though a major advance, has the drawbacks of relatively slow convergence due to the LMS-like gradient operation followed by the projection to restore the constraint, as well as the computational complexity associated with that constraint. Griffiths and Jim devised an alternative implementation that enforces the constraint up front through a "blocking matrix" operation, followed by a multichannel adaptive noise canceller, or "generalized sidelobe canceller" (GSC), that performs the minimum-variance optimization through an unconstrained LMS iterative update [25]. This algorithm still converges relatively slowly, but has a less complex implementation. It is by far the most commonly used adaptive beamforming algorithm. Figure 6.1 illustrates the GSC for a two-element array with a broadside target.

Many modifications of these MVDR beamforming algorithms have been introduced to improve their performance and robustness in various situations. For example, Greenberg's modified step-size update [24] is essential for good performance of the GSC for signals with large amplitude fluctuations (such as speech). There has also been substantial research on robust adaptive beamformers that provide more tolerance of errors in the steering vectors or in the locations, gains, and phases of the sensor elements [e.g., 14, 70].

6.2.3 Independent Component Analysis Techniques

Blind source separation (BSS) was first introduced by Jutten and Herault [32]. The basic assumption is that mixtures of multiple, statistically independent sources are received with several separate sensors; mathematically, this can be expressed as

$$\mathbf{x} = \mathbf{A}\mathbf{s}$$

where $\mathbf{s}$ is a length-N vector of source samples, $\mathbf{x}$ is a length-M vector of the received data at the M sensors, and $\mathbf{A}$ is the "mixing matrix" that expresses the gain of each source at each sensor. The goal of blind source separation is to recover $\mathbf{s}$ without prior knowledge of $\mathbf{A}$ or $\mathbf{s}$. To be possible, this requires that $\mathbf{A}$ be an invertible matrix, or that the mixing relationships from each source to the sensors be linearly independent, and that there are no more sources than sensors. Since gains and permutations of the sources can be exchanged with the mixing matrix without being visible at the sensors, blind source recovery is possible only up to unknown gains and permutations of the sources. In addition, the sources must be non-Gaussian, so

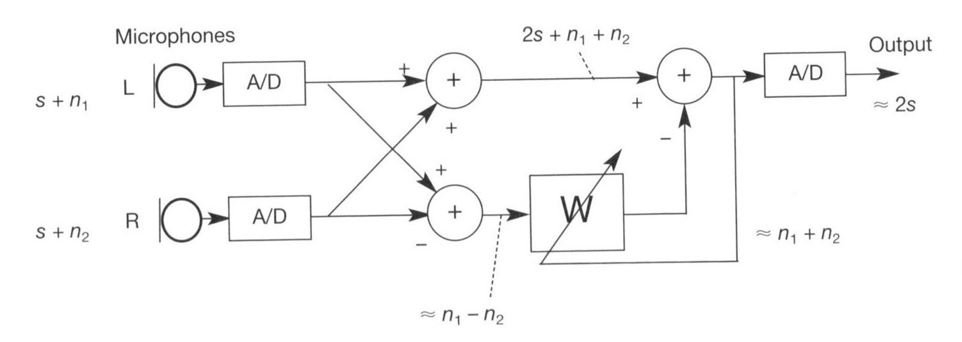

Figure 6.1 Block diagram of a two-sensors Griffiths–Jim generalized sidelobe canceller (GSC) adaptive beamformer for a broadside source (perpendicular to microphone axis). After digitization, the upper path implements a conventional delay-and-sum beam former, while the lower path implements a blocking operation that removes the target signal. This interference-only component is processed by an adaptive noise canceller to remove as much residual interference as possible.

that sufficient statistical information can be recovered to separate the sources (see [11] for an overview).

A number of algorithms have been introduced for blind source separation. Independent component analysis (ICA) is widely studied [11, 13], and a fast ICA algorithm has been developed recently [28]. Extensions have been developed for convolutive mixtures in which the relation between each source and sensor involves differing time delays or a filtering relationship [34, 50, 51], and for sources with different spectral shaping [3]. Several different approaches have been developed for nonstationary sources [4, 31, 34, 51].

Many research challenges remain to make blind source separation practical for real-world, location-based grouping. Blind separation of convolutively mixed sources remains challenging in practice, since most mixing environments introduce some form of filtering, relative time delay, and reverberation that produces a convolutive mixture (see Chapter 7). The algorithms tend to degrade badly when the real situation deviates from the idealized mathematical model. BSS also fails when there are more sources than sensors (see, however, [36] and [66]), so it does not apply in realistic and important environments such as the "cocktail party" scenario with two sensors. It remains to be seen whether blind source separation will ever play a significant practical role in location-based grouping.

6.2.4 Other Localization-Based Techniques

Other high-resolution methods for finding source locations have also been developed, most notably MUSIC [54]. MUSIC begins with an eigen-decomposition of the array correlation matrix to partition the data space into a signal subspace and a noise subspace; the separation is performed by grouping the corresponding eigenvalues into large (signal subspace) and small (assumed to be due to noise alone) groups. The inverse of the projection of steering vectors for each look direction onto the noise subspace is computed. Since the projection onto the null subspace approaches zero in the direction of sources, this creates a highly peaked response in the source directions. Although the apparent high resolution is somewhat artificial, MUSIC often yields good recovery of source locations. In our experience, when MUSIC is regularized sufficiently to yield robust results with small arrays, its performance is very similar to the MVDR beamformer. Since a null subspace is required, MUSIC can locate at most one source less than the number of sensors.

6.3 LOCATION-BASED GROUPING USING INTERAURAL TIME DIFFERENCE CUE

For most biological systems, sound localization is based on analysis of several binaural cues: interaural time differences, interaural intensity differences, and interaural spectral differences (see Chapter 5 and reviews in [71]). For pure tones, the interaural time differences (ITDs) amount to interaural phase differences, and their analysis is generally limited to low-frequency sounds due to the facts that: (1) phase coding in the auditory nerve is limited to low frequencies, and (2) encoding of interaural phase differences becomes ambiguous at high frequencies when the sound wavelength is shorter than the interaural distance. Instead, at high frequencies, the interaural time difference of the signal envelope can be used for determining the sound azimuth. In contrast to the interaural phase difference cue, analysis of interaural intensity differences is most effective for high-frequency sounds; when a quarter of the sound wavelength is shorter than the head size, the head is an obstacle to sound propagation, thus producing a head-shadowing effect. As a result, for humans, the acuity to localize pure tone sounds is greater at <1000 Hz and >4000 Hz than at 2000–3500 Hz. Finally, for living organisms with pinnae, the convolutions of each pinna produce a direction-dependent filtering of incoming sounds. The spectral shaping of one pinna differs from that of the opposite pinna, thereby creating direction-dependent interaural spectral differences.

In this and the following sections, we will cover binaural signal processing algorithms that utilize individual binaural cues. This will be followed by a description of binaural processing algorithms that make use of multiple localization cues, and a discussion of the strengths and weaknesses of the different approaches.

As described earlier, interaural differences in time/phase (ITD) confer accurate information for source azimuth, especially at low frequencies. The Jeffress model of coincidence detection [29] is well grounded anatomically and physiologically, and is a favorite of researchers in psychophysics, physiology, engineering, and computational neuroscience [2, 5, 6, 22, 38, 39, 55, 57]. The earlier models are limited to localization of one or two sources and the performance suffers when the signal is speech and there are more than two sound sources. More recently, Liu et al. [40, 41] created a binaural system that simulates Jeffress' model, utilizing interaural phase differences over the entire frequency range and a nonlinear procedure for determining source directions. This system could accurately localize four to six concurrent speech sources and characterize one or all signal sources. A schematic diagram of this system is shown in Fig. 6.2.

The system of Liu et al. assumes two inputs supplied by two identical omnidirectional microphones. Signal processing is performed in the digital frequency domain, using separate A/D converters. Once digitized, the signals are decomposed using the short-term Fourier transform. A "dual delay line" network is used for determining the directions of sound sources, at each frequency, for the left and right microphones separately. The time delays are assigned a priori such that the acoustic space in front of the two microphones is divided uniformly into 91 or 181 sectors in azimuth, and each sector is uniquely mapped to one specific location along each dual delay line. Assuming the microphones are located in free field and there is no acoustic shadowing effect, the midpoint of each dual delay line corresponds to an ITD of zero, that is, to a source located in the mid-sagittal plane. A source located on the right side of the system, corresponding to nonzero ITD, is represented on the left side of the dual delay line, and for the source further to the right its coincidence will occur further to the left of the dual delay line.

With the signals already transformed into the frequency domain, Liu et al. [40] employ a subtraction operation to determine the coincidence locations for sound

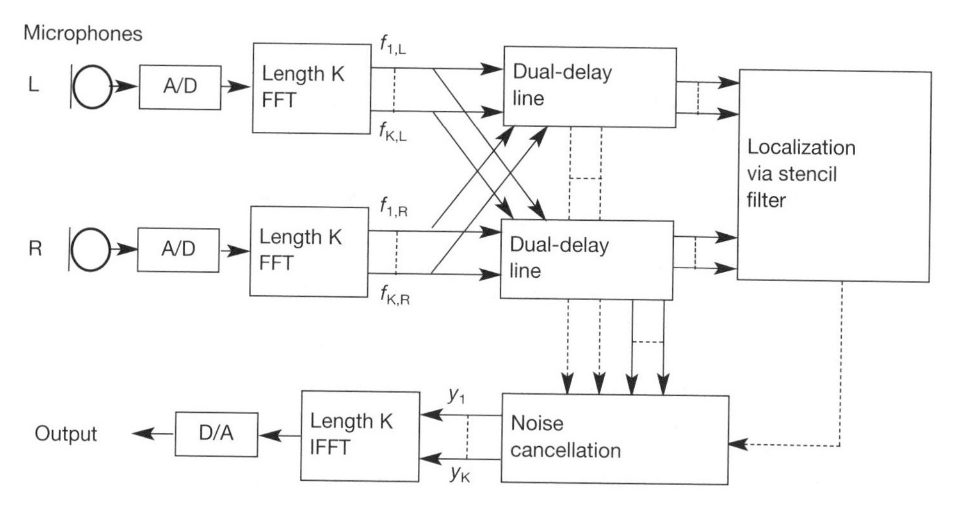

Figure 6.2 Block diagram of a localization–extraction, biologically inspired two-sensors beamformer. After digitization, each channel is transformed to the frequency domain and a dual delay line circuit determines the delay of maximal correspondence between the two inputs in a manner analogous to the interaural time-delay processing in the brain. A stencil filter fuses this information across frequency to determine all the source locations. Independent noise cancellation of the dominant interference at each time and frequency is followed by signal recovery via inverse Fourier resynthesis of the recovered signal. Reprinted with permission from Liu et al. [40]. Copyright 2000, Acoustical Society of America.

sources (in the time domain, a cross-correlation method may be used by computing a running integration over a number of critical bands [6]). A minimum between the frequency domain representations of the left and right channels, therefore, corresponds to the location of coincidence. On a frequency-by-frequency basis, the point of minimum magnitude between two channels is the same as the point of minimum phase difference; this calculation is robust against intermicrophone intensity differences and allows accurate estimation of the sound azimuth by integrating coincidence locations across frequency and time.

Temporal and spectral integrations take advantage of the broad bandwidth, and make azimuthal estimation more reliable in adverse noisy conditions. For frequency integration, the coincidence pattern over the entire frequency range is made. Because the ITD is ambiguous at high frequencies when the wavelength is shorter than the interaural distance, the ambiguity can in theory lead to erroneous estimation of the sound azimuth. To minimize the ambiguity, Liu et al. [40] create a "stencil filter" and turn the ambiguous ITD information (which is fixed for a constant interaural distance) into use for improving the estimation of the sound azimuth (Fig. 6.3). Other approaches include using an image processing technique to look for the consistency of coincidence patterns over a narrow frequency band, that is, the low-frequency range of hearing [2], or applying weights to reduce the importance of the coincidence patterns at high frequencies [57, 59]. The stencil filter approach increases the accuracy of azimuth estimation and allows the system to determine the azimuths of four to six concurrent sources in an anechoic environment. For temporal integration, Liu et al. employ a forgetting average scheme to apply weight to the system's past coincidence pattern, thereby making the system either more, or less, sensitive to the past coincidence pattern. Temporal integration provides a memory for the system, and is used to attenuate the mutual interference between concurrent signals.

The simple subtraction method, when strategically employed over the dual delay line structure, can also effectively cancel multiple interfering sound sources and consequently enhance the desired signal [41]. Noise suppression is based on knowledge of the spatial information of all sound sources. That a subtraction operation produces the proper mathematical manipulation for noise cancellation is described in their paper. We shall provide herein some intuitive examples to illustrate the strength of the algorithm.

In the simple case of having one desired source and one noise source in the ambiance, the subtraction operation applied at the location of the noise source in the dual delay line structure (determined by locating the position along the dual delay line where the minimum occurs) produces an accurate estimate of the desired signal (see Equation 24 of [41]). Conceptually, the noise cancellation scheme is equivalent to placing a null at the position of the noise source, with the constraint that the broadband gain in the target direction is kept at unity. When there are multiple interfering sources, Liu et al. take advantage of the broadband nature of the system by applying the one-null beam pattern for each frequency bin at each time instant (Fig. 6.4). In other words, at each time instant for each frequency bin, the system adaptively steers the null toward the most intense interfering source. Thus, at each time instant, for a subset of frequency bins the null is steered toward one of the interfering sources, but for other frequency bins the nulls are placed at the locations of the remaining interfering sources, while maintaining unity gain for the target. The null placements change from time to time. The system is therefore dynamic and adaptive.

Liu et al. [41] showed that the signal extraction performance is robust in real-world settings. For an anechoic condition with four concurrent talkers under different spatial configurations (speech broadcast through four separate loudspeakers from different azimuths at a fixed equal distance of 1 m from a pair of microphones), the system yields a net intelligibility-weighted gain of 8 to 9.1 dB (Table 1 of [41]). For test signals, they used both males and females speaking spondaic words; the intermicrophone distance was fixed at 14.5 cm. The gain ranges from 4.6 to 6.7 dB when the test room has a reverberation time of 0.4 s (Table 2 of [41]). By comparison, slowly adapting time-domain algorithms such as the Frost [21] beamformers are unable to track the nonstationary structure fast enough to achieve significant cancellation when there are multiple interfering sources. The binaural system of Liu et al. is computationally intensive and has yet to be implemented in a real-time system. Lockwood et al. [42] recently developed a binaural system without the computational overhead from localization of all sound sources. The algorithm is thus computationally efficient and has been implemented into a real-time

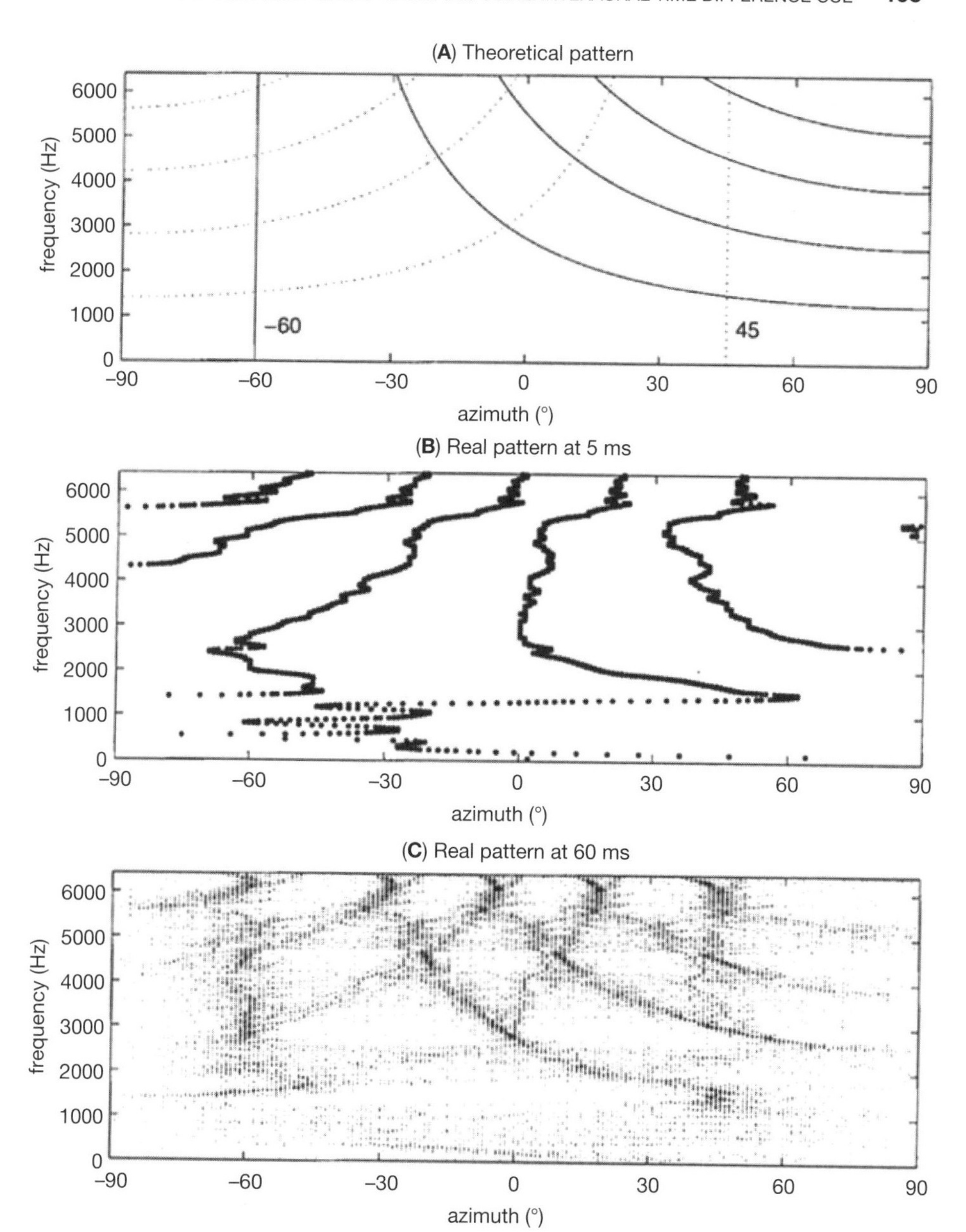

Figure 6.3 Theoretical (**A**) and actual (**B, C**) broadband coincidence patterns for a source at 60° on the left side (solid traces) and a second source at 45° on the right side (dotted traces) of a two-sensor array. Phase ambiguities at frequencies >1200 Hz are shown as curved traces; these are distinct from the unambiguous vertical traces at these azimuths. The curved traces associated with each sound azimuth are used as templates for stencil filters. With these filters, the existence of such curved traces in the actual coincidence patterns at 60 ms following the onset of two concurrent sounds at −60° and +45° (**C**) is then interpreted as having sounds at these locations. Reprinted with permission from Liu et al. [40]. Copyright 2000, Acoustical Society of America.

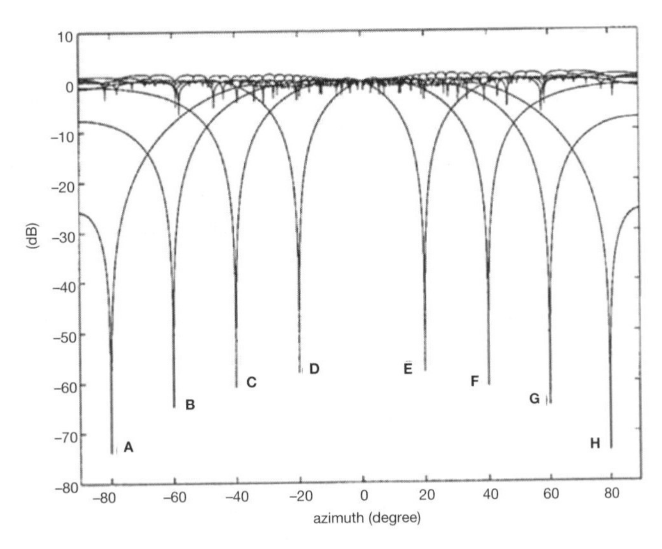

Figure 6.4 Illustration of one-null beam patterns with a target at 0° azimuth and a noise source at another azimuth: −80° (**A**), −60° (**B**), −40° (**C**), −20° (**D**), 20° (**E**), 40° (**F**), 60° (**G**), and 80° (**H**). Reprinted from Liu et al. [41]. Copyright 2001, Acoustical Society of America.

system [16]. This system assumes that the target is located directly in front of the listener (for which the interaural phase, time, and level differences are zero); in practice, the listener decides and selects any one of the acoustic sources as the target by steering the microphone array toward it.

The binaural system of Lockwood et al. [42] belongs to a class of so-called minimum variance distortionless beamformers (see Section 6.2.2; van Veen and Buckley [62]), but utilizes only two sensors. It operates in the frequency domain and thus is called the FMV algorithm (Fig. 6.5). With two microphones, minimization involves computing a 2 × 2 correlation matrix for each frequency bin. This matrix is updated every N samples, making it possible to quickly track changes in the input signals across all frequency bands. The FMV algorithm is therefore more efficient than both the time-domain correlation matrix inversion techniques and the localization-extraction binaural system of Liu et al. [40, 41]. A major consideration of the FMV system is the frequency resolution, which in practice is determined by the FFT length relative to the sampling rate. Increasing the FFT length decreases the bandwidth of each frequency bin and therefore improves the system performance (for stationary signals), as it provides a more detailed estimate of signal spectra. However, if the FFT length is too long, the performance may decrease, because the signal is not sufficiently stationary for the interval of the FFT. Also, longer FFTs require more data points, and objectionably increase the system delay.

Other frequency-domain algorithms do not perform beamforming, but rather attenuate individual frequency bands in a manner that removes interference. The algorithm of Kollmeier and colleagues [33, 52, 67], which extends an idea introduced by Allen et al. [1], uses correlation and phase and amplitude differences between channels to determine a gain for each frequency band (see Section 6.5). The idea is that bands with identical amplitude and phase are dominated by the target signal coming from straight in front of the listener and should thus be retained, whereas bands with significant amplitude or phase difference are dominated by off-axis interferers and should be suppressed. Slyh and Moses [56] describe another example of this type of algorithm. These techniques have been shown to be effective in attenuating off-axis sources using only two sensors. The caveat is that they introduce signal distortion by attenuating part of the target signal as well.

The signal extraction performance of the FMV system has been evaluated [42] under computer simulation using recordings of speech sources made with three different two-microphone arrays in three rooms with varying reverberation times (RT = 0.1, 0.37 and 0.65). For this evaluation, the three two-microphone arrays were: (1) two microphones coupled to the ear canals of a KEMAR mannequin, (2) two omnidirectional microphones in free field separated by 15 cm, and (3) two cardioid microphones in free-field separated by 15 cm. Test signals, comprising

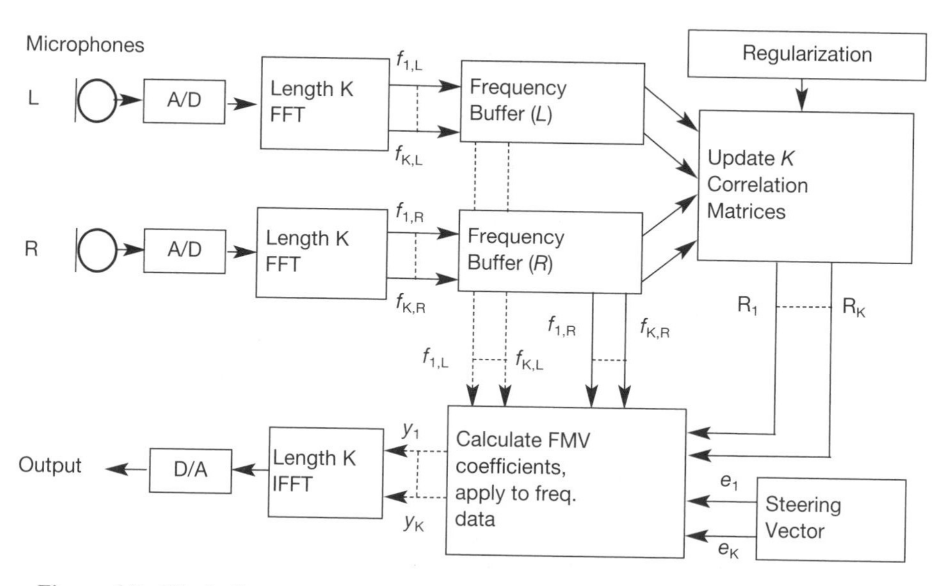

Figure 6.5 Block diagram of a frequency-domain, minimum-variance distortionless response adaptive two-sensors beamformer (FMV). After digitization, each channel is transformed to the frequency domain, and a short-time optimal narrowband adaptive beamformer is computed and applied independently to each frequency band. The desired signal is then recovered via inverse Fourier resynthesis.

two to five speech sources, were constructed for each room and array. The signals were processed with the FMV algorithm, and the signal extraction performance was quantified by calculating the signal-to-noise ratio (SNR) of the output and compared to those of two time-domain, distortionless beamformers, the Frost adaptive beamformer [21] and a version of the GSC [24, 25], as well as a frequency-domain binaural algorithm—the Peissig–Kollmeier (P–K) algorithm [33, 67]. The FMV and P–K algorithms consistently yielded higher SNR (up to 5–6 dB) when compared to the Frost and GSC algorithms for all test configurations with more than one interfering source and for all test rooms. For the one-interferer tests, the FMV and P–K algorithms sometimes performed better than the time-domain algorithms, but other times the reverse was true. Although the FMV and P–K algorithms performed similarly well in nearly all test conditions, the P–K algorithm always attenuated the frequency bands that contained interference, and, accordingly, the target distortion was usually greater (i.e., it accepted target distortion in exchange for improvement in SNR). The performance of FMV in terms of SNR gain is approximately twice that of an optimized GSC for speech signals when more than one spatially separated interferer is present, and comparable for only one inteferer.

The performance benefits of the biologically inspired Liu and FMV beamformers over conventional adaptive beamformers such as the GSC for speech sources are believed to stem from the combination of independent and rapid adaptation across frequency. Lockwood et al. [42] present evidence that these algorithms exploit the sparse, time-varying, time-frequency structure of nonstationary sources such as speech more effectively than the conventional beamformers.

The two-sensors FMV system of Lockwood et al. has been implemented in a real-time device using a DSP chip from Texas Instruments (TI-C62 fixed-point DSP) with a clock speed of 133 MHz [16]. The output of the real-time system sounds natural and is of high fidelity (examples of system outputs are available on http://www.casabook.org). Speech reception thresholds and speech intelligibility measures were obtained from listeners with normal hearing or with mild-to-moderate sensorineural hearing loss [35]. Word and sentence-length stimuli were processed through the FMV algorithm, directional microphones alone (DIR), or with a simple delay-and-sum (DAS) beamformer. Listeners' ratings of speech intelligibility, percent of words repeated correctly, and threshold for words and sentences in the presence of four competing signals showed the FMV to provide significant performance benefits across different listening environments (i.e., rooms that vary widely in size and reverberation characteristics). In one test configuration with four interfering sources, for listeners with mild-to-moderate hearing loss, FMV confers on average 30% more intelligibility and 8 dB lower speech reception threshold (SRT) than processing with the DAS, and ~50% more intelligibility and 11 dB improvement in SRT than with DIR; the figures are slightly lower for listeners with normal hearing. The results from a different test configuration also with four interferers are consistent with those described above, showing the significant benefits for the FMV when listening in challenging acoustic environments. (See examples at http://www.casabook.org.)

6.4 LOCATION-BASED GROUPING USING INTERAURAL INTENSITY DIFFERENCE CUE

As described in Section 6.3, for humans, the interaural intensity difference (IID) cue confers directional information at high frequencies. This is because when the head size is larger than a quarter of the sound wavelength it diffracts the sound and becomes an obstacle to sound propagation, thereby creating a significant IID. Many nonmammalian vertebrates with small heads (such as frogs, lizards, crocodiles, and some bird species), however, can also utilize IID to localize sound in spite of the low-frequency contents of their communication sounds [17, 18]. In these animals, the ears are acoustically coupled through Eustachian tubes (or interaural canals) and the mouth cavity. Such acoustical coupling allows the direct component of sound at the external surface of the eardrum and the indirect component at the internal surface of the eardrum to interact, forming a pressure-gradient (or pressure-difference) receiver that enhances the directionality of the ear; the two ears have asymmetrical directional patterns relative to the longitudinal body axis (see Fig. 2 of Feng and Christensen-Dahlsgaard [18]). The pressure-difference receiver, therefore, functions to amplify the amplitude (and phase) differences at the two eardrums; it generates larger IIDs than expected from the small size of the animal. Lockwood et al. [43] have utilized the same principle to create a signal extraction system using two (or more) colocated directional microphones having different directivity patterns.

Whereas most beamforming techniques for location-based signal separation and grouping exploit time differences between the sensors, some of these methods generalize easily to level-difference cues. Essentially colocated arrays of directional sensors that have been developed include underwater acoustic hydrophones for SONAR (e.g., [15, 37]), vector geophones for geophysical and seismic data acquisition (e.g., [60, 63]), colocated acoustic arrays of directional microphones [43, 45], and vector electromagnetic antennas [19].

The relative amplitude of sources at each sensor is a function of direction in such arrays, thus allowing location-based discrimination on this basis. Colocated vector sensors typically only provide two to four independent measurements; the array dimension is thus small, and fixed beamforming techniques provide limited separation, although single sources can be accurately localized [49]. The steering vectors are real-valued relative amplitudes rather than phase differences; many adaptive beamforming methods can nonetheless still be applied to locate or recover sources. MVDR methods have been extended to this case, including narrowband Capon [15] and FMV beamformers [43]. Narrowband subspace-based methods such as MUSIC and ESPRIT have also been explored [61, 68, 69].

The performance and behavior of the amplitude-based adaptive beamforming algorithms is generally comparable to that of standard spatial arrays with a similar number of elements. The FMV algorithm shows similar advantages for wideband, nonstationary signals such as speech in comparison with the generalized sidelobe canceller as with time-delay-based arrays. The algorithm is essentially identical to the binaural FMV described above, with the appropriate amplitude-based steering vectors replacing the phase-difference steering vectors used in time-difference-based arrays. For multiple speech sources, it is possible to obtain 1–2 dB higher SNR gains for two orthogonal gradient microphones than achieved by a 15 cm free-field, two-element array of omnidirectional microphones [43].

Source localization and grouping can also be accomplished using amplitude-based directional cues. As mentioned earlier, narrowband MVDR beamforming or MUSIC techniques have been extended to vector–sensor arrays. Mohan and coworkers have developed a method for localizing multiple broadband, nonstationary sources using both conventional acoustic arrays [46] and vector sensors [47], and have demonstrated its ability to locate multiple simultaneous sources in real environments. This approach begins with a short-time Fourier transform of each input channel and computation of short-time/frequency cross-correlation matrices as in FMV. Should only a single source dominate any individual time-frequency bin, the corresponding correlation matrix will have a rank-1 signal subspace, and the eigenvector corresponding to the largest eigenvalue will be the steering vector of the corresponding source. Furthermore, either amplitude-based narrowband MVDR beamforming or MUSIC can be applied to obtain a high-resolution estimate of the direction of this source. These rank-1 or "coherent" bins are identified by thresholding the ratio of the largest to the next-largest eigenvalue. The Mohan and coworkers method computes a high-resolution directional estimate for each coherent time-frequency bin using either an amplitude-based MVDR or MUSIC beamformer and fuses these estimates to produce a composite wideband location estimate in which the peaks denote the locations of multiple sources. The technique is able to locate four to six sources using only two microphones with high reliability, and detected sources are usually localized to within two degrees of their true bearing.

6.5 LOCATION-BASED GROUPING USING MULTIPLE BINAURAL CUES

Many location-based grouping algorithms utilize multiple binaural cues. For most of these models, ITD processing is based on the Jeffress' neural network as the central building block (see review by Stern and Trahiotis [58]). Lindemann [38, 39] added a contralateral-inhibition mechanism (known for processing IIDs) to a model of interaural cross-correlation to explain lateralization of pure tones with ITDs and IIDs and to explain the law of the first wavefront (see Chapter 5). Gaik [22] made additional modifications to improve the localization performance of Lindemann's model. Bodden [6, 7] and Banks [2] subsequently created binaural systems with the ability to assess the azimuth of sound incidence as well as extract a signal in the presence of spatially distributed concurrent signals. Whereas Bodden employed a time-variant optimum filter (Wiener filter), Banks utilized the so-called f-p method to separate the signal of interest from interfering signals. The performance of these systems is limited to a single interfering source, however.

To select a single voice in a complex auditory scene, Roman et al. [53] used a supervised learning approach. ITD and IID cues are used to separate a target signal from interfering sounds; an "ideal" time-frequency binaural masking paradigm (akin to human auditory masking, wherein a stronger signal masks a weaker signal for each critical band; Moore [48]) is used to select the target. Specifically, for each time-frequency unit the paradigm selects the target and assigns a mask value of 1 if it is stronger than the interference (or 0 otherwise). They employ a nonparametric classification method to determine decision regions in the ITD–IID space that correspond to an optimal estimate for an ideal mask. The performance of the binary mask approximates that of the ideal binary mask in terms of SNR (signal-to-noise ratio) and ASR (automatic speech recognition) measures. It produces an SNR gain of ~12 dB when there are two or three sources. The binaural mask model differs from the model of Bodden (and the preceding models of Lindemann and Gaik) in some respects. Namely, whereas the latter simulates the precedence effect for reverberant conditions, the former does not; on the contrary, whereas the former requires training, the latter does not.

Some beamformers, including the FMV adaptive beamformer, can be applied in the presence of simultaneous time-delay and amplitude cues, for example, in an ear-mounted two-sensors array due to the head-related transfer function (HRTF). As described in Section 6.4 for the level cue, the only difference is the change in the steering vectors, which are computed from the HRTF. Lockwood et al. report similar SNR improvements with an in-the-ear array and a free-field array [42].

The location-finding algorithm of Mohan and coworkers similarly extends with only minor modification to the multiple-cue scenario. The technique described in Section 6.4 can be applied to delay-only, amplitude-only, or multiple cues simply by applying the appropriate directional steering vectors in the directional estimation step. For the general case, these steering vectors may not have a known mathematical form and measured values must be used; for example, in the case of an ear-mounted two-sensors array, the HRTFs from many directions are measured and are then interpolated to obtain the steering vectors for all directions. Once these are known, they are applied in either the MVDR or MUSIC computation to obtain the response map as a function of direction for each coherent bin; the performance is similar to that with interelement timing and level differences.

Allen and coworkers [1], and Kollmeier and colleagues [33] who extended their method, take a different approach. These methods also compute a short-time Fourier transform of each channel and compare the corresponding time-frequency bins, but rather than beamforming, they attempt to determine whether each bin is signal or interference dominated; signal-dominated bins are retained (multiplied by a unity gain), whereas bins deemed to be interference dominated are suppressed via multiplication by a gain less than one, which may approach zero for severe interference. The modified signals at each microphone are then reconstructed through a filterbank synthesis. A potential advantage of these methods is that they produce a dichotic output that is potentially different for each channel and which may preserve some directional cues in the residual interference.

Allen and coworkers [1] argue that for a clean source, the target signal should be perfectly correlated, or coherent, at both microphones, except for possibly a delay. Reduced correlation between channels in any time-frequency bin must then be attributed to either reverberation (dereverberation was the original goal of this technique) or interference; if significant difference is observed, the energy content in that bin is attributed to reverberation or interference and reduced accordingly. Slyh and Moses [56] apply a similar idea with an emphasis on interference suppression. Kollmeier and colleagues include time-delay, level-difference, and noise-related cues to better discriminate corrupted bins [33, 52, 67]; each corresponding time-frequency bin from the two channels is compared in terms of phase (time-delay) difference, amplitude difference, and coherence. These similarity metrics are fed to a nonlinear function to produce a gain value for the bin ranging from zero to one; the output of this function should be unity for identical signals on both channels, and should approach zero for a severe mismatch. This method often produces very good interference rejection and a highly intelligible output, consistently yielding higher SNR (by up to 5–6 dB) when compared to two conventional adaptive beamforming algorithms (Frost and GSC) when there are >2 interfering sources [42]. By removing the frequency bands that contain interference, the P-K algorithm can achieve substantial noise suppression, but at the price of increasing target distortion.

6.6 DISCUSSION AND CONCLUSIONS

Location cues are among the many important features used by the human binaural hearing system to group and separate different auditory streams. Location-based grouping is perhaps the most mature and successful body of techniques in practical computational auditory scene analysis today. Conventional beamforming techniques with two sensors yield only modest gains in complex scenes, but more recent biologically inspired methods have shown substantially greater ability to separate sounds based on location and to localize multiple auditory sources and events. These successes are already significant, but localization-based grouping still has considerably more potential.

The binaural hearing system probably plays a major (perhaps dominant) role in dealing with reverberant environments, yet computational methods of dereverberation remain in their infancy, as discussed in the following chapter. The human auditory system integrates different types of cues throughout the process of auditory scene analysis, whereas current computational methods generally perform location-based processing in isolation. These and other avenues likely offer rich possibilities for further advances over the already useful state of the art.

ACKNOWLEDGMENTS

The authors thank Michael Lockwood for Figs. 6.1–6.3. Their research in this area has been variously supported by the National Institutes of Health (NIDCD), Phonak

USA, DARPA, the Mary Jane Neer Research Fund, the Charles Goodenberger Funds, and the Beckman Institute at the University of Illinois.

REFERENCES

1. J. Allen, D. A. Berkley, and J. Blauert. A multimicrophone signal-processing technique to remove room reverberation from speech signals. *Journal of the Acoustical Society of America,* 62:912–915, 1977.
2. D. Banks. Localization and separation of simultaneous voices with two microphones. *IEE Proceedings I: Communications, Speech and Vision,* 140:229–234, 1993.
3. A. Belouchrani, K. Abed-Meraim, J.-F. Cardoso, and E. Moulines. A blind source separation technique using second-order statistics. *IEEE Transactions on Signal Processing,* 45:434–444, 1997.
4. A. Belouchrani and M. G. Amin. Blind source separation based on time–frequency signal representations. *IEEE Transactions on Signal Processing,* 46:2888–2897, 1998.
5. J. Blauert. *Spatial Hearing: The Psychophysics of Human Sound Localization,* John S. Allen, translator. The MIT Press, Cambridge, MA, 1983.
6. M. Bodden. Modeling human sound source localization and the cocktail-party-effect. *Acta Acustica,* 1:43–55, 1993.
7. M. Bodden. Auditory demonstration of a cocktail-party-processor. *Acta Acustica,* 82:356–357, 1996.
8. A. S. Bregman. *Auditory Scene Analysis.* The MIT Press, Cambridge, MA, 1990.
9. A. W. Bronkhorst and R. Plomp. Effect of multiple speechlike maskers on binaural speech recognition in normal and impaired hearing. *Journal of the Acoustical Society of America,* 92:3132–3139, 1992.
10. J. Capon. High-resolution frequency-wavenumber spectrum analysis. *Proceedings of the IEEE,* 57(8):1408–1419, 1969.
11. J.-F. Cardoso. Blind signal separation: statistical principles. *Proceedings of the IEEE,* 86(10):2009–2025, 1998.
12. E. C. Cherry. Some experiments on the recognition of speech, with one and with two ears. *Journal of the Acoustical Society of America,* 25:975–979, 1953.
13. P. Comon. Independent component analysis—A new concept? *Signal Processing,* 36:287–314, 1994.
14. H. Cox, R. M. Zeskind, and M. M. Owen. Robust adaptive beamforming. *IEEE Transactions on Acoustics, Speech, and Signal Processing, ASSP*-35, 10:1365–1376, 1987.
15. G. L. D'Spain, W. S. Hodgkiss, G. L. Edmonds, J. C. Nickles, F. H. Fisher, and R. A. Harriss. Initial analysis of the data from the vertical DIFAR array. In *Proceedings of OCEANS '92,* pp. 346–351, 1992.
16. M. E. Elledge, M. E. Lockwood, R. C. Bilger, A. S. Feng, D. L. Jones, C. R. Lansing, W. D. O'Brien, and B. C. Wheeler. Real-time implementation of a frequency-domain beamformer on the TI C62X EVM. In *10th Annual DSP Technology Education Research Conference,* Houston, TX, August 2–4, 2000.

17. R. R. Fay, and A. S. Feng. Mechanisms for directional hearing among nonmammalian vertebrates. In W. A. Yost and G. Gourevitch, G., editors, *Directional Hearing,* pp. 179–213. Springer-Verlag, New York, 1987.

18. A. S. Feng and J. Christensen-Dahlsgaard. Interconnections between the ears in non-mammalian vertebrates. In P. Dallos and D. Oertel, D. editors, *Handbook of the Senses: Audition,* Elsevier, San Diego, 2006.

19. E. R. Ferrara, Jr. and T. M. Parks. Direction finding with an array of antennas having diverse polarizations. *IIEEE Transactions on Antennas and Propagation,* AP-31(2):231–236, 1983.

20. J. L. Flanagan, J. D. Johnston, R. Zahn, and G. W. Elko. Computer steered microphone arrays for sound transduction in large rooms. *Journal of the Acoustical Society of America,* 78:1508–1518, 1985.

21. O. L. Frost. An algorithm for linearly constrained adaptive array processing. *Proceedings of the IEEE,* 60:926–935, 1972.

22. W. Gaik. Combined evaluation of interaural time and intensity differences: Psychoacoustic results and computer modeling. *Journal of the Acoustical Society of America,* 94:98–110, 1993.

23. J. W. Grabke and J. Blauert. Cocktail party processors based on binaural models. In D. Rosenthal and H. G. Okuno, editors, *Computational Auditory Scene Analysis,* pp. 243–255. Lawrence Erlbaum, Mahwah, NJ, 1998.

24. J. E. Greenberg. Modified LMS algorithms for speech processing with an adaptive noise canceller. *IEEE Transactions on Speech and Audio Processing,* 6:338–351, 1998.

25. L. J. Griffiths and C. W. Jim. An alternative approach to linearly constrained adaptive beamforming. *IEEE Transactions on Antennas and Propagation,* AP-30:27–34, 1982.

26. M. L. Hawley, R. Y. Litovsky, and J. F. Culling. The benefit of binaural hearing in a cocktail party: Effect of location and type of interferer. *Journal of the Acoustical Society of America,* 115:833–843, 2004.

27. S. Haykin. *Adaptive Filter Theory.* Prentice-Hall, Upper Saddle River, NJ, 2002.

28. A. Hyvärinen and E. Oja. A fast fixed-point algorithm for independent component analysis. *Neural Computation,* 9(7):1483–1492, 1997.

29. L. A. Jeffress. A place theory of sound localization. *Journal of Comparative Physiology and Psychololog*y, 41:35–39, 1948.

30. D. H. Johnson and D. E. Dudgeon. *Array Signal Processing: Concepts and Techniques.* Prentice-Hall, Englewood Cliffs, NJ, 1993.

31. D. L. Jones. A new method for blind source separation of nonstationary signals. *Proceedings of the IEEE International Conference on Acoustics, Speech and Signal Processing,* 5:2893–2896, 1999.

32. C. Jutten and J. Herault. Blind separation of sources, Part I: An adaptive algorithm based on neuromimetic architecture. *Signal Processing,* 24:1–10, 1991.

33. B. Kollmeier, J. Peissig, and V. Hohmann. Real-time multiband dynamic compression and noise reduction for binaural hearing aids. *Journal of Rehabilitation Research and Development,* 30(1):82–94, 1993.

34. B. S. Krongold and D. L. Jones. Blind source separation of nonstationary convolutively

mixed signals. In *Proceedings of 10th IEEE Workshop on Statistical Signal and Array Processing,* pp. 53–57, 2000.

35. J. Larsen, C. R. Lansing, R. C. Bilger, B. C. Wheeler, S. A. Phatak, M. E. Lockwood, W. D. O'Brien, and A. S. Feng. Human speech perceptual performance in noise with a two-sensor FMV beamforming algorithm. *Acoustics Research Letters Online,* 5(3): 100–105, 2004.
36. T. W. Lee, M. S. Lewicki, M. Girolami, and T. J. Sejnowski. Blind source separation of more sources than mixtures using overcomplete representations. *IEEE Signal Processing Letters,* 6:87–90, 1999.
37. C. B. Leslie, J. M. Kendall, and J. L. Jones. Hydrophone for measuring particle velocity. *Journal of the Acoustical Society of America,* 28:711–715, 1956.
38. W. Lindemann. Extension of a binaural cross-correlation model by contralateral inhibition. I. Simulation of lateralization for stationary signals. *Journal of the Acoustical Society of America,* 80:1608–1622, 1986.
39. W. Lindemann. Extension of a binaural cross-correlation model by contralateral inhibition. II. The law of the first wave front. *Journal of the Acoustical Society of America,* 80:1623–1637, 1986.
40. C. Liu, B. C. Wheeler, W. D. O'Brien, Jr., R. C. Bilger, C. R. Lansing, and A. S. Feng. Localization of multiple sound sources with two microphones. *Journal of the Acoustical Society of America,* 108:1888–1905, 2000.
41. C. Liu, B. C. Wheeler, W. D. O'Brien, Jr., C. R. Lansing, R. C. Bilger, D. L. Jones, and A. S. Feng. A two-microphone dual delay-line approach for extraction of a speech sound in the presence of multiple interferers. *Journal of the Acoustical Society of America,* 110:3218–3231, 2001.
42. M. E. Lockwood, D. L. Jones, R. C. Bilger, C. R. Lansing, W. D. O'Brien, Jr., B. C. Wheeler, and A. S. Feng. Performance of time- and frequency-domain binaural beamformers based on recorded signals from real rooms. *Journal of the Acoustical Society of America,* 115:379– 391, 2004.
43. M. E. Lockwood, D. L. Jones, Q. Su, and R. N. Miles. Beamforming with collocated microphone arrays. *Journal of the Acoustical Society of America,* 114:2451, 2003.
44. R. C. McDonough. Application of the maximum-likelihood method and the maximum-entropy method to array processing. In S. Haykin, editor, *Nonlinear Methods of Spectral Analysis: Topics in Applied Physics,* pp. 181–243. Springer-Verlag, New York, 1979.
45. R. N. Miles, S. Sundermurthy, C. Gibbons, R. Hoy, and D. Robert. Differential Microphone. US Patent no. 6,788,796, Sept. 7, 2004.
46. S. Mohan, M. L. Kramer, B. C. Wheeler, and D. L. Jones. Localization of nonstationary sources using a coherence test. In *Proceedings of IEEE Workshop on Statistical Signal Processing,* pp. 470–473, 2003.
47. S. Mohan, M. E. Lockwood, D. L. Jones, Q. Su, and R. N. Miles. Sound source localization with a gradient array using a coherence test. *Journal of the Acoustical Society of America,* 114:2451. 2003.
48. B. C. J. Moore. *An Introduction to the Psychology of Hearing.* Academic Press, San Diego, 1997.
49. A. Nehorai and E. Paldi. Acoustic vector-sensor array processing. *IEEE Transactions on Signal Processing,* 42(9):2481–2491, 1994.

50. H. L. Nguyen Thi and C. Jutten. Blind source separation for convolutive mixtures. *Signal Processing,* 45:209–229, 1995.

51. L. Parra and C. Spence. Convolutive blind separation of non-stationary sources, *IEEE Transactions on Speech and Audio Processing,* 8(3):320–327, 2000.

52. J. Peissig and B. Kollmeier. Directivity of binaural noise reduction in spatial multiple noise-source arrangements for normal and impaired listeners. *Journal of the Acoustical Society of America,* 101, 1660–1670, 1997.

53. N. Roman, D. L. Wang, and G. J. Brown. Speech segregation based on sound localization. *Journal of the Acoustical Society of America,* 114:2236–2252, 2003.

54. R. Schmidt. Multiple emitter location and signal parameter estimation. *IEEE Transactions on Antennas and Propagation,* AP-34(3):276–280, 1986.

55. S. A. Shamma, N. Shen, and P. Gopalaswamy. Stereausis: Binaural processing without neural delays. *Journal of the Acoustical Society of America,* 86:989–1006, 1989.

56. R. E. Slyh and R. L. Moses. Microphone array speech enhancement in overdetermined signal scenarios. *International Conference on Acoustics, Speech and Signal Processing,* 2:347–350, 1993.

57. R. M. Stern and C. Trahiotis. The role of consistency of interaural time over frequency in binaural lateralization. In Y. Cazals, L. Demany, and K. Horner, editors, *Auditory Physiology and Perception, Proceedings of the 9th International Symposium on Hearing,* pp. 547–554. Pergamon Press, Oxford, 1992.

58. R. M. Stern and C. Trahiotis. Models of binaural interaction. In B. C. J. Moore, editor, *Hearing.* Academic Press, New York, 1995.

59. R. M. Stern A. S. Zeiberg, and C. Trahiotis. Lateralization of complex binaural stimuli: A weighted-image model. *Journal of the Acoustical Society of America,* 84:156–165, 1988.

60. R. R. Stewart, J. E. Gaiser, R. J. Brown, and D. C. Lawton. Converted-wave seismic exploration: Methods. *Geophysics,* 67(5):1348–1363, 2002.

61. P. Tichavsky, K. T. Wong, and M. D. Zoltowski. Near-field/far-field azimuth and elevation angle estimation using a single vector hydrophone. *IEEE Transactions on Signal Processing,* 49(11):2498–2510, 2001.

62. B. D. van Veen and K. M. Buckley. Beamforming: A versatile approach to spatial filtering. *IEEE Signal Processing,* 5:4–24, 1988.

63. J. E. White. Directional sound detection. US patent no. 2,982,942, 1961.

64. B. Widrow and M. E. Hoff, Jr. Adaptive switching circuits. *IRE WESCON Conference Record,* Pt. 4:96–104, 1960.

65. B. Widrow, P. E. Mantey, L. J. Griffiths, and B. B. Goode. Adaptive antenna systems, *Proceedings of the IEEE,* 55:2143–2159, 1967.

66. S. Winter, H. Sawada, S. Araki, and S. Makino. Overcomplete BSS for convolutive mixtures based on hierarchical clustering. In *Proceedings of ICA,* pp. 652–660, 2004.

67. T. Wittkop, S. Albani, V. Hohmann, J. Peissig, W. S. Woods, and B.Kollmeier. Speech processing for hearing aids: Noise reduction motivated by models of binaural interaction. *Acustica,* 83:684–699, 1997.

68. K. T. Wong and M. D. Zoltowsky. Uni-vector-sensor ESPRIT for multisource azimuth, elevation, and polarization estimation. *IEEE Transactions on Antennas and Propagation,* AP-45(10):1467–1474, 1997.

69. K. T. Wong and M. D. Zoltowsky. Root-MUSIC-based azimuth-elevation angle-of-arrival estimation with uniformly spaced but arbitrarily oriented velocity hydrophones. *IEEE Transactions on Signal Processing,* 47(12):3250–3260, 1999.

70. S. Q. Wu, and J. Y. Zhang. A new robust beamforming method with antennae calibration errors, In *Proceedings of IEEE Wireless Communications and Networking Conference,* 2:869–872, 1999.

71. W. A. Yost and G. Gourevitch. *Directional Hearing.* Springer-Verlag, New York, 1987.

CHAPTER 7

Reverberation

GUY J. BROWN and KALLE J. PALOMÄKI

7.1 INTRODUCTION

In real-world acoustic environments, sound reaches a receiver (such as a microphone or a human listener) directly from the source and after reflection from various surfaces. These attenuated, time-delayed reflections of the original sound combine additively to form reverberation. Most often, reverberation is associated with enclosed spaces such as meeting rooms and concert halls. However, reverberation can also occur in outdoor environments such as forests [52]. As a result, auditory mechanisms have evolved to combat the deleterious effects of reverberation upon sound localization and communication signals such as speech. These mechanisms are still only partially understood, and creating computational systems that match the robustness of human listeners in reverberant environments remains a challenging problem for computational auditory scene analysis (CASA) and automatic speech recognition (ASR).

Consider an enclosed environment in which there is a loudspeaker and a microphone. The room impulse response (RIR) completely characterizes the properties of the room for the specific configuration of the sound source and receiver. Typically, the RIR is measured by recording a continuous wideband signal in the room of interest; the response to an idealized Dirac impulse is then obtained by deconvolving the original test signal from the recorded signal. An idealized RIR is shown in Fig. 7.1. The first event in the RIR is a pulse due to the direct sound (the "first wavefront"). This is followed by a set of sparse early reflections for the next 50 msec or so (depending upon the size of the room), and then by dense late reflections that form a decaying "tail."

The most important objective measure for describing the acoustical properties of a room is the reverberation time, T_{60}. This is defined as the time required (in seconds) for the sound level of a steady-state source to drop by 60 dB after it is abruptly turned off. The RIR shown in Fig. 7.1 is characterized by a T_{60} of approximately 0.25 s. Another useful measure is the direct sound-to-reverberation

Computational Auditory Scene Analysis. Edited by DeLiang Wang and Guy J. Brown

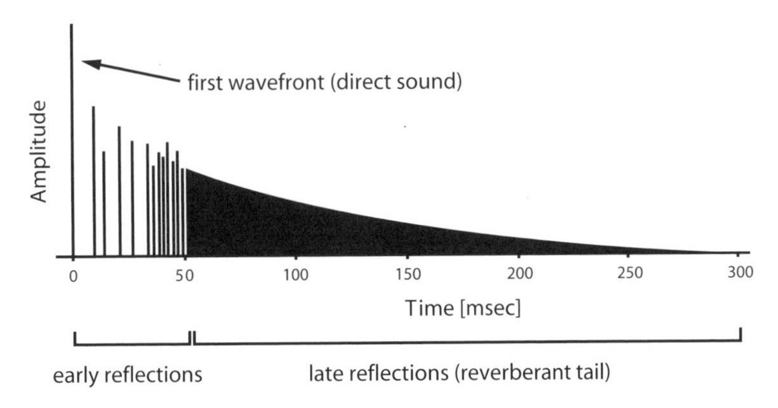

Figure 7.1 The anatomy of an idealized room impulse response. Sparse early reflections are followed by a dense reverberant tail. This impulse response has a reverberation time T_{60} of about 0.25 s.

ratio (DRR), which is the ratio of the direct sound energy to reverberant energy expressed in dB.

Again, consider a scenario in which a single sound source is recorded by a single microphone in an enclosed room. A number of factors determine the amount of reverberation received at the microphone, and its characteristics. First, the reverberation becomes more apparent the further the microphone is moved away from the sound source. This is because sound amplitude decays as a function of the distance travelled, so that larger distances between the source and receiver give rise to a smaller DRR. Second, reverberation is influenced by the volume of the room. For example, a small office may have a T_{60} of 0.3 s or so, whereas a concert hall typically has a T_{60} of 1.5 s or more. Third, the pattern of reflections associated with a reverberant space is influenced by the proximity of the sound source and the microphone to reflective surfaces. The time delay between direct sound and strong early reflections introduces nulls in the frequency response of the room; this results in comb filtering that causes spectral distortion ("coloration") of the original signal. Furthermore, echoes of the direct sound are spectrally distorted because room materials absorb sound energy in a frequency-dependent manner.

In order to gain insight into the problems that reverberation poses for CASA, and their possible solutions, this chapter covers a relatively broad range of related material. First, we review the effects of reverberation on human listeners and machines (including ASR systems) in Sections 7.2 and 7.3. We then discuss the mechanisms that are thought to underlie the robustness of human hearing in reverberant environments in Section 7.4, and compare these with current approaches to reverberation-robust acoustic processing in Section 7.5. CASA approaches that have been evaluated in reverberant conditions are then reviewed in Section 7.6. We conclude with a discussion in Section 7.7, and make some suggestions for future work in this challenging area.

7.2 EFFECTS OF REVERBERATION ON LISTENERS

Reverberation presents both a problem and an opportunity for auditory perception. On the one hand, reverberation has a negative impact on speech perception, sound localization and the ability to use auditory grouping cues. However, it also provides a useful cue to the distance of a sound source, and allows listeners to create a spatial impression of an acoustic environment.

7.2.1 Speech Perception

When an utterance is produced in a reverberant environment, the direct sound from the talker combines additively with scaled and delayed versions of the speech that arise from room reflections. As a result, the speech signal is smeared over time and dips in the temporal envelope become filled by energy from reflections. This decreases the intelligibility of the speech. For example, Gelfand and Silman [36] have shown that phoneme recognition is poorer in reverberant conditions than in quiet. They found that place of articulation and stop and frication information were most affected by reverberation. Additionally, phonemes that occurred in the initial position of spoken words were less affected than those in the final position.

These findings may be explained in terms of self-masking and overlap masking of reverberated speech [12, 60]. In self-masking, reverberation due to the early part of a phoneme masks later parts of the same phoneme. This leads to the degradation of phonetic information, such as the blurring of formant transitions and the smearing of the onsets and decays of transient phonemes. In overlap masking, echoes of previous phonemes mask later phonemes. In particular, strong vowels can mask weaker consonants that follow them. This is believed to be the larger effect of the two, consistent with the observation that speech intelligibility in reverberation can be improved by slowing the speaking rate [12].

The effects of reverberation on speech are illustrated in Fig. 7.2, which shows a cochleagram for the utterance "you took me by surprise" spoken by a male talker (see Chapter 1, Section 1.3.2). In panel **A**, the speech has been recorded in an anechoic environment, and in panel **B** the same signal has been convolved with a room impulse response with a T_{60} of 0.5 s. The cochleagram of the reverberated speech is clearly smeared over time, and a number of specific effects are also visible (see also [5]). First, formant transitions are less distinct and appear flattened. Second, silent intervals caused by vocal tract closure during the production of stop consonants are filled by energy from reflections. For example, a closure in the word "took" (0.4–0.45 s) is apparent in the anechoic speech but is abolished in the reverberant case. Third, reverberation blurs temporal distinctions such as voice onset time (i.e., the time between a stop release and the subsequent onset of voicing). This is apparent in the word "by" between 0.6 and 0.75 s. Fourth, noise bursts associated with fricatives and affricates persist longer; in panel **B** of the figure, the initial /s/ of "surprise" (between 0.75 and 0.8 s) is much extended. Finally, amplitude modulation associated with the fundamental frequency of

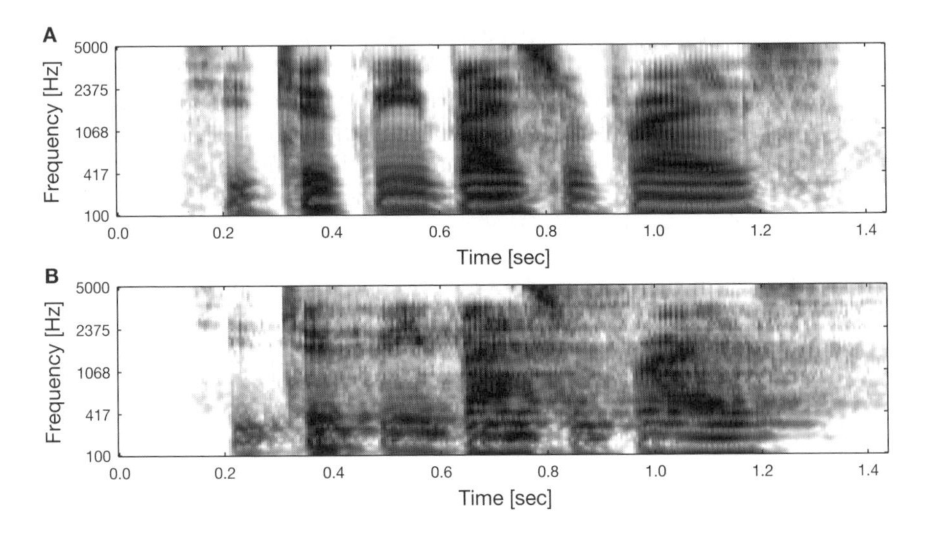

Figure 7.2 The effect of reverberation on speech. (**A**) Cochleagram of the utterance "you took me by surprise" spoken by a male talker in an anechoic environment. Dark pixels indicate regions of high energy, and light pixels indicate regions of low energy. (**B**) Cochleagram of the same utterance convolved with the impulse response of a reverberant room. The T_{60} reverberation time was 0.5 sec, and the DRR was –3 dB.

voiced speech is reduced. In panel **A** of the figure, amplitude modulation is apparent as a regular pattern of vertical bands, such as in the voiced part of the word "surprise" between 0.95 and 1.15 s. This structure is much reduced in the reverberated speech shown in panel **B**.

Recall from Section 7.1 that the impulse response of a reverberant room consists of sparse early reflections followed by dense late reverberation. These two components of the impulse response have different effects on speech intelligibility. Early reflections are highly correlated with the original speech, and can actually assist the intelligibility of speech by increasing its loudness. Late reflections are poorly correlated with the original speech and, therefore, behave more like additive noise. This observation has led to the concept of the ratio of useful/detrimental sound energy, which was introduced by Lochner and Burger [64] as a means of predicting speech intelligibility in reverberant environments. Such a measure is essentially an early-to-late-arriving energy ratio, such as the one defined by European standard BS EN ISO 3382:2000 [68]:

$$C_{te} = 10 \log_{10}\left(\int_0^{te} p^2(t)\, dt \Big/ \int_{te}^{\infty} p^2(t)\, dt \right) \text{dB} \tag{7.1}$$

Here, $p(t)$ is the instantaneous sound pressure of the room impulse response measured at time t and C_{te} is termed the early-to-late index (see also [13]). The standard [68] suggests that the value of the early time limit te should be 50 ms for predicting speech intelligibility, and 80 ms for predicting the "clarity" of music (for details of the latter, see [82]).

Finally, we note that the characterization of early reflections as "useful" for speech perception is something of a simplification. Although early reflections help to boost the loudness of speech above the noise floor introduced by late reverberation, they also distort the spectral envelope of the speech signal. These distortions arise from comb filtering caused by successive reflections, and from the frequency-dependent absorption of sound by surfaces in the room. Apparently, listeners are able to compensate perceptually for the spectral distortion of speech caused by reverberation; we return to this point in Section 7.4.4.

7.2.2 Sound Localization

Human listeners predominantly use comparisons between the two ears in order to localize sounds in space (see Chapter 5). Sound energy emitted from a source that lies to one side of a listener will be more intense in the ear nearest to the source, principally because sound reaching the farther ear is attenuated by the head (the so-called "head shadow"). Sound will also reach the nearer ear first, and the farther ear a short time later. Hence, interaural intensity difference (IID) and interaural time difference (ITD) cues are available for judging the direction of a sound. Additionally, the head and pinnae act as a direction-dependent filter that imposes particular spectral characteristics upon incoming sounds, depending on their angle of incidence with the head. Listeners are able to use these spectral changes to judge the location of a sound source (particularly its elevation).

In reverberant conditions, the efficacy of these cues is reduced by room reflections. In particular, the distribution of observed IIDs and ITDs is broadened by reverberation, because sound energy arrives from reflections as well as from the direct path between the source and the listener's ears. This is illustrated in Fig. 7.3. Panel **A** of the figure shows the distribution of IID and ITD for a mixture of two male speech signals. Both cues were computed for an auditory filter channel with a center frequency of 500 Hz, using the approach described by Faller and Merimaa [34] (see also Chapter 1 and Chapter 5). The sound sources were spatialized at azimuths of –15 degrees and +25 degrees in an anechoic virtual room, using the ROOMSIM package [17]. Note that IID and ITD estimates cluster into two groups and, therefore, provide an unambiguous cue to the location of the two sources [85]. However, if the sources are spatialized in a reverberant room (panel **B**), the situation is less clear. The distributions of IID and ITD are now broader, and they do not peak at the true locations of the sound sources.

In practice, the extent to which localization performance is degraded by reverberation depends upon a number of factors, including the position of the listener in the reverberant room and the characteristics of the sound source. For example, Hartmann [44] found that steady broadband noise, which was easily localized in an dry acoustic environment, was much less accurately localized in the presence of reverberation (see also [37]). He also found that localization performance was related to the spectral density of the sound source; spectrally sparse complex tones were lo-

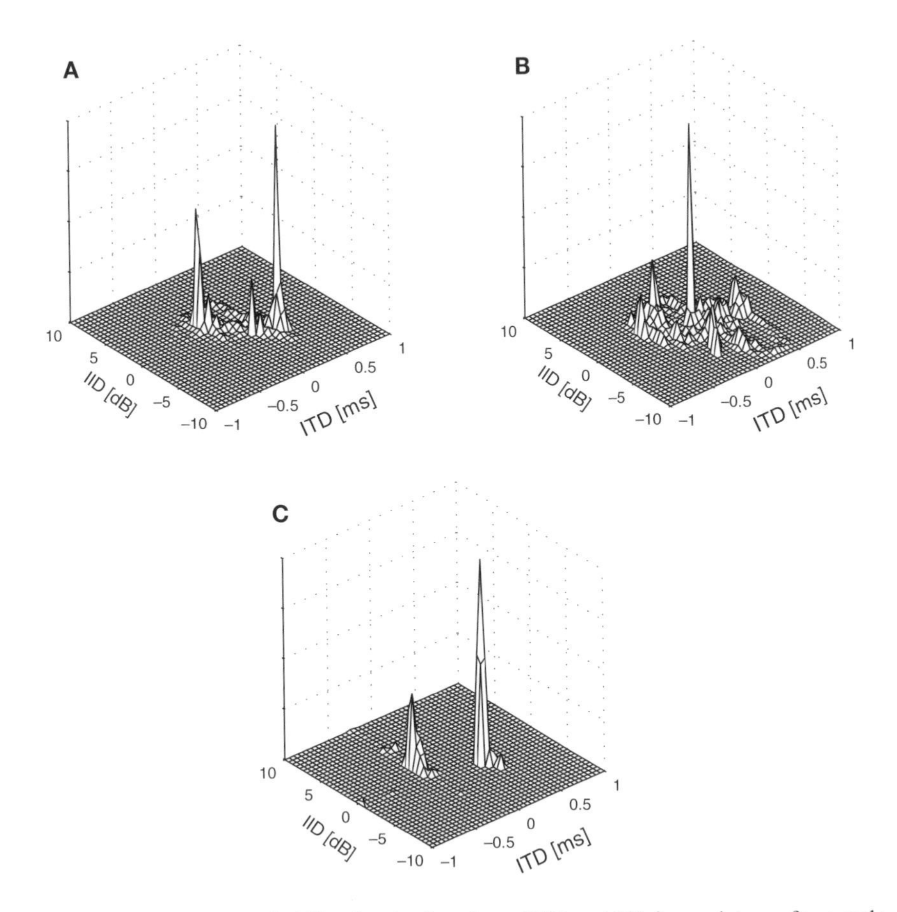

Figure 7.3 The joint probability density function of ITD and IID for a mixture of two male speech signals, spatialized at azimuths of –15 degrees and +25 degrees. ITD and IID estimates were computed every 0.05 ms for an auditory filter channel with a center frequency of 500 Hz using the approach described by Faller and Merimaa [34]. Sources were spatialized and reverberated using the ROOMSIM package [17]. (**A**) Sources spatialized in an anechoic room. (**B**) Sources spatialized in a simulated reverberant room of size 6.25 m × 3.75 m × 2.5 m, for which the walls had absorption characteristics typical of acoustic plaster. The reverberation time T_{60} was 0.5 s. (**C**) Sources spatialized in a reverberant room as in the previous panel, but only the subset of ITD and IID estimates are plotted for which the interaural coherence was at least 0.99.

calized less well than broadband noise, and it was almost impossible to localize a steady low-frequency tone in reverberant conditions. For continuous broadband noise, localization accuracy fell as the reverberation time increased.

However, Hartmann [44] also reported that the ability of listeners to localize sounds with strong attack transients was independent of reverberation time. This is consistent with the notion that a "precedence effect" underlies the ability of listeners to localize sound in reverberant spaces [95]. Apparently, listeners are able to selectively weight incoming directional information, such that a greater importance is attributed to cues carried by direct sound from the source than cues associated with later reflections. The precedence effect is triggered by rapid acoustic onsets, which explains why transient sounds can be accurately localized in reverberant environments but steady-state sounds cannot. The mechanisms underlying the precedence effect are further discussed in Section 7.4.3.

7.2.3 Source Separation and Signal Detection

Although much research has considered the effect of reverberation on single sources such as speech, fewer studies have investigated its effect on multisource listening environments (and specifically the effect of reverberation on auditory grouping cues). In this section, we consider the effect of reverberation on some of the perceptual grouping cues introduced in Chapter 1, including harmonicity, spatial cues, onsets, and offsets.

It is well known that listeners are able to use harmonic relationships to segregate concurrent sounds (see Chapter 3). For example, listeners are better able to identify two concurrent vowel sounds ("double vowels") when there is a difference in fundamental frequency ($\Delta F0$) between them, relative to the case in which they have the same $F0$. However, in reverberant conditions, the effectiveness of grouping by harmonicity depends on the extent to which the $F0$ fluctuates [22]. Consider a sound composed of stationary harmonics, such as a synthetic vowel. Reverberation causes several echoes of each harmonic to occur, which differ in amplitude and phase from the direct sound and combine additively with it. The amplitude of each harmonic is, therefore, perturbed by a random amount, but harmonic relationships remain unchanged. However, if the $F0$ of the harmonic sound fluctuates (as it does in natural speech), the direct sound and delayed echoes have different harmonic frequencies. As a result, harmonicity is reduced.

This point is illustrated by Fig. 7.4, which shows a summary correlogram for an utterance spoken by a male talker in anechoic and reverberant conditions (see Chapters 1 and 2 for details of how this representation is computed). In panel **A** of the figure, the $F0$ of the speech is clearly evident in the form of dark horizontal bands that occur in the region of 100 Hz. The effects of moderate reverberation ($T_{60} = 0.5$ s.) on this $F0$ information are shown in panel **B** of the figure. Note that periodicity is apparent in regions of the reverberated speech whereas there was none in the anechoic signal, due to the influence of echoes. This is particularly apparent in the region between 1.2 and 1.3 s. Also, the $F0$ due to the brief vowel at about 0.85 s has been corrupted by echoes from the more intense diphthong that precedes it.

Following this reasoning, Culling et al. [22] predicted that reverberation would reduce the ability of listeners to use a $\Delta F0$ to separate concurrent sounds with fluctuating $F0$s, but not sounds with static $F0$s. They confirmed this using double vowel stimuli, to which reverberation was added using a physical model of small-room acoustics. When the target and masking vowels had static $F0$s, the masked identifi-

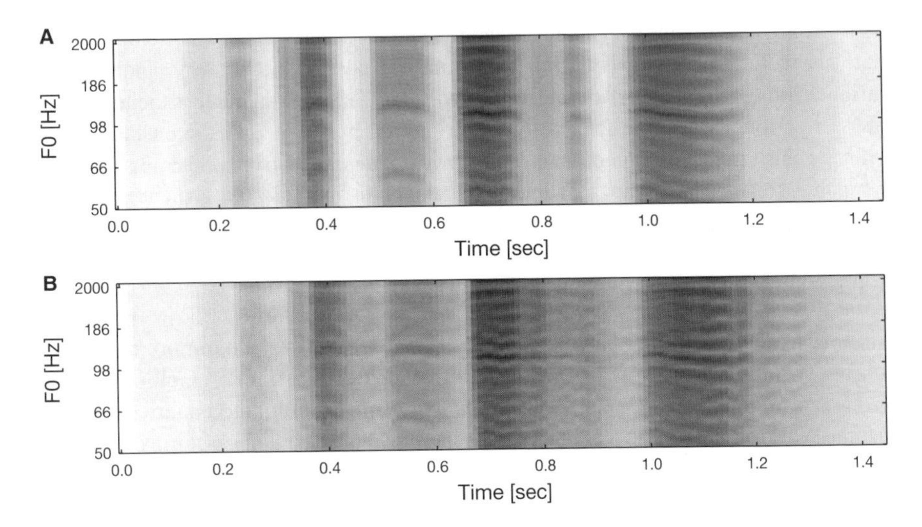

Figure 7.4 Effect of reverberation on pitch information. (**A**) Summary correlogram for the utterance "you took me by surprise" spoken by a male speaker. The dark bands in the region of 100 Hz correspond to the fundamental frequency. (**B**) Summary correlogram for the same utterance convolved with the impulse response of a reverberant room. The T_{60} reverberation time was 0.5 s, and the DRR was –3 dB.

cation threshold was lower when a $\Delta F0$ was introduced, and this advantage was not affected by reverberation. However, when the $F0$s of the target and masking vowels were sinusoidally modulated, the advantage of the $\Delta F0$ was abolished in the presence of reverberation.

In a subsequent study, Culling et al. [20] extended their experimental paradigm to running speech. Specifically, they measured the speech reception threshold (SRT) for a male talker in the presence of an interfering female talker, using conditions in which the two voices had natural $F0$ contours or were resynthesized on a monotone. Again, they found that reverberation disrupted the ability of listeners to exploit a difference between the natural $F0$ contours of the two voices. However, the monotonous speech was no more intelligible than naturally intonated speech in reverberant conditions. Culling et al. predicted from their previous study [22] that monotonous speech should be more intelligible in reverberant conditions than intonated speech. However, this was based on the assumption that monotonous and intonated speech are equally intelligible in anechoic conditions. This is not the case for continuous speech, which is more intelligible when naturally intonated. Hence, although the SRT was degraded more due to reverberation for intonated speech than monotonous speech, the intelligibility of monotonous and naturally intonated speech was about the same in the reverberated condition.

A related experiment by Darwin and Hukin [24] investigated the grouping of two simultaneous target words with an attended sentence and a simultaneous unattended sentence. They found that listeners had a weak tendency to group target words with the attended sentence when they both had the same monotonous $F0$, but that this preference was eliminated by reverberation. However, natural pitch cues were more robust, and a difference in vocal tract size between the attended and unattended voices provided a cue that was very resistant to the effects of reverberation. Darwin and Hukin also found that vocal tract differences and prosodic cues for speech segregation were more resistant to reverberation than a difference in ITD between the two talkers. The advantage of the latter was abolished by relatively small amounts of reverberation.

Darwin and Hukin's findings are consistent with previous studies, which show that reverberation seriously disrupts the ability of listeners to use spatial cues to segregate concurrent voices. For example, Plomp [80] measured the intelligibility of speech in the presence of a spatially separated speech masker in anechoic and reverberant conditions. In anechoic conditions, a gain in intelligibility of 4–5 dB was obtained when the target speech and masker emanated from different azimuths. However, even moderate reverberation (T_{60} = 0.4 s) reduced this intelligibility gain to only 2–3 dB. Also, the advantage of a spatial separation between the target speech and the masker further diminished as the T_{60} was increased. Similarly, Culling et al. [20] have reported that listeners can only exploit a difference in azimuth between two competing voices in anechoic conditions. Also, it appears that ITD only provides a weak cue for simultaneous (across-frequency) grouping of sounds that originate from different azimuths [21, 28]. It seems likely that listeners do not adopt such a strategy because it would fail in reverberant conditions, due to the disruptive effects of reverberation on interaural coherence. However, ITD may play a role in sequential grouping, by providing a cue that directs auditory attention to a specific spatial location [23].

It was noted in Chapter 1 that listeners tend to perceptually group acoustic components that appear and disappear at the same time (i.e., they use principles of grouping by "common onset" and "common offset"). Figure 7.5 illustrates the likely effect of reverberation on such grouping cues. Panel **A** shows the acoustic onsets and offsets that occur in an utterance spoken by a male talker in anechoic conditions. This representation was computed from a cochleagram with a frame rate of 10 ms, in which the usual compression stage was omitted. Black pixels indicate time-frequency regions in which the energy in a frequency channel has increased by at least 5 dB compared to the previous time frame (i.e., an onset). Similarly, white pixels indicate offsets in which the energy in a channel has fallen by at least 5 dB compared to the previous time frame. Panel **B** shows the same representation for speech that has been reverberated by a room impulse response with a T_{60} of 0.5 s. Note that the activity in panel **B** occurs a little later than that in panel **A**, because of the delay introduced by convolution with the room impulse response. Strong onsets at the start of the utterance still persist in the reverberated speech, but weaker onsets (e.g., at 0.46 and 0.95 s) are abolished because dips in the spectrum are filled by energy from reflections. The acoustic offsets evident in panel **A** are almost completely abolished by reverberation according to the 5 dB criterion used here. This analysis suggests that offsets are unlikely to be a reliable grouping cue in reverberant conditions, whereas onsets will often be preserved (although this depends on their magnitude and temporal context).

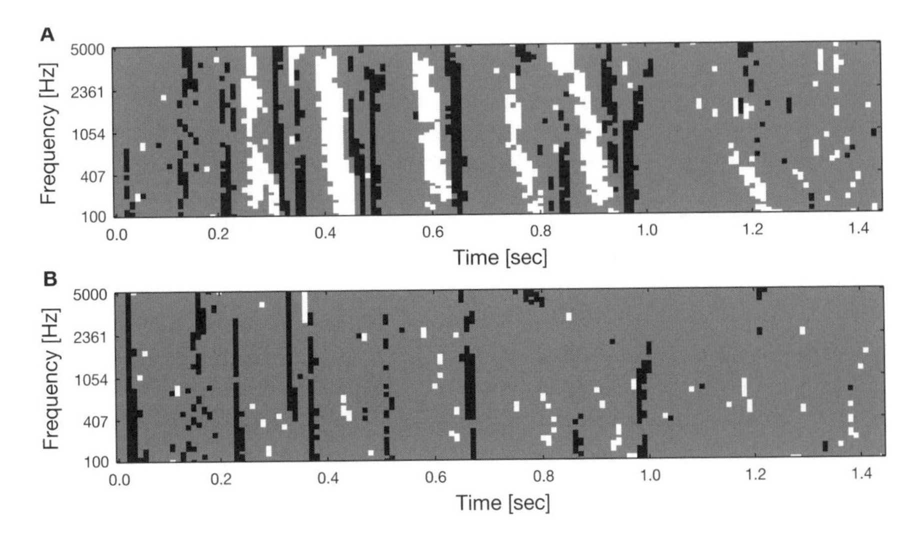

Figure 7.5 Effect of reverberation on onsets and offsets. (**A**) Onsets and offsets for the utterance "you took me by surprise" spoken by a male talker in an anechoic environment. Onsets are marked with black pixels, and offsets are marked with white pixels. (**B**) Onsets and offsets for the same utterance convolved with the impulse response of a reverberant room. The T_{60} reverberation time was 0.5 sec, and the DRR was –3 dB. Note that weak onsets are eliminated and no substantial offsets now occur.

Relatively few studies have investigated the effect of reverberation on the ability of listeners to detect a target signal that has been masked by an interferer. Such signal detection experiments are distinct from those discussed above, because the former do not concern speech intelligibility, only the ability to detect a masked signal. Zurek et al. [105] have investigated the detection of a brief one-third-octave band of noise masked by continuous broadband noise in anechoic and reverberant conditions, using one and two ears. Impulse responses for reverberant conditions were synthesized using the image method [2]. They found that monaural detection performance was generally predicted by the target-to-masker power ratio, which varied according to head-diffraction effects and reverberation. Binaural detection was generally well predicted by Durlach's equalization–cancellation (EC) model [32]. Although originally proposed to explain binaural masking differences due only to masking sounds, the EC model is equally successful at predicting the effects of reverberation because it is sensitive to the interaural decorrelation caused by reverberation. Zurek et al. [105] found that reverberation had a greater effect on monaural than binaural detection thresholds, leading to a substantial binaural advantage.

In summary, reverberation has a varying degree of impact on auditory grouping cues. Prosodic cues and those relating to vocal tract size are more robust to reverberation than ITD. Acoustic onsets are often preserved, so that grouping by common onset may still be a useful cue in reverberant conditions. Finally, binaural cues appear to contribute little to across-frequency grouping when reverberation is present, but they still contribute to signal detection in reverberant conditions. In the latter case, the binaural advantage is gained due to unmasking of the target in a frequency-independent manner, which is well predicted by the EC model.

7.2.4 Distance Perception

The role of reverberation in determining the perceived distance of a sound source was first noted by Békésy [93]. Specifically, he showed that the perceived distance of a sound source in an enclosed environment depends upon the ratio of the energies of direct and reflected sound, and also on the time delay between direct and reflected sound. Subsequent studies have confirmed this finding. For example, Mershon and Bowers [70] have shown that listeners can make absolute and relative judgments of source distance on the basis of the ratio of direct to reflected sound. Furthermore, the use of this cue does not depend on prior experience; listeners are able to judge the distance of a novel sound presented for the first time in an unfamiliar environment.

Bronkhorst and Houtgast [16] have investigated the role of reverberation in distance perception using virtual sources, and have proposed a quantitative model based on their findings. They spatialized sound sources using head-related transfer functions (HRTFs), and presented them to listeners over headphones. The authors found that perceived distance was determined by the distance of the virtual source and the number of reflections and their relative sound levels. They also confirmed the existence of an "auditory horizon" effect [70]; for small distances, perceived distance is approximately proportional to source distance, but for distances greater than 2 meters or so, the perceived distance grows at a slower rate than source distance. They explain their findings in terms of a modified direct-to-reverberant energy ratio in which a 6 ms window is used to estimate the energy of the direct sound. The authors note that such a window might be a general property of spatial hearing, since a similar integration time is implicated in the precedence effect.

It should be noted that perceived distance depends on a number of cues in addition to those related to reverberation; for example, sound intensity provides a relative cue to changes in distance [71]. Furthermore, cues to the distance of a sound source are combined in complex ways. For instance, Zahorik [103] reports that the perceptual weight assigned to intensity and direct-to-reverberation energy ratio depends on the angular direction of the sound source and its type (noise or speech).

7.2.5 Auditory Spatial Impression

In addition to their role in distance perception, room reflections also determine the overall spatial impression associated with an acoustic source and its surrounding environment. Here, we address the topic of spatial impression only briefly; the reader is referred to the review articles of Rumsey [86], Blauert [10] (Section 3.3) and Griesinger [42] for a full account.

Auditory spatial impression can be characterized by a variety of dimensions, but perhaps the most rigorously defined are apparent source width (ASW) and listener envelopment (LEV). ASW relates to the perceived width of an acoustic source, and is primarily determined by the relative level of early lateral reflections that arrive within 80 ms of the direct sound [66, 7, 8, 14]. LEV relates to the subjective impression of being immersed in a reverberant sound field, and is determined by the level, temporal distribution, and direction of arrival of late reflections (i.e., those arriving more than 80 ms after the direct sound). Bradley and Soulodre [14] show that listeners are able to distinguish between ASW and LEV, and argue that they represent different aspects of spatial impression. However, Merimaa and Hess [69] suggest that the concepts of ASW and LEV are not clear for naïve listeners. They investigated the ability of listeners to use ASW and LEV to characterize the spatial impression of concert halls, and found that listeners only developed consistent criteria for judging ASW and LEV after extensive training. They also found that ASW and LEV depended on the type of stimuli used (noise, cello, or snare drum) in addition to the reverberation time of the hall.

Bradley and Soulodre [14] suggest that the origin of ASW and LEV can be explained in terms of the precedence effect (see Section 7.4.3). During the precedence effect, direct sound and early reflections are temporally and spatially fused. Early lateral reflections increase the level of the fused event and cause some ambiguity in its perceived direction, which can be quantified in terms of the ASW. Later-arriving reflections are not temporally or spatially fused with the direct sound, leading to a more diffuse spatial effect that is perceived as envelopment.

7.3 EFFECTS OF REVERBERATION ON MACHINES

We now consider the effects of reverberation on the performance of automatic speech recognition (ASR) and CASA systems. The former is reviewed here for two reasons. First, the effects of reverberation on ASR performance are well documented, and provide a good example of the difficulties involved in developing robust algorithms for reverberant environments. Second, ASR is often paired with CASA in the same application, and the two may be tightly integrated (see Chapter 4). A detailed discussion of specific CASA processing for reverberant conditions is given in Section 7.6.

The current generation of statistical automatic speech recognition systems work by matching the acoustic input against an acoustic model for each class of speech sound, such as a phoneme or word (see Chapters 4 and 9). A key assumption underlying this approach is that speech encountered during testing will closely resemble the acoustic model, which is usually trained using a corpus of "clean" speech. In this respect, reverberated speech presents a problem because it is spectrally distorted by early reflections, and a noise floor is introduced by the combination of late reflections with the direct sound. Both of these factors reduce the similarity between the target speech and the acoustic model, causing a reduction in recognition accuracy.

One approach to this problem is to train the acoustic model on reverberated speech [19]. However, it must then be assumed that the characteristics of the reverberation are known a priori, or that there is a mechanism for selecting an appropriate acoustic model given particular reverberant conditions. Other ways of combating the problem, by dereverberating the input speech or using robust acoustic features, are discussed in Section 7.5.

It is instructive to consider the differing effects of reverberation time and source–receiver distance on ASR performance. Results from numerous studies indicate that ASR accuracy declines monotonically with increasing T_{60}. This is illustrated by panel **A** of Fig. 7.6, which shows the performance of several ASR systems in various reverberant conditions. It should be noted that this figure does not provide meaningful information about the relative performance of the systems, because they vary in their recognition architecture, training corpus, and testing conditions. To emphasize this point, recognition accuracy has been normalized to 100% in the $T_{60} = 0$ (anechoic) condition. However, in all cases the trend of decreasing recognition accuracy with increasing T_{60} is very clear.

Further insights into the effect of T_{60} on recognizer accuracy have been reported by Gillespie and Atlas [38]. They convolved speech with an impulse response consisting of a direct path and a single strong echo, which was delayed by a variable

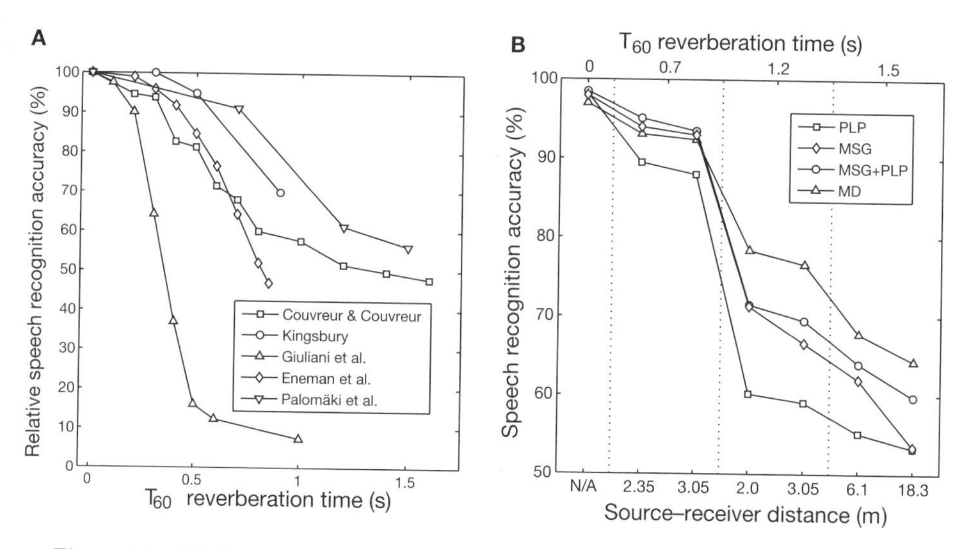

Figure 7.6 (**A**) The dependence of automatic speech recognition accuracy on T_{60} reverberation time, as reported by five studies. Accuracy is expressed as a percentage relative to performance in the $T_{60} = 0$ (anechoic) condition. Data from Couvreur and Couvreur [19], Kingsbury [54], Giuliani et al. [39], Eneman et al. [33], and Palomäki et al. [78]. (**B**) Speech recognition accuracy from the study of Palomäki et al. [78], showing the effects of T_{60} and source–receiver distance. Points delimited by vertical dotted lines indicate the same T_{60} condition. Note that accuracy falls more abruptly when the T_{60} is increased than it does when the source–receiver distance is increased. Data are shown for speech recognition with perceptual linear prediction features (PLP), modulation-filtered spectrogram features (MSG), a combination of both (MSG+PLP), and missing-data ASR with spectral features (MD).

number of milliseconds. Recognition performance fell monotonically as the echo delay time (and hence the T_{60}) was increased. The authors note that the performance curve they obtained has a knee at 25 ms; for echo delay times below this, performance was less affected by reverberation. They further note that 25 ms is approximately equal to the frame duration used for acoustic feature analysis in ASR systems.

In related work, Gölzer and Kleinschmidt [40] have undertaken a detailed study of the effects of early and late reflections on ASR; specifically, they describe techniques for deriving a value of the early time limit, *te*, after which reflections will have a detrimental effect on an ASR system (recall Eq. 7.1). This was achieved by convolving speech with a room impulse response in which certain portions had been zeroed. The reverberated speech was then recognized by a HMM-based system which was trained on unreverberated speech using mel-frequency cepstral coefficient (MFCC) or relative spectral–perceptual linear prediction (RASTA-PLP) features (see Section 7.5.2). Their results broadly mirror those of human perceptual studies; although the highest ASR accuracy was obtained using the direct sound only, it was found that early reflections did not substantially effect performance, whereas late reverberation had an adverse effect. The early time limit depended upon the characteristics of the room impulse response and the manner in which it was manipulated, but lay in a range between 25 ms and 50 ms. Again, this is consistent with predictions from human performance. This pattern of results did not depend on the acoustic features used; qualitatively similar results were obtained for the recognizers using MFCC and RASTA-PLP features.

A number of studies have also shown that speech recognition performance falls when the T_{60} is held constant and the DRR is decreased (for example, see [78]). However, Gillespie and Atlas [38] suggest that long T_{60} reverberation times have a much more detrimental effect on recognition accuracy than a low DRR. This view is supported by panel **B** of Fig. 7.6, which demonstrates that an increase in source–receiver distance (and hence a decrease in DRR) generally causes a smaller drop in recognition accuracy than an increase in T_{60}.

In summary, these experiments suggest that late reflections present a particular problem for ASR in reverberant environments, because they have a temporal smoothing effect that extends over several analysis frames. Early reflections are less problematic, because they usefully increase the level of the speech and introduce spectral distortions that can be addressed using within-frame processing schemes (such as cepstral mean normalization; see Section 7.5.2).

The effect of reverberation on CASA systems is much less well documented than its effect on ASR. Indeed, the majority of monaural CASA systems have not been evaluated in reverberant conditions, presumably because of the difficulty of the task. Figure 7.7 illustrates the effect of reverberation on pitch tracking, which is a component of many CASA systems (see Chapters 2 and 3). When compared with Fig. 7.4, it can be seen that the pitch tracker—the YIN algorithm of de Cheveigné and Kawahara [26]—provides an accurate estimate of the $F0$ contour in the anechoic condition (panel **A**). However, when the speech is reverberated (panel **B**), sections of the pitch contour are lost (at about 0.4 s and 1.1 s) and some

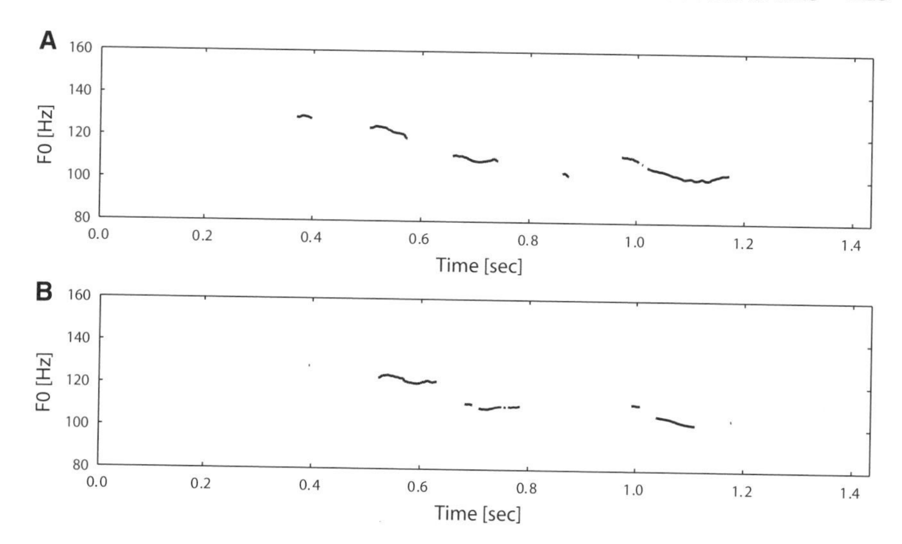

Figure 7.7 The effect of reverberation on pitch tracking (compare with Fig. 7.4). (**A**) Pitch track for the utterance "you took me by surprise" spoken by a male speaker in an anechoic environment. (**B**) Pitch track for the same utterance convolved with the impulse response of a reverberant room. The T_{60} reverberation time was 0.5 sec, and the DDR was –3 dB. Both pitch tracks were obtained using the YIN pitch tracker [26].

tracks become extended, because the algorithm follows the $F0$ of the reflected sound. Multipitch tracking in reverberant conditions represents an even greater challenge.

Recent work by Roman and Wang [84] has attempted to address this issue by combining dereverberation and pitch-based sound segregation in a monaural CASA system. Initially, their system estimates an inverse filter that equalizes the room response corresponding to the location of the target speaker (see Section 7.5.1). This serves to partially restore the harmonicity of the target speaker, while further smearing the interfering signals that originate from other directions. After this preprocessing step, pitch-based speech segregation is performed using a variant of Hu and Wang's [50] approach. Roman and Wang report that their system achieves a substantial improvement in SNR across a range of different test conditions.

A number of workers have also proposed binaural CASA systems that attempt to segregate a target speaker from spatially separated noise, and have evaluated them in reverberant conditions (e.g., [65, 11, 62, 79]). In such systems, binaural cues provide a mechanism for grouping acoustic components that originate from the same location in space, and therefore allow direct sound to be segregated from reflected sound. Generally speaking, binaural CASA systems offer some degree of robustness in reverberant conditions, but their performance degrades with increasing T_{60} and DRR in much the same way that ASR accuracy does. A detailed discussion of binaural approaches to CASA in reverberant environments is given in Section 7.6.

7.4 MECHANISMS UNDERLYING ROBUSTNESS TO REVERBERATION IN HUMAN LISTENERS

It was established in Section 7.2 that reverberation has a disruptive effect on many aspects of auditory perception. That said, human listeners still have a remarkable tolerance to reverberation. For example, Nabelek and Robinson [74] have reported the performance of human listeners in a single-talker speech recognition task, in anechoic and reverberant conditions. Speech recognition accuracy in anechoic conditions was 99.7%, and this degraded to 97.0%, 92.5%, and 88.7% for T_{60} reverberation times of 0.4, 0.8, and 1.2 s, respectively. These figures compare very favorably with the performance of ASR systems in similar reverberant conditions, as shown in Fig. 7.6. In this section, we consider the factors that underlie the robustness of human listeners to reverberation.

7.4.1 The Role of Slow Temporal Modulations in Speech Perception

The remarkable tolerance of human speech perception to reverberation has been attributed to the ability of human hearing to exploit information about the temporal modulation characteristics of speech [49, 30, 29]. Temporal modulations in speech occur at modulation frequencies between 1 and 16 Hz, with a strong component at approximately 4 Hz that reflects the syllable rate [49]. Whereas fast temporal fluctuations in speech are smoothed out by reverberation (see Fig. 7.2), these slow modulations remain relatively robust in reverberant environments.

Early evidence for the important role of slow temporal modulations in speech perception came from Homer Dudley's experiments with the channel vocoder at Bell Laboratories [31]. The channel vocoder generates speech by filtering a source signal (such as periodic buzz in the case of voiced speech) by a filter model of the vocal tract. Dudley found that highly intelligible speech could be generated by using slowly varying filter parameters with a cutoff at 25 Hz and dominant frequencies below 10 Hz. More recently, filtered speech has been used to establish that slow modulations are important for speech intelligibility. For example, Drullman et al. [30, 29] split speech into a series of frequency bands and applied either low- or high-pass filtering to the temporal envelope in each band. Reducing the low-pass cutoff frequency below 16 Hz reduced the intelligibility of the speech, whereas increasing it above 16 Hz did not significantly affect intelligibility [30]. Similarly, intelligibility was unaffected by high-pass filtering so long as envelope modulations above 4 Hz were preserved [29]. Further evidence for the role of slow modulations in speech perception has come from an analysis of American English telephone conversations by Greenberg et al. [41]. They show that stressed and unstressed syllables are reflected in the lower (< 6 Hz) and upper (6–20 Hz) regions of the modulation spectrum, respectively. In particular, they note that the low-frequency region of the modulation spectrum is crucial for speech understanding since stressed syllables tend to be associated with words that have a high informational content.

Observations about the role of slow temporal modulations in speech perception have led to measures of speech intelligibility based on the concept of the modulation transfer function (MTF) [49]. Such approaches characterize the effects of a signal transmission path by assessing the extent to which modulations in the talker's speech are preserved at the receiver. To assess the MTF, a temporally modulated signal is emitted from a talker's position and recorded at a listener's position. The MTF is then defined as the reduction in the modulation index of the intensity envelope of the recorded test signal, relative to the original, as a function of modulation frequency. Modulation frequencies in the range 0.63–12.5 Hz are usually considered, and the MTF is computed within seven octave bands. As such, a full MTF analysis yields a matrix that describes the frequency-dependent effects of acoustic disturbances (including noise and reverberation) on a signal transmission path. Furthermore, the MTF can be transformed into a single index (called the speech transmission index, STI). To do so, it is assumed that regardless of the actual acoustic disturbance present (e.g., reverberation), each element in the matrix of MTF values can be treated as an equivalent signal-to-noise ratio (SNR), as though the disturbance was due to interfering noise only. The STI is then obtained by a weighted mean of the equivalent SNR over all modulation frequency bands and carrier frequency bands. Numerous studies have shown that the STI is an accurate predictor of speech intelligibility in reverberant environments.

Interestingly, Singh and Theunissen [89] have presented evidence that slow temporal modulations may be a general characteristic of behaviorally relevant sounds. They observe that in addition to speech, most other natural sounds (such as animal vocalizations) have their modulation energy concentrated at low spectral and temporal modulation frequencies. Furthermore, they postulate that the auditory system exploits the statistical properties of natural sounds in order to achieve effective representations of behaviorally relevant signals. There is some support for this view from physiological studies, which report a concentration of neurons that are tuned to low temporal and spectral modulation frequencies in the mammalian auditory thalamus and cortex [72].

7.4.2 The Binaural Advantage

It is well known that binaural hearing can reduce the effects of reverberation. One of the earliest illustrations of this was provided by Koenig [57], who noted that the room reverberation apparent in single-microphone listening is suppressed when listening dichotically via a pair of spatially separated microphones.

The advantage of binaural hearing in reverberant environments is most often explained in terms of mechanisms that attend to a specific location (see Section 7.2.3) or which compare the signals at the two ears. In the latter respect, one of the most important factors is the precedence effect (see below). Additionally, there is evidence that listeners employ an approach that treats reverberation as uncorrelated noise and cancels it using binaural mechanisms. Libbey and Rogers [60] argue that this binaural overlap–masking release contributes to the binaural intelligibility advantage in reverberant conditions, but is insufficient to explain it completely. They show that speech intelligibility is greater with real reverberation than with reverberation-like noise, in which localization cues have been removed by phase randomization. It therefore appears that reverberation is not simply cancelled, because it provides some useful information for speech identification. One explanation for this finding is that reverberation contains cues to location that are used by listeners.

It should also be noted that we benefit from listening with two ears because the SNR at any particular time is usually favorable in one ear. In small, highly reverberant environments, this advantage may be reduced because echoes are integrated at each ear, leading to a reduction in the IID. In general, however, it is possible that listeners obtain some advantage by using the "better ear." Devore and Shinn-Cunningham [27] note that in reverberant conditions, the better ear is not necessarily the one nearest to the target sound source; the SNR may be better in the left or right ear at any particular time, depending on the arrival of echoes. Listeners may therefore employ a mechanism that dynamically selects the "better ear" based on short-term estimates of the SNR.

7.4.3 The Precedence Effect

In Section 7.2.2, it was noted that listeners retain the ability to localize sound in reverberant environments based on the direct sound, even though reflections reach the subject from many other directions. The term *precedence effect* is used to describe a number of related phenomena that are thought to underlie this ability [95, 104, 10, 61]. Apparently, listeners apply a perceptual weighting to directional cues such that those associated with the direct sound (the "first wave front") are emphasized, and those associated with later reflections are suppressed (recall Fig. 7.1).

A number of workers have investigated the precedence effect using lead-lag sound pairs (i.e., a direct sound followed by a single echo) in which the directions of the direct sound and the echo differ. It appears that such stimuli are interpreted differently by the binaural system depending on the time delay between the direct sound and the echo [10]. If the lead–lag difference is less than 1 ms, then *summing localization* occurs and listeners hear a single fused event whose direction lies between that of the lead and lag sounds. Fusion still occurs for lead–lag differences greater than 1 ms, but in such cases the direction of the fused event corresponds to the direction of the first arriving wave front. Hence the precedence effect plays an active role, and the lead sound is said to have *localization dominance* over the lag sound [61]. Furthermore, if the time between the lead and lag sounds exceeds a critical delay called the *echo threshold,* then the lead–lag pair is no longer perceived as a single fused event [10]. Rather, it is heard as two separate events that are localized at the directions of the lead and lag sounds. The echo threshold is about 5 ms for single clicks, but for sounds with a more complex character (such as speech or music) it can be as long as 50 ms [61, 10]. A further point is that the precedence effect is strongest for identical lead–lag pairs, but is still invoked to some extent even if the lag sound is not an exact replica of the lead [61]. This makes sense from an ecological point of view, because room reflections are usually spectrally distorted copies of the direct sound.

It has also been found that the precedence effect is subject to a "buildup" phenomenon, such that sound pairs with a constant lead–lag interval are initially heard as separate events, but are perceived as fused after a number of repetitions [91, 35].

If the directions of the lead and lag sounds are changed abruptly in such experiments, then the previously fused lead–lag sound event decomposes into two events that originate from their own directions [18]. Hence, it appears that the precedence effect is influenced by complex adaptation processes.

The precedence effect is often assumed to involve a mechanism in which the first wavefront generates an inhibitory neural signal, which then suppresses later neural activity associated with reflections [104, 67, 51]. A diagram of such a scheme is shown in Fig. 7.8. Alternatively, Faller and Merimaa [34] suggest that some aspects of the precedence effect can be modeled by an approach based on interaural coherence. More specifically, they suggest that the binaural system only uses IID and ITD cues if they are coherent, i.e., if they correspond to the true direction of a sound source. They propose a computational model in which an interaural coherence function $c(t)$ is computed within each frequency channel of an auditory model, where t represents time and

$$c(t) = \max_{\tau} r(t, \tau) \tag{7.2}$$

Here, $r(t, \tau)$ is a running normalized cross-correlation function (see Chapters 1 and 5) and τ is the cross-correlation lag (which corresponds to the ITD). When $c(t) \approx 1$, the interaural coherence is high and the corresponding IID and ITD cues are taken to be reliable evidence for the direction of a sound source. In terms of the precedence effect, a high interaural coherence is therefore indicative of the "first wave front." Similarly, when $c(t) \ll 1$, the IID and ITD cues are discarded, since they are likely to be due to a mixture of direct sound and room reflections. We note that Faller and Merimaa's approach is closely related to an earlier system described by Allen et al. [1], which employs a gain control in each frequency band to remove uncorrelated signals and pass correlated signals.

Output from Faller and Merimaa's model is shown in Fig. 7.3. Panel **A** of the figure shows the distribution of IID and ITD for a mixture of two male speech signals that are spatialized at different azimuths in an anechoic environment. Peaks in the distribution correspond to the location of each source. In panel **B**, the same ut-

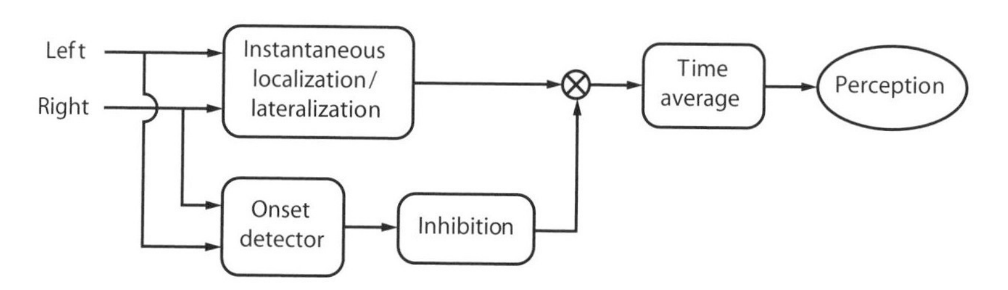

Figure 7.8 General structure for a model of the precedence effect. Inhibition triggered by an onset detector suppresses localization cues that arrive shortly after the direct sound. Adapted from Zurek [104] (see also [67]), with permission of Springer Science and Business Media.

terances are spatialized in a reverberant environment and the peaks disappear. The reverberant case is also shown in panel C, but the ITD and IID cues for time t are only included if the corresponding interaural coherence $c(t) \geq 0.99$. Note that peaks in the distribution are now visible, and that they correspond closely to the peak positions (and hence the source directions) shown in the anechoic case of panel **A**. The figure therefore confirms that ITD and IID cues carried by the first wavefront offer an effective means for determining the direction of sound sources, in accordance with the precedence effect.

7.4.4 Perceptual Compensation for Spectral Envelope Distortion

It was noted in Section 7.2.1 that early reflections (i.e., those arriving within 50 ms or so of the direct sound) distort the spectral envelope of speech sounds. Human listeners are able to compensate for such distortions, but are the mechanisms underlying this compensation central or peripheral in origin? To address this question, Watkins [96] investigated the perception of speech sounds that were preceded by "carrier" sounds whose spectral envelopes were altered. The speech sounds were drawn from a continuum between "itch" and "etch," so that perceptual compensation induced by the carrier could be quantified in terms of a phoneme boundary shift (i.e., the greater the perceptual compensation, the greater the boundary shift). Phoneme boundary shifts were smaller when the carrier and test sounds were presented to different ears, and speech carriers gave larger boundary shifts than noise carriers. Perceptual compensation was also maintained when the stimuli suggested a change of talker, but was reduced when there was evidence for a change in the characteristics of the distorting channel. These results suggest that the mechanisms underlying perceptual compensation for spectral envelope distortion are central, rather than peripheral, in origin. Watkins suggests that the auditory system might exploit information about the rate of spectral change in order to separate sound-source characteristics from channel characteristics, since speech contains rapid changes in spectral content, whereas the spectral changes induced by a communication channel or room reverberation are relatively slowly varying.

In a subsequent study, Watkins [97] asked whether mechanisms similar to those of the precedence effect operate in the perception of spectral envelope; specifically, is the percept of a vowel dominated by the spectral shape conveyed during the first wavefront? Watkins used vowels in which the first part (corresponding to direct sound) was unfiltered and the remainder was filtered to simulate the spectral distortion caused by reverberation. The ITD of the direct sound and remainder were also manipulated separately. Watkins confirmed that the lateral position of the whole sound was largely determined by the ITD of the direct sound, in accordance with the precedence effect. However, no evidence was found for the involvement of similar mechanisms in the perception of spectral envelope; the spectrally distorted ("reverberated") section of the vowel had a substantial influence on its perceived identity, even when the ITD of the reverberated part differed from that of the direct sound. This seems intuitively reasonable; whereas reflections contain misleading directional information that should be suppressed, they also contain useful information about speech (to the extent that speech can be recognized from reflections only [60]).

7.5 REVERBERATION-ROBUST ACOUSTIC PROCESSING

Generally speaking, three approaches to dealing with the problem of reverberation have been described in the literature. First, acoustic processing can be used to remove the echoes introduced by room reflections, in an attempt to retrieve a "clean" signal from a reverberated one. This is called *dereverberation.* A second line of research concerns the development of acoustic features that are robust to reverberation. This approach has mainly been applied in ASR, in order to obtain representations of speech that are relatively invariant across a range of acoustic environments. Finally, a recently proposed approach is *reverberation masking,* in which the parts of the signal that are least contaminated by reverberation are identified, and given a higher weight in subsequent processing. For example, automatic recognition of reverberated speech can be based on a decoder that treats contaminated and uncontaminated regions of the speech differently during decoding.

It should be noted that all of the schemes reviewed below involve the processing of speech that has been produced in a reverberant environment. An alternative approach is to modify speech prior to transmission in a reverberant environment, so that it is more robust to the effects of reverberation. Arai et al. [4] suggest such an approach. They note that steady-state regions of the speech signal (such as vowels) are relatively redundant, and also contribute substantially to overlap masking because they tend to be relatively high in energy. Accordingly, they describe a preprocessing technique in which the steady-state portions of speech are suppressed prior to reverberation. They find that speech intelligibility is improved by this preprocessing under some reverberation conditions.

7.5.1 Dereverberation

Broadly, dereverberation techniques can be classified into three categories: spatial filtering, feature-based techniques, and inverse filtering. These differ in regard to the number of sensors (microphones) required and the degree to which the characteristics of the target source and acoustic environment are exploited.

Spatial Filtering. In spatial filtering, sound energy that originates from a desired direction is enhanced, whereas sound energy from other directions is suppressed. Hence, the technique can be used to enhance the direct sound from a talker while suppressing room reflections (and other acoustic sources) that originate from different directions. Usually, the data for spatial filtering is obtained through discrete spatial sampling by an array of sensors (i.e., two or more microphones), in which case the approach is termed beamforming [94]. There is a substantial literature on the use of beamforming for sound separation and dereverberation; the reader is referred to Chapter 6 for details.

Feature-Based Techniques. A number of workers have described feature-based dereverberation techniques that exploit the properties of speech. For example, Yegnanarayana and Murthy [102] note that the DRR of reverberated speech varies within each glottal cycle, in addition to longer-scale changes due to overlap masking (see Section 7.2.1). Consequently, they propose a dereverberation scheme in which high-DRR regions are enhanced within short (2 ms) and longer (20 ms) segments. This is achieved by selectively weighting the residual signal obtained from a linear prediction (LP) analysis. The appropriate weights are derived by using the statistics of the LP residual signal for reverberant speech. More specifically, the samples of the LP residual have a Gaussian-like probability density function for reverberant tails in which the DRR is low, but have a skewed distribution in voiced regions for which the DRR is high. The authors report that speech reconstructed from the weighted LP residual has much reduced reverberation without significant loss in quality.

Another characteristic of speech is the presence of low-frequency temporal envelope modulation, which is known to be important for speech intelligibility (see Section 7.4.1). Accordingly, Avendano and Hermansky [6] propose a dereverberation scheme that attempts to design an inverse modulation transfer function (IMTF), which reverses the effect of reverberation on the modulation spectrum of speech. They achieve this by splitting the speech signal into frequency bands, and learning the coefficients of a filter for each band that minimizes the distance between the filtered temporal envelope of the reverberated speech and the temporal envelope of clean speech. They find that filters learned in this way bear some resemblance to idealized IMTFs. Furthermore, reverberated speech had modulation characteristics that were closer to those of clean speech after processing by the filters. However, this approach did not yield reverberation-free speech, and the processing introduced audible artefacts.

The harmonic structure of speech can also be exploited in reverberation-robust acoustic processing. For example, Brandstein [15] presents a technique for estimating the time delay of a speech signal received at a pair of spatially separated microphones that is robust in the presence of reverberation. The technique is based on the assumption that parts of speech that exhibit a clean harmonic structure have been relatively unaffected by reverberation; they should therefore be highly weighted in the time-delay computation, since they will convey accurate information about the source location. Using similar logic, Wu and Wang [99] describe an approach for speech dereverberation in which the correlogram is used to estimate the T_{60}. Their approach is based on the observation that reverberation corrupts the harmonic structure of speech and, hence, that pitch strength is inversely related to the reverberation time. Their measure of pitch strength is derived from a correlogram-based system for $F0$ tracking [101] (see Chapter 2). A histogram is formed of the relative time lags between the pitch period found at each time frame, and the closest peak in each channel of the corresponding correlogram. For anechoic speech, this distribution is centered on zero and has a sharp peak. However, reverberation corrupts the harmonic structure of speech, leading to a broadening of the distribution. Indeed, Wu and Wang show that the T_{60} can be estimated directly from a measure of the spread of the relative time lag distribution. Having estimated the reverberation time in this way, the speech is then enhanced by a subsequent processing stage in which echoes are identified and subtracted.

Inverse Filtering. Consider a source signal $s(n)$ that is emitted in a reverberant room with an impulse response $h(n)$. The received reverberated signal is then given by $x(n) = h(n) * s(n)$, where the symbol $*$ represents convolution. The inverse filtering approach to dereverberation requires a filter $w(n)$ that reverses the effect of the room response; then an estimate of the "clean" signal $y(n)$ can be recovered by $y(n) = w(n) * x(n)$.

If the room impulse response $h(n)$ is known, then in principle it can be inverted to give $w(n)$. However, this approach is limited in practice because it requires that the impulse response be minimum phase (this ensures that an inverse filter exists that is causal and stable). This condition is not met in most real acoustic environments. However, Miyoshi and Kaneda [73] show that exact inverse filtering of acoustic spaces with a nonminimum phase characteristic can be achieved by using multiple microphones. This approach is known as the multiple-input/output theorem (MINT). More recently, Hatziantoniou and Mourjopoulos [45] have shown that smoothing of room transfer functions can result in inverse filters that are more robust for practical applications, that is, they yield fewer artifacts and are less sensitive to changes in listening position.

In many situations, $s(n)$ and $h(n)$ are unknown, so that the inverse filter $w(n)$ cannot be determined directly and must be estimated. This estimation problem is known as blind deconvolution [46] or, more specifically, blind dereverberation. Various approaches to blind dereverberation have been suggested in the literature, including those based on the complex cepstrum [92] and independent component analysis [9].

Recently, an interesting approach to blind dereverberation has been proposed by Nakatani et al. [76], which they term harmonic dereverberation (HERB). This may be regarded as a hybrid method, since it performs blind inverse filtering but is also feature based, in that it exploits information about the harmonic structure of speech. More specifically, periodic sections of the speech signal are used to estimate an inverse filter, which can then be used to dereverberate the signal (including unvoiced regions). The basis of HERB is the enhancement of periodicity in local time regions, during which the $F0$ is assumed to be constant. As noted previously, reverberation causes a degradation in periodicity because the $F0$ of direct sound is mixed with different $F0$s associated with echoes. It follows that dereverberation can be achieved by designing a filter $w(n)$ that makes $y(n)$ a periodic signal in every local time region; if a single periodicity is present, then reverberant components must have been eliminated.

In one approach described by Nakatani et al. [76], $w(n)$ takes the form of an average filter function that transforms $x(n)$ into a quasiperiodic signal. Specifically, $w(n)$ is formed in the frequency domain as an average of the ratio of the output of an adaptive harmonic filter and the observed reverberated signal. This technique is able to learn $w(n)$ from a reverberant speech signal, and applying the resulting filter leads to high quality dereverberated speech. However, approximately 5000 utterances were required during training; most real-world applications would require that $w(n)$ be estimated from a smaller amount of speech data. Also, dereverberation performance was best when the reverberation time was relatively long.

A further problem with HERB is that it is more effective for female speakers than for male speakers. This is because the low $F0$ of male speech yields harmonics that are close together in frequency, making them difficult to distinguish from reverberant energy with an adaptive harmonic filter. In a subsequent paper [75], this issue is addressed and the constraint that the $F0$ should be constant within each analysis window is relaxed. This is achieved by using dynamic time warping to obtain a signal that has an approximately constant $F0$ in a warped time domain. The temporally warped signal is then dereverberated, and overlap–add synthesis is used to reconstruct the dereverberated signal on its original time scale. The authors report substantial improvements in dereverberation performance using this approach, as illustrated in Fig. 7.9.

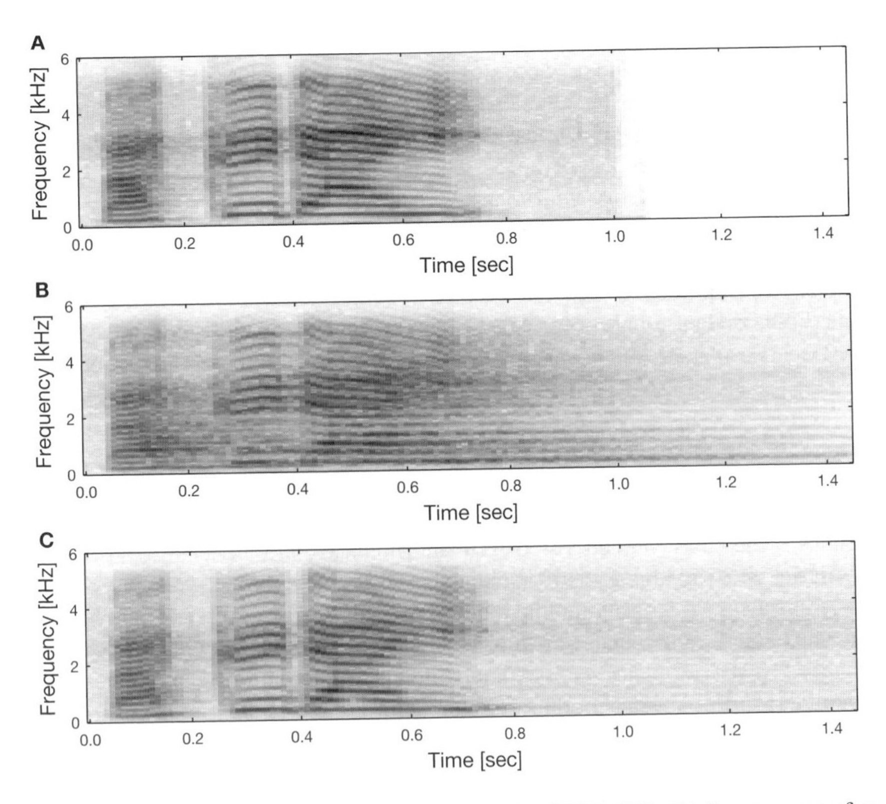

Figure 7.9 Dereverberation of a female utterance by HERB [75]. (**A**) Spectrogram of an anechoic speech signal. (**B**) Spectrogram of reverberated speech, generated by convolving the anechoic signal with a room impulse response ($T_{60} = 1.0$ s). (**C**) Spectrogram of dereverberated speech. The dereverberated signal was obtained by convolving the reverberated speech with an inverse filter estimated by HERB. Note that the long reverberant tail has been removed, and the harmonic structure in the region of 1–2 kHz has been restored.

Wu and Wang [100] have also proposed an algorithm for one-microphone reverberant speech enhancement in which inverse filtering plays a key role. In the first stage of their system, an inverse filter is estimated and applied, in order to reduce coloration of the speech and increase the DRR. Following this, late reverberation is suppressed by subtracting the power spectrum of late-reflection components from that of the inverse filtered speech. Wu and Wang show that this approach offers substantially better enhancement than the system of Yegnanarayana and Murthy [102] (see Section 7.5.1).

7.5.2 Reverberation-Robust Acoustic Features

Another approach to dealing with reverberation is to represent the acoustic signal in a way that is relatively invariant in both anechoic and reverberant environments. Such "reverberation-robust" acoustic features have principally been applied in ASR, as a means of reducing the mismatch between anechoic speech used for training the recognizer and reverberated speech encountered during testing.

Typically, acoustic features for ASR are computed at intervals of 10 ms or so, over a short time frame of approximately 20 ms. Such frame-based features tend to perform poorly in reverberant conditions, because reverberation smears energy over a number of consecutive frames. For example, cepstral mean normalization (CMN) [63] is effective in dealing with convolutional distortions with a short impulse response, but does not yield a significant gain in ASR accuracy in strongly reverberant conditions (for example, see [78]).

This suggests that the key to reducing the mismatch between anechoic and reverberated speech may lie in representations that emphasize the temporal structure of speech signals. As discussed in Section 7.4.1, modulation frequencies that reflect the syllable rate of speech are important for human speech intelligibility. Similarly, the importance of such low-frequency envelope modulations for speech enhancement and ASR has been highlighted by many studies. Modulation filtering has been applied as a method for enhancement of speech contaminated by additive noise or reverberation, resulting in improvements in intelligibility [59, 87]. Similarly, a number of workers have shown that ASR can be made more robust to reverberation and slowly varying noise by applying temporal modulation filtering to a time-frequency representation of speech [48, 54, 55, 47]. For example, Hermansky and Morgan [47] describe RASTA-PLP features for ASR, which exploit low-frequency speech modulations in the range 0.26 Hz to 12.8 Hz. Similarly, Kandera et al. [53] compare the ASR performance for narrow modulation frequency bands, showing that the modulation filter bandwidth increases logarithmically with increasing centre frequency. The authors show that the most important modulation frequency range is between 2 and 4 Hz, and that modulation frequencies above about 16 Hz are not important for ASR performance. These findings are compatible with results obtained from speech intelligibility tests using human subjects.

Variants of RASTA have proven successful for spectral distortions due to transmission lines and additive noise [47]. However, the RASTA method was found to be rather unsuccessful with reverberated speech [54]. To address this, Kingsbury et al. [55] present an extension of the RASTA-PLP approach which they term the modulation spectrogram. The modulation spectrogram is derived from an initial spectral decomposition using a bank of bandpass filters, which roughly approximate a cochlear frequency analysis. The output of each filter is then lowpass filtered to 28 Hz, downsampled and normalized by its long-term average. A modulation filter is then applied, which passes significant energy in the range 0–8 Hz. The log magnitudes are then plotted in spectrographic format, as shown in Fig. 7.10. Note that the modulation spectrograms in the figure appear to be relatively invariant in anechoic and reverberant conditions when compared with the cochleagrams shown in Fig. 7.2. Kingsbury demonstrates that these features provide a robust representation of speech for ASR in reverberant conditions, which outperforms RASTA-PLP in reverberant conditions (see also [54]). In his subsequent work, Kingsbury [54] shows improved results with a system that uses two sets of modulation-filtered spectral features, one of which is low-pass filtered at 8 Hz and another that is band-pass filtered between 8 and 16 Hz.

A related approach is reported by Kleinschmidt et al. [56], who employ a model of auditory perception (PEMO) as a front end for robust ASR in noise and reverberation (see also [25]). Like Kingsbury's approach, their system emphasizes slow modulations in the acoustic signal, in this case by low-pass filtering the output of a gammatone filterbank that has been compressed by a nonlinear adaptation mechanism. The modulation transfer function of PEMO is band pass with a peak at 4 Hz, which provides a good match to the modulation spectrum of speech. However, the performance of their ASR system was substantially poorer in reverberant condi-

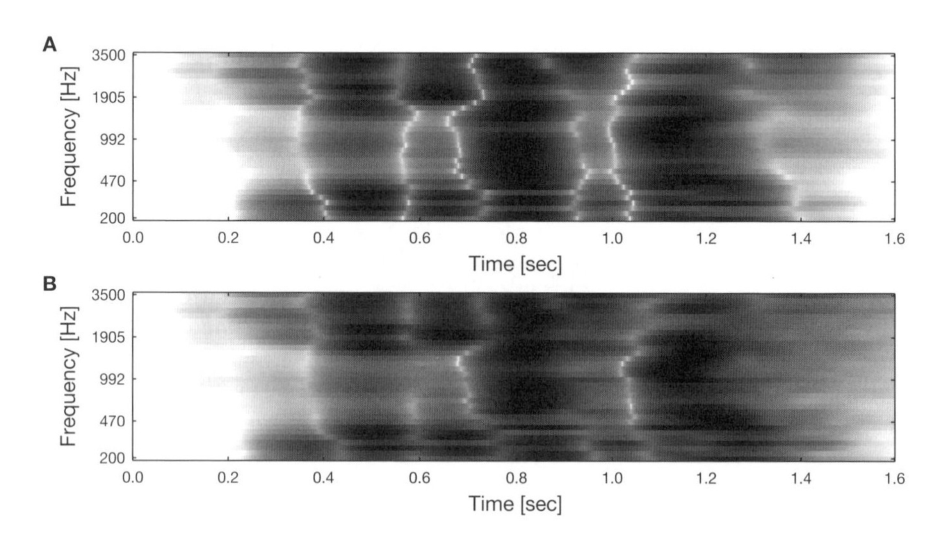

Figure 7.10 Modulation spectrograms for the utterance "you took me by surprise" spoken by a male speaker. Dark pixels indicate regions of high energy, and light pixels indicate regions of low energy. (**A**) Utterance spoken in anechoic conditions. (**B**) The same utterance convolved with the impulse response of a reverberant room (T_{60} = 0.5 s, DRR = –3 dB).

tions than in anechoic conditions, even though the recognizer was trained on PEMO features derived from reverberated speech.

Finally, we note that the studies described above illustrate a general problem with reverberation-robust acoustic features; namely, they may not generalize well to acoustic conditions other than those to which the features have been "tuned." For example, Kingsbury et al. [55] report that the modulation spectrogram gave worse performance on clean (anechoic) speech than conventional perceptual linear prediction (PLP) features. One solution to this problem may be to combine reverberation-robust features with conventional features. Indeed, Kingsbury et al. show that a recognizer that uses a combination of modulation spectrum and PLP features during decoding performs better in clean and reverberant conditions than recognizers trained on either feature alone. Another possible solution is an approach based on reverberation masking, as discussed in the following section.

7.5.3 Reverberation Masking

Palomäki et al. [78] (see also [77]) propose an alternative approach to dealing with reverberation that is based on the "missing data" framework for ASR. In this approach, the ASR system is provided with acoustic features and a time-frequency mask that indicates whether each feature constitutes reliable evidence for the target speech signal or not (see Chapter 9). More specifically, Palomäki et al. use modulation filtering to identify spectrotemporal regions of the acoustic input that contain strong speech energy (i.e., those that are least contaminated by reverberation). This information is used to derive a binary reverberation mask in which a value of 1 indicates a reliable speech feature and a value of 0 represents a feature that is likely to be corrupted by reverberation. The acoustic features and the reverberation mask are then passed to the ASR system for decoding.

In the first stage of this system, the envelope is extracted from each channel of a gammatone filterbank and processed with a finite impulse response (FIR) filter, which consists of a linear phase low-pass component and a differentiator. The low-pass component serves to detect and smooth modulations in the speech range, whereas the differentiator emphasizes onsets that are likely to correspond to early reflections and direct sound. The resulting modulation filter is band pass, with 3 dB cutoff frequencies at 1.5 Hz and 8.2 Hz. Subsequently, the reverberation mask is set to 1 if the corresponding value of the modulation-filtered envelope lies above a threshold value, and to 0 otherwise. This threshold is set automatically according to a "blurredness" metric, which quantifies the degree to which the spectrum is smoothed by reverberation. The blurredness metric is based on a ratio of the average-to-maximum values of the temporal envelope in filterbank channels. In reverberant conditions, dips in the speech envelope are filled by energy originating from reflections, leading to an increase in the blurredness metric and a concomitant increase in the threshold for reliable speech regions.

The missing data approach was originally conceived as a way of handling additive noise in ASR, and the assumption underlying the system of Palomäki et al. is that late reverberation can be treated in a similar way. However, early reflections also spectrally distort the speech features (see Section 7.2.1). Hence, the final stage of the system is a spectral normalization process, in which a normalization factor is computed from spectral features that are denoted as reliable in the reverberation mask.

The reverberation mask estimation process is illustrated in Fig. 7.11. Anechoic and reverberated speech are shown in panels **A** and **B**, respectively. Panel **C** shows an a priori mask, in which white pixels denote "reliable" time-frequency regions whose values did not change appreciably between the anechoic and reverberant conditions. The a priori mask may be regarded as a "ground truth," which indicates the time-frequency regions that have been least affected by reverberation. Panel **D** shows a reverberation mask estimated from the reverberated speech signal only, which is in good agreement with the a priori mask.

Palomäki et al. evaluate their system on a connected digit recognition task in a range of reverberant conditions. Its performance is similar to that of Kingsbury's [54] system, although, as noted above, the reverberation masking approach may have some advantages. Specifically, assumptions about the acoustic environment are restricted to the mask estimation process, which can be adapted dynamically to changing conditions. In contrast, Kingsbury's system (and others that use features that generalize over many noise or reverberation conditions) may be suboptimal for a particular acoustic environment.

The reverberation masking approach is proposed as an engineering solution to the problem of reverberation, rather than a model of auditory function. However, the approach is to some extent supported by psychophysical studies of human

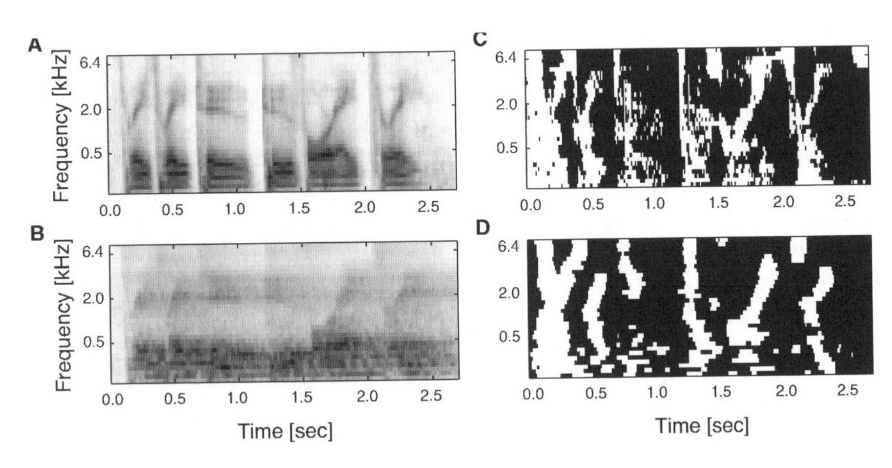

Figure 7.11 An illustration of reverberation masking. (**A**) Cochleagram for connected digits spoken by a male talker in anechoic conditions. (**B**) Cochleagram for the same utterance convolved with a room impulse response (T_{60} = 1.7 s, DRR = –10 dB). (**C**) Reverberation mask derived from a priori knowledge of the anechoic and reverberated signals. (**D**) Reverberation mask derived from the reverberated signal only, using the system described by Palomäki et al. [77]. From [77]. Copyright © 2002 IEEE.

speech perception. For example, Libbey and Rogers [60] report that binaural overlap-masking release makes some contribution to the binaural advantage in reverberant conditions (see Section 7.4.2). The system of Palomäki et al. [78] assumes a similar mechanism (i.e., one in which reverberation is assumed to act like masking noise), although it is based on monaural rather than binaural processing.

7.6 CASA AND REVERBERATION

Relatively few CASA systems have been evaluated in reverberant conditions, and even fewer include mechanisms that specifically address the problem of room reverberation. Those that do are usually based on a principle of directional filtering, using two or more microphones. Such systems aim to preserve sounds that emanate from a desired direction, while suppressing other sounds (including room reflections) that originate from other directions. The general processing scheme of such models involves the extraction of binaural cues from each frequency band of an auditory model, followed by selective weighting of those bands whose binaural cues are consistent with the desired source direction. Such systems are reviewed here with an emphasis on their performance in reverberant conditions; a more general discussion of binaural and multiple-microphone CASA systems can be found in Chapters 5 and 6.

7.6.1 Systems Based on Directional Filtering

One of the earliest attempts to employ directional filtering in a CASA system is the work of Lyon [65], who investigated the performance of a binaural sound separation system on reverberated mixtures of speech and an interfering sound. In his system, acoustic inputs for the left and right ear are processed by a cochlear model, and the resulting outputs for each frequency band are cross correlated. The time delay at which the largest peak occurs in each cross correlation function is interpreted as the ITD of a sound source. In a second pathway, cochleagrams are also computed for each ear. These are filtered using a time-variable gain determined by the ITD in each frequency band, in order to produce an estimate of the cochleagram for each constituent sound source in the acoustic input. Lyon considered the specific problem of separating and dereverberating a mixture of two sources, and used eight time-variable gains in order to produce cochleagrams for the left and right direct sound, and for the left and right reverberant sound (echoes). The system was only evaluated informally, but apparently separated the speech and noise sources, and their respective echoes, with some success. However, Lyon notes that the separation performance deteriorates in the most reverberant parts of the test signal, since the occurrence of multiple echoes prevents an accurate estimate of source direction.

A more sophisticated variant of Lyon's approach is described by Bodden [11]. In the first stage of his system, the cross-correlation used by Lyon is extended by algorithms that apply contralateral inhibition and adaptation to head-related transfer functions. As a result, location estimates from the cross-correlation are influenced both by ITD and by IID. Subsequent stages of Bodden's system estimate the azimuth of each sound source, and this information is used to determine the transfer function of a time-variant optimum filter (Wiener filter), which selectively enhances the target speech signal. However, the system performed poorly in the presence of reverberation. Even when tracking a single sound source, "ghost" sound sources were identified due to strong reflections that were mistaken for a second source (see Fig. 9 of [11]). Bodden suggests that the inclusion of mechanisms related to the precedence effect, particularly those concerned with the role of onsets in sound localization, should improve the performance of his system in reverberant conditions.

Subsequent work has adopted a similar approach to the Lyon and Bodden systems, but with various enhancements. For example, the system of Peissig (reviewed in [98]) computes an interaural coherence function for each frequency channel. This is used to determine whether the channel is excited by direct sound (which causes a strong interaural coherence) or diffuse reflections. Aoki and Furuya [3] introduce two further innovations: the use of information about the phase and amplitude differences associated with particular source directions, and a time-weighted average of binaural cues that avoids undue emphasis of time-frequency regions that are momentarily masked by room reflections. Aoki and Furuya report that their system performs well in reverberant conditions, but they only compare its performance against their own previous work.

A sophisticated CASA system that employs directional filtering is the two-stage binaural processor described by Liu et al. [62]. In the first stage of their system, sources are localized by processing the inputs from two microphones through a dual delay line and coincidence detection mechanism. Subsequently, noise sources at nontarget locations are cancelled. This is achieved by subtraction of the two input signals, such that a nulling pattern is generated for each point on the dual delay line. The nulling pattern is configured so that a null occurs in the direction of the noise source, and unity gain is maintained in the direction of the target source. Nulls are steered independently in different frequency bands, so that the noise source that contributes the most energy to a band is cancelled. Hence, their system is able to cancel multiple interfering sounds, provided that their locations can be accurately determined. Liu et al. show that for four talkers in an anechoic environment, their system is able to extract a target speaker with high quality while canceling the three interfering talkers by 3–11 dB. However, in a moderately reverberant environment ($T_{60} = 0.4$ s) the total noise cancellation fell by about 2 dB. The authors note that the reduction in cancellation performance caused by reverberation is likely to limit the usefulness of their algorithm in real-world environments.

In the CASA systems discussed so far, directional filtering has been used to cancel interfering sound sources while preserving the target speech signal. A different approach is proposed by Roman and Wang [83], who describe a binaural system intended for multisource reverberant environments. In their system, adaptive filtering is used to cancel the *target* speech source. The authors note that a correlation exists between the amount of cancellation that occurs in a time-frequency region and the relative level of the target and interference in that region. Their approach is based on the estimation of an output-to-input energy ratio $R(t,f)$, which is given by

$$R(t,f) = \frac{|Z(t,f)|^2}{|Y(t,f)|^2} \tag{7.3}$$

Here, $Z(t,f)$ represents the residue in the time-frequency region denoted by (t,f) after cancellation of the target by an adaptive filter, and $Y(t,f)$ represents the corresponding acoustic input. Note that if the time-frequency region contains no noise, then the target will be cancelled effectively and $R(t,f) \rightarrow 0$. Similarly, regions that are dominated by noise will leave a substantial residue after cancellation of the target, such that $R(t,f) \gg 0$. A simple threshold can, therefore, be applied to $R(t,f)$ in order to estimate an ideal binary mask for the target source. Roman and Wang evaluate their approach for cases of one or four interfering speakers in reverberant conditions, using a connected digit recognition task. The recognizer was trained and tested on moderately reverberated speech with a T_{60} of 0.35 s. They report ASR performance that is very close to that obtained with ideal binary masks for SNRs above 0 dB.

7.6.2 CASA for Robust ASR in Reverberant Conditions

It was noted in Section 7.3 that reverberation presents a serious problem for current ASR systems. Given the well-established benefits that binaural hearing brings to human speech perception in reverberant conditions, it seems reasonable to ask whether ASR can benefit in a similar way by using two or more microphones. A number of workers have pursued this approach by combining directional filtering with ASR.

Sullivan and Stern [90] provide an early example of this approach. They describe a system in which the discrete cosine transform is used to convert a cross-correlation function into cepstral coefficients, which can then be used as the acoustic features for an ASR system. The cross-correlation function was derived from signals recorded by either two or four spatially separated microphones. The ASR system was trained on speech recorded in a quiet office, but was tested on speech recorded in the presence of additive noise in a much more reverberant space. Cepstral normalization was, therefore, used to compensate (at least to some extent) for the difference in acoustic conditions between the training and test data. The authors found that better ASR performance was obtained by using more microphones.

A different approach is adopted by Palomäki et al. [79]. Rather than using directional filtering to obtain features for a conventional ASR system, they use it to obtain a time-frequency mask for a missing-data speech recognizer (see Chapters 1 and 9). Another notable feature of their system is that it models certain aspects of the precedence effect, and is therefore tailored specifically for reverberant conditions. A schematic diagram of their system is shown in Fig. 7.12. Acoustic inputs to the system are convolved with binaural impulse responses, which are generated by a model of small-room acoustics. The binaural impulse responses also include the

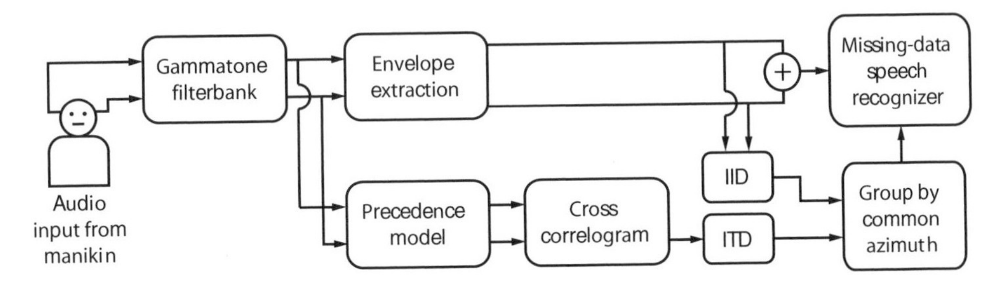

Figure 7.12 Schematic diagram of the CASA system of Palomäki et al. [79], which includes a simple model of the precedence effect to improve source localization in reverberant conditions. Reprinted from [79]. Copyright © 2004, with permission from Elsevier.

effects of head-related transfer functions (recorded from a binaural manikin) so that realistic ITD and IID cues are available.

In the first stage of the system, the left and right inputs are processed by a gammatone filterbank. The output of each filter is then processed in two ways. First, the envelope is extracted from each filter channel and smoothed to form features for the missing data speech recognizer. Second, the filter response is rectified, compressed, and processed by a simple model of the precedence effect. This model is based on the phenomenological account of Zurek [104], in which acoustic energy due to room reflections is suppressed by an inhibitory mechanism. Specifically, an inhibitory signal is generated by low-pass filtering the envelope of the filter channel response, and this is subtracted from the (rectified and compressed) filter output. As a result, sustained activity is suppressed and transients (which are more likely to be due to direct sound) are preserved. The processed filter output is then cross-correlated in order to determine the direction of the target source. Frequency bands that are dominated by the target talker are then identified, and the corresponding elements in the missing data mask are set to unity (indicating that they contain reliable evidence for the target). The remaining mask elements are assumed to be unreliable, and are set to zero. Also, mask elements are only marked as reliable if the observed IID and ITD are consistent with one another. The acoustic features and the mask are then passed to the missing-data ASR system for decoding. The authors evaluate their system in moderately reverberant conditions ($T_{60} = 0.45$ s), using mixtures of a target speaker and a spatially separated interferer. The speech recognizer was trained on clean speech and, hence, a spectral normalization process was used to compensate for the spectral distortion caused by reverberation of the test signals. In most conditions, the performance of their system was substantially higher than that obtained for a conventional ASR system trained on MFCC features. However, their results suggest that the reverberation robustness of the system could be much improved in situations in which there is substantial interference from other sources; even slight reverberation ($T_{60} = 0.3$ s) caused a dramatic fall in ASR performance when the SNR was low.

Palomäki et al. [79] present a separate evaluation of the precedence effect component of their system, and though it appears to help localization in reverberant environments, its performance is still far from optimal. Further gains in performance might be obtained by using a more sophisticated model of the precedence effect, which is more closely based on physiological mechanisms. Such models are described by Huang et al. [51] and Wittkop et al. [98]. In the latter case, the model is based on the physiology of the barn owl, and includes a neural feedback circuit that suppresses echoes. Wittkop et al. do not assess the noise reduction ability of this system, but claim that it is able to reliably track the location of four concurrent sound sources in reverberant conditions.

Harding et al. [43] further develop the idea of using directional filtering to estimate a mask for missing data ASR in reverberant conditions. They consider stimulus configurations in which a target talker and an interfering talker are present in moderately reverberant rooms (T_{60} times of 0.34 s and 0.51 s were used). Their approach differs in two main respects from the study of Palomäki et al. First, the acoustic models for speech recognition were trained on reverberated (rather than clean) speech. Second, time-frequency masks derived from binaural cues were subject to a number of postprocessing operations that aimed to improve their quality. Specifically, erosion and dilation were employed to remove noise from the masks, and artificial neural networks were used to learn a mapping between the estimated masks and ideal masks (see Chapter 1). The latter approach allowed the estimated masks to be transformed into masks that were more similar to the ideal masks. The best ASR performance was obtained with the learned mapping, and it was found that a mapping that was learned for a particular reverberation condition generalized well to different reverberation conditions.

7.6.3 Systems that Use Multiple Cues

It was noted in Section 7.2.2 that ITD and IID become unreliable cues for source direction as the reverberation time increases, because the interaural coherence of sound sources is disrupted by echoes. This suggests that CASA systems must use additional cues if they are to work successfully in reverberant environments.

Following this reasoning, Shamsoddini and Denbigh [88] describe a binaural CASA system that combines harmonicity and localization cues. In their system, a cross correlation is computed between two input signals derived from spatially separated microphones, and is used to determine the direction of the dominant sound source at each time instant. The spectrum of the target voice (which is assumed to be at zero degrees azimuth) is then estimated, and cancelled from the two inputs in order to give a crude estimate of the spectrum of the interference. The spectrum of the interference is then subtracted from the spectrum of the target voice. Frequency coefficients of the target that have lower energy than the corresponding coefficients of the interference are completely removed, since they are assumed to be corrupted by noise or reverberation. The $F0$ of the target and interference are then tracked. If the target speech is voiced, then the harmonics of its $F0$ are enhanced. Similarly, the harmonics of a voiced interferer are suppressed. Finally, the target voice is reconstructed using overlap–add. Shamsoddini and Denbigh evaluate their system by testing the intelligibility of reconstructed target speech, for configurations with one and two spatially separated interferers in a reverberant room. Substantial improvements in intelligibility were obtained in both conditions. The authors also demonstrate that processing by the algorithm leads to improved automatic recognition of target speech under similar conditions. However, they do not indicate how the performance of their system varies as a function of reverberation time.

Kollmeier and Koch [58] (see also [98]) describe a separation system based on the assumption that there is an interaction between modulation detection and binaural grouping mechanisms in the auditory system. In their system, acoustic inputs for the left and right ears are analyzed into frequency bands using the discrete Fourier transform (DFT). A second DFT is then performed on the envelope in each frequency band in order to obtain a complex modulation spectrum. Binaural cues are then extracted by computing the ratio of the left and right modulation spectra and taking the real and imaginary parts, which correspond to the interaural level and phase differences, respectively. A weighting function is derived from these binaural cues so that modulation components that arise from a target location are passed unchanged, whereas those originating from other directions are suppressed. A reconstructed signal is then obtained from the weighted modulation spectrum by taking the inverse DFT (twice) and overlap adding. Kollmeier and Koch evaluated their algorithm by testing the intelligibility of a target talker in the presence of one, two, or four spatially separated interferers. In anechoic conditions, processing by the algorithm gave an improvement corresponding to an increase in signal-to-noise ratio of approximately 2 dB. Importantly, a similar advantage was also obtained in a reverberant environment ($T_{60} = 1.33$ s). As such, their study illustrates that information known to be important for the robustness of human speech perception in reverberant conditions (i.e., envelope modulation) can be successfully exploited in a computational system.

7.7 DISCUSSION AND CONCLUSIONS

Robustness in reverberant conditions remains a key problem for CASA systems. Currently, the performance profile of CASA for reverberated signals is similar to that of ASR; the systems work to some extent, but performance degrades sharply as the reverberation time increases. Also, reverberation further complicates the already difficult problem of source separation; even a modest amount of room reverberation can cause a substantial fall in performance if an interfering sound source is also present. The magnitude of the task facing CASA in this respect is well illustrated by the analysis of Plomp [81] for human listeners. He shows that in typical "cocktail party" situations in rooms with highly absorbent walls and ceiling, the SNR is only about 2 dB above the level required for almost complete understanding of a nearby talker. Even slight reverberation erodes this small margin, and strong reflections from the walls or ceiling will completely eliminate it. We are therefore aiming to produce CASA systems that work reliably in conditions that push the human auditory system to its very limits.

What, then, should be the research goal for CASA in terms of reverberation? In Section 1.2.1, CASA was described as the challenge of constructing a machine system that achieves human performance in ASA. In regard to reverberation, this implies that CASA systems should match human performance in reverberant conditions, including the graceful degradation in performance associated with more difficult reverberant, multisource environments. However, it should be noted that the criteria for success need not be perfect dereverberation of the target signal. Although this may be desirable for some applications (such as hearing aids, in which the target signal must be reconstructed and delivered to the listener after processing), it may not be necessary for applications such as ASR and automatic music transcription. For example, Palomäki et al. [78] have shown that reverberation can be handled within the missing-data framework for ASR, by treating reverberant and nonreverberant regions of the speech spectrum differently during decoding.

A review of the factors thought to underlie the robustness of human hearing to reverberation suggests a number of specific directions for future CASA research. First, although binaural hearing confers robustness to reverberation in respect to sound localization and signal detection, binaural grouping cues (common ITD and IID) are severely degraded by reverberation. As a result, human listeners are largely unable to exploit a difference in spatial location between concurrent sound sources in reverberant environments (see Section 7.2.3). In contrast, CASA systems that have been evaluated in reverberant conditions tend to use binaural grouping cues exclusively (see, e.g., [65, 11, 85]). The combination of binaural cues with other (monaural) cues may, therefore, be required in order to obtain robustness in reverberant conditions. In particular, the psychophysical studies reviewed in Section 7.2.3 suggest that grouping strategies based upon prosodic cues, common onset, and vocal tract size may be more robust in the presence of reverberation.

Second, relatively few binaural CASA systems employ models of the precedence effect. A notable exception is the binaural separation system described by Palomäki et al. [79], but even in this case the model is crude and requires further development. Given the well-established role of the precedence effect in human sound localization in reverberant conditions (see Section 7.4.3), one would expect binaural CASA systems to benefit from more sophisticated models of precedence-like processing.

Also, recent advances in dereverberation and reverberation-robust speech recognition suggest that much can be gained by exploiting the structure inherent in speech signals (particularly in terms of low-frequency envelope modulation [54], glottal pulsing [102], and harmonic structure [76]). A notable example of this in the CASA literature is the system described by Kollmeier and Koch [58], which combines spatial filtering with modulation filtering in the speech range. The reverberation masking approach proposed by Palomäki et al. also exploits speech modulation, and could be integrated into CASA systems that employ time-frequency masks.

A related point is that the majority of CASA systems focus on data-driven approaches that are primarily motivated by peripheral auditory function. However, psychophysical studies indicate that central auditory processing plays an important role in robustness to reverberation, by compensating for the spectral distortion of speech (see Section 7.4.4). Furthermore, higher neural centers can exploit binaural information in speech identification to overcome the effects of reverberation. Indeed, Libbey and Rogers [60] speculate that two pathways may be involved in which direct sound and reverberated sound are treated differently, the former via a single-channel pathway and the latter via a binaural pathway. It is unlikely that CASA systems will match human performance until these central mechanisms are fully understood and incorporated in models.

We also note that CASA algorithms often focus on maximum noise rejection, but do so at the expense of preserving speech quality. This is particularly problematic in reverberant environments, in which little improvement in intelligibility is observed even though the reverberation is effectively suppressed because of the distortions introduced by processing. In the case of algorithms for hearing-impaired listeners, Kollmeier and Koch [58] note that effective noise suppression in favorable acoustic conditions is less important than ensuring that speech quality is not degraded by processing in adverse acoustic conditions.

A final point concerns the general approach that should be taken to the problem of reverberation—should reverberation be modeled, or just treated as noise? The reverberation-masking approach, in which late reverberation is regarded as additive noise, has proved to be quite effective. However, there is evidence that such masking mechanisms do not play a major role in human perception [60]. Also, learning plays a role in sound localization in reverberant environments, and in auditory spatial impression (see Section 7.2). This suggests that listeners are able to build an internal model of multisource reverberant environments, an approach that could bring benefits to CASA.

In conclusion, reverberation is a hard problem for CASA. However, our understanding of the mechanisms that underlie the robustness of human listeners to reverberation is growing, and this represents an important and largely untapped resource for future CASA research. We hope that this chapter will serve to stimulate further work in this challenging and important area.

ACKNOWLEDGMENTS

The room impulse responses that we used for our simulations were originally collected by Jim West, Michael Gatlin, and Carlos Avendano. The authors would like to thank Tomohiro Nakatani for providing the signals on which Fig. 7.9 is based. Guy J. Brown was supported by EPSRC grant GR/R47400/01. The work of Kalle J. Palomäki was done at the University of Sheffield, Department of Computer Science (EC IHP HOARSE project) and at the Helsinki University of Technology, Laboratory of Acoustics and Audio Signal Processing (grant 1277811 from the Academy of Finland).

REFERENCES

1. J. B. Allen, D. A. Berkeley, and J. Blauert. Multimicrophone signal-processing technique to remove room reverberation from speech signals. *Journal of the Acoustical Society of America,* 62(4):912–915, 1977.

2. J. B. Allen and D. A. Berkley. Image method for efficiently simulating small-room acoustics. *Journal of the Acoustical Society of America,* 65(4):943–950, 1979.
3. M. Aoki and K. Furuya. Real-time source separation based on sound localization in a reverberant environment. In *Proceedings of the 12th IEEE Workshop on Neural Networks for Signal Processing,* pp. 475–484, Martigny, September 2002.
4. T. Arai, K. Kinoshita, N. Hodoshima, A. Kusumoto, and T. Kitamura. Effects of suppressing steady-state portions of speech on intelligibility in reverberant conditions. *Acoustical Science and Technology,* 23(4):229–232, 2002.
5. P. Assmann and Q. Summerfield. The perception of speech under adverse acoustic conditions. In S. Greenberg, W. A. Ainsworth, A. N. Popper, and R. R. Fay, editors, *Speech Processing in the Auditory System,* volume 18 of *Springer Handbook of Auditory Research.* Springer-Verlag, Berlin, 2004.
6. C. Avendano and H. Hermansky. Study on the dereverberation of speech based on temporal envelope filtering. In *Proceedings of the International Conference on Spoken Language Processing,* pp. 889–892, Philadelphia, October 1996.
7. M. Barron. The subjective effects of first reflections in concert halls—The need for lateral reflections. *Journal of Sound and Vibration,* 15(4):475–494, 1971.
8. M. Barron and A. H. Marshall. Spatial impression due to early lateral reflections in concert halls: The derivation of a physical measure. *Journal of Sound and Vibration,* 77(2):211–232, 1981.
9. A. J. Bell and T. J. Sejnowski. An information-maximisation approach to blind separation and blind deconvolution. *Neural Computation,* 7:1129–1159, 1995.
10. J. Blauert. *Spatial Hearing,* rev. ed. MIT Press, Cambridge, MA, 1997.
11. M. Bodden. Modelling human sound-source localization and the cocktail party effect. *Acta Acustica,* 1:43–55, 1993.
12. R. H. Bolt and A. D. MacDonald. Theory of speech masking by reverberation. *Journal of the Acoustical Society of America,* 21(6):577–580, 1949.
13. J. S. Bradley. Predictors of speech intelligibility in rooms. *Journal of the Acoustical Society of America,* 80(3):837–845, 1986.
14. J. S. Bradley and G. A. Soulodre. The influence of late arriving energy on spatial impression. *Journal of the Acoustical Society of America,* 97(4):2263–2271, 1995.
15. M. S. Brandstein. Time-delay estimation of reverberated speech exploiting harmonic structure. *Journal of the Acoustical Society of America,* 105(5):2914–2919, 1999.
16. A. W. Bronkhorst and T. Houtgast. Auditory distance perception in rooms. *Nature,* 397:517–520, February 1999.
17. D. R. Campbell. *The ROOMSIM User Guide (V3.3),* 2004. Available at http://media.paisley.ac.uk/~campbell/Roomsim/.
18. R. K. Clifton. Breakdown of echo suppression in the precedence effect. *Journal of the Acoustical Society of America,* 82:1834–1835, 1987.
19. L. Couvreur and C. Couvreur. Blind model selection for automatic speech recognition in reverberant environments. *Journal of VLSI Signal Procesing,* 36:189–203, 2004.
20. J. F. Culling, K. I. Hodder, and C. Y. Toh. Effects of reverberation on perceptual segregation of competing voices. *Journal of the Acoustical Society of America,* 114(5):2871–2876, 2003.
21. J. F. Culling and Q. Summerfield. Perceptual separation of concurrent speech sounds:

Absence of across-frequency grouping by common interaural delay. *Journal of the Acoustical Society of America,* 98(2):785–797, 1995.

22. J. F. Culling, Q. Summerfield, and D. H. Marshall. Effects of simulated reverberation on the use of binaural cues and fundamental frequency differences for separating concurrent vowels. *Speech Communication,* 14:71–95, 1994.
23. C. J. Darwin and R. W. Hukin. Perceptual segregation of a harmonic from a vowel by interaural time difference and frequency proximity. *Journal of the Acoustical Society of America,* 102(4):2316–2324, 1997.
24. C. J. Darwin and R. W. Hukin. Effects of reverberation on spatial, prosodic, and vocal-tract size cues to selective attention. *Journal of the Acoustical Society of America,* 108(1):335–342, 2000.
25. T. Dau, D. Püschel, and A. Kohlrausch. A quantitative model of the "effective" signal processing in the auditory system. I. Model structure. *Journal of the Acoustical Society of America,* 99(6):3615–3622, 1996.
26. A. de Cheveigné and H. Kawahara. YIN, a fundamental frequency estimator for speech and music. *Journal of the Acoustical Society of America,* 111(4):1917–1930, 2002.
27. S. Devore and B. G. Shinn-Cunningham. Perceptual consequences of including reverberation in spatial auditory displays. In *Proceedings of the International Conference on Auditory Display,* pp. 75–78, Boston, MA, July 2003.
28. W. R. Drennan, S. Gatehouse, and C. Lever. Perceptual segregation of competing speech sounds: The role of spatial location. *Journal of the Acoustical Society of America,* 114(4):2178–2189, 2003.
29. R. Drullman, J. M. Festen, and R. Plomp. Effect of reducing slow temporal modulations on speech reception. *Journal of the Acoustical Society of America,* 95:2670–2680, 1994.
30. R. Drullman, J. M. Festen, and R. Plomp. Effects of temporal envelope smearing on speech reception. *Journal of the Acoustical Society of America,* 95:1053–1064, 1994.
31. H. Dudley. Remaking speech. *Journal of the Acoustical Society of America,* 11(2): 169–177, 1939.
32. N. I. Durlach. Equalization and cancellation theory of binaural masking level differences. *Journal of the Acoustical Society of America,* 35(8):1206–1218, 1963.
33. K. Eneman, J. Duchateau, M. Moonen, D. Van Campernolle, and H. Van Hamme. Assessment of dereverberation algorithms for large vocabulary speech recognition systems. In *Proceedings of Eurospeech,* pp. 2689–2692, Geneva, September 2003.
34. C. Faller and J. Merimaa. Source localization in complex listening situations: Selection of binaural cues based on interaural coherence. *Journal of the Acoustical Society of America,* 116(5):3075–3089, 2004.
35. R. L. Freyman, R. K. Clifton, and R. Y. Litovsky. Dynamic effects in the precedence effect. *Journal of the Acoustical Society of America,* 90:874–884, 1991.
36. S. A. Gelfand and S. Silman. Effects of small room reverberation upon the recognition of some consonant features. *Journal of the Acoustical Society of America,* 66(1):22–29, 1979.
37. C. Giguère and S. M. Abel. Sound localization: Effects of reverberation time, speaker array, stimulus frequency, and stimulus rise/decay. *Journal of the Acoustical Society of America,* 94(2):769–776, 1983.
38. B. W. Gillespie and L. E. Atlas. Acoustic diversity for improved speech recognition in

reverberant environments. In *Proceedings of the International Conference on Acoustics, Speech and Signal Processing,* pp. 557–560, Orlando, May 2002.

39. D. Giuliani, M. Omologo, and P. Svaizer. Experiments of speech recognition in a noisy and reverberant environment using a microphone array and HMM adaptation. In *Proceedings of the International Conference on Spoken Language Processing,* pp. 1329–1332, Philadelphia, October 1996.
40. H. Gölzer and M. Kleinschmidt. Importance of early and late reflections for automatic speech recognition in reverberant environments. In *Elektronische Sprachsignalverarbeitung (ESSV),* 2003.
41. S. Greenberg, H. Carvey, L. Hitchcock, and S. Chang. Temporal properties of spontaneous speech—A syllable-centric perspective. *Journal of Phonetics,* 31:465–485, 2003.
42. D. Griesinger. The psychoacoustics of apparent source width, spatiousness and envelopment in performance spaces. *Acta Acustica,* 83:721–731, 1997.
43. S. Harding, J. Barker, and G. J. Brown. Mask estimation based on sound localization for missing data speech recognition. In *Proceedings of the International Conference on Acoustics, Speech and Signal Processing,* Philadelphia, March 2005, Volume 1, pp. 537–540.
44. W. M. Hartmann. Localization of sound in rooms. *Journal of the Acoustical Society of America,* 74(5):1380–1391, 1983.
45. P. Hatziantoniou and J. Mourjopoulos. Real-time room equalization based on complex smoothing: Robustness results. In *Proceedings of the 116th Convention of the Audio Engineering Society,* Berlin, May 2004.
46. S. Haykin, editor. *Blind Deconvolution.* Prentice-Hall, Upper Saddle River, NJ, 1994.
47. H. Hermansky and N. Morgan. RASTA processing of speech. *IEEE Transactions on Speech and Audio Processing,* 2:578–589, 1994.
48. H. G. Hirsch, P. Meyer, and H. W. Ruehl. Improved speech recognition using high-pass filtering of subband envelopes. In *Proceedings of Eurospeech,* pp. 413–416, Genoa, September 2001.
49. T. Houtgast and H. J. M. Steeneken. A review of the MTF concept in room acoustics and its use for estimating speech intelligibility in auditoria. *Journal of the Acoustical Society of America,* 77:1069–1077, 1985.
50. G. Hu and D. L. Wang. Monaural speech segregation based on pitch tracking and amplitude modulation. *IEEE Transactions on Neural Networks,* 15(5):1135–1150, 2004.
51. J. Huang, N. Ohnishi, X. Guo, and N. Sugie. Echo avoidance in a computational model of the precedence effect. *Speech Communication,* 27:223–233, 1999.
52. W. H. T. Huisman and K. Attenborough. Reverberation and attenuation in a pine forest. *Journal of the Acoustical Society of America,* 90(5):2664–2677, 1991.
53. N. Kandera, T. Arai, H. Hermansky, and M. Pavel. On the relative importance of various components of the modulation spectrum for automatic speech recognition. *Speech Communication,* 28:43–55, 1999.
54. B. E. D. Kingsbury. *Perceptually Inspired Signal-Processing Strategies for Robust Speech Recognition in Reverberant Environments.* PhD thesis, University of California, Berkeley, 1998.
55. B. E. D. Kingsbury, N. Morgan, and S. Greenberg. Robust speech recognition using the modulation spectrogram. *Speech Communication,* 25(1–3):117–132, 1998.

56. M. Kleinschmidt, J. Tchorz, and B. Kollmeier. Combining speech enhancement and auditory feature extraction for robust speech recognition. *Speech Communication,* 34(1–2):75–91, 2001.
57. W. Koenig. Subjective effects in binaural hearing. *Journal of the Acoustical Society of America,* 22(1):61–62, 1950.
58. B. Kollmeier and R. Koch. Speech enhancement based on physiological and psychoacoustical models of modulation perception and binaural interaction. *Journal of the Acoustical Society of America,* 95(3):1593–1602, 1994.
59. T. Langhans and H. W. Strube. Speech enhancement by nonlinear multiband envelope filtering. In *Proceedings of the International Conference on Acoustics, Speech and Signal Processing,* pp. 156–159, Paris, May 1982.
60. B. Libbey and P. H. Rogers. The effect of overlap-masking on binaural reverberant word intelligibility. *Journal of the Acoustical Society of America,* 115:3141–3151, 2004.
61. R. Y. Litovsky, S. H. Colburn, W. A. Yost, and S. J. Guzman. The precedence effect. *Journal of the Acoustical Society of America,* 106:1633–1654, 1999.
62. C. Liu, B. C. Wheeler, W. D. O'Brien, C. R. Lansing, R. C. Bilger, D. L. Jones, and A. S. Feng. A two-microphone dual delay-line approach for extraction of a speech sound in the presence of multiple interferers. *Journal of the Acoustical Society of America,* 110(6):3218–3231, 2001.
63. F. H. Liu, R. M. Stern, X. Huang, and A. Acero. Efficient cepstral normalization for robust speech recognition. In *Proceedings of the Sixth ARPA Workshop on Human Language Technology,* Princeton, NJ, March 1993. Morgan Kaufmann.
64. J. P. A. Lochner and J. F. Burger. The influence of reflections on auditorium acoustics. *Journal of Sound and Vibration,* 1(4):426–454, 1964.
65. R. F. Lyon. A computational model of binaural localization and separation. In *Proceedings of the International Conference on Acoustics, Speech and Signal Processing,* pp. 1148–1151, 1983.
66. A. H. Marshall. A note on the importance of room cross-section in concert halls. *Journal of Sound and Vibration,* 5(1):100–112, 1997.
67. K. D. Martin. Echo suppression in a computational model of the precedence effect. In *Proceedings of the IEEE Workshop on Applications of Signal Processing to Acoustics and Audio,* Mohonk, NY, 1997.
68. BS EN ISO 3382. Measurement of the reverberation time of rooms with reference to other acoustical parameters, 2000.
69. J. Merimaa and W. Hess. Training of listeners for evaluation of spatial attributes of sound. In *Proceedings of the 117th Convention of the Audio Engineering Society,* San Francisco, October 2004.
70. D. H. Mershon and J. N. Bowers. Absolute and relative cues for the auditory perception of egocentric distance. *Perception,* 8:311–322, 1979.
71. D. H. Mershon and E. King. Intensity and reverberation as factors in the auditory perception of egocentric distance. *Perception and Psychophysics,* 18:409–415, 1975.
72. L. M. Miller, M. A. Escabí, H. L. Read, and C. E. Schreiner. Spectrotemporal receptive fields in the lemniscal auditory thalamus and cortex. *Journal of Neurophysiology,* 87:516–527, 2002.
73. M. Miyoshi and Y. Kaneda. Inverse filtering of room acoutics. *IEEE Transactions on Acoustics, Speech and Signal Processing,* 36(2):145–152, 1988.

74. A. K. Nabelek and P. K. Robinson. Monaural and binaural speech perception in reverberation for listeners of various ages. *Journal of the Acoustical Society of America,* 71:1242–1248, 1982.

75. T. Nakatani, K. Kinoshita, M. Miyoshi, and P. S. Zolfaghari. Harmonicity based monaural speech dereverberation with time warping and F0 adaptive window. In *Proceedings of the International Conference of Spoken Language Processing,* pp. 873–876, Jeju Island, Korea, 2004.

76. T. Nakatani, M. Miyoshi, and K. Kinoshita. One microphone blind dereverberation based on quasi-periodicity of speech signals. In S. Thrun, L. K. Saul, and B. Schölkopf, editors, *Advances in Neural Information Processing Systems 16,* pp. 1417–1424. MIT Press, Cambridge, MA, 2004.

77. K. J. Palomäki, G. J. Brown, and J. Barker. Missing data speech recognition in reverberant conditions. In *Proceedings of the International Conference on Acoustics, Speech and Signal Processing,* pp. 65–68, Orlando, May 2002.

78. K. J. Palomäki, G. J. Brown, and J. Barker. Techniques for handling convolutional distortion with "missing data" automatic speech recognition. *Speech Communication,* 43(1–2):123–142, 2004.

79. K. J. Palomäki, G. J. Brown, and D. L. Wang. A binaural processor for missing data speech recognition in the presence of noise and small-room reverberation. *Speech Communication,* 43(4):361–378, 2004.

80. R. Plomp. Binaural and monaural speech intelligibility of connected discourse in reverberation as a function of azimuth of a single competing sound source (speech or noise). *Acustica,* 34:200– 211, 1976.

81. R. Plomp. Acoustical aspects of cocktail parties. *Acustica,* 38:186–191, 1977.

82. W. Reichardt, O. Abdel Alim, and W. Schmidt. Abhängigkeit der grenzen zwischen brauchbarer und unbrauchbarer durchsichtigkeit von der art des musikmotives, der nachhallzeit und der nachhalleinsatzzeit. *Applied Acoustics,* 7(4):243–264, 1974.

83. N. Roman and D. L. Wang. Binaural sound segregation for multisource reverberant environments. In *Proceedings of the International Conference on Acoustics, Speech and Signal Processing,* pp. 373–376, Montreal, May 2004.

84. N. Roman and D. L. Wang. A pitch-based model for separation of reverberant speech. In *Proceedings of Interspeech,* pp. 2109–2112, Lisbon, September 2005.

85. N. Roman, D. L. Wang, and G. J. Brown. Speech segregation based on sound localization. *Journal of the Acoustical Society of America,* 114(4):2236–2252, 2003.

86. F. Rumsey. Spatial quality evaluation for reproduced sound: Terminolgy, meaning, and a scene-based paradigm. *Journal of the Audio Engineering Society,* 50(9):651–666, 2002.

87. M. Schlang. An auditory based approach for echo compensation with modulation filtering. In *Proceedings of Eurospeech,* pp. 661–664, Paris, September 1989.

88. A. Shamsoddini and P. N. Denbigh. A sound segregation algorithm for reverberant conditions. *Speech Communication,* 33(3):179–196, 2001.

89. N. C. Singh and F. E. Theunissen. Modulation spectra of natural sounds and ethological theories of auditory processing. *Journal of the Acoustical Society of America,* 114(6):3394–3411, 2003.

90. T. M. Sullivan and R. M. Stern. Multi-microphone correlation-based processing for robust speech recognition. In *Proceedings of the International Conference on Acoustics, Speech and Signal Processing,* pp. 91–94, Minneapolis, 1993.

91. W. R. Thurlow and T. E. Parks. Precedence-suppression effects for two click sources. *Perceptual and Motor Skills,* 13:7–12, 1961.

92. M. Tohyama, R. H. Lyon, and T. Koike. Source waveform recovery in a reverberant space by cepstrum dereverberation. In *Proceedings of the IEEE International Conference on Acoustics, Speech and Signal Processing,* pp. 157–160, Minneapolis, April 1993.

93. G. v. Békésy. Über die entstehung der entfernungsempfindung beimhören. *Akustische Zeitschrift,* 3:21–31, 1938.

94. B. D. van Veen and K. M. Buckley. Beamforming: A versatile approach to spatial filtering. *IEEE ASSP Magazine,* pp. 4–24, April 1988.

95. H. W. Wallach, E. B. Newman, and M. R. Rosenzweig. The precedence effect in sound localization. *American Journal of Psychology,* 62:315–337, 1949.

96. A. J. Watkins. Central, auditory mechanisms of perceptual compensation for spectral-envelope distortion. *Journal of the Acoustical Society of America,* 90(6):2942–2955, 1991.

97. A. J. Watkins. The influence of early reflections on the identification and lateralization of vowels. *Journal of the Acoustical Society of America,* 106:2933–2944, 1999.

98. T. Wittkop, S. Albani, V. Hohmann, J. Peissig, W. S. Woods, and B. Kollmeier. Speech processing for hearing aids: Noise reduction motivated by models of binaural interaction. *Acustica,* 83:684– 699, 1997.

99. M. Wu and D. L. Wang. A one-microphone algorithm for reverberant speech enhancement. In *Proceedings of the International Conference on Acoustics, Speech and Signal Processing,* pp. 844–847, Hong Kong, April 2003.

100. M. Wu and D. L. Wang. A two-stage algorithm for one-microphone reverberant speech enhancement. *IEEE Transactions on Audio, Speech and Language Processing,* 14: 774–784, 2006.

101. M. Wu, D. L. Wang, and G. J. Brown. A multipitch tracking algorithm for noisy speech. *IEEE Transactions on Speech and Audio Processing,* 11(3):229–241, 2003.

102. B. Yegnanarayana and P. S. Murthy. Enhancement of reverberant speech using LP residual signal. *IEEE Transactions on Speech and Audio Processing,* 8(3):267–281, 2000.

103. P. Zahorik. Assessing auditory distance perception using virtual acoustics. *Journal of the Acoustical Society of America,* 111(4):1832–1846, 2002.

104. P. M. Zurek. The precedence effect. In W. A. Yost and G. Gourevitch, editors, *Directional Hearing,* pp. 85–105. Springer-Verlag, New York, 1987.

105. P. M. Zurek, R. L. Freyman, and U. Balakrishnan. Auditory target detection in reverberation. *Journal of the Acoustical Society of America,* 115(4):1609–1620, 2004.

CHAPTER 8

Analysis of Musical Audio Signals

MASATAKA GOTO

8.1 INTRODUCTION

This chapter introduces research approaches that aim to build a computational model that allows musical audio signals to be understood through the description of music scenes. A typical research approach to computational auditory scene analysis (CASA) is sound-source separation—the extraction of each individual audio signal that constitutes a sound mixture. Human listeners can obviously understand various properties of sound mixtures they hear in a real-world environment and this suggests that they detect the existence of certain auditory objects in sound mixtures and obtain a description of them. This understanding, however, is not necessarily evidence that the human auditory system extracts the individual audio signal corresponding to each auditory object. This is because separation is not a necessary condition for understanding: Even if a mixture of two components cannot be separated, it can be understood from their salient features that the mixture includes them. In fact, from the viewpoint of auditory psychology, it has been pointed out that perceptual sound-source segregation is different from signal-level separation. For example, Bregman noted that "there is evidence that the human brain does not completely separate sounds" [5]. The approach of developing a computational model of monaural or binaural sound-source separation might deal with a hard problem that is not solved by any mechanism in this world (not solved even by the human brain), although sound-source separation is valuable from an engineering viewpoint.

In the context of CASA, it is therefore essential to build a computational model that can obtain a certain description of auditory scenes from sound mixtures. To emphasize its difference from sound-source separation and restoration, this approach was named "auditory scene description" [41, 38]. In modeling auditory scene description, it is important to discuss what constitutes an appropriate description of audio signals. An easy way of specifying the description is to borrow the terminology of existing discrete symbol systems, such as musical scores consisting of musical notes or speech transcriptions consisting of text characters. Those symbols,

Computational Auditory Scene Analysis. Edited by DeLiang Wang and Guy J. Brown

however, fail to express nonsymbolic properties such as the expressiveness of a musical performance and the prosody of spontaneous speech. To take such properties into account, it is necessary to introduce a subsymbolic description represented as continuous quantitative values. At the same time, it is also necessary to choose an appropriate level of abstraction for the description, because descriptions such as raw waveforms and spectra are too concrete even though they have continuous values. This appropriateness of the abstraction level depends on the purpose of the description and on the use to which it will be put.

The focus of this chapter is the problem of "music scene description" [41, 35, 38], that is, auditory scene description in music, for complex real-world audio signals like those recorded on commercially distributed compact discs (CDs). The audio signals are thus assumed to contain simultaneous sounds of various instruments. In music scene description, "scenes" that occur within a musical performance—such as melody, bass, beat structure, and chorus and phrase repetition—are described. The following sections introduce this music-scene-description research approach that obtains those scene descriptions that are intuitively meaningful to musically untrained listeners, without trying to extract every musical note from musical audio signals.

8.2 MUSIC SCENE DESCRIPTION

The goal of music scene description is to enable a computer to understand musical audio signals at the level of untrained human listeners. People listening to music can easily hum the melody, clap hands in time to the musical beat, notice a phrase being repeated, and find chorus sections. The brain mechanisms underlying these abilities, however, are not well understood. In addition, it has been difficult to implement these abilities on a computer system, although a system with them is useful in various applications such as music information retrieval, music production/editing, and music interfaces. It is therefore an important challenge to build a music-scene-description system that can understand complex real-world music signals like those from CD recordings.

This music-scene-description approach emphasizes methods that can obtain a certain description of a music scene from sound mixtures of various musical instruments in a musical piece. Since various levels of abstraction for the description are possible, it is necessary to consider which level is an appropriate first step toward the ultimate description in human brains. Musical scores are inadequate for this because, as pointed out by Goto [44, 45, 41] and Scheirer [85], untrained listeners understand music to some extent without mentally representing audio signals as musical scores. For example, listeners who cannot identify the name and constituent notes of a chord can nevertheless feel the harmony and chord changes. This suggests that a chord is perceived as a combined whole sound (tone color) without reducing it to its constituent notes (as in reductionism). Furthermore, even if it is possible to derive separated signals and musical notes, it would still be difficult to obtain high-level music descriptions like melody lines and chorus sections. Music transcription, identifying the names (symbols) of all musical notes and chords, is in fact a skill mastered only by trained musicians. To consider appropriate descriptions instead of musical scores, Goto [41, 38] proposed the following three viewpoints:

1. An intuitive description that can be easily obtained by untrained listeners.
2. A basic description that trained musicians can use as a basis for higher-level music understanding.
3. A useful description to facilitate the development of various practical applications.

8.2.1 Music Scene Descriptions

According to the three viewpoints, the following local and global descriptions (Figure 8.1) have been proposed:

(1) Melody and bass lines (local frequency structure). Melody and bass lines represent the temporal trajectory of the melody and bass. The melody is a series of single tones and is heard more distinctly than the rest. The bass is a series of single tones and is the lowest part in polyphonic music. Note that a melody or bass line here is not represented as a series of musical notes; it is a continuous representation of fundamental frequency ($F0$, perceived as pitch) and power transitions.

(2) Hierarchical beat structure (local temporal structure). The hierarchical beat structure represents the fundamental temporal structure of music and comprises the quarter-note and measure levels, that is, the positions of quarter-note beats and bar lines.

(3) Chorus sections and repeated sections (global music structure). Chorus sections represent the most representative, uplifting, and prominent thematic sections in the structure of a musical piece (especially in popular music). Since chorus sections are usually repeated, they are represented as a list of the start and end points of every chorus section. Repeated sections represent the repetition of temporal regions with various lengths.

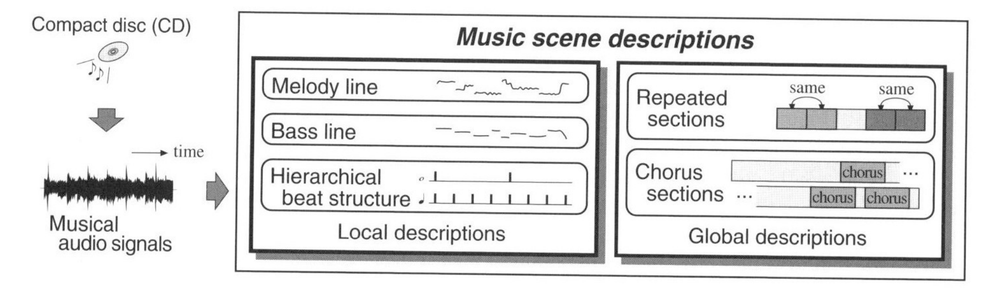

Figure 8.1 Music scene descriptions.

The idea behind these descriptions came from introspective observation of how untrained listeners listen to music.

In this chapter, the following sections introduce methods for producing these descriptions from music signals, such as CD recordings, which contain simultaneous sounds of various instruments (with or without drum sounds). Section 8.3 describes methods for estimating melody and bass lines, Section 8.4 describes methods for estimating beat structure, and Section 8.5 describes methods for estimating chorus sections and repeated sections.

Although this chapter focuses on methods used to estimate the above descriptions, these methods are relevant to and influenced by various methods developed in more traditional approaches: automatic music transcription and sound source separation. Automatic music transcription as a research theme has a long history going back to the 1970s, and has progressed steadily as the difficulty of the target music has increased from monophonic sounds of melodies to polyphonic sounds from a single instrument and a mixture of sounds from several instruments. This progression has been accompanied by a shift toward more specialized research topics, namely, sound source separation and multiple $F0$ estimation (see Section 2.4).

Research related to the understanding of musical audio signals in both traditional and music-scene-description approaches has developed significantly since the mid-1990s. The major development regarding the understanding of musical audio signals has been supported by advances in hardware and in techniques for processing audio signals. Before the mid-1990s, it was difficult to calculate a fast Fourier transform (FFT) in real time; it can now be done so quickly that the time required for its computation can essentially be ignored. This jump in processing performance has let researchers devise computationally intensive approaches that could not be considered in the past, and has also promoted the use of a wide range of statistical techniques.

For example, techniques based on probabilistic models such as the hidden Markov model (HMM) and various techniques making use of maximum likelihood estimation and Bayesian estimation have been proposed. In 1994, Kashino et al. [56, 58] proposed a method based on a probabilistic model called OPTIMA (Organized Processing toward Intelligent Music Scene Analysis). This method was novel in its use of a graphical model to describe the hierarchical structure of frequency components, musical notes, and chords, and in determining the most likely interpretation based on this hierarchical relationship. Then, in 1999, Goto [41, 38] proposed a predominant-$F0$ estimation method (PreFEst) that does not assume the number of sound sources. This method prepares probability distributions that represent the shape of harmonic structures for all possible $F0$s, and models input frequency components as a mixture (weighted sum) of those distributions. It then estimates the parameters of this model—the amplitude (weight) of each component sound in the input sound mixture and the shape of its harmonic structure—by using maximum a posteriori probability (MAP) estimation (the details of this method are described in Section 8.3). At the beginning of this century, many probabilistic or iterative methods were proposed. Klapuri [59] proposed a method for sequentially determining the components in a sound mixture by repeatedly estimating the predominant $F0$ and removing its harmonic components. Davy and Godsill [16] proposed a method for estimating model parameters such as the number of simultaneous sounds, the number of frequency components making up each sound, $F0$s, and amplitude by modeling the signal as a weighted sum of sound waveforms in the time domain and applying the Markov chain Monte Carlo (MCMC) algorithm. Cemgil, Kappen, and Barber [7] proposed a method for estimating notes, tempo, and waveforms by associating them with a graphical model that models the waveform generation process when performing a musical score at a certain (local) tempo. Kameoka, Nishimoto, and Sagayama [54] proposed a method that formalizes the problem as the clustering of frequency components under harmonic-structure constraints and determines the number of clusters (sound sources) that minimizes the Akaike Information Criterion (AIC) so as to estimate the median ($F0$) and weight (amplitude) of each cluster. They also proposed a method that extends the basis concept of the harmonic-structure modeling of the PreFEst to temporal modeling and estimates, by using MAP estimation, model parameters consisting of the temporal curve of a power envelope, onset time, and duration, as well as $F0$, amplitude, and the shape of the harmonic structure of each auditory object [55].

8.2.2 Difficulties Associated with Musical Audio Signals

Every music-scene-description method assumes specific properties of the input musical audio signals because the difficulty of estimating music scene descriptions depends on various properties of musical signals. These include the number of channels (monaural or stereo), the number of simultaneous sounds in the mixture (monophonic or polyphonic), and the music genre.

In general, most methods deal with monaural audio signals because stereo signals on CDs can be easily converted to monaural signals by averaging the left and right channels. Although a method depending on stereo information cannot be applied to monaural signals, a method assuming monaural signals can be applied to stereo signals and be considered essential to music understanding, since human listeners have no difficulty understanding the above descriptions even from monaural signals. In particular, the difficulty of estimating melody and bass lines depends on the number of channels; the estimation for stereo audio signals is easier than the estimation for monaural audio signals because the sounds of those lines are usually panned to the center of stereo recordings and the localization information can help the estimation.

Most music-scene-description methods also assume polyphonic sound mixtures because CD recordings usually contain simultaneous sounds of various instruments. A method that works only with monophonic signals, which is much simpler and easier, cannot be applied to polyphonic signals in general because frequency components of several sounds often overlap and rhythms and music structures are more complex in polyphonic music. In fact, even state-of-the-art technologies cannot separate sound sources and transcribe musical scores from polyphonic CD recordings. Most music-scene-description methods, therefore, do not rely on separated sounds and scores but directly extract intuitively meaningful descriptions from polyphonic music.

Since unique difficulties arise with each music genre, there is no universal method that can deal with all genres. For example, methods of estimating melody and bass lines usually deal only with music that has distinct melody and bass lines, such as popular songs. The difficulty of estimating the beat structure depends on how explicitly the beat structure is expressed in the target music; it depends on temporal properties such as tempo changes and deviations, rhythmic complexity, and the presence of drum sounds (e.g., beat tracking for expressive classical works is more difficult than that for popular music with drum sounds). Most methods of estimating chorus sections deal with only popular music, which has distinct repeated choruses.

8.3 ESTIMATING MELODY AND BASS LINES

The detection of melody and bass lines is important because the melody forms the core of Western music and is very influential in the identity of a musical piece, whereas the bass is closely related to the tonality. These lines are fundamental to the perception of music by both trained and untrained listeners. They are also useful in various applications such as automatic music indexing for query by humming (QBH) [53, 31, 92, 70, 91, 76, 93, 89, 51, 11], which enables a user to retrieve a musical piece by humming or singing its melody; computer participation in live human performances; musical performance analysis of outstanding recorded performances; and automatic production of accompaniment tracks for karaoke using CDs.

It is difficult to estimate the fundamental frequency ($F0$) of melody and bass lines in monaural sound mixtures from CD recordings. Most previous $F0$ estimation methods cannot be applied to this estimation because they require that the input audio signal contains just a single-pitch sound with aperiodic noise, or that the number of simultaneous sounds be known beforehand. The main reason $F0$ estimation in sound mixtures is difficult is that in the time–frequency domain the frequency components of one sound often overlap the frequency components of simultaneous sounds. In popular music, for example, part of the voice's harmonic structure is often overlapped by harmonics of the keyboard instrument or guitar, by higher harmonics of the bass guitar, and by noisy inharmonic frequency components of the snare drum. A simple method of locally tracing a frequency component is, therefore, neither reliable nor stable. Moreover, sophisticated $F0$ estimation methods relying on the existence of the $F0$'s frequency component (the frequency component corresponding to the $F0$) not only cannot handle the missing fundamental, but are also unreliable when the $F0$'s frequency component is smeared by the harmonics of simultaneous sounds.

The $F0$ estimation of melody and bass lines in CD recordings was first achieved in 1999 by Goto [41, 38]. Goto proposed a real-time method called PreFEst (<u>Pre</u>dominant-<u>F</u>0 <u>Est</u>imation method), which detects the melody and bass lines in monaural sound mixtures. Unlike previous $F0$ estimation methods, PreFEst does not assume the number of sound sources, locally trace frequency components, or even rely on the existence of the $F0$'s frequency component. PreFEst basically estimates the $F0$ of the most predominant harmonic structure—the most predominant $F0$ corresponding to the melody or bass line—within an intentionally limited frequency range of the input mixture. It simultaneously takes into consideration all possibilities for the $F0$ and treats the input mixture as if it contains all possible harmonic structures with different weights (amplitudes). It regards a probability density function (PDF) of the input frequency components as a weighted mixture of harmonic-structure tone models (represented by PDFs) of all possible $F0$s and simultaneously estimates both their weights corresponding to the relative dominance of every possible harmonic structure, and the shape of the tone models by maximum a posteriori probability (MAP) estimation considering their prior distribution. It then considers the maximum-weight model as the most predominant harmonic structure and obtains its $F0$. The method also considers the $F0$'s temporal continuity by using a multiple-agent architecture.

The following sections first explain the PreFEst method in detail and then introduce other methods for detecting the melody line developed by Paiva, Mendes, and Cardoso [74, 73], Marolt [67, 68], Eggink and Brown [25], and Li and Wang [64], and a method for detecting the bass line developed by Hainsworth and Macleod [49]. Figure 8.2 shows an overview of PreFEst. PreFEst consists of three compo-

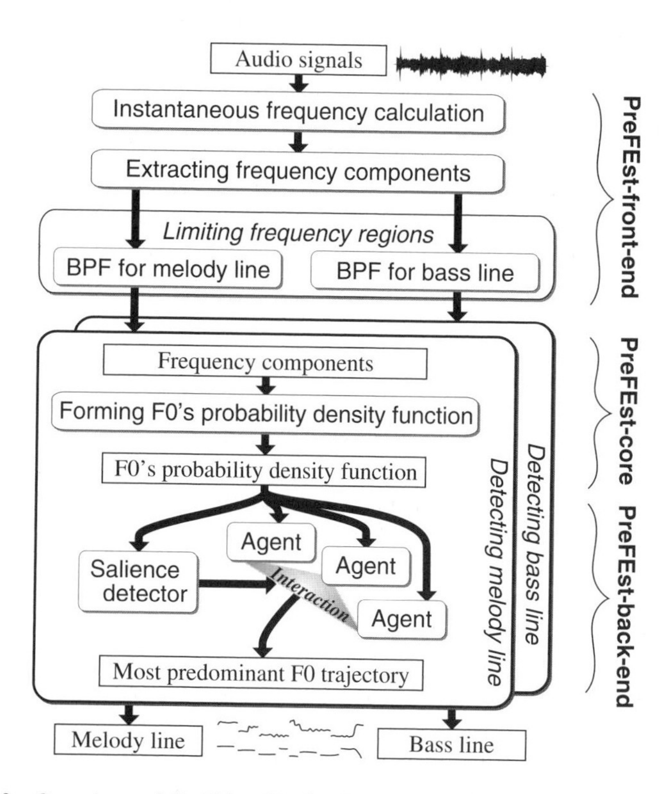

Figure 8.2 Overview of PreFEst (Predominant-F0 Estimation method) for detecting melody and base lines in CD recordings.

nents, the PreFEst-front-end for frequency analysis, the PreFEst-core to estimate the predominant $F0$, and the PreFEst-back-end to evaluate the temporal continuity of the $F0$. Since the melody line tends to have the most predominant harmonic structure in middle- and high-frequency regions and the bass line tends to have the most predominant harmonic structure in a low-frequency region, the $F0$s of the melody and bass lines can be estimated by applying the PreFEst-core with appropriate frequency-range limitation.

8.3.1 PreFEst-front-end: Forming the Observed Probability Density Functions

The PreFEst-front-end first uses a multirate filterbank [38] to obtain adequate time–frequency resolution under a real-time constraint. It then extracts frequency components by using an instantaneous-frequency-related measure and obtains two sets of band-pass-filtered frequency components, one for the melody line (261.6–4186 Hz) and the other for the bass line (32.7–261.6 Hz). To enable the application of statistical methods, each set of the band-pass-filtered components is represented as a probability density function (PDF), called an *observed PDF*. The observed PDF is denoted by $p_\Psi^{(t)}(x)$, where t is the time measured in units of frame shifts (10 ms), and x is the log-scale frequency measured in units of cents (a musical-interval measurement). Here, frequency f_{Hz} in hertz is converted to frequency f_{cent} in cents so that there are 100 cents to a tempered semitone and 1200 to an octave:

$$f_{\text{cent}} = 1200 \log_2 \frac{f_{\text{Hz}}}{440 \times 2^{(3/12)-5}} \tag{8.1}$$

8.3.2 PreFEst-core: Estimating the *F0*'s Probability Density Function

For each set of filtered frequency components represented as an observed PDF $p_\Psi^{(t)}(x)$, the PreFEst-core forms a probability density function of the $F0$, called the $F0$'s PDF, $p_{F0}^{(t)}(F)$, where F is the log-scale frequency in cents. The PreFEst-core considers each observed PDF to have been generated from a weighted-mixture model of the tone models of all possible $F0$s; the tone model is the PDF corresponding to a typical harmonic structure and indicates where the harmonics of the $F0$ tend to occur. Because the weights of tone models represent the relative dominance of every possible harmonic structure, these weights can be regarded as the $F0$'s PDF; the more dominant a tone model is in the mixture, the higher the probability of the $F0$ of its model.

8.3.2.1 *Weighted-Mixture Model of Adaptive Tone Models.* To deal with diversity of the harmonic structure, the PreFEst-core can use several types of harmonic-structure tone models. The PDF of the mth tone model for each $F0$ F is denoted by $p[x|F, m, \mu^{(t)}(F, m)]$ (Fig. 8.3), where the model parameter $\mu^{(t)}(F, m)$ represents the shape of the tone model. The number of tone models is M_i $(m = 1, \ldots, \text{M}_i)$, where

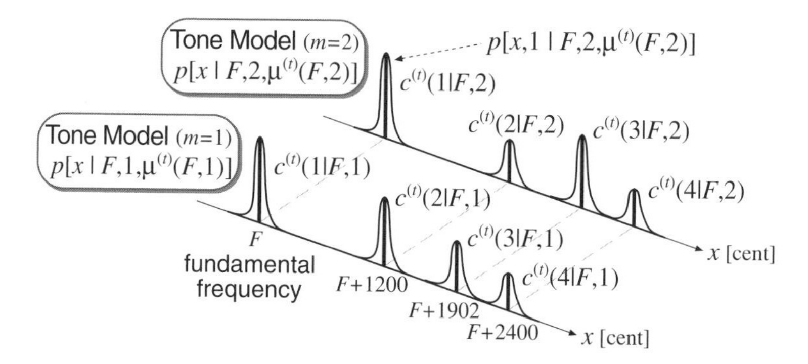

Figure 8.3 Model parameters of multiple adaptive tone models $p[x|F, m, \mu^{(t)}(F, m)]$.

i denotes the melody line (i = m) or the bass line (i = b). Each tone model is defined by

$$p[x|F, m, \mu^{(t)}(F, m)] = \sum_{h=1}^{H_i} p[x, h|F, m, \mu^{(t)}(F, m)] \tag{8.2}$$

$$p[x, h|F, m, \mu^{(t)}(F, m)] = c^{(t)}(h|F, m)\, G(x; F + 1200 \log_2 h, W_i) \tag{8.3}$$

$$\mu^{(t)}(F, m) = \{c^{(t)}(h|F, m) \mid h = 1, \ldots, H_i\} \tag{8.4}$$

$$G(x; x_0, \sigma) = \frac{1}{\sqrt{2\pi\sigma^2}}\, e^{-(x-x_0)^2/2\sigma^2} \tag{8.5}$$

where H_i is the number of harmonics considered, W_i is the standard deviation σ of the Gaussian distribution $G(x; x_0, \sigma)$, and $c^{(t)}(h|F, m)$ determines the relative amplitude of the hth harmonic component (the shape of the tone model) and satisfies

$$\sum_{h=1}^{H_i} c^{(t)}(h|F, m) = 1 \tag{8.6}$$

In short, this tone model places a weighted Gaussian distribution at the position of each harmonic component.*

The PreFEst-core then considers the observed PDF $p_{\Psi}^{(t)}(x)$ to have been generated from the following model $p(x|\theta^{(t)})$, which is a weighted mixture of all possible tone models $p[x|F, m, \mu^{(t)}(F, m)]$:

*Although only harmonic-structure tone models are described here, PreFEst can be applied to any weighted mixture of arbitrary tone models (even if their components are inharmonic) by simply replacing Eq. 8.3 with $p[x, h|F, m, \mu^{(t)}(F, m)] = c^{(t)}(h|F, m)p_{\text{arbit}}(x; F, h, m)$, where $p_{\text{arbit}}(x; F, h, m)$ is an arbitrary PDF (h is merely the component number in this case). Even with this general tone model, in theory the F0's PDF can be estimated by using the same Eqs. 8.19 and 8.20. Both any harmonic- and inharmonic-structure tone models can also be used together.

$$p(x|\theta^{(t)}) = \int_{\mathrm{Fl}_i}^{\mathrm{Fh}_i} \sum_{m=1}^{\mathrm{M}_i} w^{(t)}(F, m)\, p[x|F, m, \mu^{(t)}(F, m)]\, dF \tag{8.7}$$

$$\theta^{(t)} = \{w^{(t)}, \mu^{(t)}\} \tag{8.8}$$

$$w^{(t)} = \{w^{(t)}(F, m) \mid Fl_i \leq F \leq Fh_i, m = 1, \ldots, \mathrm{M}_i\} \tag{8.9}$$

$$\mu^{(t)} = \{\mu^{(t)}(F, m) \mid Fl_i \leq F \leq Fh_i, m = 1, \ldots, \mathrm{M}_i\} \tag{8.10}$$

where Fl_i and Fh_i denote the lower and upper limits of the possible (allowable) $F0$ range and $w^{(t)}(F, m)$ is the weight of a tone model $p[x|F, m, \mu^{(t)}(F, m)]$ that satisfies

$$\int_{Fl_i}^{Fh_i} \sum_{m=1}^{\mathrm{M}_i} w^{(t)}(F, m)\, dF = 1 \tag{8.11}$$

Because the number of sound sources cannot be known a priori, it is important to simultaneously take into consideration all $F0$ possibilities as expressed in Eq. 8.7. If it is possible to estimate the model parameter $\theta^{(t)}$ such that the observed PDF $p_\Psi^{(t)}(x)$ is likely to have been generated from the model $p(x|\theta^{(t)})$, the weight $w^{(t)}(F, m)$ can be interpreted as the $F0$'s PDF $p_{F0}^{(t)}(F)$:

$$p_{F0}^{(t)}(F) = \sum_{m=1}^{\mathrm{M}_i} w^{(t)}(F, m) \qquad (\mathrm{Fl}_i \leq F \leq \mathrm{Fh}_i) \tag{8.12}$$

8.3.2.2 Introducing a Prior Distribution. To use prior knowledge about $F0$ estimates and the tone-model shapes, a prior distribution $p_{0i}(\theta^{(t)})$ of $\theta^{(t)}$ is defined as follows:

$$p_{0i}(\theta^{(t)}) = p_{0i}(w^{(t)}) p_{0i}(\mu^{(t)}) \tag{8.13}$$

$$p_{0i}(w^{(t)}) = \frac{1}{Z_w}\, e^{-\beta_{wi}^{(t)} D_w(w_{0i}^{(t)}; w^{(t)})} \tag{8.14}$$

$$p_{0i}(\mu^{(t)}) = \frac{1}{Z_\mu}\, e^{-\int_{\mathrm{Fl}_i}^{\mathrm{Fh}_i} \sum_{m=1}^{\mathrm{M}_i} \beta_{\mu i}^{(t)}(F,m) D_\mu[\mu_{0i}^{(t)}(F,m); \mu^{(t)}(F,m)] dF} \tag{8.15}$$

Here, $p_{0i}(w^{(t)})$ and $p_{0i}(\mu^{(t)})$ are unimodal distributions: $p_{0i}(w^{(t)})$ takes its maximum value at $w_{0i}^{(t)}(F, m)$ and $p_{0i}(\mu^{(t)})$ takes its maximum value at $\mu_{0i}^{(t)}(F, m)$, where $w_{0i}^{(t)}(F, m)$ and $\mu_{0i}^{(t)}(F, m)$ [$c_{0i}^{(t)}(h|F, m)$] are the most probable parameters. Z_w and Z_μ are normalization factors, and $\beta_{wi}^{(t)}$ and $\beta_{\mu i}^{(t)}(F, m)$ are parameters determining how much emphasis is put on the maximum value. The prior distribution is uniform when $\beta_{wi}^{(t)}$ and $\beta_{\mu i}^{(t)}(F, m)$ are 0. In Eqs. 8.14 and 8.15, $D_w(w_{0i}^{(t)}; w^{(t)})$ and $D_\mu[\mu_{0i}^{(t)}(F, m); \mu^{(t)}(F, m)]$ are the following Kullback–Leibler information:

$$D_w(w_{0i}^{(t)}; w^{(t)}) = \int_{\mathrm{Fl}_i}^{\mathrm{Fh}_i} \sum_{m=1}^{\mathrm{M}_i} w_{0i}^{(t)}(F, m) \log \frac{w_{0i}^{(t)}(F, m)}{w^{(t)}(F, m)} \, dF \tag{8.16}$$

$$D_\mu[\mu_{0i}^{(t)}(F, m); \mu^{(t)}(F, m)] = \sum_{h=1}^{\mathrm{H}_i} c_{0i}^{(t)}(h|F, m) \log \frac{c_{0i}^{(t)}(h|F, m)}{c^{(t)}(h|F, m)} \tag{8.17}$$

8.3.2.3 MAP Estimation Using the EM Algorithm. The problem to be solved is to estimate the model parameter $\theta^{(t)}$, taking into account the prior distribution $p_{0i}(\theta^{(t)})$, when $p_\Psi^{(t)}(x)$ is observed. The MAP estimator of $\theta^{(t)}$ is obtained by maximizing

$$\int_{-\infty}^{\infty} p_\Psi^{(t)}(x) \, [\log p(x|\theta^{(t)}) + \log p_{0i}(\theta^{(t)})] \, dx \tag{8.18}$$

Because this maximization problem is too difficult to solve analytically, the PreFEst-core uses the Expectation–Maximization (EM) algorithm [17], which is an algorithm in which two steps—the expectation step (E-step) and the maximization step (M-step)—are iteratively applied to compute MAP estimates from incomplete observed data [i.e., from $p_\Psi^{(t)}(x)$]. With respect to $\theta^{(t)}$, each iteration updates the old estimate $\theta'^{(t)} = \{w'^{(t)}, \mu'^{(t)}\}$ to obtain a new (improved) estimate $\overline{\theta^{(t)}} = \{\overline{w^{(t)}}, \overline{\mu^{(t)}}\}$ by using the following equations [38]:

$$\overline{w^{(t)}(F, m)} = \frac{\overline{w_{\mathrm{ML}}^{(t)}(F, m)} + \beta_{wi}^{(t)} w_{0i}^{(t)}(F, m)}{1 + \beta_{wi}^{(t)}} \tag{8.19}$$

$$\overline{c^{(t)}(h|F, m)} = \frac{\overline{w_{\mathrm{ML}}^{(t)}(F, m)}\; \overline{c_{\mathrm{ML}}^{(t)}(h|F, m)} + \beta_{\mu i}^{(t)}(F, m) c_{0i}^{(t)}(h|F, m)}{\overline{w_{\mathrm{ML}}^{(t)}(F, m)} + \beta_{\mu i}^{(t)}(F, m)} \tag{8.20}$$

where $\overline{w_{\mathrm{ML}}^{(t)}(F, m)}$ and $\overline{c_{\mathrm{ML}}^{(t)}(h|F, m)}$ are, when the noninformative prior distribution [$\beta_{wi}^{(t)} = 0$ and $\beta_{\mu i}^{(t)}(F, m) = 0$] is given, the following maximum likelihood estimates:

$$\overline{w_{\mathrm{ML}}^{(t)}(F, m)} = \int_{-\infty}^{\infty} p_\Psi^{(t)}(x) \frac{w'^{(t)}(F, m) \, p[x|F, m, \mu'^{(t)}(F, m)]}{\int_{\mathrm{Fl}_i}^{\mathrm{Fh}_i} \sum_{\nu=1}^{\mathrm{M}_i} w'^{(t)}(\eta, \nu) \, p[x|\eta, \nu, \mu'^{(t)}(\eta, \nu)] \, d\eta} \, dx \tag{8.21}$$

$$\overline{c_{\mathrm{ML}}^{(t)}(h|F, m)} = \frac{1}{\overline{w_{\mathrm{ML}}^{(t)}(F, m)}} \int_{-\infty}^{\infty} p_\Psi^{(t)}(x) \frac{w'^{(t)}(F, m) \, p[x, h|F, m, \mu'^{(t)}(F, m)]}{\int_{\mathrm{Fl}_i}^{\mathrm{Fh}_i} \sum_{\nu=1}^{\mathrm{M}_i} w'^{(t)}(\eta, \nu) \, p[x|\eta, \nu, \mu'^{(t)}(\eta, \nu)] \, d\eta} \, dx \tag{8.22}$$

For an intuitive explanation of Eq. 8.21,

$$\frac{w'^{(t)}(F, m)\, p[x|F, m, \mu'^{(t)}(F, m)]}{\int_{Fl_i}^{Fh_i} \sum_{m=1}^{M_i} w'^{(t)}(\eta, \nu)\, p[x|\eta, \nu, \mu'^{(t)}(\eta, \nu)]\, d\eta}$$

is called the decomposition filter. For the integrand on the right side of Eq. 8.21, it can be considered that, because of this filter, the value of $p_{\Psi}^{(t)}(x)$ at frequency x is decomposed into (is distributed among) all possible tone models $p[x|F, m, \mu'^{(t)}(F, m)]$ ($\mathrm{Fl}_i \leq F \leq \mathrm{Fh}_i$, $1 \leq m \leq \mathrm{M}_i$) in proportion to the numerator of the decomposition filter at x. The higher the weight $w'^{(t)}(F, m)$, the larger the decomposed value given to the corresponding tone model. Note that the value of $p_{\Psi}^{(t)}(x)$ at different x is also decomposed according to a different ratio in proportion to the numerator of the decomposition filter at that x. Finally, the updated weight $\overline{w_{\mathrm{ML}}^{(t)}(F, m)}$ is obtained by integrating all the decomposed values given to the corresponding mth tone model for the $F0$ F.

This decomposition behavior is the advantage of the weighted-mixture modeling. Comb-filter-based or autocorrelation-based multiple $F0$ estimation methods (see Section 2.4.2) cannot easily support the decomposition of an overlapping frequency component (overtone) shared by several simultaneous tones and tend to have difficulty distinguishing sounds with overlapping overtones. In addition, PreFEst can simultaneously estimate all the weights $\overline{w_{\mathrm{ML}}^{(t)}(F, m)}$ (for all the range of F) so that these weights can be optimally balanced; it does not determine the weight at F after determining the weight at another F. This simultaneous estimation of all the weights is an advantage of PreFEst compared to traditional recursive-subtraction-based multiple $F0$ estimation methods (see Chapter 2, Section 2.4.1) where components of the most dominant harmonic structure identified are subtracted from a mixture and then this is recursively done again starting from the residue of the previous subtraction. In these recursive-subtraction-based methods, once inappropriate identification or subtraction occurs, the following recursions starting from the wrong residue become unreliable.

After the above iterative computation of Eqs. 8.19 and 8.20, the $F0$'s PDF $p_{F0}^{(t)}(F)$ can be obtained from $w^{(t)}(F, m)$ according to Eq. 8.12. The tone-model shape $c^{(t)}(h|F, m)$, which is the relative amplitude of each harmonic component of all types of tone models $p[x|F, m, \mu^{(t)}(F, m)]$, can also be obtained.

8.3.3 PreFEst-back-end: Sequential *F0* Tracking by Multiple-Agent Architecture

A simple way to identify the most predominant $F0$ is to find the frequency that maximizes the $F0$'s PDF. This result is not always stable, however, because peaks corresponding to the $F0$s of simultaneous sounds sometimes compete in the $F0$'s PDF for a moment and are transiently selected, one after another, as the maximum.

The PreFEst-back-end therefore considers the global temporal continuity of the $F0$ by using a multiple-agent architecture [41] in which agents track different temporal trajectories of the $F0$. The final $F0$ output is determined on the basis of the most dominant and stable $F0$ trajectory. Figure 8.4 shows an example of the final output.

8.3.4 Other Methods

Although the PreFEst method resulted from pioneering research regarding melody and bass detection and weighted-mixture modeling for $F0$ estimation, many issues still need to be resolved. For example, if an application requires MIDI-level note sequences of the melody line, the $F0$ trajectory should be segmented and organized into notes. Note that the PreFEst method does not deal with the problem of detecting the absence of melody and bass lines; it simply estimates the predominant $F0$ without discriminating between sound sources. In addition, since the melody and bass lines are generated from a process that is statistically biased rather than random, this bias can also be incorporated into their detection. This section introduces other recent approaches that deal with these issues in describing polyphonic audio signals.

Paiva, Mendes, and Cardoso [74, 73] proposed a method of obtaining the melody note sequence by using a model of the human auditory system described in [90] as a frequency analysis front end and applying MIDI-level note tracking, segmentation, and elimination techniques. Although the techniques used differ from the PreFEst method, the basic idea that "the melody generally clearly stands out of the background" [73] is the same as the basic PreFEst concept that the $F0$ of the most predominant harmonic structure is considered the melody. The advantage of this method is that MIDI-level note sequences of the melody line are generated, whereas the output of PreFEst is a simple temporal trajectory of the $F0$. The method first estimates predominant $F0$ candidates by using correlograms that represent the periodicities in a cochleagram (auditory nerve responses of an ear model). It then forms the temporal trajectories of $F0$ candidates. It quantizes their frequencies to the closest MIDI note numbers and then tracks them according to their frequency proximity, where only one-semitone transition is considered continuous. After this tracking, $F0$ trajectories are segmented into MIDI-level note candidates by finding a sufficiently long trajectory having the same note number and by dividing it at clear local minima of its amplitude envelope. Because there still remain many inappropriate notes, it eliminates notes whose amplitude is too low, whose duration is too short, or which have harmonically related $F0$s and almost same onset and offset times. Finally, the melody note sequence is obtained by selecting the most predominant notes according to heuristic rules. Since simultaneous notes are not allowed, the method eliminates simultaneous notes that are less dominant and not in a middle frequency range.

Marolt [67, 68] proposed a method of detecting the melody line by representing it as a set of short vocal fragments of $F0$ trajectories. This method is based on the PreFEst method with some modifications; it uses the PreFEst-core to estimate

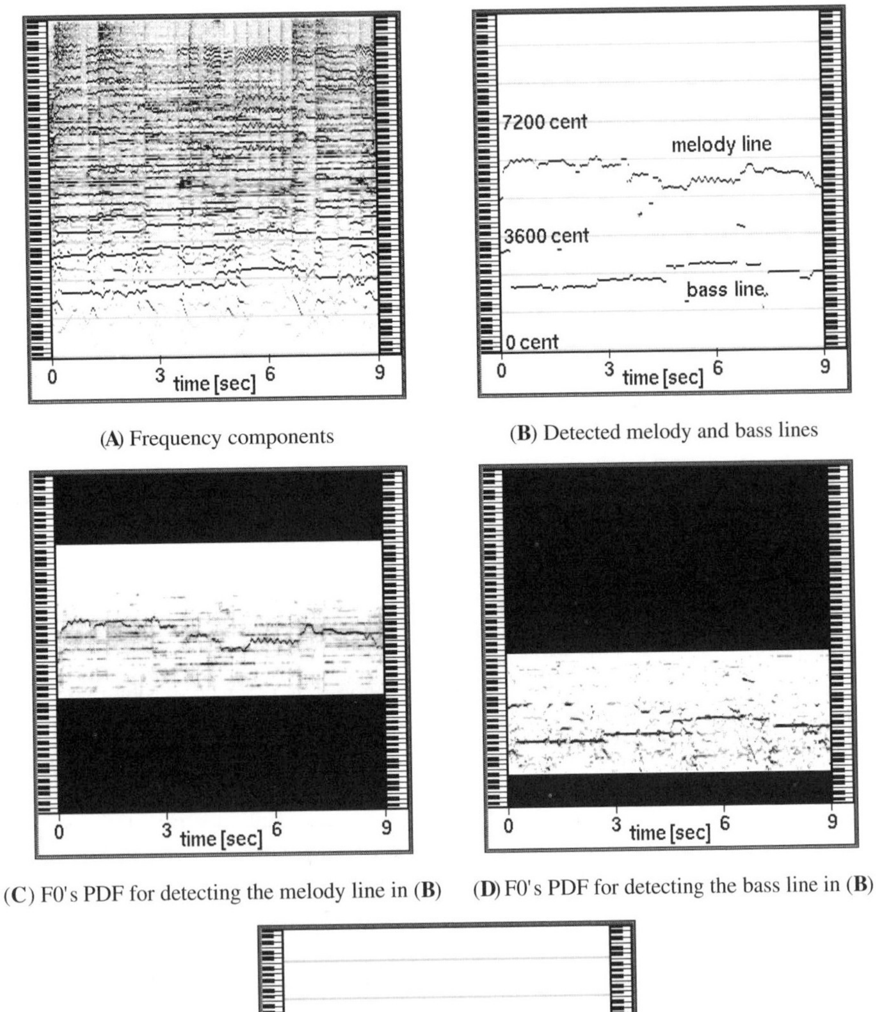

(**A**) Frequency components

(**B**) Detected melody and bass lines

(**C**) F0's PDF for detecting the melody line in (**B**)

(**D**) F0's PDF for detecting the bass line in (**B**)

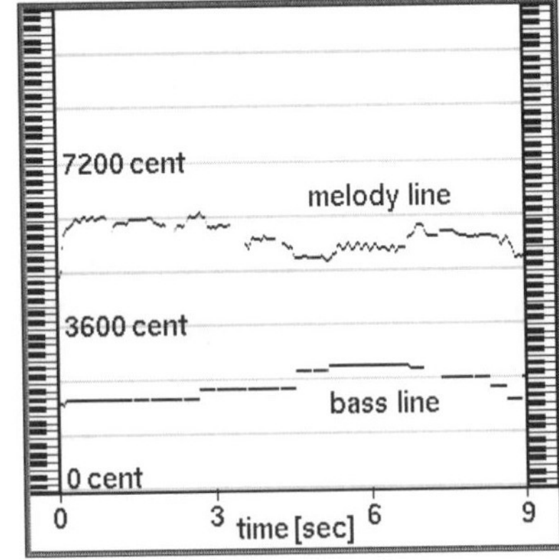

(**E**) Correct melody and bass lines hand labeled for the evaluation

Figure 8.4 Audio-synchronized real-time graphics output for a popular-music excerpt with drum sounds: (**A**) frequency components, (**B**) the corresponding melody and bass lines detected (final output), (**C**) the corresponding $F0$'s PDF estimated when detecting the melody line, (**D**) the corresponding $F0$'s PDF estimated when detecting the bass line, and (**E**) the correct $F0$'s hand labeled for evaluating the PreFEst method. These interlocking windows have the same vertical axis of log-scale frequency.

predominant $F0$ candidates, but uses spectral modeling synthesis (SMS) analysis instead of the PreFEst-front-end. The advantage of this method is that the $F0$ candidates are tracked and grouped into melodic fragments and these fragments are then clustered into the melody line. The melodic fragments denote reasonably segmented signal regions that exhibit strong and stable $F0$ and are formed by tracking temporal trajectories of the $F0$ candidates. This tracking is similar to the PreFEst-back-end. Because the fragments belong to not only the melody (lead vocal), but also to different parts of the accompaniment, they are clustered to find the melody cluster by using Gaussian mixture models (GMMs) according to their five properties:

1. Dominance (average weight of a tone model estimated by the EM algorithm)
2. Pitch (centroid of the $F0$s within the fragment)
3. Loudness (average loudness of harmonics belonging to the fragment)
4. Pitch stability (average change of $F0$s during the fragment)
5. Onset steepness (steepness of overall loudness change during the first 50 ms of the fragment)

Eggink and Brown [25] proposed a method of detecting the melody line with the emphasis on using various knowledge sources to choose the most likely succession of $F0$s as the melody line. Unlike other methods, this method is specialized for a classical sonata or concerto, where a solo melody instrument can span the whole pitch range, ranging from the low tones of a cello to a high-pitched flute, so the frequency-range limitation used in the PreFEst method is not feasible. In addition, because the solo instrument does not always have the most predominant $F0$, additional knowledge sources are necessary to extract the melody line. The main advantage of this method is the leverage provided by knowledge sources, including local knowledge about an instrument recognition module and temporal knowledge about tone durations and interval transitions, which are integrated in a probabilistic search. Those sources can both help to choose the correct $F0$ among multiple concurrent $F0$ candidates and to determine sections in which the solo instrument is actually present. The knowledge sources consist of two categories: local knowledge and temporal knowledge. The local knowledge concerning $F0$ candidates obtained by picking peaks in the spectrum includes

- $F0$ strength (the stronger the spectral peak, the higher its likelihood of being the melody)
- Instrument-dependent $F0$ likelihood (the likelihood values of an $F0$ candidate in terms of its frequency and the pitch range of each solo instrument, which are evaluated by counting the frequency of its $F0$ occurrence in different standard MIDI files)
- Instrument likelihood (the likelihood values of an $F0$ candidate being produced by each solo instrument, which are evaluated by the instrument recognition module)

The instrument recognition module uses trained Gaussian classifiers of the frequency and power of the first ten harmonic components, their deltas, and their delta-deltas [26]. On the other hand, the temporal knowledge concerning tone candidates obtained by connecting $F0$ candidates includes

- Instrument-dependent interval likelihood (the likelihood values of an interval transition between succession tones, which are evaluated by counting the frequency of its interval occurrence in different standard MIDI files)
- Relative tone usage (measures related to tone durations between successive tones, which are used to penalize overlapped tones)

These knowledge sources are combined to find the most likely "path" of the melody through the space of all $F0$ candidates in time. Since the melody path occasionally follows the accompaniment, additional post processing is done to eliminate sections where the solo instrument is actually silent.

Li and Wang [64] proposed a method of detecting the melody line of a singing voice, which extended a multiple $F0$ estimation method for noisy speech developed by Wu, Wang, and Brown [100]. This method first uses a 128 channel gammatone filterbank to simulate the cochlear filtering. The point of this method is to split these channels into two subbands: low-frequency channels below 800 Hz and high-frequency channels above 800 Hz. To extract periodicity information, an autocorrelation is then calculated on the filter output of each channel in the low-frequency channels while this is done on the output envelope of each channel in the high-frequency channels to exploit the beating phenomenon. The method then selects plausible peaks in the autocorrelation results; it retains the maximum value of the autocorrelation in the low-frequency channels when it is high enough and within an allowed $F0$ range, and retains the first peak of the autocorrelation in the high-frequency channels. Li and Wang reported that the use of all of the high-frequency channels is important in distinguishing different harmonic structures. These retained peaks are used to score a collection of $F0$ hypotheses corresponding to 0, 1, and 2 $F0$s, that is, hypotheses in three i-dimensional subspaces ($i = 0, 1, 2$), for each channel, and these scores are then integrated across the channels. The method finally tracks the most probable temporal trajectory of the predominant $F0$ in the integrated scored $F0$ hypotheses by using a trained HMM that models the $F0$ generation process.

Although the above methods deal with the melody line, Hainsworth and Macleod [49] proposed a method of obtaining the bass note sequence by maintaining multiple hypotheses. The method first extracts the onset times of bass notes by picking peaks of a smoothed temporal envelope of a total power below 200 Hz. It then generates hypotheses regarding the $F0$ of each extracted note; the $F0$ and amplitude of each hypothesis are estimated by fitting a quadratic polynomial to a large amplitude peak and subtracting it from the spectrum. The first four harmonic components of those hypotheses are tracked over time by using a comb-filter-like analysis. Finally, the method selects the most likely hypothesis for each onset on the basis of its duration and the amplitude of harmonic components, and further tidies up these hypotheses by removing inappropriate overlaps and relatively low amplitude notes.

8.4 ESTIMATING BEAT STRUCTURE

In this section, beat tracking (including measure or bar-line estimation) is defined as the process of organizing musical audio signals into a hierarchical beat structure. It is also an important initial step in the computational modeling of music understanding because the beat is fundamental, for both trained and untrained listeners, to the perception of Western music. There are many applications, such as music-synchronized computer graphics, stage-lighting control, video/audio editing, intelligent computer accompaniment, and human–computer improvisation in live ensembles.

Beat tracking and measure estimation were researched in the 1980s using MIDI signals or score-level information [14, 18, 19, 1, 22, 80, 81, 82, 62], and Goto and Muraoka began a series of studies [43, 32, 45, 33] on audio-based, real-time beat tracking for CD recordings in 1994. They developed a system that can track beats by leveraging musical knowledge about chord changes and drum patterns as well as onset times under the assumption of a nearly constant tempo. Then, taking a hint from these studies, Scheirer [84] proposed an audio-based beat-tracking method that could accommodate changes in tempo by using comb-filter resonators without detecting onset times. Todd and Brown [96] also proposed a bottom-up beat-tracking method that visualizes a hierarchical rhythmic structure by using a multiscale smoothing model applied to onsets that were detected by a model of human auditory periphery. At the beginning of this century, a variety of methods having even fewer restrictions were proposed [21, 63, 86, 98, 47, 46, 52, 48, 50, 2, 15, 61, 60]. Methods targeting MIDI signals [79, 6, 94, 95, 75] have also progressed significantly.

As depicted in Figure 8.1, the hierarchical beat structure comprises the *quarter-note level* (the tactus level represented as almost regularly spaced beat times) and the *measure level* (bar lines).* The basic nature of tracking the quarter-note level is represented by two parameters: period and phase. The period is the temporal difference between the times of two successive beats and is also called the *interbeat interval (IBI)*. Hence, it is inversely proportional to the tempo, which is defined as the number of beats per minute. The phase corresponds to actual beat positions and equals zero at beat times. The measure level is defined on beat times because the beat structure is hierarchical: the beginnings of measures (bar-line positions) are labeled on beat times. The goal of beat tracking is to determine all beat times and the beginnings of measures while estimating the time-varying period and phase of the quarter-note level.

The main problems in recognizing the beat structure in CD recordings can be summarized as:

- **Problem 1. Estimating the period and phase by using cues in audio sig-**

*Although music scene description does not rely on score representation, for convenience score-representation terminology is used here. In this formulation, the quarter-note level indicates the temporal basic unit that a human feels in music and that usually corresponds to quarter notes in scores.

nals. It is necessary to extract beat-tracking cues from the input audio signals by using signal-processing techniques and then estimate the period and phase of the quarter-note level. Although various cues such as onset times, notes, melodies, chords, and repetitive note patterns have been used in score-based or MIDI-based beat tracking [14, 18, 19, 1, 22, 80, 81, 82, 62], most of those cues are hard to precisely detect in polyphonic audio signals.

- **Problem 2. Dealing with ambiguity of interpretation.** The intrinsic reason that beat tracking is difficult is due to the problem of inferring an original beat structure that is not expressed explicitly. This leads to various ambiguous situations, such as ones in which different periods (IBIs) seem plausible and where several onset times obtained by frequency analysis may correspond to a beat. Therefore, it must be taken into account that multiple interpretations of beats are possible at any given time.
- **Problem 3. Using musical knowledge to make musical decisions.** Higher-level processing using musical knowledge is necessary to determine the measure level of the hierarchical beat structure. Musical knowledge is also useful for selecting the best hypothesis in the above ambiguous situations.

The following sections describe solutions to these problems by introducing various beat-tracking approaches.

8.4.1 Estimating Period and Phase

The basic approach is to detect onset times and use them as cues for estimating the period and phase. Many methods rely on the fact that a frequent *interonset interval (IOI),* the temporal difference between two onset times, is likely to be the beat period and onset times tend to coincide with beat times (i.e., sounds are likely to occur on beats).

To estimate the beat period, a simple technique is to calculate the histogram of IOIs between two adjacent onset times [43, 47, 52] or cluster the IOIs [21] and pick out the maximum peak or the top-ranked cluster within an appropriate tempo range. This does not necessarily correspond to the beat period, though, because the temporal difference between adjacent onset times could correspond to eighth-note or sixteenth-note intervals. Although this can be improved by considering the IOI between alternate (every other) or every third onset times, this approach is somewhat ad hoc. A more sophisticated technique is to calculate a windowed autocorrelation function of an onset-time sequence [44, 45, 33], power envelope of the input signal [46], or continuous onset representation with peaks at onset positions [15, 2], and pick out peaks in the result. This can be considered an extended version of the IOI histogram because it naturally takes into account various temporal differences such as those between adjacent, alternate, and every third onset times. Another sophisticated technique is to apply a set of comb-filter resonators, each tuned to a possible period, to the time-varying degree of musical accentuation [84]. Since the degree of musical accentuation can be measured by a half-wave rectified differential of the power envelope of the input signal, this technique does not require onset-time detection.

For audio-based beat tracking, it is essential to split the full frequency band of the input audio signal into several frequency subbands and calculate periodicities in each subband. Goto and Muraoka [44, 45, 33] proposed a method in which the beat-period analysis is first performed within logarithmically equally spaced subbands and those results are then combined across the subbands by using a weighted sum. They introduced a seven-dimensional onset-time vector whose dimensions correspond to the onset times of seven subbands as shown in Figure 8.5. Each onset time is obtained by picking peaks of the summation of the degree of onset within each subband. The degree of onset is the speed of a local increase in the magnitude spectrum, calculated by considering the magnitude present in nearby time–frequency regions. The result of autocorrelation of the onset times within each dimension is finally combined across all the dimensions. Scheirer [84] also used the idea of this subband-based beat tracking and confirmed its effectiveness

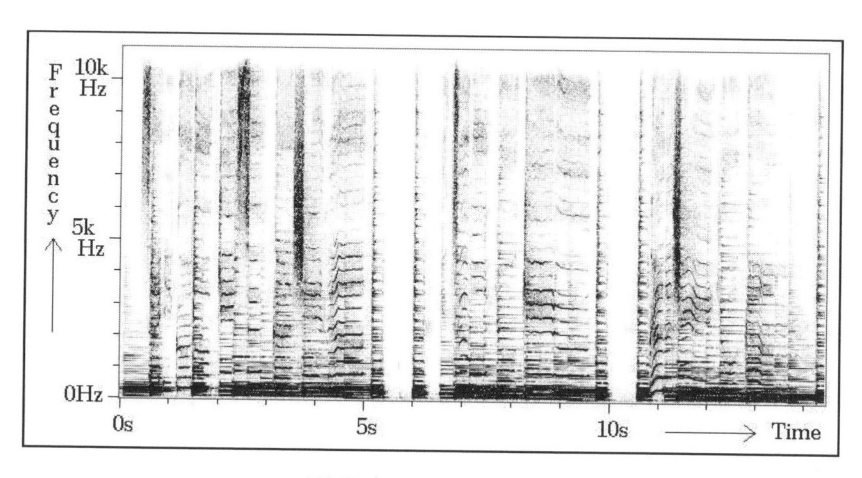

(**A**) Frequency spectrum

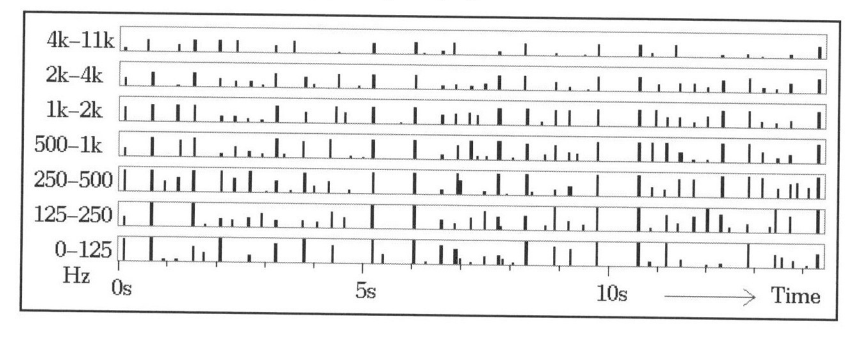

(**B**) Onset-time vector sequence

Figure 8.5 Example of a frequency spectrum and a seven-dimensional onset-time vector sequence, which represents onset times detected within each frequency subband.

through a psychoacoustic demonstration using amplitude modulated noise. Instead of detecting onset times, Scheirer applied a set of comb-filter resonators to the degrees of musical accentuation (half-wave rectified differentials of power envelopes) of six subbands to find the most resonant period. To locate periodicity in subband signals, Sethares and Staley [86] used a periodicity transform instead of using comb-filter resonators. The usefulness of using subbands has also been confirmed by other studies [98, 46]. Klapuri, Eronen, and Astola [60] combine the advantages of both subband-based approaches of Goto and Muraoka [44, 45, 33] and Scheirer [84]. The speed of a local increase in the magnitude spectrum is evaluated by using the degrees of musical accentuation obtained using 36 narrow-band subbands and a dynamic power-envelope compression technique, and then linearly summing them to obtain the degrees of musical accentuation for nine subbands, which are then used for the comb-filtering resonators. Klapuri, Eronen, and Astola [60] pointed out that the key problems are in measuring the degree of musical accentuation and in modeling higher-level musical knowledge, not in finding the exactly correct period estimator.

After estimating the beat period, the phase should be estimated. When onset times are used to estimate the period, a windowed cross-correlation function is applied between an onset-time sequence and a tentative beat-time sequence whose interval is the estimated period. The result can be used to predict the next beat in a real-time beat-tracking system [44, 45, 33]. A similar cross-correlation function has also been applied to continuous onset representation with peaks at onset positions [15, 2]. On the other hand, when the degrees of musical accentuation are used, the internal state of the delays of comb-filter resonators that have lattices of delay-and-hold stages can be used to determine the phase [84].

8.4.2 Dealing with Ambiguity

There are variations in how ambiguous situations in determining the beat structure are managed. A traditional approach is to maintain multiple hypotheses, each having a different possible set of period and phase. For MIDI-based beat tracking, a beam search technique [1, 80, 81] was proposed to maintain multiple hypotheses. For audio-based beam tracking, multiple-agent architectures [43, 45, 33, 21] have been proposed to maintain hypotheses in more active and efficient ways. As illustrated in Figure 8.6, multiple agents examine various hypotheses of the beat structure in parallel according to different strategies. Each agent estimates the beat period and phase described in Section 8.4.1 as well as the measure level while interacting with another agent to track beats cooperatively and adapting to the current situation by adjusting its strategy. Even if some agents lose track of beats, correct beats can be tracked as long as other agents have the correct hypothesis. The reliability of each hypothesis is evaluated and the final beat-tracking result is determined based on the most reliable hypothesis. While the number of hypotheses is fixed in a multiple-agent architecture developed by Goto and Muraoka [45, 33], it is not fixed and hypotheses are branched in a multiple-agent architecture developed by Dixon [21].

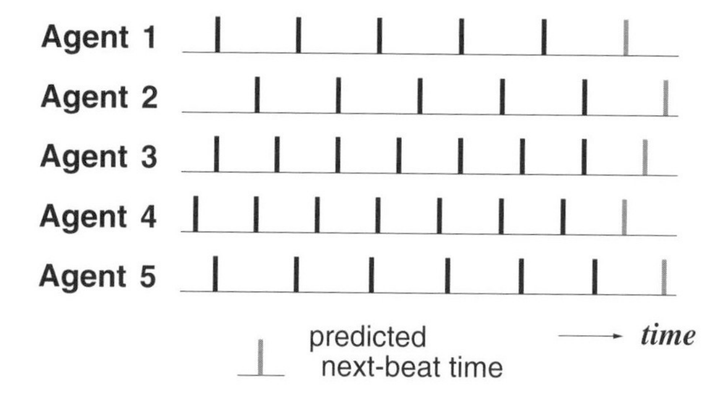

Figure 8.6 Multiple hypotheses maintained by multiple agents.

A more advanced, computationally intensive approach to examine multiple hypotheses is to use probabilistic generative models and estimate their parameters. Probabilistic approaches with maximum likelihood estimation, MAP estimation, and Bayesian estimation could maintain distributions of all parameters, such as the beat period and phase, and find the best hypothesis as if all possible pairs of the period and phase were evaluated simultaneously. For example, Hainsworth and Macleod [48, 50] explored the use of particle filters for audio-based beat tracking on the basis of MIDI-based methods by Cemgil and Kappen [6]. They made use of Markov chain Monte Carlo (MCMC) algorithms and sequential Monte Carlo algorithms (particle filters) to estimate model parameters, such as the period and phase. Lang and de Freitas [61] estimated the period and phase of a simple graphical model by using MAP estimation. In addition to the period and phase, Laroche [63] added a "swing" parameter representing a slight delay of the second and fourth quarter beats* and simultaneously estimated them by using maximum likelihood estimation with a simple model representing onset positions in a measure.

8.4.3 Using Musical Knowledge

Because the reliability of each hypothesis and the estimation of the measure level depend on the musical context, various kinds of musical knowledge have been studied. Goto and Muraoka [43, 45, 33] used musical knowledge concerning chord changes and drum patterns with a focus on popular music. On the other hand, Klapuri, Eronen, and Astola [60] used musical knowledge concerning temporal relationship among different levels of the hierarchical beat structure and encoded this prior knowledge in HMMs that could jointly estimate periods at different hierarchical levels and then separately estimate their phases.

The use of musical knowledge regarding chord changes and drum patterns requires the estimation of chord changes and drum patterns, but this estimation is dif-

*Here, one beat is subdivided into four equal quarter beats without a swing parameter.

ficult when using only a bottom-up frequency analysis without referring to the beat structure. A real-time beat-tracking system developed by Goto and Muraoka [43, 45, 33] addressed this issue by leveraging the integration of top-down and bottom-up processes as depicted in Figure 8.7. The system first obtains multiple hypotheses of provisional beat times (a quarter-note-level beat structure) on the basis of onset times without using musical knowledge regarding chord changes or drum patterns. The system makes use of the provisional beat times of each hypothesis as top-down information to detect chord changes in a frequency spectrum obtained by a bottom-up frequency analysis, without identifying musical notes or chords by name. The frequency spectrum is sliced into strips at the beat times and the dominant frequencies of each strip are estimated by using a histogram of frequency components in the strip. Chords are considered to be changed when the dominant frequencies change between adjacent strips as shown in Figure 8.8 [44, 45]. The idea for the above method corresponds to the observation that a listener who cannot identify chord names can nevertheless perceive chord changes. On the other hand, the sys-

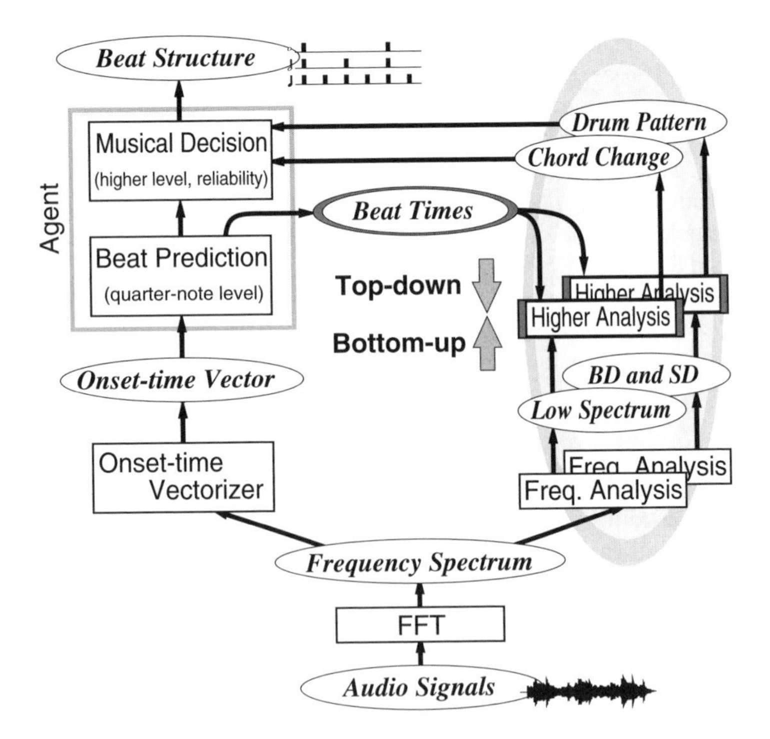

Figure 8.7 Synergy between the estimation of the hierarchical beat structure, chord changes, and drum patterns. Chord changes and drum patterns are obtained, at "Higher Analysis" in the figure, by using provisional beat times as top-down information. The hierarchical beat structure is then estimated at "Musical Decision" by using the chord changes and drum patterns.

tem makes use of the provisional beat times as top-down information to form the onset times of drums into drum patterns whose grid is aligned with the beat times as shown in Figure 8.9. The onset times of the sounds of bass drum (BD) and snare drum (SD) can be detected through bottom-up frequency analyses [105, 27, 103, 83, 20] such as those using onset and noise components in the frequency spectrum [43, 33].

After the chord changes and drum patterns are obtained, components of the higher-level beat structure, such as the measure level, can be estimated by using musical knowledge about them. The following musical knowledge about chord changes can be used under the assumption that the time signature of an input song is 4/4: chords are more likely to change on beat times than at other positions between two successive correct beats, change more on half-note times than at other positions of beat times, and change more at the beginnings of measures than at other positions of half-note times. For music with drum sounds, several drum patterns, like those illustrated in Figure 8.10, are prestored, and the beginning of each pattern corresponds to the beginning of a measure or a half measure (half-note time). When an input drum pattern that is currently detected in the input signal closely matches one of the prestored drum patterns, the following knowledge can be used: the beginning of the input drum

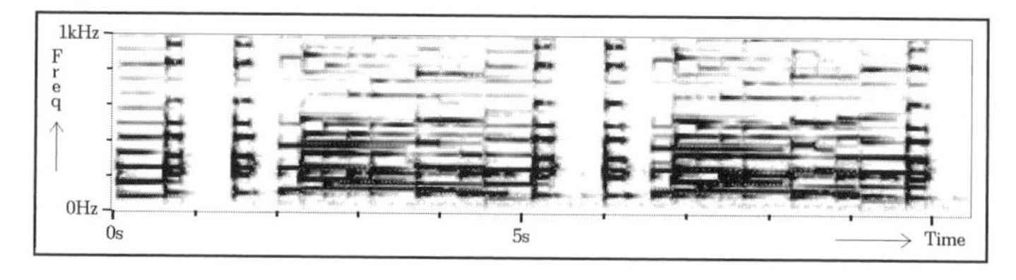

(**A**) Frequency spectrum

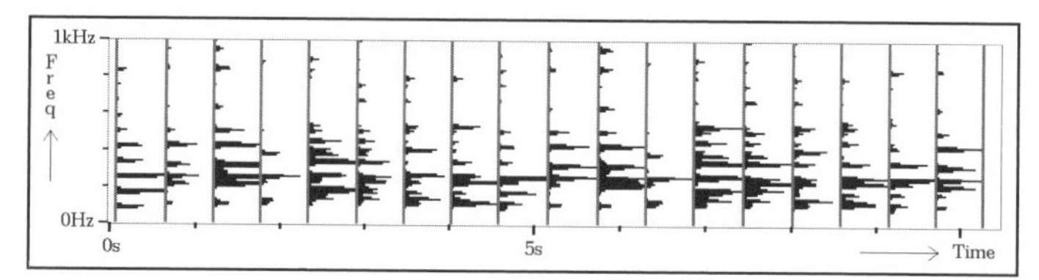

(**B**) Histograms of frequency components in spectrum strips sliced at beat times

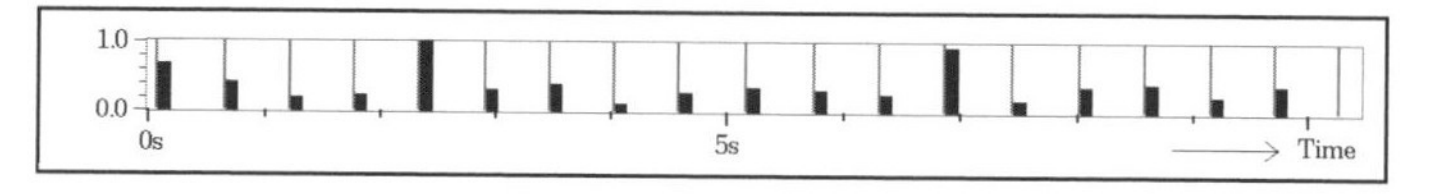

(**C**) Quarter-note chord-change possibilities

Figure 8.8 Example of detecting chord changes based on provisional beat times. A chord change is represented by a chord-change possibility that indicates how much the dominant frequency components included in chord tones and their harmonic overtones change in a frequency spectrum.

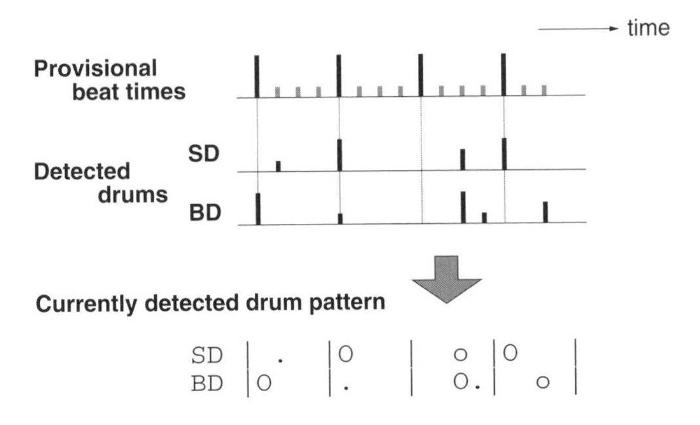

Figure 8.9 Example of forming a drum pattern by making use of provisional beat times. A drum pattern is represented by the temporal pattern of a bass drum and a snare drum.

pattern corresponds to the beginning of a half measure (half-note time). Figure 8.11 shows a sketch of how the half-note and measure times are determined based on the chord changes and drum patterns. The reliability of each hypothesis can then be evaluated according to how well the chord changes and drum patterns obtained using the provisional beat times of a hypothesis conform to the musical knowledge.

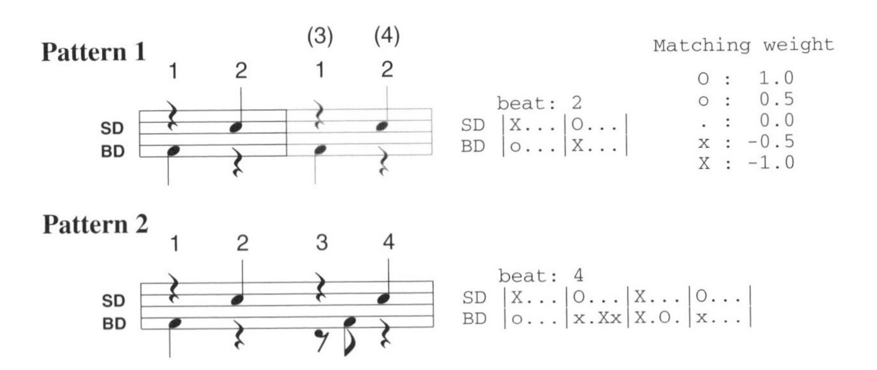

Figure 8.10 Examples of prestored drum patterns. These represent the ways drum sounds are typically used in popular music. The matching score is calculated by comparing each drum pattern with the detected onset times of bass drum (BD) and snare drum (SD). The score is weighted by the product of a matching weight indicated in a prestored pattern and the reliability of detected drum onset (see Figure 8.9). A drum sound detected at a negative matching weight is given a penalty.

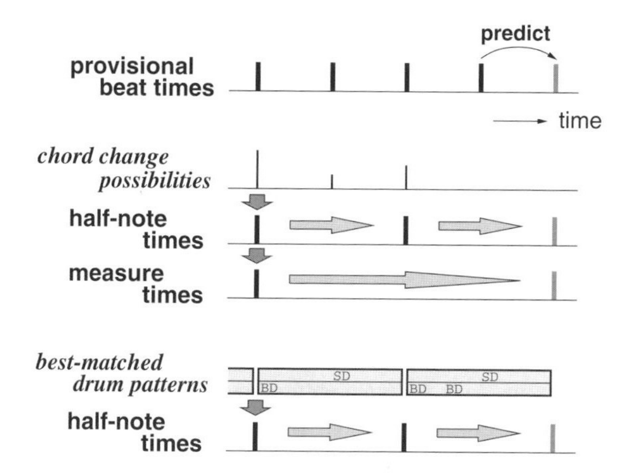

Figure 8.11 The hierarchical beat structure is estimated by using musical knowledge regarding chord changes and drum pattern.

8.5 ESTIMATING CHORUS SECTIONS AND REPEATED SECTIONS

Chorus ("hook" or refrain) sections of popular music are the most representative, uplifting, and prominent thematic sections in the music structure of a song, and human listeners can easily understand where the chorus sections are because these sections are the most repeated and memorable portions of a song. Automatic detection of chorus sections is essential for the computational modeling of music understanding and is useful in various practical applications. In music browsers or music retrieval systems, it enables a listener to quickly preview a chorus section as a "music thumbnail" (a musical equivalent of an image thumbnail) to find a desired song. It can also increase the efficiency and precision of music retrieval systems by enabling them to match a query with only the chorus sections. A novel music listening interface called SmartMusicKIOSK, which enables users to jump and listen to the chorus with just a push of a button, has already been developed [36].

To detect chorus sections, typical approaches do not rely on prior information regarding acoustic features unique to choruses but focus on the fact that chorus sections are usually the most repeated sections of a song. They thus adopt the following basic strategy: detect similar sections that repeat within a musical piece (such as a repeating phrase) and output those that appear most often. At the beginning of this century, this strategy led to methods for extracting a single segment from several chorus sections by detecting a repeated section of a designated length as the most representative part of a musical piece [65, 4, 9]; methods for segmenting music, discovering repeated structures, or summarizing a musical piece through bottom-up analyses without assuming the output segment length [12, 13, 77, 78, 3, 28, 10, 8, 99, 66]; and a method for exhaustively detecting all chorus sections by determining the start and end points of every chorus section [34, 36].

Although this basic strategy is simple and effective, it is difficult for a computer to judge repetition because it is rare for repeated sections to be exactly the same. The following summarizes the main problems that must be addressed in finding music repetition and determining chorus sections.

- **Problem 1. Extracting acoustic features and calculating their similarity.** Whether a section is a repetition of another must be judged on the basis of the similarity between the acoustic features obtained from each frame or section. In this process, the similarity must be high between acoustic features even if the accompaniment or melody line changes somewhat in the repeated section (e.g., the absence of accompaniment on bass and/or drums after repetition). That is, it is necessary to use features that capture useful and invariant properties.
- **Problem 2. Finding repeated sections.** A pair of repeated sections can be found by detecting contiguous temporal regions having high similarity. However, the criterion establishing how high similarity must be to indicate repetition depends on the song. For a song containing many repeated accompaniment phrases, for example, only a section with very high similarity should be considered the chorus-section repetition. For a song containing a chorus section with accompaniments changed after repetition, on the other hand, a section with somewhat lower similarity can be considered the chorus-section repetition. This criterion can be easily set for a small number of specific songs by manual means. For a large open song set, however, the criterion should be automatically modified based on the song being processed.
- **Problem 3. Grouping repeated sections.** Even if many pairs of repeated sections with various lengths are obtained, it is not obvious how many times and where a section is repeated. It is therefore necessary to organize repeated sections that have common sections into a group. Both ends (the beginning and end points) of repeated sections must also be estimated by examining the mutual relationships among various repeated sections. For example, given a song having the structure (A B C B C C), the long repetition corresponding to (B C) would be obtained by a simple repetition search. Both ends of the C section in (B C) could be inferred, however, from the information obtained regarding the final repetition of C in this structure.
- **Problem 4. Detecting modulated repetition.** Because the acoustic features of a section generally undergo a significant change after modulation (key change), similarity with the section before modulation is low, making it difficult to judge repetition. The detection of modulated repetition is important since modulation sometimes occurs in chorus repetitions, especially in the latter half of a song.*
- **Problem 5. Selecting chorus sections.** Because various levels of repetition can be found in a musical piece, it is necessary to select a group of repeat-

*Masataka Goto's survey of Japan's popular-music hit chart (top 20 singles ranked weekly from fiscal 2000 to fiscal 2003) showed that modulation occurred in chorus repetitions in 152 songs (10.3%) out of 1481.

ed sections corresponding to chorus sections. A simple selection of the most repeated sections is not always appropriate though. For example, another section such as verse A is occasionally repeated more often than chorus sections.

Regarding the above repetition-based methods, the following sections mainly describe a method called *RefraiD* (Refrain Detection method) [34, 36] and briefly introduce techniques used in the other methods in each relevant section and Section 8.5.6. Since the RefraiD method addresses all of the above problems and detects all chorus sections in a popular-music song regardless of whether a key change occurs, it is suitable for music scene description. Figure 8.12 shows the process flow of the RefraiD method. First, a 12-dimensional feature vector called a chroma vector, which is robust with respect to changes of accompaniments, is extracted from each frame of an input audio signal and then the similarity between these vectors is calculated (solution to Problem 1). Each element of the chroma vector corresponds to one of the 12 pitch classes (C, C#, D, . . . , B) and is the sum of the magnitude spectrum at frequencies of its pitch class over six octaves. Pairs of repeated sections are then listed (found) using an adaptive repetition-judgment criterion which is configured by an automatic threshold selection method based on a discriminant criterion (solution to Problem 2). To organize common repeated sections into groups and to identify both ends of each section, the pairs of repeated sections are integrated (grouped) by analyzing their relationships over the whole song (solution to Problem 3). Because each element of a chroma vector corresponds to a different pitch class, a before-modulation chroma vector is close to the after-modulation chorus vector

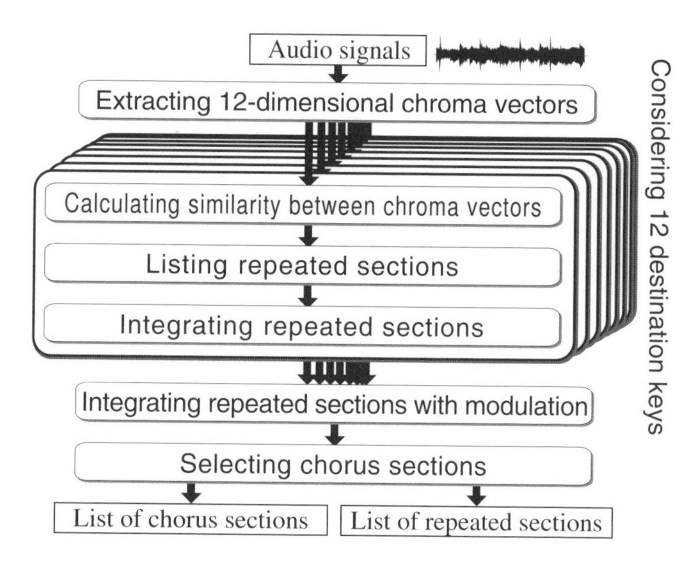

Figure 8.12 Overview of RefraiD (Refrain Detection method) for detecting all chorus sections with their beginning and end points while considering modulations (key changes).

whose elements are shifted (exchanged) by the pitch difference of the key change. By considering 12 kinds of shift (pitch differences), 12 sets of the similarity between nonshifted and shifted chroma vectors are then calculated, pairs of repeated sections from those sets are listed, and all of them are integrated (solution to Problem 4). Finally, the *chorus measure,* which is the possibility of being chorus sections for each group, is evaluated (solution to Problem 5), and the group of chorus sections with the highest chorus measure as well as other groups of repeated sections are output (Figure 8.13).

8.5.1 Extracting Acoustic Features and Calculating Their Similarity

The following acoustic features, which capture pitch and timbral features of audio signals in different ways, were used in various methods: chroma vectors [34, 36, 4, 12, 13], mel-frequency cepstral coefficients (MFCC) [65, 9, 3, 28, 10], (dimension reduced) spectral coefficients [9, 28, 10, 8, 99], pitch representations using $F0$ estimation or constant-Q filterbanks [12, 13, 8, 66], and dynamic features obtained by supervised learning [77, 78].

8.5.1.1 Pitch Feature: Chroma Vector. The chroma vector is a perceptually motivated feature vector using the concept of *chroma* in Shepard's helix representation of musical pitch perception [88]. According to Shepard [88], the perception of pitch with respect to a musical context can be graphically represented by using a

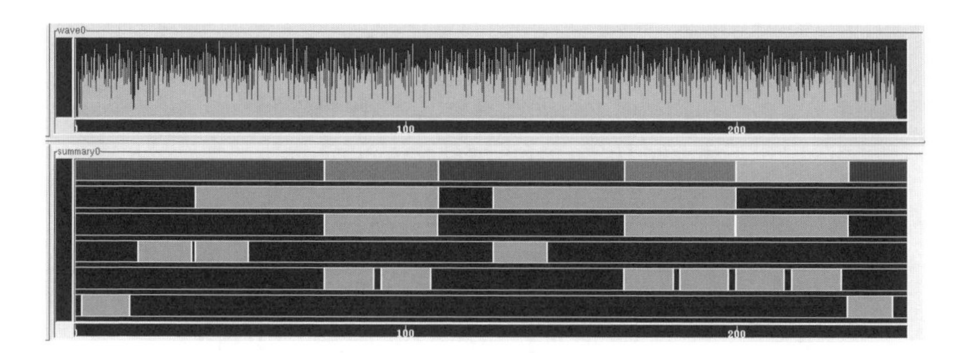

Figure 8.13 An example of chorus sections and repeated sections detected by the RefraiD method. The horizontal axis is the time axis (in seconds) covering the entire song. The upper window shows the power. On each row in the lower window, gray sections indicate similar (repeated) sections. The top row shows the list of the detected chorus sections, which were correct for this song (RWC-MDB-P-2001 No. 18 of the *RWC Music Database* [39, 37]) and the last of which was modulated. The bottom five rows show the list of various repeated sections (only the five longest repeated sections are shown). For example, the second row from the top indicates the structural repetition of "verse A → verse → chorus"; the bottom row with two short gray sections indicates the similarity between the "intro" and "ending."

continually cyclic helix that has two dimensions, chroma and height, as shown at the right of Figure 8.14. Chroma refers to the position of a musical pitch within an octave that corresponds to a cycle of the helix; that is, it refers to the position on the circumference of the helix seen from directly above. On the other hand, height refers to the vertical position of the helix seen from the side (the position of an octave).

Figure 8.14 shows an overview of calculating the chroma vector used in the RefraiD method [34, 36]. This represents magnitude distribution on the chroma that is discretized into twelve pitch classes within an octave. The 12-dimensional chroma vector $\vec{v}(t)$ is extracted from the magnitude spectrum, $\Psi_p(f, t)$ at the log-scale frequency f at time t, calculated by using the short-time Fourier transform (STFT). Each element of $\vec{v}(t)$ corresponds to a pitch class c ($c = 1, 2, \ldots, 12$) in the equal temperament and is represented as $v_c(t)$:

$$v_c(t) = \sum_{h=\mathrm{Oct_L}}^{\mathrm{Oct_H}} \int_{-\infty}^{\infty} BPF_{c,h}(f)\, \Psi_p(f, t)\, df \tag{8.23}$$

The $BPF_{c,h}(f)$ is a band-pass filter that passes the signal at the log-scale frequency $F_{c,h}$ (in cents) of pitch class c (chroma) in octave position h (height), where

$$F_{c,h} = 1200h + 100(c - 1) \tag{8.24}$$

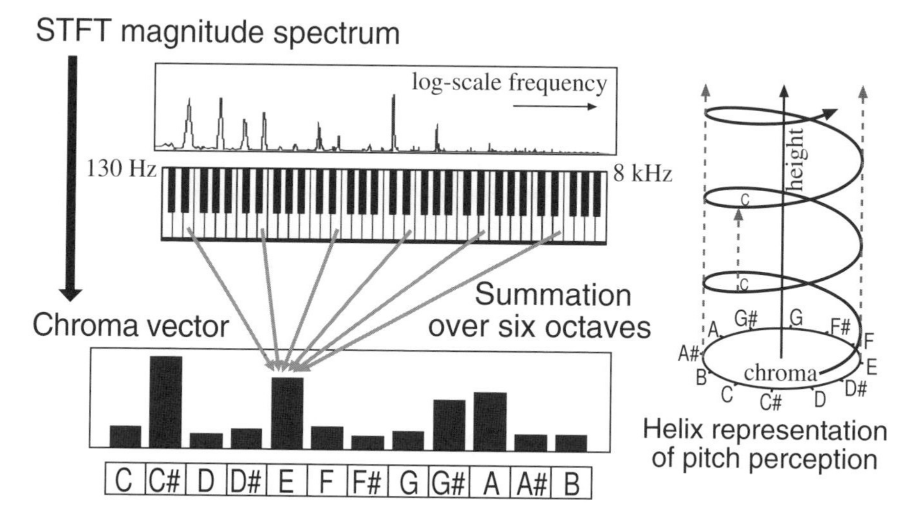

Figure 8.14 Overview of calculating a 12-dimensional chroma vector. The magnitude at six different octaves is summed into just one octave that is divided into 12 log-spaced divisions corresponding to pitch classes. The Shepard's helix representation of musical pitch perception [88] is shown at the right.

The $BPF_{c,h}(f)$ is defined using a Hanning window as follows:

$$BPF_{c,h}(f) = \frac{1}{2}\left(1 - \cos\frac{2\pi[f - (F_{c,h} - 100)]}{200}\right) \tag{8.25}$$

This filter is applied to octaves from Oct_L to Oct_H. In Goto's implementation [34, 36], an STFT with a 256 ms Hanning window shifted by 80 ms is calculated for audio signals sampled at 16 kHz, and the Oct_L and Oct_H are respectively 3 and 8, covering six octaves (130 Hz to 8 kHz).

There are variations in how the chroma vector is calculated. For example, Bartsch and Wakefield [4] developed a technique in which each STFT bin of the log-magnitude spectrum is mapped directly to the most appropriate pitch class, and Dannenberg and Hu [12, 13] also used this technique. A similar indiscrete concept was called the chroma spectrum [97].

There are several advantages to using the chroma vector. Because it captures the overall harmony (pitch-class distribution), it can be similar even if accompaniments or melody lines are changed to some degree after repetition. In fact, the chroma vector is effective for identifying chord names [30, 101, 102, 87, 104]. The chroma vector also enables modulated repetition to be detected as described in Section 8.5.4.

8.5.1.2 *Timbral Feature: MFCC and Dynamic Features.* Although the chroma vectors capture pitch-related content, the MFCCs typically used in speech recognition capture spectral content and general pitch range, and are useful for finding timbral or "texture" repetitions. Dynamic features [77, 78] are more adaptive spectral features designed for music structure discovery. Through a supervised learning method, they are selected from the spectral coefficients of a filterbank output by maximizing the mutual information between the selected features and hand-labeled music structures. The dynamic features are beneficial in that they reduce the size of the results when calculating similarity (i.e., the size of the similarity matrix described in Section 8.5.1.3) because the frame shift can be longer (e.g., 1 s) than for other features.

8.5.1.3 *Calculating Similarity.* Given a feature vector such as the chroma vector or MFCC at every frame, the next step is to calculate the similarity between feature vectors. Various distance or similarity measures, such as the Euclidean distance and the cosine angle (scalar product), can be used for this. Before calculating the similarity, feature vectors are usually normalized, for example, to a mean of zero and a standard deviation of one or to a maximum element of one.

In the RefraiD method [34, 36], the similarity $r(t, l)$ between the feature vectors (chroma vectors) $\vec{v}(t)$ and $\vec{v}(t - l)$ is defined as

$$r(t, l) = 1 - \frac{\left| \dfrac{\vec{v}(t)}{\max_c v_c(t)} - \dfrac{\vec{v}(t-l)}{\max_c v_c(t-l)} \right|}{\sqrt{12}} \tag{8.26}$$

where l is the lag. Since the denominator $\sqrt{12}$ is the length of the diagonal line of a 12-dimensional hypercube with edge length 1, $r(t, l)$ satisfies $0 \leq r(t, l) \leq 1$.

For the given feature vectors of a constant-Q filterbank output with center frequencies at 36 tempered semitones in three octaves, Lu, Wang, and Zhang [66] introduced a structure-based distance measure to emphasize pitch similarity over timbral similarity. This measure does not depend on the norm of the difference between feature vectors, but on the structure of it, that is, how the peak intervals in the difference conform to harmonic relationships.

8.5.2 Finding Repeated Sections

By using the same similarity $r(t, l)$, two equivalent representations can be obtained: a similarity matrix [9, 12, 13, 28, 10, 99] and a time-lag triangle (or time-lag matrix) [34, 36, 4, 78, 66], as shown in Figure 8.15. For the similarity matrix, the similarity $s(t, u)$ between feature vectors $\vec{v}(t)$ and $\vec{v}(u)$,

$$s(t, u) = r(t, t - u) \tag{8.27}$$

is drawn within a square in the two-dimensional $(t–u)$ space. For the time-lag triangle, the similarity $r(t, l)$ between feature vectors $\vec{v}(t)$ and $\vec{v}(t - l)$ is drawn within a right-angled isosceles triangle in the two-dimensional time-lag $(t–l)$ space. If a nearly constant tempo can be assumed, each pair of similar sections is represented by two noncentral diagonal line segments in the similarity matrix or a horizontal line segment in the time-lag triangle. Because the actual $r(t, l)$ obtained from a musical

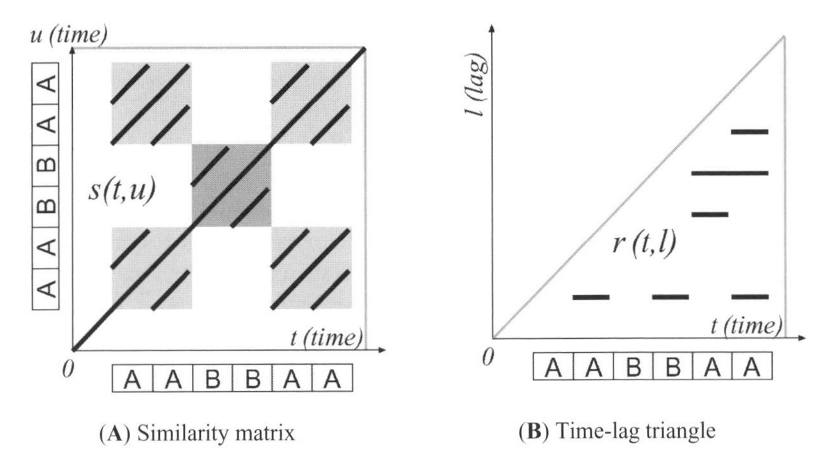

(**A**) Similarity matrix (**B**) Time-lag triangle

Figure 8.15 An idealized example of a similarity matrix and time-lag triangle drawn from the same feature vectors of a musical piece consisting of four "A" sections and two "B" sections. The diagonal line segments in the similarity matrix or horizontal line segments in the time-lag triangle, which represent similar sections, appear when short-time pitch features like chroma vectors are used.

piece is noisy and ambiguous, it is not a straightforward task to detect these line segments.

The RefraiD method [34, 36] finds all horizontal line segments [contiguous regions with high $r(t, l)$] in the time-lag triangle by evaluating $R_{all}(t, l)$, the possibility of containing line segments at the lag l at the current time t (e.g., at the end of a song) as follows (Figure 8.16):*

$$R_{all}(t, l) = \frac{1}{t - l + 1} \sum_{\tau=l}^{t} r(\tau, l) \tag{8.28}$$

Before this calculation, $r(t, l)$ is normalized by subtracting a local mean value while removing noise and emphasizing horizontal lines.

The method then picks up each peak in $R_{all}(t, l)$ along the lag l after smoothing $R_{all}(t, l)$ with a moving average filter along the lag and removing the global drift caused by cumulative noise in $r(t, l)$ from $R_{all}(t, l)$. The method next selects only high peaks above a threshold to search the line segments. Because this threshold is closely related to the repetition-judgment criterion which should be adjusted for each song, an automatic threshold-selection method based on a discriminant criterion [72] is used. When dichotomizing the peak heights into two classes by a threshold, the optimal threshold is obtained by maximizing the discriminant criterion measure defined by the following between-class variance:

$$\sigma_B^2 = \omega_1 \omega_2 (\mu_1 - \mu_2)^2 \tag{8.29}$$

where ω_1 and ω_2 are the probabilities of class occurrence (number of peaks in each class/total number of peaks), and μ_1 and μ_2 are the means of the peak heights in each class.

For each detected high peak with lag $L1$, the line segments are finally searched on the one-dimensional function $r(\tau, L1)$ ($L1 \leq \tau \leq t$). After smoothing $r(\tau, L1)$ using a moving-average filter, the method obtains line segments on which the smoothed $r(\tau, L1)$ is above a threshold (Figure 8.17). This threshold is also adjusted through the automatic threshold-selection method [72].

Instead of using the similarity matrix and time-lag triangle, there are other approaches that do not explicitly find repeated sections. To segment music, represent music as a succession of states (labels), and obtain a music thumbnail or summary, these approaches segment and label (i.e., categorize) contiguous frames (feature vectors) by using clustering techniques [65] or ergodic HMMs [65, 77, 78].

8.5.3 Grouping Repeated Sections

Since each line segment in the time-lag triangle indicates just a pair of repeated sections, it is necessary to organize into a group the line segments that have common

*This can be considered the Hough transform, in which only horizontal lines are detected; the parameter (voting) space $R_{all}(t, l)$ is therefore simply one dimensional along l.

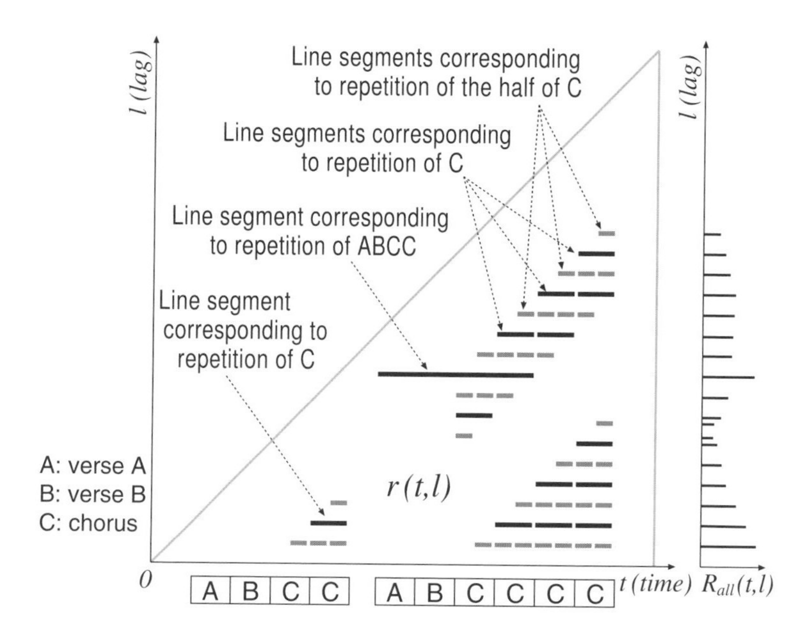

Figure 8.16 A sketch of line segments, the similarity $r(t, l)$ in the time-lag triangle, and the possibility $R_{\text{all}}(t, l)$ of containing line segments at lag l.

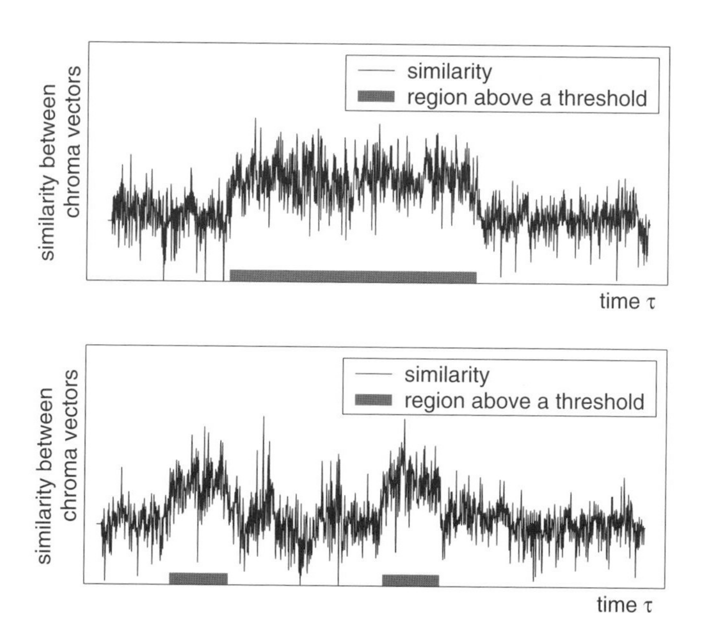

Figure 8.17 Examples of the similarity $r(\tau, L1)$ at high-peak lags $L1$. The bottom horizontal bars indicate the regions above an automatically adjusted threshold, which means they correspond to line segments.

sections. When a section is repeated n times ($n \geq 3$), the number of line segments to be grouped together should theoretically be $n(n-1)/2$ if all of them are found in the time-lag triangle.

Aiming to exhaustively detect all the repeated (chorus) sections appearing in a song, the RefraiD method groups line segments having almost the same section while redetecting some missing (hidden) line segments not found in the bottom-up detection process (described in Section 8.5.2) through top-down processing using information on other detected line segments. In Figure 8.16, for example, two line segments corresponding to the repetition of the first and third C and the repetition of the second and fourth C, which overlap with the long line segment corresponding to the repetition of ABCC, can be found even if they were hard to find in the bottom-up process. The method also appropriately adjusts the start and end times of line segments in each group because they are sometimes inconsistent in the bottom-up line-segment detection.

8.5.4 Detecting Modulated Repetition

The processes described above do not deal with modulation (key change), but they can easily be extended to it. A modulation can be represented by the pitch difference of its key change, ζ (0, 1, . . . , 11), which denotes the number of tempered semitones. For example, $\zeta = 9$ means the modulation of nine semitones upward or the modulation of three semitones downward. One of the advantages of the 12-dimensional chroma vector $\vec{v}(t)$ is that a transposition amount ζ of the modulation can naturally correspond to the amount by which its 12 elements are shifted (rotated). When $\vec{v}(t)$ is the chroma vector of a certain performance and $\vec{v}(t)'$ is the chroma vector of the performance that is modulated by ζ semitones upward from the original performance, they tend to satisfy

$$\vec{v}(t) \approx S^{\zeta}\vec{v}(t)' \tag{8.30}$$

where S^{ζ} is the ζth power of S, and S is a 12-by-12 shift matrix defined by

$$S = \begin{pmatrix} 0 & 1 & 0 & \cdots & \cdots & 0 \\ 0 & 0 & 1 & 0 & \cdots & 0 \\ \vdots & & \ddots & \ddots & \ddots & \vdots \\ 0 & \cdots & \cdots & 0 & 1 & 0 \\ 0 & \cdots & \cdots & \cdots & 0 & 1 \\ 1 & 0 & \cdots & \cdots & \cdots & 0 \end{pmatrix} \tag{8.31}$$

To detect modulated repetition by using this feature of chroma vectors and considering 12 destination keys, the RefraiD method [34, 36] calculates 12 kinds of extended similarity as follows:

$$r_{\zeta}(t, l) = 1 - \frac{\left| \dfrac{S^{\zeta}\vec{v}(t)}{\max_c v_c(t)} - \dfrac{\vec{v}(t-l)}{\max_c v_c(t-l)} \right|}{\sqrt{12}} \tag{8.32}$$

Starting from each $r_\zeta(t, l)$, the processes of finding and grouping the repeated sections are performed again. Nonmodulated and modulated repeated sections are then grouped if they share the same section.

8.5.5 Selecting Chorus Sections

A group corresponding to the chorus sections is finally selected from groups of repeated sections (line segments). In general, a group that has many and long repeated sections tends to be the chorus section. In addition to this property, the RefraiD method evaluates the chorus measure, which is the possibility of being chorus sections for each group, by considering the following three heuristic rules with a focus on popular music:

1. The length of the chorus has an appropriate, allowed range (7.7 to 4 s in Goto's implementation).
2. When there is a repeated section that is long enough to likely correspond to the repetition of a long section like (verse A $\rightarrow$ verse B $\rightarrow$ chorus) $\times$ 2, the chorus section is likely to be at the end of that repeated section.
3. Because a chorus section tends to have two half-length repeated subsections within its section, a section having those subsections is likely to be the chorus section.

The group that maximizes the chorus measure is finally selected as the chorus section.

8.5.6 Other Methods

Since the above sections mainly describe the RefraiD method [34, 36] with the focus on detecting all chorus sections, this section briefly introduces other methods that aim at music thumbnailing, music segmentation, structure discovery, or music summarization.

Several methods of detecting the most representative part of a song for use as a music thumbnail have been studied. Logan and Chu [65] developed a method using clustering techniques and HMMs to categorize short segments (1 s) in terms of their acoustic features, where the most frequent category is then regarded as a chorus. Bartsch and Wakefield [4] developed a method that calculates the similarity between acoustic features of beat-length segments obtained by beat tracking and finds the given-length segment with the highest similarity averaged over its segment. Cooper and Foote [9] developed a method that calculates a similarity matrix of acoustic features of short frames (100 ms) and finds the given-length segment with the highest similarity between it and the whole song. Note that these methods assume that the output segment length is given and do not identify both ends of a repeated section.

Music segmentation or structure discovery methods in which the output segment length is not assumed have also been studied. Dannenberg and Hu [12, 13] developed a structure discovery method of clustering pairs of similar segments obtained by several techniques such as efficient dynamic programming or iterative greedy algorithms. This method finds, groups, and removes similar pairs from the beginning to group all the pairs. Peeters, La Burthe, and Rodet [77, 78] developed a supervised learning method of modeling dynamic features and studied two structure discovery approaches: the sequence approach of obtaining repetitions of patterns and the state approach of obtaining a succession of states. The dynamic features are selected from the spectrum of a filterbank output by maximizing the mutual information between the selected features and hand-labeled music structures. Aucouturier and Sandler [3] developed two methods of finding repeated patterns in a succession of states (texture labels) obtained by HMMs. They used two image-processing techniques, the kernel convolution and Hough transform, to detect line segments in the similarity matrix between the states. Foote and Cooper [28, 10] developed a method of segmenting music by correlating a kernel along the diagonal of the similarity matrix, and clustering the obtained segments on the basis of the self-similarity of their statistics. Chai and Vercoe [8] developed a method of detecting segment repetitions by using dynamic programming, clustering the obtained segments, and labeling the segments based on heuristic rules such as the rule of first labeling the most frequent segments, removing them, and repeating the labeling process. Wellhausen and Crysandt [99] studied the similarity matrix of spectral-envelope features defined in the MPEG-7 descriptors and a technique of detecting noncentral diagonal line segments. Lu, Wang, and Zhang [66] developed a method of analyzing all repeated sections by using a structure-based distance measure that emphasizes pitch similarity over timbral similarity. Their method also estimates the tempo of a song and discriminates between vocal and instrumental sections to facilitate music structure analysis.

8.6 DISCUSSION AND CONCLUSIONS

This chapter has described the music-scene-description research approach toward developing a system that understands real-world musical audio signals without deriving musical scores or separating signals. In particular, it has focused on methods describing melody, bass, beat structure, and chorus and phrase repetition in polyphonic CD recordings. The following sections discuss the importance of this approach, evaluation issues, and directions and suggestions for future work.

8.6.1 Importance

The music-scene-description approach is important from a CASA viewpoint because it explores what is essential for understanding audio signals in a human-like fashion. The ideas and techniques are expected to be extended not only to music signals, but also to general audio signals including music, speech, environmental sounds, and mixtures of these. Traditional speech-recognition frameworks have been developed for dealing with only isolated speech signals or a single-pitch sound with background noise, which should be removed or suppressed without considering their relationship. Research on understanding musical audio signals is complementary to speech-recognition research and is a good starting point for creating a new framework for understanding general audio signals, because music is polyphonic, temporally structured, and complex, yet still well organized. In particular, relationships between various simultaneous or successive sounds are important and unique to music. Music-scene-description research will contribute to such a general framework.*

This approach is also important from industrial or application viewpoints since end users can now easily "rip" audio signals from CDs, compress and store them on a personal computer, load a huge number of songs onto a portable music player, and listen to them anywhere and anytime. These users want to retrieve and listen to their favorite music or a portion of a musical piece in a convenient and flexible way. Reflecting these demands, the target of processing has expanded from the internal content of individual musical pieces to entire musical pieces and even sets of musical pieces [42]. While the primary target of music scene description is the internal content of a piece, the obtained descriptions are useful for dealing with sets of musical pieces; for example, various music scene descriptions facilitate the computation of similarity between musical pieces. Such similarity enables a user to use musical pieces themselves as search keys to retrieve other musical pieces having a similar feeling, and can also be used to automatically classify musical pieces into genres or music styles. The more accurate and detailed we can make the obtained music scene descriptions, the more advanced and intelligent music applications and interfaces will become.

8.6.2 Evaluation Issues

To evaluate automatic music-scene-description methods, it is necessary to label musical pieces in an adequate-size music database with their correct descriptions (metadata). This labeling task is time consuming and troublesome. More seriously, there was no available common music database with correct metadata since most musical pieces used by researchers are generally copyrighted and cannot be shared by other researchers. To overcome this situation and establish benchmarks (evaluation frameworks) for music scene description, Goto et al. [39, 40, 37] developed the *RWC (Real World Computing) Music Database,* a copyright-cleared music database that is available to researchers as a common foundation for research. It contains six original collections: the *Popular Music Database* (100 pieces), *Royalty-Free Music Database* (15 pieces), *Classical Music Database* (50 pieces), *Jazz Music Database* (50 pieces), *Music Genre Database* (100 pieces), and *Musical Instrument Sound Database* (50 instruments). For all 315 musical pieces, audio signals, standard MIDI files, and text files of lyrics were prepared. For the 50 instruments, individual sounds at half-tone intervals were captured. This database has

*In fact, Masuda-Katsuse [69] has extended the PreFEst method described in Section 8.3.2 and demonstrated its effectiveness for speech recognition in realistic noisy environments.

been distributed to researchers around the world and has already been widely used. For musical instrument sounds, there are other databases released for public use: the McGill University Master Samples [71] and the University of Iowa Musical Instrument Samples [29]. Musical pieces licensed under a Creative Commons license can also be used for evaluation purposes.

After developing this database, Goto [35] also developed a multipurpose music scene labeling editor (metadata editor) that enables a user to hand-label a musical piece with music scene descriptions shown in Figure 8.1. The editor can deal with both audio files and standard MIDI files and supports interactive audio/MIDI playback while editing. Along a wave or MIDI piano roll display it shows subwindows in which any selected descriptions can be displayed and edited. To facilitate the support of various descriptions, its architecture is based on a plug-in system in which an external module for editing each description is installed as plug-in software. As a first step, the RefraiD method was evaluated by using the chorus-section metadata for 100 songs of the *RWC Music Database: Popular Music* (80 of the 100 songs were correctly detected) [34, 36].

8.6.3 Future Directions

Although various methods for detecting melody and bass lines, tracking beats, and finding chorus sections have been developed and successful results have been achieved to some extent, there is considerable room for improving these methods and developing new ones. The following are examples of such future directions:

- Other music scene descriptions as well as those described in this chapter should be studied in depth. For example, musical instrument identification [57, 24, 23, 26] for recognizing the musical instrument name of each component sound in a polyphonic sound mixture simultaneously with $F0$ estimation, and drum-sound detection [43, 33, 105, 27, 103, 83, 20] for estimating the onset times of drum sounds in CD recordings are also important issues.
- While this chapter has focused on research on audio-level music understanding, research on symbol-level music understanding dealing with musical scores and MIDI has also been progressing well. The fusion of symbol processing and audio signal processing is still far from sufficient and will be an important target in the future. Such fusion will bridge a gap between these forms of processing and enable symbol processing to be based on proper symbol grounding and audio signal processing to cover abstract semantic computing, eventually enabling music understanding that reflects the manifold meaning of music.
- The integration of top-down and bottom-up audio processing is an important and interesting issue. There are various ways of integration in which top-down processing using musical knowledge and hypothetical analysis can cooperate with bottom-up processing based on audio signal processing. As described in Section 8.4.3, for example, in a real-time beat-tracking system

developed by Goto and Muraoka [43, 45, 33], the integration of top-down hypothetical analysis and bottom-up acoustical analysis was able to exploit the synergy between the estimation of the hierarchical beat structure, drum patterns, and chord changes.

- An integrated method exploiting the relationships between music scene descriptions will be a promising next step because each method has generally been studied independently and implemented separately. For example, the result of describing the melody and bass lines could help to find the repetition in those lines when describing the chorus and phrase repetition, and this result could help to describe the melody line, bass line, and beat structure because the repeated sections are likely to have similar melody and bass lines and the same beat structure. The result of describing the beat structure could also help to determine section boundaries when describing the chorus and phrase repetition.
- The integration of various forms of CASA research with the music-scene-description research is also expected. Music scene description is not a pure signal-processing problem. Even if the primary research goal is derived from an engineering viewpoint, it will be fruitful to understand human auditory processing in more detail, and learn from auditory psychology and physiology as well as other CASA-related contributions.

REFERENCES

1. P. E. Allen and R. B. Dannenberg. Tracking musical beats in real time. In *Proceedings of the 1990 International Computer Music Conference,* pp. 140–143, 1990.
2. M. Alonso, B. David, and G. Richard. Tempo and beat estimation of musical signals. In *Proceedings of International Symposium on Music Information Retrieval 2004,* pp. 158–163, 2004.
3. J.-J. Aucouturier and M. Sandler. Finding repeating patterns in acoustic musical signals: Applications for audio thumbnailing. In *Proceedings of AES 22nd International Conference on Virtual, Synthetic and Entertainment Audio,* pp. 412–421, 2002.
4. M. A. Bartsch and G. H. Wakefield. To catch a chorus: Using chroma-based representations for audio thumbnailing. In *Proceedings of IEEE Workshop on Applications of Signal Processing to Audio and Acoustics (WASPAA '01),* pp. 15–18, 2001.
5. A. S. Bregman. Constraints on computational models of auditory scene analysis, as derived from human perception. *The Journal of the Acoustical Society of Japan (E),* 16(3):133–136, May 1995.
6. A. T. Cemgil and B. Kappen. Monte Carlo methods for tempo tracking and rhythm quantization. *Journal of Artificial Intelligence Research,* 18:45–81, 2003.
7. A. T. Cemgil, B. Kappen, and D. Barber. Generative model based polyphonic music transcription. In *Proceedings of Workshop on Applications of Signal Processing to Audio and Acoustics 2003,* pp. 181–184, 2003.
8. W. Chai and B. Vercoe. Structural analysis of musical signals for indexing and thumb-

nailing. In *Proceedings of ACM/IEEE Joint Conference on Digital Libraries,* pp. 27–34, 2003.

9. M. Cooper and J. Foote. Automatic music summarization via similarity analysis. In *Proceedings of International Symposium on Music Information Retrieval 2002,* pp. 81–85, 2002.
10. M. Cooper and J. Foote. Summarizing popular music via structural similarity analysis. In *Proceedings of Workshop on Applications of Signal Processing to Audio and Acoustics 2003,* pp. 127–130, 2003.
11. R. B. Dannenberg, W. P. Birmingham, G. Tzanetakis, C. Meek, N. Hu, and B. Pardo. The MUSART testbed for query-by-humming evaluation. In *Proceedings of International Symposium on Music Information Retrieval 2003,* pp. 41–47, 2003.
12. R. B. Dannenberg and N. Hu. Discovering musical structure in audio recordings. In *Proceedings of the International Conference on Music and Artificial Intelligence 2002,* pp. 43–57, 2002.
13. R. B. Dannenberg and N. Hu. Pattern discovery techniques for music audio. *Journal of New Music Research,* 32(2):153–163, 2003.
14. R. B. Dannenberg and B. Mont-Reynaud. Following an improvisation in real time. In *Proceedings of the 1987 International Computer Music Conference,* pp. 241–248, 1987.
15. M. E. P. Davies and M. D. Plumbley. Causal tempo tracking of audio. In *Proceedings of International Symposium on Music Information Retrieval 2004,* pp. 164–169, 2004.
16. M. Davy and S. J. Godsill. Bayesian harmonic models for musical signal analysis. In *Bayesian Statistics* 7, pp. 105–124, 2003.
17. A. P. Dempster, N. M. Laird, and D. B. Rubin. Maximum likelihood from incomplete data via the EM algorithm. *Journal of the Royal Statistical Society B,* 39(1):1–38, 1977.
18. P. Desain and H. Honing. The quantization of musical time: A connectionist approach. *Computer Music Journal,* 13(3):56–66, 1989.
19. P. Desain and H. Honing. Advanced issues in beat induction modeling: syncopation, tempo and timing. In *Proceedings of the 1994 International Computer Music Conference,* pp. 92–94, 1994.
20. C. Dittmar and C. Uhle. Further steps towards drum transcription of polyphonic music. In *Proceedings of the 116th Audio Engineering Society Convention (AES116),* 2004.
21. S. Dixon. Automatic extraction of tempo and beat from expressive performances. *Journal of New Music Research,* 30(1):39–58, 2001.
22. A. Driesse. Real-time tempo tracking using rules to analyze rhythmic qualities. In *Proceedings of the 1991 International Computer Music Conference,* pp. 578–581, 1991.
23. J. Eggink and G. J. Brown. Application of missing feature theory to the recognition of musical instruments in polyphonic audio. In *Proceedings of International Symposium on Music Information Retrieval 2003,* pp. 125–131, 2003.
24. J. Eggink and G. J. Brown. A missing feature approach to instrument recognition in polyphonic music. In *Proceedings of International Conference on Acoustics, Speech and Signal Processing 2003,* volume V, pp. 553–556, 2003.
25. J. Eggink and G. J. Brown. Extracting melody lines from complex audio. In *Proceedings of International Symposium on Music Information Retrieval 2004,* pp. 84–91, 2004.
26. J. Eggink and G. J. Brown. Instrument recognition in accompanied sonatas and concer-

tos. In *Proceedings of International Conference on Acoustics, Speech and Signal Processing 2004,* pp. IV–217–220, 2004.

27. D. FitzGerald, B. Lawlor, and E. Coyle. Drum transcription in the presence of pitched instruments using prior subspace analysis. In *Proceedings of Irish Signals and Systems Conference (ISSC),* pp. 202–206, 2003.
28. J. T. Foote and M. L. Cooper. Media segmentation using self-similarity decomposition. In *Proceedings of SPIE Storage and Retrieval for Media Databases 2003,* volume 5021, pp. 167–175, 2003.
29. L. Fritts. University of Iowa Musical Instrument Samples. Available at http://theremin.music.uiowa.edu/MIS.html.
30. T. Fujishima. Realtime chord recognition of musical sound: A system using common lisp music. In *Proceedings of International Computer Music Conference,* pp. 464–467, 1999.
31. A. Ghias, J. Logan, D. Chamberlin, and B. C. Smith. Query by humming: Musical information retrieval in an audio database. In *Proceedings of ACM Multimedia 95,* pp. 231–236, 1995.
32. M. Goto. *A Study of Real-time Beat Tracking for Musical Audio Signals* (in Japanese). PhD thesis, Waseda University, 1998.
33. M. Goto. An audio-based real-time beat tracking system for music with or without drum sounds. *Journal of New Music Research,* 30(2):159–171, 2001.
34. M. Goto. A chorus-section detecting method for musical audio signals. In *Proceedings of International Conference on Acoustics, Speech and Signal Processing 2003,* volume V, pp. 437–440, April 2003.
35. M. Goto. Music scene description project: Toward audio-based real-time music understanding. In *Proceedings of International Symposium on Music Information Retrieval 2003,* pp. 231–232, 2003.
36. M. Goto. SmartMusicKIOSK: Music listening station with chorus-search function. In *Proceedings of the ACM Symposium on User Interface Software and Technology, 2003,* pp. 31–40, 2003.
37. M. Goto. Development of the RWC music database. In *Proceedings of ICA 2004,* pp. I–553– 556, 2004.
38. M. Goto. A real-time music scene description system: Predominant-F0 estimation for detecting melody and bass lines in real-world audio signals. *Speech Communication,* 43(4):311– 329, 2004.
39. M. Goto, H. Hashiguchi, T. Nishimura, and R. Oka. RWC music database: Popular, classical, and jazz music databases. In *Proceedings of International Symposium on Music Information Retrieval 2002,* pp. 287–288, 2002.
40. M. Goto, H. Hashiguchi, T. Nishimura, and R. Oka. RWC music database: Music genre database and musical instrument sound database. In *Proceedings of International Symposium on Music Information Retrieval 2003,* pp. 229–230, 2003.
41. M. Goto and S. Hayamizu. A real-time music scene description system: Detecting melody and bass lines in audio signals. In *Working Notes of the IJCAI-99 Workshop on Computational Auditory Scene Analysis,* pp. 31–40, 1999.
42. M. Goto and K. Hirata. Invited review, "recent studies on music information processing." *Acoustical Science and Technology* (edited by the Acoustical Society of Japan), 25(6):419–425, November 2004.
43. M. Goto and Y. Muraoka. A beat tracking system for acoustic signals of music. In *Pro-*

ceedings of the Second ACM International Conference on Multimedia, pp. 365–372, 1994.

44. M. Goto and Y. Muraoka. Real-time rhythm tracking for drumless audio signals—chord change detection for musical decisions. In *Working Notes of the IJCAI-97 Workshop on Computational Auditory Scene Analysis*, pp. 135–144, 1997.
45. M. Goto and Y. Muraoka. Real-time beat tracking for drumless audio signals: Chord change detection for musical decisions. *Speech Communication*, 27(3–4):311–335, 1999.
46. F. Gouyon and P. Herrera. A beat induction method for musical audio signals. In *Proceedings of the International Workshop on Image Analysis for Multimedia Interactive Services, 2003*, 2003.
47. F. Gouyon, P. Herrera, and P. Cano. Pulse-dependent analyses of percussive music. In *Proceedings of AES 22nd International Conference on Virtual, Synthetic and Entertainment Audio*, 2002.
48. S. Hainsworth and M. Macleod. Beat tracking with particle filtering algorithms. In *Proceedings of Workshop on Applications of Signal Processing to Audio and Acoustics 2003*, pp. 91–94, 2003.
49. S. W. Hainsworth and M. D. Macleod. Automatic bass line transcription from polyphonic music. In *Proceedings of International Computer Music Conference 2001*, pp. 431–434, 2001.
50. S. W. Hainsworth and M. D. Macleod. Particle filtering applied to musical tempo tracking. *EURASIP Journal on Applied Signal Processing*, 15:2385–2395, 2004.
51. N. Hu and R. B. Dannenberg. A comparison of melodic database retrieval techniques using sung queries. In *Proceedings of the Joint Conference on Digital Libraries 2002*, pp. 301–307, 2002.
52. K. Jensen and T. H. Andersen. Beat estimation on the beat. In *Proceedings of Workshop on Applications of Signal Processing to Audio and Acoustics 2003*, pp. 87–90, 2003.
53. T. Kageyama, K. Mochizuki, and Y. Takashima. Melody retrieval with humming. In *Proceedings of International Computer Music Conference 1993*, pp. 349–351, 1993.
54. H. Kameoka, T. Nishimoto, and S. Sagayama. Extraction of multiple fundamental frequencies from polyphonic music using harmonic clustering. In *Proceedings of the International Conference on Acoustics 2004*, pp. I-59–62, 2004.
55. H. Kameoka, T. Nishimoto, and S. Sagayama. Audio stream segregation of multi-pitch music signal based on time-space clustering using gaussian kernel 2-dimensional model. In *Proceedings of International Conference on Acoustics, Speech and Signal Processing 2005*, pp. III-5–8, 2005.
56. K. Kashino. *Computational Auditory Scene Analysis for Music Signals* (in Japanese). PhD thesis, University of Tokyo, 1994.
57. K. Kashino and H. Murase. A sound source identification system for ensemble music based on template adaptation and music stream extraction. *Speech Communication*, 27(3– 4):337–349, 1999.
58. K. Kashino, K. Nakadai, T. Kinoshita, and H. Tanaka. Organization of hierarchical perceptual sounds: Music scene analysis with autonomous processing modules and a quantitative information integration mechanism. In *Proceedings of the International Joint Conference on Artificial Intelligence*, pp. 158–164, 1995.

59. A. P. Klapuri. Multiple fundamental frequency estimation based on harmonicity and spectral smoothness. *IEEE Transactions on Speech and Audio Processing,* 11(6): 804–816, 2003.

60. A. P. Klapuri, Antti J. Eronen, and Jaakko T. Astola. Analysis of the meter of acoustic musical signals. *IEEE Transactions on Speech and Audio Processing,* 14(1):342–355, 2006.

61. D. Lang and N. de Freitas. Beat tracking the graphical model way. In *Proceedings of Neural Information Processing Systems, 2004,* 2004.

62. E. W. Large. Beat tracking with a nonlinear oscillator. In *Working Notes of the IJCAI-95 Workshop on Artificial Intelligence and Music,* pp. 24–31, 1995.

63. J. Laroche. Estimating tempo, swing and beat locations in audio recordings. In *Proceedings of Workshop on Applications of Signal Processing to Audio and Acoustics 2001,* pp. 135–139, 2001.

64. Y. Li and D. Wang. Detecting pitch of singing voice in polyphonic audio. In *Proceedings of International Conference on Acoustics, Speech and Signal Processing 2005,* pp. III-17–20, 2005.

65. B. Logan and S. Chu. Music summarization using key phrases. In *Proceedings of International Conference on Acoustics, Speech and Signal Processing 2000,* pp. II-749–752, 2000.

66. L. Lu, M. Wang, and H.-J. Zhang. Repeating pattern discovery and structure analysis from acoustic music data. In *Proceedings of the 6th ACM SIGMM International Workshop on Multimedia Information Retrieval,* pp. 275–282, 2004.

67. M. Marolt. Gaussian mixture models for extraction of melodic lines from audio recordings. In *Proceedings of International Symposium on Music Information Retrieval 2004,* pp. 80–83, 2004.

68. M. Marolt. On finding melodic lines in audio recordings. In *Proceedings of the 7th International Conference on Digital Audio Effects (DAFx'04),* 2004.

69. I. Masuda-Katsuse. A new method for speech recognition in the presence of non-stationary, unpredictable and high-level noise. In *Proceedings Eurospeech 2001,* pp. 1119–1122, 2001.

70. T. Nishimura, H. Hashiguchi, J. Takita, J. Xin Zhang, M. Goto, and R. Oka. Music signal spotting retrieval by a humming query using start frame feature dependent continuous dynamic programming. In *Proceedings of the 2nd Annual International Symposium on Music Information Retrieval (International Symposium on Music Information Retrieval 2001),* pp. 211–218, October 2001.

71. F. Opolko and J. Wapnick. McGill University Master Samples (CDs), 1987.

72. N. Otsu. A threshold selection method from gray-level histograms. *IEEE Transactions on Systems, Man and Cybernetics,* SMC-9(1):62–66, 1979.

73. R. P. Paiva, T. Mendes, and A. Cardoso. An auditory model based approach for melody detection in polyphonic musical recordings. In *Proceedings of the the 2nd International Symposium on Computer Music Modeling and Retrieval (CMMR 2004),* 2004.

74. R. P. Paiva, T. Mendes, and A. Cardoso. A methodology for detection of melody in polyphonic musical signals. In *Proceedings of the 116th Audio Engineering Society Convention (AES116),* 2004.

75. B. Pardo. Tempo tracking with a single oscillator. In *Proceedings of International Symposium on Music Information Retrieval 2004,* pp. 154–157, 2004.

76. S. Pauws. CubyHum: A fully operational query by humming system. In *Proceedings of International Symposium on Music Information Retrieval 2002,* pp. 187–196, October 2002.
77. G. Peeters, A. LaBurthe, and X. Rodet. Toward automatic music audio summary generation from signal analysis. In *Proceedings of International Symposium on Music Information Retrieval 2002,* pp. 94–100, 2002.
78. G. Peeters and X. Rodet. Signal-based music structure discovery for music audio summary generation. In *Proceedings of International Computer Music Conference 2003,* pp. 15–22, 2003.
79. C. Raphael. Automated rhythm transcription. In *Proceedings of International Symposium on Music Information Retrieval 2001,* pp. 99–107, 2001.
80. D. Rosenthal. Emulation of human rhythm perception. *Computer Music Journal,* 16(1):64– 76, 1992.
81. D. Rosenthal. *Machine Rhythm: Computer Emulation of Human Rhythm Perception.* PhD thesis, Massachusetts Institute of Technology, 1992.
82. R. Rowe. *Interactive Music Systems.* MIT Press, Cambridge, MA, 1993.
83. V. Sandvold, F. Gouyon, and P. Herrera. Percussion classification in polyphonic audio recordings using localized sound models. In *Proceedings of International Symposium on Music Information Retrieval 2004,* pp. 537–540, 2004.
84. E. D. Scheirer. Tempo and beat analysis of acoustic musical signals. *Journal of the Acoustical Society of America,* 103(1):588–601, 1998.
85. E. D. Scheirer. *Music-Listening Systems.* PhD thesis, MIT, 2000.
86. W. A. Sethares and Thomas W. Staley. Meter and periodicity in musical performance. *Journal of New Music Research,* 30(2):149–158, 2001.
87. A. Sheh and D. P.W. Ellis. Chord segmentation and recognition using EM-trained hidden Markov models. In *Proceedings of International Symposium on Music Information Retrieval 2003,* pp. 183–189, 2003.
88. R. N. Shepard. Circularity in judgments of relative pitch. *Journal of the Acoustical Society of America,* 36(12):2346– 2353, 1964.
89. J. Shifrin, B. Pardo, C. Meek, and W. Birmingham. HMM-based musical query retrieval. In *Proceedings of the Joint Conference on Digital Libraries 2002,* pp. 295–300, 2002.
90. M. Slaney and R. F. Lyon. On the importance of time—a temporal representation of sound. In M. Cooke, S. Beet, and M. Crawford, editors, *Visual Representations of Speech Signals,* pp. 95–116. Wiley, 1993.
91. J. Song, S. Y. Bae, and K. Yoon. Mid-level music melody representation of polyphonic audio for query-by-humming system. In *Proceedings of International Symposium on Music Information Retrieval 2002,* pp. 133–139, October 2002.
92. T. Sonoda, M. Goto, and Y. Muraoka. A WWW-based melody retrieval system. In *Proceedings of the International Computer Music Conference,* pp. 349–352, 1998.
93. T. Sonoda, T. Ikenaga, K. Shimizu, and Y. Muraoka. The design method of a melody retrieval system on parallelized computers. In *Proceedings of International Conference on Web Delivering of Music (WEDELMUSIC) 2002,* pp. 66–73, 2002.
94. H. Takeda, T. Nishimoto, and S. Sagayama. Automatic rhythm transcription of multiphonic MIDI signals. In *Proceedings of International Symposium on Music Information Retrieval 2003,* pp. 263–264, 2003.

95. H. Takeda, T. Nishimoto, and S. Sagayama. Rhythm and tempo recognition of music performance from a probabilistic approach. In *Proceedings of International Symposium on Music Information Retrieval 2004,* pp. 357–364, 2004.

96. N. P. M. Todd and G. J. Brown. Visualization of rhythm, time and metre. *Artificial Intelligence Review,* 10:253–273, 1996.

97. G. H. Wakefield. Mathematical representation of joint time-chroma distributions. *SPIE'99,* pp. 637–645, 1999.

98. Y. Wang and M. Vilermo. A compressed domain beat detector using MP3 audio bitstreams. In *Proceedings of the Ninth ACM International Conference on Multimedia,* pp. 194–202, 2001.

99. J. Wellhausen and H. Crysandt. Temporal audio segmentation using MPEG-7 descriptors. In *Proceedings of SPIE Storage and Retrieval for Media Databases 2003,* volume 5021, pp. 380–387, 2003.

100. M. Wu, D. Wang, and G. J. Brown. A multipitch tracking algorithm for noisy speech. *IEEE Transactions on Speech and Audio Processing,* 11(3):229–241, 2003.

101. H. Yamada, M. Goto, H. Saruwatari, and K. Shikano. Multi-timbre chord classification for musical audio signals (in Japanese). In *Proceeding of the 2002 Autumn Meeting of the Acoustical Society of Japan,* pp. 641–642, September 2002.

102. H. Yamada, M. Goto, H. Saruwatari, and K. Shikano. Multi-timbre chord classification method for musical audio signals: Application to musical pieces (in Japanese). In *Proceeding of the 2003 Spring Meeting of the Acoustical Society of Japan,* pp. 835–836, March 2003.

103. K. Yoshii, M. Goto, and H. G. Okuno. Automatic drum sound description for real-world music using template adaptation and matching methods. In *Proceedings of International Symposium on Music Information Retrieval 2004,* pp. 184–191, 2004.

104. T. Yoshioka, T. Kitahara, K. Komatani, T. Ogata, and H. G. Okuno. Automatic chord transcription with concurrent recognition of chord symbols and boundaries. In *Proceedings of International Symposium on Music Information Retrieval 2004,* pp. 100–105, 2004.

105. A. Zils, F. Pachet, O. Delerue, and F. Gouyon. Automatic extraction of drum tracks from polyphonic music signals. In *Proceedings of International Conference on Web Delivering of Music (WEDELMUSIC) 2002,* pp. 179–183, 2002.

CHAPTER 9

Robust Automatic Speech Recognition

JON BARKER

9.1 INTRODUCTION

Automatic speech recognition (ASR) has been an active research area for over 30 years and in this time great strides have been made. The research focus has progressed from isolated words to connected speech, and from systems tuned to a given speaker to those able to handle unknown speakers. The development of sophisticated search algorithms and faster computers has lead to ever-increasing vocabulary sizes; from early systems with vocabularies containing less than one hundred words, to the systems of today, which may employ vocabularies of 50,000 words or more. Now, the research emphasis is moving toward increasingly natural speech, progressing from people speaking artificially prepared corpora (TIMIT [49]), to people reading newspapers (*Wall Street Journal* [83]), to transcribing radio newscasters (Broadcast News [12]), and most recently, dealing with real conversations involving two people (Switchboard [54]) or even several speakers (ICSI and AMI meeting projects [64, 92]). Today's state-of-the-art recognizers are highly sophisticated systems incorporating detailed knowledge of acoustics, grammar, and discourse.

ASR has come a long way but, disappointingly, it has failed to become the pervasive technology that many foresaw. Deploying ASR "in the wild" has proved surprisingly challenging. For example, we are still unable to design something as seemingly simple as a television set with a reliable speech interface. Despite all the progress that has been made, we are still little closer to understanding how to recognize speech in uncontrolled environments, and, crucially, there is still no general solution to the problem of recognizing speech in the presence of competing sound sources. In this regard, speech recognition remains a brittle technology.

In the early days of speech research, dealing with the variability in the speech signal itself was a sufficient challenge; nobody wished to make the problem harder by mixing the speech signal with other sounds. In recent years, the desire for broader deployment of speech technology has brought the robustness issue increasingly

Computational Auditory Scene Analysis. Edited by DeLiang Wang and Guy J. Brown

to the forefront. However, given the complexity of speech recognition systems, the favored approaches usually involve "bolting on" some extra component that will work with minimal disturbance to the existing recognition machinery.

The development of ASR in machines stands in marked contrast to the development of speech perception in humans. Our early ancestors were presumably listening to their environment long before the evolution of speech. As has been discussed in this book, the task of listening is primarily about forming descriptions of separate sound sources from the complex mixtures that arrive at the ear (see Chapter 1). As speech developed, it was just another sound source in this mixture to which the existing well-established listening systems could be applied. So, in evolutionary terms, speech perception could more reasonably be described as something that was bolted onto a system for processing the noisy environment, rather than vice versa. As our speech and listening apparatus have further coevolved as part of the same communication tool, the processes have become closely coupled.

So we might argue that the robustness of speech recognition in humans is something that arises out of a deeper, more primitive, auditory-scene analysis system, and, therefore, computational auditory scene analysis (CASA) is an obvious solution to the current brittleness of ASR technology. However, understanding the relationship between auditory scene analysis (ASA) and speech perception is a challenging problem. It is not obvious how scene analysis and speech recognition can be effectively combined. The simplest model would be to employ CASA as a "front end" that can feed acoustic features directly into an ASR system. However, as we will see later in this chapter, there are good reasons why such a simple approach cannot work well.

This chapter will look at both CASA and ASR and examine the ways in which they may be combined. However, before considering CASA as a solution to the problems encountered by ASR, there needs to be proper justification for the CASA approach. To this end, Section 9.2 will commence by presenting a detailed examination of the evidence that ASA is an active component of speech perception. Much of this evidence accrues from classic experiments first conducted some twenty or thirty years ago. However, it is worth reexamining this body of work, because if ASA does not underpin speech perception in humans, then there is little motivation for attempting to ground robust ASR systems on computational models of ASA. Furthermore, an understanding of the way in which perceptual organization and speech perception interact will inform the design of the CASA-based ASR systems discussed in later sections.

Having examined the perceptual foundations of the CASA/ASR approach, Section 9.3 will formalize the speech recognition problem. The section will briefly explain the standard statistical ASR framework. It will then examine some traditional approaches to dealing with noisy speech data and contrast these with CASA-driven approaches. The chapter will then proceed to examine in detail some of the more recent CASA-driven approaches to ASR. Section 9.4 will examine systems that attempt to use primitive ASA grouping principles to first segregate the speech and the background in the spectrotemporal domain, and then base ASR on the partial description of speech recovered from the mixture. By tackling the problem in two independent steps, these systems effectively decouple segregation and recognition. This is a simple strategy, but ultimately limited. Section 9.5 will consider more sophisticated systems in which these processes are not decoupled. In these systems, the segregation will be a function of both primitive processes and the speech models that govern recognition. Finally, Section 9.6 concludes with a discussion of the major unanswered questions that need to be addressed before the full potential of CASA-based ASR systems can be realized.

9.2 ASA AND SPEECH PERCEPTION IN HUMANS

Before proposing CASA as a solution to the problem of achieving robust ASR in machines, it is worth reviewing the evidence that ASA can account for the perception of speech in humans. After all, if there is no evidence that humans exploit ASA to benefit speech perception then there would seem little point in striving to engineer ASA into our ASR systems.

Although auditory scene analysis was initially proposed by Bregman as a general account of auditory perceptual organization, the account was built up from evidence gained from experiments employing a narrow range of artificial stimuli. Whether or not the primitive grouping principles that apply to simple stimuli, can be said to apply equally to complex stimuli such as speech has been the subject of much debate. Researchers trying to answer this question have extended Bregman's work using digital signal processing and synthesis techniques that have allowed complex artificial stimuli to be carefully controlled. This level of control has allowed ASA principles to be directly tested on speech-like stimuli. This section will review some of this work, identifying key experiments that test the hypothesis that speech perception is governed by ASA principles. The section has been organized to separately consider simultaneous grouping, sequential grouping, and schema-driven processes. Although these grouping principles have already been discussed in Chapter 1, the current section focuses exclusively on the extent to which they can be applied to the speech signal. The section concludes by reviewing a specific challenge to the ASA account, which argues that it does not adequately account for the perceived coherence of speech.

9.2.1 Speech Perception and Simultaneous Grouping

Simultaneous grouping is one of the cornerstones of the ASA account. It describes how coexistent elements may be grouped across frequency due to a similarity of their characteristics. Bregman was able to demonstrate a broad range of such grouping effects using simple stimuli. For example, components will show a tendency to group together if they share a common fundamental frequency, or if their onsets occur at the same time, or if they share a common pattern of amplitude modulation. However, anyone who has briefly studied a speech spectrogram may question whether such simple rules may be usefully applied to a pattern as complex as speech. Below we review the experimental evidence that suggest that the perceptual organization of speech is at least partially governed by simultaneous grouping principles.

Grouping by Harmonicity. The periodic vibration of the vocal chords during the production of vowels and voiced consonants (such as /d/, /g/ and /b/) give the speech signal a clear harmonic structure. The frequency of vibration, termed the fundamental frequency ($F0$), is perceived as the speech's pitch and varies smoothly during speech, with typical $F0$ values ranging between 50 and 500 Hz. In multiple-speaker situations, it is unlikely that two speakers will share the same pitch at the same time. So competing speakers will produce separate sets of regularly spaced harmonics, with the harmonics of each speaker spaced at a distinct interval. Given that this structure exists, it easy to imagine how it may be exploited to separate and group components of voiced speech signals. Indeed, there are many computational models that separate speech in this way (for a review see Chapter 2).

Grouping by harmonicity appears to be a strong cue for the perceptual organisation of speech. As long ago as 1957, Broadbent and Ladefoged [18] demonstrated that if synthetic formants are delivered to opposite ears, listeners will report hearing a single voice if the formants are synthesized with the same fundamental frequency. If the same formants are synthesized at different $F0$s, listeners report hearing more than one voice. This result clearly suggests that a common $F0$ can act to group formants into a common stream. Darwin [35] supports this finding using an ambiguous, synthetic four-formant stimulus that may be heard as either the syllable /ru/ or the syllable /li/ depending on how the formants are grouped. Specifically, four formants were arranged such that when produced with a common $F0$ the stimulus was heard as the syllable /ru/. In contrast, when the second formant (F2) was mistuned, its contribution to the perceived syllable was reduced and the stimulus was heard as /li/, that is, as though F2 had been removed from the complex. In further studies, Gardner et al. [52] reported that large $F0$ differences were necessary to fully exclude the influence of F2. With smaller differences a "duplex perception" was formed whereby the second formant both formed a separate stream and contributed to the syllable perception.

These results demonstrate that a mixture of two sounds with different $F0$s tends to segregate into separate perceptual streams. In contrast, studies employing simultaneous speech sources have also shown how the possession of a common fundamental can result in perceptual fusion. Much of this work has involved the presentation of simultaneous vowels ("double vowels"). In the double vowel paradigm, originally used by Scheffers [99], subjects are played two simultaneous vowel sounds and are asked to identify them both. Scheffers showed that listeners could typically identify about 45% of the vowels when they were presented with the same $F0$ but this score significantly improved, to around 62%, when a very small $F0$ difference was introduced. It would appear that these small differences in $F0$ are necessary for listeners to perceptually segregate the two vowels and, hence, identify them more easily. This effect is very robust and has been replicated many times, see, for example, Zwicker [119], Assmann and Summerfield [2, 3], and Culling [29, 30] (see also Chapter 2).

Results of double vowel studies have been interpreted in terms of a harmonic sieve. A mechanism acting like a sieve lets through components that occur at the harmonics of a particular $F0$, and blocks components that do not. Only the components admitted by the sieve will contribute to the vowel percept. Numerous auditory models have been produced that mimic this effect. Most are based on autocorrelation techniques that apply an autocorrelation function to the output of the cochlear filters and use the peaks due to the fundamental as a basis for grouping energy across different frequency regions. Examples of these include Weintraub [114], Slaney and Lyon [101], Meddis and Hewitt [74], and Brown and Cooke [20]. There are other models, like that of Meyer and Berthommier [75], that operate by performing a spectral analysis of the filter outputs—an analysis of envelope modulation (see Chapter 3)—and which, again, group components on the basis of peaks due to the fundamental frequency. Leaving questions of mechanism aside, double-vowel studies all support the notion that speech perception can be influenced by a gestalt-like, data-driven mechanism that can group energy across frequency bands on the basis of common $F0$. Grouping by harmonic relation is one of the most robust organization cues and, as we shall see, has been employed at the core of most successful CASA-motivated, robust ASR systems.

Grouping by Common Onset and Offset. Effects of onset grouping can be demonstrated by using similar experiments to those described in the previous section. Employing the four-formant /ru/–/li/ stimulus, Darwin [35] demonstrated that if the second formant was given a lead of 300 ms, its contribution to the perception of the complex was reduced. However, the effect observed is very weak and the asynchrony used is many times greater than that needed to promote streaming in simple repetitive tone complexes of the kind employed by Bregman. More convincing demonstrations of common onset can be found by misaligning individual harmonics within a synthetic vowel [39]. In this case, onsetting a harmonic as little as 30 ms sooner than the rest of the formant can change the perceived vowel identity. Possible explanations based on peripheral adaptation to the early harmonic are ruled out by further experiments in which the effect is shown to also occur when the harmonic is delayed [39, 95]. Darwin and Sutherland then went on to show that the effect of an early onset harmonic can be reversed by adding a "capture" tone at the octave of the misaligned harmonic that onsets synchronously with it and offsets when the rest of the vowel commences. Darwin and Sutherland conclude that their findings are evidence of the influence of grouping mechanisms, but that coincidence of onset and offset times constitutes neither a necessary nor sufficient condition for grouping the harmonics of a single voice. They suggest that other principles must be employed in conjunction with onset and offset grouping to ensure that the rapidly modulated harmonics of normal speech are bound together.

Darwin and Sutherland's observation that onset and offset cues are neither necessary nor sufficient to initiate grouping is a specific case of the more general observation that grouping rules are not absolute. For example, onset (and offset) cues have different efficacy depending on the nature of the stimulus employed. There are larger tolerances for onset differences in speech perception studies than in pitch perception studies. Generally, listening experiments involving highly learned patterns, and in particular those employing speech, appear to indicate that grouping rules may be more accurately described as grouping "suggestions." They influence, rather than dictate, the final perceptual organization of the input. We will return to this point later when discussing schemas. We shall also see that the statistical nature of primitive grouping is something that needs to be better understood before CASA and ASR can be integrated in a principled manner.

9.2.2 Speech Perception and Sequential Grouping

Having considered how simultaneous components of a speech signal may be grouped across frequency, we shall now examine the mechanisms that group components arriving sequentially through time. Bregman describes how tones and glides group sequentially following the principle of "good continuation"; that is, tones will group together if their pitches form a sufficiently smooth contour. Are there similar forms of good continuation that can be applied to explain the coherence of speech?

If speech synthesis is performed simply by splicing together examples of naturally spoken words, the resulting speech is perceived to be badly garbled and intelligibility breaks down even at moderately low speaking rates. Even if the words spliced are all originally recorded from the same speaker, listeners report hearing them as originating from different speakers or from different directions [76]. This effect may possibly be due to the lack of good continuation at the word boundaries, leading to poor sequential streaming. Each new word starts a new stream and is effectively interpreted as coming from a new physical source.

Considering the classic source/filter model of speech production [48] there are two classes of possible discontinuity that occur in this primitive form of synthesis. The first is discontinuities in the source characteristics, for example, discontinuities in fundamental frequency during voiced speech. The second is discontinuities in the nature of the filter, for example, the position and bandwith of spectral resonances. To what degree does the successful sequential streaming of speech rely on continuity in these domains?

F0 Continuity. Through the use of a word-shadowing paradigm, Darwin [34] was able to demonstrate how sequential streaming is dependent on good $F0$ continuity. The experiment was inspired by an earlier "speaker shadowing" study by Treisman [105]. In this study, subjects were presented with a monologue spoken by a single speaker in one ear (the target ear), and another monologue spoken by a different speaker in the other ear (the distractor ear). They were asked to repeat whatever they heard at the target ear while ignoring the distractor ear. At some point, the two speakers were swapped between the two ears. It was found that at this point subjects would repeat a few words from the distractor ear, despite having been instructed to shadow only the target ear. It was initially concluded that attention was being drawn by high-level linguistic properties. Darwin however, noted that when the voices were switched, there were also low-level discontinuities, such as discontinuities in pitch, occurring at each ear. To isolate these effects, he repeated the experiment but with two additional conditions. In one, the "content" of the dialogue was swapped between ears without any pitch shift. In the other, through the use of resynthesis techniques, the pitch was swapped between ears but there was no break in the semantic content of the dialogues. He found that words from the nontarget ear were only repeated under the condition in which the pitch was switched.

Continuity of fundamental frequency has also been shown to influence speech perception at the phonetic level. Darwin and Bethell-Fox [36] studied the effect of pitch-contour discontinuities on the perception of alternating vowel sequences. They employed a repeating formant pattern that alternated between two steady-state values joined via linear formant transitions. Each pattern had as one of its steady-state values formants appropriate to the vowel /a/. The formant values for the other steady-state region were chosen so that the formant transitions on either side of the /a/ vowel formed patterns appropriate to one of the three voiced stop consonants /b/, /d/, and /g/. The steady states and the transitions each had a duration of 60 ms. In one condition, the stimuli were synthesized with a constant $F0$ of 136 Hz. In another condition, the $F0$ alternated between 101 Hz and 178 Hz, changing at the midpoint of each of the formant transition regions. Subjects were asked to identify the consonant immediately preceding the /a/ vowel, choosing from either /b/, /d/, /g/, /w/, /l/, or /y/. It was found that when the fundamental frequency was held constant, subjects heard mainly /w/, /r/, and /l/, whereas when the fundamental frequency alternated, subjects heard the stop consonants /b/, /d/, and /g/. This effect was explained by assuming that the two different components of the $F0$ contour form two perceptually distinct streams. This segregation has the effect of producing illusory silences in each stream during the portions of the acoustic signal attributed to the other stream. These silences coupled with the appropriate formant transitions indicate the presence of stop consonants.

Continuity of $F0$ can act to sequentially group temporally disjoint vowel sounds by the same principle that the continuity of frequency can group dislocated components in simple tone sequences. This was demonstrated by Nooteboom et al. [76] using sequences of synthetic vowel sounds. From a set of nine vowel sounds, short sequences were constructed, with each constituent vowel being 100 ms in duration and separated by a short silence. The fundamental frequency of alternate vowels and the length of the separation were systematically varied and listeners were required to report the order in which the vowels occurred. The ability to make reliable judgments on the temporal order of the vowels and therefore correctly report the sequence order was taken as an indication that the vowels were being grouped into the same perceptual stream. It was found that sequential streaming was strongest for large temporal separations or small $F0$ differences, with a trade-off being demonstrated such that as the $F0$ difference increased, larger gaps were needed to produce the same strength of streaming.

The ability of $F0$ continuity to group temporally disjoint components could go some way toward explaining the perceptual coherence of speech containing abrupt transitions that might otherwise cue segregation. For example, consider stop consonants, such as /t/ in the word "butter." Before the plosive release of the /t/ there is the short stop period which is essentially silence. How then is the "bu" occurring before the silent period perceptually linked to the "ter" occurring after it? Could this occur due to $F0$ continuity?

This question has been addressed by Dorman [41]. A 50 ms silence was inserted before the /ʃ/ occurring in the phrase, "please say shop." With the silence inserted, listeners reported hearing, "please say chop." The dip in energy acts to signal /ch/ rather than /ʃ/. However, when the phrase "please say" spoken by a *female* speaker was spliced to the word "shop" spoken by a *male* speaker, with an identical 50 ms silence between, listeners heard the stimulus as, "please say shop." In this case, the large $F0$ difference prevents the two halves of the stimulus from sequentially grouping, and as they now form separate streams there is no within-stream energy dip to alter the interpretation of /ʃ/.

Spectral Continuity. The previous section illustrated that $F0$ continuity can string together subsequent components of a speech signal to form a coherent percept. In the last example, it was shown how this continuity can act "at a distance," joining syllables on the basis of $F0$ similarity of the vowels they contain. However, what is not explained simply by $F0$ is the coherence of these syllables themselves. Considering that syllables are typically composed of a mixture of voiced and unvoiced phonemes, where is the continuity that binds them? For example, in the "say shop" above, it was seen how $F0$ can bind "say" to "shop," but what binds the /ʃ/, /a/, and /p/ components of the word "shop"?

During voiced speech, air is forced through vibrating vocal chords and the resulting quasiperiodic pressure wave propagates through a system of cavities formed by the throat, nose, and mouth before being emitted as sound. The cavities through which the wave passes can be modeled as a filter acting to amplify some frequency components and suppress others. The strength and position of the resonances, that is, formants, are dependent on the precise shape of the cavities and, hence, articulatory "controls" such as tongue position, mouth openness, the tension in muscles in the throat and so on. These articulators must all have continuous trajectories and, hence, formant trajectories in voiced speech must also be continuous.

Unfortunately, when considering speech composed of mixtures of voiced and nonvoiced segments the model is not so simple. Only the cavities through which the excitation wave must pass will act to filter the signal. This means, for examples, that the formant structure imposed on the /ʃ/ of "shop," generated by air flowing turbulently between the tongue and the palette, will not be the same as that imposed on the vowel /a/. This effect is also illustrated in the spectrogram of "she sells sea shells" shown in Fig. 9.1. However, considering a word like "house," the aspirant /h/, being generated below the cavities of the nose and mouth, will have the same filtering applied to it as the vowel. It is possible that this spectral continuity could form a useful sequential grouping cue.

The effects of this form of spectral continuity were demonstrated in an experiment by Verbrugge and Rakerd [108]. They used the same "please say shop" stimulus as used by Dorman to show the effects of $F0$ continuity. Dorman had shown how a shift in $F0$ prevented a silence from being incorporated into the phonetic

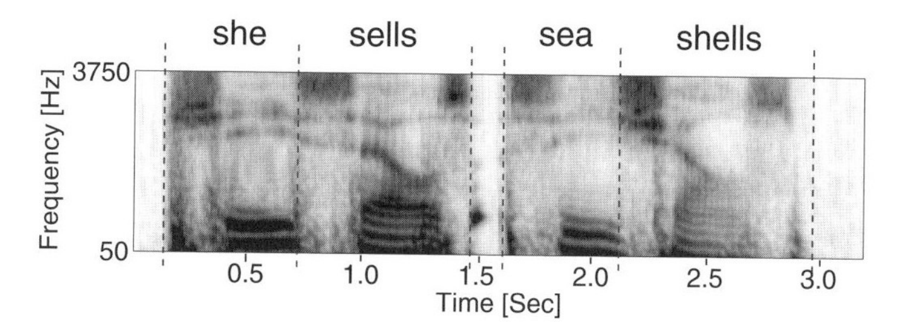

Figure 9.1 A cochleagram (see Chapter 1) of the utterance "she sells sea shells." The utterance is heard as a single coherent stream despite containing discontinuities at the vowel consonant boundaries that might be presumed to promote segregation.

stream and being interpreted as a stop consonant closure. However, it was found that the silence would retain its effectiveness as a cue, despite the $F0$ shift, as long as the general articulatory pattern was made to be continuous across the two speakers. This confirms an independent sequential grouping effect of spectral continuity.

The speech spectrum also provides evidence of invariant filter properties, such as vocal tract size. Although these properties may be hard to estimate reliably over a short subword time interval, reliable estimates can be made over longer time intervals. Darwin and Hukin [37] showed that vocal-tract size cues could override spatial cues in determining which of two simultaneous target words belongs in an attended sentence. A follow-up study [38] showed that the effects of apparent vocal tract length are robust to reverberation. We can speculate that other higher-level invariant properties, such as accent or speaker identity, can also help to group elements of speech into a common stream.

As mentioned earlier, there are events during speech production that can produce apparent spectral discontinuities. For example, sudden shifts in the point at which the excitation is generated lead to sudden changes in the spectral resonances of the speech stream. How is coherence maintained across these boundaries? Consider a word such as, "essay." The /s/ has very different spectral characteristics from those of the surrounding vowels, and yet it does not perceptually segregate. The fact that the /s/ commences at precisely the time that the /eh/ terminates is not sufficient reason for them to be sequentially grouped. If it were, then sequential streaming mechanisms could be easily fooled into incorporating extraneous noises into the speech stream. For instance, consider the effect of imposing an intense wideband click on a segment of speech. Ladefoged [69] demonstrated that listeners presented with such a stimulus may hear the speech as uninterrupted and will have great difficulty in identifying the precise point during the speech at which the click occurred. The click has clearly been segregated into a separate perceptual stream.

Why, then, does the /s/ in "essay" get perceptually integrated into the speech stream whereas a noise superimposed on top of a word does not? What is the essential difference between these two cases? This can perhaps be answered by studies of the induction effect on simple tones. When a noise band is employed to mask a tone, listeners report hearing the tone as continuing through the noise, even if the masked portion of the tone had actually been removed before the noise was inserted. This effect is known as *induction*. It has been shown that there are several strict conditions that must be met for an induced tone to be heard. One of these is that the tone must not change in anyway as it enters or leaves the noise band [16]. The shortest silent interval inserted in between the tone and the noise, or a small deflection in the tone's trajectory, will cancel the induction effect. Either of these conditions weakens the interpretation that the tone is continuing under the noise. Considering the example of "essay," the /eh/ and /ay/ can probably be fairly reliably grouped by $F0$ and spectral continuity as discussed earlier. So now there are two possible interpretations: the /s/ is part of the speech (hence filling the gap) or the /s/ is some other sound masking the /eh/ component. Now consider the spectral properties at the boundaries of the /s/. Even if there is an apparent spectral discontinuity, at a microscopic level there will be rapid transitions in the vowel sounds as the articulators move into and out of the required configuration for producing a sibilant. These transitions would not occur if the /s/ were simply a superimposed noise. So, the most consistent interpretation must be to integrate the /s/ into the gap in the /eh/ component.

We have reviewed evidence that suggests that both $F0$ and spectral continuity promote sequential grouping that binds components of the speech signal across time. However, these experiments say little about the mechanism underlying these effects. To what extent can grouping by spectral continuity, say, be regarded as a "primitive" process? An alternative view would be that the effect is a product of learned schema-driven grouping processes. The /s/ is incorporated into "essay" simply because we have a model for how the word "essay" should sound, and if the observed acoustics fit that model, then a coherent percept will be formed. This balance between the extent to which grouping is governed by generic data-driven processes, and the extent to which it is governed by specific learned patterns, is a theme we shall return to later in the context of CASA-motivated ASR techniques.

9.2.3 Speech Schemas

Listening experiments with speech-like stimuli, and attempts to build computational models of auditory scene analysis, have demonstrated that primitive grouping processes of the type discussed so far are insufficient in themselves to account for auditory organization. Bregman describes a higher level of auditory processing that he terms "schema-driven" processes. These processes can be envisaged as "looking down" on the output of earlier processes to find support for learned models of commonly occurring sounds. For example, the schema-driven processes may apply to the output of the primitive grouping stage, and act to bind together local source fragments into a coherent whole. Alternatively, in some models, schema-driven processes that operate directly on the signal are invoked to explain the perception of mixed sources in situations in which there are no apparent primitive cues for separation, for example, simultaneous vowels with equal $F0$ as described later in this section.

Schema-driven processes are presumed to play a central role in explaining the perceptual coherence of speech. Despite the presence of $F0$ and spectral continuity, it has been argued that the speech signal remains insufficiently continuous for it to cohere on the basis of primitive processes alone [91]. What prevents tongue twisters such as "she sells sea shells" from disassociating into separate streams of consonants and vowels? The /ʃ/ and /s/ sounds are similar and can be imagined as grouping together on this basis, but there appears to be little reason for this unvoiced group to adhere to the acoustically distinct /ee/ and /eh/ vowels sounds (see Fig. 9.1). (However, as we have seen, even at these boundaries there is some spectral continuity due to the smoothness in the trajectories of the vocal articulators.) Indeed, a meaningless cycle of alternating hiss and buzz sounds *will* segregate into two streams [112]. Bregman argues that it is learned speech schemas that promote perceptual coherence despite apparent discontinuity. A compelling demonstration of this idea can be found by listening to the group of African languages that employ a "click" sound that is wholly unfamiliar as a speech sound to speakers of languages outside this group. Untrained listeners experience these languages as decomposing into two perceptual streams, with one stream sounding like someone speaking a foreign language, and the other sounding like a rapid sequence of clicks. The two streams appear totally unrelated and it is very hard to imagine them as a coherent whole. Africans who have grown up listening to these languages will hear the clicks as being as well integrated into the speech stream as more familiar plosives sound to us. It is presumed they have learned a model of the sound sequences that occur in their language, which acts to pull the clicks into the speech stream as long as they appear in linguistically appropriate places. Pushing this theme further, we may speculate about the experience of someone who had somehow managed to grow up never having heard any form of speech and who is then suddenly introduced to speech for the first time. Would such a person experience speech as a coherent stream, or as a confusing jumble of interlaced sound sources?

Indirect evidence of a source segregation role for speech schemas has come from double-vowel experiments. These experiments have been commonly employed to demonstrate the role of primitive grouping cues such as harmonicity (see Chapter 2). However, it has also been observed that even when there is no $F0$ difference, subjects are able to achieve recognition scores well above chance levels. For example, in a much-repeated study using eight vowel sounds synthesized with identical $F0$s, Scheffers reports that both vowels were recognized correctly 45% of the time compared with a chance level of 3.6% [99]. Even when the formants were excited using noise, mimicking whispered vowels, both could be recognised correctly 26% of the time. In such conditions, successful recognition occurs despite the lack of any low-level cues to promote primitive segregation. One explanation of these results would be that listeners have learned models of the vowel pair combinations, and that the double-vowel stimulus is being identified directly without segregation per se. However, listeners are able to gain results above chance without any specific training. An alternative explanation is that some form of schema-induced segregation is occurring, allowing the vowel sounds to be recognized in much the same way that the vowels separated by primitive mechanisms are recognized.

Less controversially, speech schemas appear to be powerful mechanisms for fusing elements of the speech stream. There are many examples in which schemas appear to override primitive processes that promote segregation. For example, Cutting [33] demonstrated such a case by taking the first two formants of diphones such as /da/, /ba/, and /ga/ and playing them to opposing ears. Even when the formants were synthesized using separate $F0$s, listeners were able to perceive the correct phonetic category, a perception that could only arise if the formants were being fused.

It might then be asked, if such a powerful system of schema-driven perceptual organisation exists, why is there any need for the primitive processes at all? Is it possible that auditory organization can be driven in a totally "top-down"' fashion by this learned knowledge? Indeed, some people may take this strong view. However, an opposing view is that without the primitive processes there would be nothing to "bootstrap" the schemas. Schemas work by using knowledge of learned familiar patterns, but it is perhaps necessary to have access to primitive mechanisms so that coherent patterns can become familiar in the first place. Even once the schemas are fully developed, there remains an advantage to having access to primitive processes, for it is these that act to constrain the organizational hypotheses that the schema-driven processes generate. As Bregman points out, "If a schema-based integration could become entirely independent of primitive scene analysis, then camouflage would be impossible" [16, p. 404], as camouflage only works by minimizing the cues that allow primitive processes to segregate a body from its background. Once the hypothesis that the object is there has been tried, then the camouflage fails to work. So primitive processes perhaps limit the space over which schema-driven processes need to "search" in order to arrive at the correct hypothesis. This is an idea that can be directly translated into an ASR system.

Significant differences have been observed in the way in which primitive processes and schema-driven processes organize the available acoustic information. Primitive grouping seems to behave in a symmetrical manner, in that one grouping of components defines a background consisting of the components it excludes. In contrast, schema-driven grouping behaves asymmetrically; the figure does not define the ground. For example, Dowling [42] demonstrated that increasing listeners' familiarity with a tune could help them in segregating it from a set of distractor tones, but teaching them the pattern of the distractor tones did not help them to hear out the tune. This is a key point, and clearly limits the space of plausible models. This may suggest that although a CASA-based ASR system requires detailed models of the target source, it perhaps does not need detailed models of the background.

Finally, it may be noted that the concept of a schema is something with which most speech technologists would feel very comfortable. As we will see, automatic speech recognition is based on a "top-down" model-driven search. Typically, models of utterances are built up from models of words, which are in turn constructed from models of subphonetic speech units. The effectiveness of an ASR system is a function of how well it can model the speech acoustics, and how efficiently it can search over the space of word sequences for the sequence most likely to have generated the observed signal. Although the models employed by ASR are, for many reasons, openly acknowledged to be imperfect representations of speech, they are probably the closest thing we have to workable speech schema.

9.2.4 Challenges to the ASA Account of Speech Perception

The previous sections have reviewed long-standing accumulated evidence that the perceptual organization of speech is mediated by schema-driven processes acting on a primitive layer of gestalt-grouping mechanisms. Despite the strength of this evidence, the ASA account of speech perception is not universally accepted. Objections arise from the "speech is special" school of thought, which can be broadly characterized as believing that speech perception requires mechanisms that are fundamentally different from those required for the perception of all other sound sources. One version of this idea is that speech perception operates in a preemptive fashion that is isolated from mechanisms that describe the perception of other signals [91]. It is easy to believe this is true. The speech signal has properties that make it unique amongst all the sound sources received by the ear. First, it communicates a complex message. Whereas most environmental sounds have a fairly simple structure, the speech signal is necessarily complicated. It is composed of rapid sequences of segments with highly contrasting properties, and with each segment seamlessly blended with the next. Second, speech is being received by a system that, at some level, "understands" how the signal was constructed. Speech production and speech perception are complementary tasks, and it is not hard to imagine that they may be linked by common mechanisms (this idea is central to Liberman's "motor theory" of speech perception [71]).

In 1994, Remez at al. [91] published a critique of auditory scene analysis that focused on ASA's seeming inability to account for the coherence of speech. They argued that speech is perceptually coherent despite violating the gestalt-grouping principles that are the foundation of ASA. Further, they argued that as ASA is posited as a general account, its failure to handle one particular stimulus falsified the entire account. Their arguments were supported by a series of experiments using an artificial speech stimulus, known as sine-wave speech (SWS). This speech-like stimulus was generated using just three time-varying sinusoids whose frequency and amplitude mimic those of the first three speech formants. Listeners hear this stimulus as a perceptual whole, rather than as individual sinusoids, and can, with some practice, transcribe sine-wave sentences with a high degree of accuracy. Remez et al. argued that there was nothing in Bregman's account to explain the grouping of the SWS sinusoids. Low-level grouping cues cannot be engaged as the sinusoids are not harmonically related, nor is their frequency or amplitude variation obviously correlated. They also argued that speech schemas would not be invoked as the SWS stimuli are sufficiently unspeech-like to naive listeners. In place of ASA, they suggested that phonetic perceptual organization is achieved "by specific sensitivity to the acoustic modulations characteristic of speech signals."

The coherence of sine-wave speech despite an apparent lack of grouping cues poses an interesting challenge for ASA. However, SWS is significantly less intelligible than natural speech and there are many listeners who find it wholly unintelligible [91]. Carrell and Opie [22] studied the effect of "reinserting" grouping cues into the SWS stimulus by coherently amplitude modulating the three sinusoids at a frequency of 100 Hz. Common amplitude modulation is well documented as a powerful grouping cue [17, 57], and when applied to SWS it becomes very difficult to "hear out" the individual sinusoids. Carrell and Opie showed that this modulation dramatically improved the intelligibility of the stimulus. Crucially, a second condition, in which different modulation frequencies were applied to each of the three simultaneous sinusoids had no effect on intelligibility.

Further, Remez et al.'s claim that SWS is devoid of grouping cues is perhaps overstated. The sinusoids have common onsets. This initial grouping coupled with a partial triggering of speech schemas could account for the loosely coherent percept that is generated. Treating a stimulus as a single source is in some respects the "default perception," and Bregman has shown that an acoustic mixture will not fragment into multiple streams until sufficient evidence for a more complicated interpretation has accumulated [15]. If SWS is to demonstrate that speech perception is independent of primitive grouping, it should be possible to "hear out" and transcribe a single speaker in a simultaneous SWS utterance condition. However, intelligibility in this condition is severely reduced [4]. Listeners do not perceive two speech streams, but are occasionally able to hear isolated words of one or other of the sources. According to Bregman's account, if an organizational cue were reintroduced to these mixtures—for example, if the two speakers were commodulated at different frequencies—then the perception of two speech streams should be restored and intelligibility should increase. To date, this key experiment does not appear to have been attempted.

Sine wave speech remains an interesting test of CASA-motivated ASR models. The fact that humans are willing to search for linguistic meaning in SWS and other similarly highly distorted speech signals says much about the difference between ASR and human speech perception. The human acoustic model seems able to see beneath the surface properties of the signal and recognize the commonality between SWS and speech. Acoustic models employed with ASR are essentially superficial and will be broken by the slightest spectral distortion. It is likely that until we develop a more sophisticated model of speech acoustics, attempts to use our understanding of ASA to enable ASR will be met with limited success.

9.2.5 Interim Summary

We have seen in this section that the ASA account of perceptual organization can provide the basis for understanding the results of a wide range of speech perception studies. There is evidence that elements may be grouped together across frequency if they are harmonically related, or if they share a common time of onset. Elements may be sequentially grouped through time if grouping them produces a continuous $F0$ track or a smoothly varying spectral shape. However, it has also been seen that primitive grouping alone cannot explain the coherence of the speech signal. Instead, it would appear that the primitive processing would encourage speech to fracture into a number of locally distinct vowel and consonantal fragments. The coherence of speech is at least partly dependent on learned schema-driven processes that pull these fragments back into a common stream.

We have looked at arguments against the ASA account that claim that it fails to support speech perception. These arguments may be persuasive, but they attack an extreme and simplified caricature of the ASA account in which primitive grouping makes firm decisions, and then schema-driven processes have to correct the mistakes that have been made. It is generally understood that top-down and bottom-up ASA processes are not "decoupled" in this manner. The true picture is likely to be far more complex. The primitive grouping processes may be more realistically modeled as "suggesting" groupings rather than dictating them. These suggestions may carry more or less weight but they are never absolute. Then primitive processes and schema-driven processes may be imagined as being intimately coupled by a system that operates to mutually satisfy bottom-up and top-down constraints. Computational models of ASA have often leaned toward a simpler interpretation of Bregman's ideas. This simplicity has no doubt been driven by a desire to keep models clearly compartmentalized and therefore easy to engineer. However, we shall see in the remainder of this chapter that more sophisticated models are starting to emerge.

9.3 SPEECH RECOGNITION BY MACHINE

Having discussed speech perception in some detail, we will now proceed to introduce the statistical framework that is the basis of all currently competitive ASR technology. The problems of deploying this technology in an unpredictable environment will be discussed, and the limitations of various mainstream solutions will be briefly examined. As a detailed account of these topics is outside the scope of this chapter, readers with no prior knowledge of ASR, are referred to the many good books dedicated to the subject [65, 61, 86]. For comprehensive reviews of robust ASR see [66, 55, 68, 10]. This section ends with a brief overview of some of the earlier attempts that have been made to couple CASA and ASR. Particular techniques will be discussed in detail in the sections that follow.

9.3.1 The Statistical Basis of ASR

The first stage of speech recognition is to analyze the acoustic signal and extract a temporal sequence of acoustic feature vectors that describe the speech. The goal of the recognizer can then be stated as finding the word sequence $\hat{W} = \{w_1, w_2, \ldots, w_{nw}\}$ that has the maximum a posteriori probability given the observed sequence of acoustic feature vectors, $X = \{x_1, x_2, \ldots, x_{nx}\}$:

$$\hat{W} = \operatorname*{argmax}_{W} P(W|X) \tag{9.1}$$

This equation may be rearranged using Bayes' theorem to give

$$\hat{W} = \underset{W}{\operatorname{argmax}} \frac{P(X|W)P(W)}{P(X)} \tag{9.2}$$

As $P(X)$ is independent of W, it can be dropped from the maximization, and we are left with

$$\hat{W} = \underset{W}{\operatorname{argmax}}\, P(X|W)P(W) \tag{9.3}$$

Here, $P(W)$ is the prior probability of a particular word sequence and is estimated using a language model. $P(X|W)$ is the distribution of the feature vector sequence for a given sequence of words—the acoustic model. In most successful speech recognizers, the acoustic model is represented using a hidden Markov model (HMM), which typcially models each word sequence as a sequence of stationary states, $Q = \{q_1, q_2, \ldots, q_{n_X}\}$, where each state in the sequence represents some very small segment of a word (typically the beginning, middle, or end portion of a single phoneme). The problem of finding the best word sequence is then equivalent to that of finding the best state sequence and Eq. 9.3 can be rewritten as

$$\hat{Q} = \underset{Q}{\operatorname{argmax}}\, P(X|Q)P(Q) = \underset{Q}{\operatorname{argmax}}\, P(x_1, x_2, \ldots, x_n|q_1, q_2, \ldots, q_n)P(q_1, q_2, \ldots, q_n) \tag{9.4}$$

where n is being used to represent n_x, the number of feature vectors in the observed sequence.

The model is simplified by making a couple of approximations. First, it is assumed that the state sequence can be described as being generated by a first-order Markov process (i.e., one in which the probability of the next state depends only on the identity of the current state). Second, it is assumed that the observation, x_t, depends only on the current state, q_t. Under these assumptions Eq. 9.4 becomes

$$\hat{Q} = \underset{Q}{\operatorname{argmax}}\, P(X|Q)P(Q) = \underset{Q}{\operatorname{argmax}} \prod_{i=1}^{n} P(x_i|q_i) \prod_{i=1}^{n} P(q_i|q_{i-1})P(q_1) \tag{9.5}$$

Note, this is a hidden Markov model in which the observed acoustics, X, depend on an unobserved (i.e., "hidden") sequence of states, Q. This model can be turned into a practical speech recognition system by taking advantage of two important algorithms. The first is the Baum–Welch algorithm [8] (a case of the more general expectation–maximization algorithm [40]), which allows one to take a corpus of transcribed speech data and learn the parameters of the HMM that best fit the model to the data. For a typical large vocabulary speech recognition system, tens or even hundreds of hours worth of transcribed training data will be employed. The second is the Viterbi algorithm, which takes a trained HMM and a sequence of acoustic observation vectors and efficiently solves Eq. 9.5 to return the most likely sequence of states, and hence words. Details of these algorithms are outside the scope of this chapter, but the interested reader is referred to Rabiner's very accessible HMM tutorial [87].

Clearly, for good recognition performance, the distributions, $P(x|q)$, that are estimated from the training data need to be a good model of the data that is to be recognized. For example, if all the data in the training set has been collected from male speakers, then recognition performance will probably be poor for female speakers. If the training data has been collected in a nonreverberant environment, then it will not generally produce models suitable for deploying in a reverberant environment (see Chapter 7). And most crucially, if the training data has been collected in the absence of additive noise, the speech recognizer will not work well when competing sound sources are present. Much research in recent years has been devoted to finding solutions to this problem.

9.3.2 Traditional Approaches to Robust ASR

Speech recognition technology can be remarkably fragile. It is often designed to work well in a very narrow range of operating conditions, but with performance that deteriorates significantly if the operating environment is not carefully controlled. There has been much research devoted to engineering ways to make speech recognition more "robust." All robust strategies must deal with the central problem—error caused by mismatch between the acoustics of the training data and the acoustics experienced at the time of operation. There are many techniques that have been explored, but they can be roughly characterized under one of the following three general strategies.

The first strategy is to design a speech parameterization that is insensitive to changes in the environment. An example of this approach is the use of RASTA features [60], which essentially filter out parts of the signal that do not share the characteristic 4 to 50 Hz range of envelope modulation frequencies typical of speech. The parameterization is essentially "deaf" to near-stationary and rapidly varying additive noise components. JRASTA generalizes the idea to accommodate nonstationary convolutive noise, that is, effects caused by changes in the communication channel. Although these techniques can be very effective in some situations, they have clear drawbacks. The most serious problem is the reliance on an easily characterized difference between the target utterance and the noise sources. If the target and masker have similar properties, for example, if the masker is in fact a second speech source, then neither RASTA nor any similar strategy will be able to confer an advantage.

The second strategy is to reduce mismatch by adapting the speech models to match the statistics of the noisy signal. If the noise conditions are known in advance, then simply training on noisy examples of the speech can be advantageous. Systems trained on a range of noise conditions (multicondition training) are generally more robust than systems that are trained more narrowly [84]. However, in the absence of prior knowledge of the noise, the design of the training set is problematic. Multiband systems [14] that filter the speech signal into a small number of overlapping frequency bands and analyze each band separately can be trained with broadband additive noise in such a way as to produce systems that generalize well over a range of noise conditions [44]. However, such systems cannot in general handle nonstationary noises at low signal-to-noise ratios (SNRs), and perform poorly in multispeaker conditions. Online model adaptation, which changes the model parameters at recognition time, can work well for small mismatches [115]. These techniques are typically used to adapt models to the characteristics of a given speaker that has not been encountered in the training data. A more sophisticated strategy for adapting models to deal with additive noise is to combine models of the clean speech signal with models for the noise. Again, this presumes that sufficient is known about the noise to make an adequate model. Techniques that employ this strategy, such as HMM decomposition [107] and parallel model combination [51], have been shown to produce very good results when the noise can be adequately modeled with a simple structure. However, it is hard to see how these techniques can be applied in truly unpredictable environments in which a library of many different noise models would be required to describe the scene. The factorial nature of HMM model combination leads to an explosion in the size of the state space when more than a few models are present. For example, a task as simple as recognizing connected digits may require as many as 200 states for a gender-dependent implementation. Describing two simultaneous speakers would require 40,000 states, and three simultaneous speakers would imply 8,000,000 states. Curiously, systems based on this approach are able to simultaneously recognize each source present; for two speakers, two text transcriptions are produced. Humans are not capable of this task, which may suggest that our auditory system does not model complex noisy signals in this way.

As a final strategy, rather than attempting to model the noisy data, one may attempt to filter out the noise so that the signal better fits the models of clean speech. These approaches are termed speech enhancement. Speech enhancement techniques have been reviewed by Ephraim et al. [47], and the approach has been compared with CASA in Chapter 1 of this book. Briefly, unlike CASA systems, these techniques are mostly designed to work in very specific situations in which the characteristics of the noise are well understood. Nevertheless, this sort of system can have useful application, for example, for cockpit or in-car speech recognition. If the noise is quasistationary, then good results can be achieved by simply estimating its spectrum during periods of the signal when there is judged to be no speech present, and then subtracting this estimate from the spectrum of the acoustic mixture [13]. There have been many variants of this technique, which differ in detail according to how they compute the noise spectrum estimate and in how they remove this estimate from the mixture (e.g., [73]). The noise can also be successfully filtered out in situations in which there are multiple microphones present (see Chapter 6). If there are at least as many microphones as there are noise sources, then techniques based on independent components analysis (ICA) can be employed to estimate the speech signal as a linear mixture of the multiple observations [9]. However, these techniques are disrupted by reverberation and are obviously incompatible with single-channel applications.

In summary, robust ASR draws on a broad spectrum of techniques, each of which can perform very well if certain assumptions about the noise happen to be met. However, real world noise is essentially unpredictable. If we want our ASR systems to operate in the wide range of noisy environments with which humans regularly cope then we need to find more general techniques.

9.3.3 CASA-Driven Approaches to ASR

The appeal of CASA as a technique is that it is based on a very general account of sound organization. The only assumptions it makes are assumptions that arise from the physics of sound, and these, therefore, apply equally well to all sound sources. Although these loose assumptions may not lead us to a noise-free speech signal in a single step, they can be developed into a framework that provides a potential solution to the robust ASR problem.

Weintraub (1985) [114] was the first to explicitly pair CASA and ASR. His model attempted to apply ASA principles to the separation of two simultaneous speakers with the hope of improving automatic speech recognition results. He employed a pitch-based separation technique that followed the neural autocoincidence model of Licklider [72]. The speech signals were labeled as voiced, unvoiced, transitional or silent and a Markov model was used to track the states of each speaker. The state sequence controlled a dual-pitch tracking algorithm, and the pitch estimates were used to divide the mixed speech energy to form spectral estimates for each source. Judged in terms of speech recognition performance, Weintraub's system was disappointing. However, considering that twenty years ago the statistical foundations of ASR were still being laid, it is perhaps not surprising that the project met with limited success. Weintraub himself was fully aware of the deficiencies of his approach, and had sufficient insight to realize that the two-speaker problem could not be solved without the influence of top-down model-driven factors that were not implemented in his system.

Following the difficulties experienced by Weintraub, researchers backed away from using ASR as a CASA performance metric. Instead, focus turned to systems that attempted to reconstruct the separated signals, and which measured performance in terms of improvements in signal-to-noise ratio. This more general metric has the advantage of being applicable to nonspeech sounds, and could be applied to truly general CASA systems such as those developed in the early 1990s by Cooke [25] and Brown [19] and, later, Ellis [45].

CASA systems that are able to enhance speech through segregation and resynthesis have obvious applications. In recent years there has been increased interest in CASA-motivated speech-enhancement systems (e.g., [77, 11, 104, 98]). However, signals reconstructed for human ears are not necessarily well suited for machine recognition. An enhanced speech signal may contain artifacts that prove little more than a minor distraction to the human speech perception system, but which are disastrous for a standard ASR system. This difference in goals has lead to a divergence in techniques.

From the mid-1990s, fresh attempts were made to link CASA and ASR. This more recent work has taken advantage of the progress made in ASR since Weintraub's initial efforts. A more advanced statistical framework has meant that it is no longer necessary to resynthesize masked speech, but that missing information can be handled in a principled manner. Cooke et al.'s "missing data" approach [24] has demonstrated that it is possible to obtain good performance in some circumstances while still using the same basic left-to-right, data-driven processing employed by Weintraub. We shall examine these systems in some depth in the next section.

Ultimately, however, when dealing with highly unpredictable noise sources, purely bottom-up systems have failed to segregate sound sources reliably. When interpreting complex, potentially ambiguous, acoustic scenes, it appears necessary that the source segregation mechanism have access to top-down information. This realization spurred the development of top-down mediated systems such as Ellis' "prediction-driven CASA" [45]. However, although these earlier systems, have proved capable of producing convincing descriptions of complex mixtures of environmental sounds, they have not provided the statistical framework necessary for compatibility with modern ASR. Recently, systems have started to emerge that address this issue by describing bottom-up and top-down processing as components of a common statistical model. Section 9.5 will examine the development of one such technique, "speech fragment decoding" [5], which employs missing-data techniques at its core to evaluate multiple bottom-up organizations of the acoustic mixture.

9.4 PRIMITIVE CASA AND ASR

We saw in the previous section that attempts to resynthesize the output of a CASA front end and pass the result to traditional ASR systems have proved disappointing. A more sophisticated approach is to perform ASR directly using the representations employed by the CASA front end without intermediate resynthesis. For this approach to work, several problems have to be addressed, chief among which are, (i) CASA systems produce partial representations of sources in which information has been lost due to masking, and (ii) the representations employed by CASA systems do not fit the simple statistical models preferred by ASR systems. This section will examine solutions to these problems that have enabled CASA-based ASR systems to produce robust recognition results on tasks with small-to-medium-sized vocabularies.

9.4.1 Speech and Time-Frequency Masking

CASA techniques operating in the spectral domain are able to exploit the fact that most sound sources are sparsely represented in time and frequency. Highly nonstationary, structured sources, such as speech, have their energy concentrated narrowly in time and/or frequency. The ear is able to form a spectrotemporal representation in which sound sources, which may be totally overlapping in time, are only partially overlapping in the time-frequency plane. By analyzing the signal at each point in time and frequency in terms of features such as interaural time difference (ITD), interaural intensity difference (IID), and amplitude modulation (AM), sufficient evidence can be gathered to group time-frequency points and recover partial descriptions of the various sources present. Inevitably, sources will still overlap to some degree. In some frequency channels, multiple sources will have significant energy at the same time, and weaker sources will be dominated by the most energetic source. In this case, the weaker sources are effectively "masked" and information may be lost. However, despite this overlap, partial descriptions of each source can be seen in the spectrotemporal gaps left by the competing sources.

Consider the case of two simultaneous speakers. Figure 9.2 displays an auditory filterbank representation (henceforth referred to as a cochleagram) of a male speaker (panel **A**). The center panel shows a cochleagram of the same utterance but with the addition of a simultaneous female speaker mixed at an SNR of 0 dB. What is striking is that many of the key features of the male speech are clearly visible in the

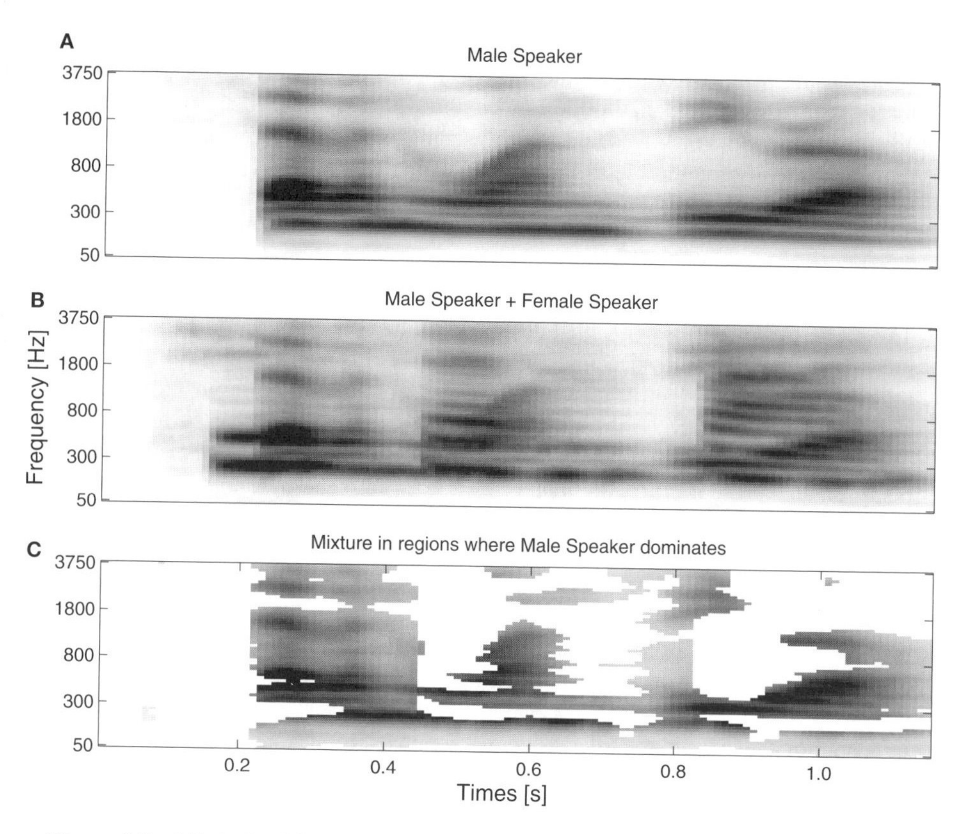

Figure 9.2 (**A**) A "cochleagram" of the utterance "seven one zero" spoken by a male speaker. (**B**) A female speaker saying "two seven five" has been added to the male utterance at 0 dB SNR (i.e., both sources have equal energy). (**C**) The spectrotemporal regions of the mixture in the central panel where the contribution of energy from the male source is greater than that from the female source. Note that in these regions the pattern of male speech energy is mostly unaffected by the presence of the female speech.

mixture and appear relatively undistorted. Panel **C** shows the spectrotemporal regions where gaps in the female speaker leave the male source virtually unchanged. It is clear that though some of the energy of the male source is effectively masked, most of the regions of significant energy remain unmasked. So by "listening in the right places" we can receive an incomplete but otherwise undistorted rendition of either of the speech sources present. Similar results would be expected for mixtures of other nonstationary sources.

Although information is lost due to masking, fortunately the speech signal is highly redundant. Information is multiply encoded and cues for identifying speech units are spread both across frequency and time. This redundancy is demonstrated by intelligibility tests using speech from which information has been artificially removed. For example, if conversational speech is filtered to remove all frequencies above 1800 Hz, it remains highly intelligible [50, 1]. However, if these low frequencies are removed, and the frequencies above 1800 Hz are retained, then the speech is equally intelligible. In fact, speech remains intelligible even after band-pass filtering with pass bands as narrow as one-third of an octave [113]. Similar effects occur in the time domain. Experiments in which the speech signal is gated on and off in a periodic fashion so that half the signal is removed, show that the speech remains perfectly intelligible as long as the gating function has a period not much larger than the average phoneme rate.

9.4.2 The Missing-Data Approach to ASR

The properties of the speech signal described in the previous section suggest a simple two-step approach to speech recognition. In the first step, primitive ASA grouping processes act to identify regions of the spectrotemporal representation that are energetically dominated by the speech source. This information can be represented as a binary spectrotemporal "mask," in which each time-frequency unit is labeled as being either "present" (i.e., dominated by the target speech source) or "missing" (i.e., dominated by one of the competing masking sound sources). In the second step, this partial representation is matched against a model of noise-free (i.e., "clean") speech. This straightforward approach is the basis of the "missing data" speech recognition technique [24].

The missing-data approach is consistent with Bregman's principle of "exclusive allocation." A binary mask allocates time-frequency regions as belonging exclusively to either the target source or one of the masker sources. To use a visual analogy, auditory scenes are regarded as being composed of opaque objects; parts of objects are either visible (in which case they are undisturbed by the other objects present), or they are masked (in which case they are totally hidden by the occluding object). This is an unusual manner in which to model sound. When sound sources combine, all sources present will contribute to the energy observed at every time and frequency point. So is it reasonable to build ASR on a front end that treats sound in this peculiar way?

In modeling sound as "opaque," we expect that when speech is mixed with noise, the energy in time-frequency regions where the speech source dominates is similar to the energy that would be observed if the speech had been present in isolation. This turns out to be a very good approximation in most cases. The sparsity of the speech encoding means that when speech is mixed with other signals with similar average energy, in the local regions where the speech energy is concentrated, it typically dominates the background energy by several orders of magnitude. When using a representation that compresses the energy (for example, via a log or cube root function), the background energy has no appreciable effect on the observed level of the speech energy, that is, the observed log energy of the mixture is very close to the log energy of the most energetic source (this observation has become known as the "max approximation" [107]; see also Chapter 1). The "max approximation" is illustrated in Fig. 9.3. The panel **A** shows a log-scaled cochleagram of a simultaneous male and female utterance. The lower panel shows a cochleagram generated according to the max approximation by first computing cochleagrams of the two sources in isolation, and then combining the cochleagrams by selecting each time-frequency point from the cochleagram with the maximum value. Note that the gray-scale images appear almost identical. The bottom panel compares a spectral slice taken across the cochleagram of the mixture with one taken across the max-approximation cochleagram. If the source energies combined in a strictly additive fashion, then the maximum theoretical difference between the mixture and the max approximation would be about 3 dB. In practice, the error rarely exceeds 4 dB, which is small in comparison with the large dynamic range of the speech signal (see Fig. 9.3**C**). These larger errors occur when the two sources have equal energy, but because of the sparsity of the sources, this local equal energy condition rarely occurs, even when the speakers are mixed at a global SNR of 0 dB.

For missing-data speech recognition to be implemented, two problems need to be solved. First, how can we reliably identify the regions of the spectrotemporal representation that are not masked? This step requires something akin to CASA. It remains a challenging problem, but, as we shall see later, it turns out that for some tasks, surprisingly good recognition results can be obtained with crude approximations of the unmasked regions. Second, how can we compute the probability of a word sequence given a sequence of spectral-feature vectors in which some features are marked as missing? This problem of dealing with incomplete observations is one that occurs frequently in statistical analysis (e.g., dealing with unanswered questions in surveys) and solutions to it are well understood. If classification decisions need to be based on incomplete data, there are essentially two strategies:

1. **Marginalization,** which amounts to considering all possible values of the missing features
2. **Imputation,** first estimate values for the missing features, and then classify on the basis of the reconstructed data

Missing-data speech recognition systems can be designed using either of these two approaches. Each presents its own particular advantages and disadvantages, which will be discussed in detail in the following sections.

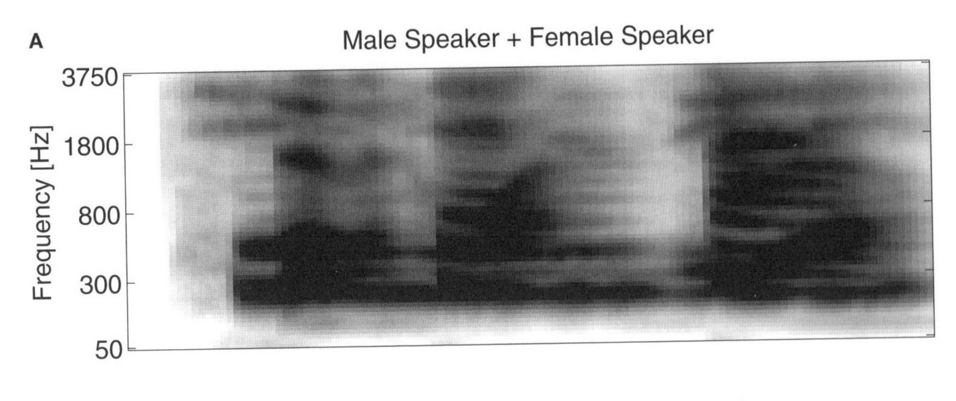

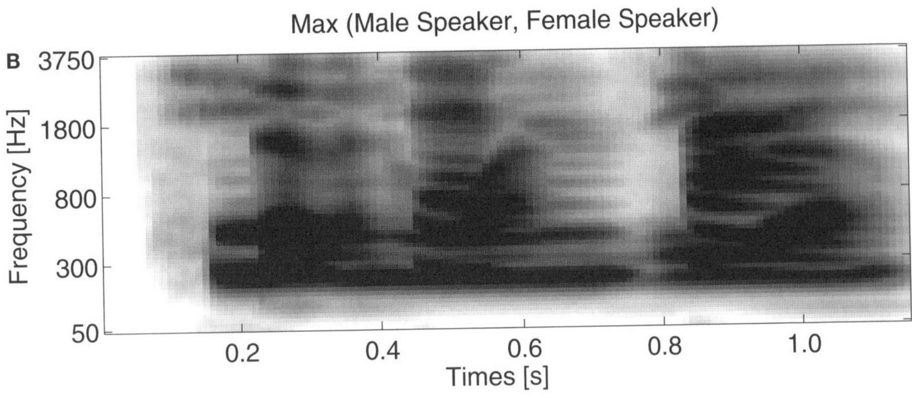

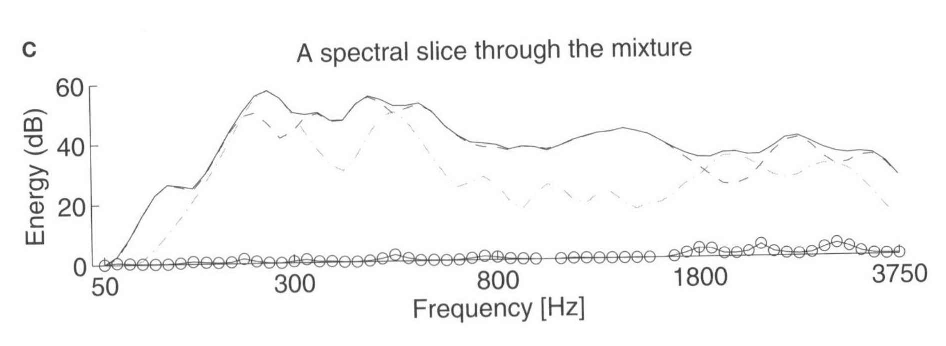

Figure 9.3 (**A**) A cochleagram of simultaneous male and female digit sequence utterances ("seven one zero" and "two seven five") mixed so that each has equal energy. (**B**) max(M_{tf}, F_{tf}) where M_{tf} and F_{tf} are cochleagrams of the male and female utterances prior to mixing. Note how similar this is to the cochleagram of the mixture (top panel). (**C**) A spectral slice taken from the cochleagram of the male plus female speech mixture at frame 30. The spectrum is always close to the maximum of the spectra for the two individual sources (dashed lines). The difference between the two, that is, the error produced by the "max approximation," is plotted as the line marked with the circles.

9.4.3 Marginalization-Based Missing-Data ASR Systems

Many of the marginalization-based missing-data ASR techniques discussed in the following sections were originally developed by Cooke et al. The reader is referred to [24] for more detailed analysis.

Marginalization. Let us consider the spectral feature vector x as being segregated into "present features," x_p, and "missing features," x_m. In the arguments that follow, the order of the components of the vectors x can be interchanged without loss of generality, so we can safely write $x = \{x_p, x_m\}$. As we saw in Section 9.3.1, in a standard ASR system we are searching for the state sequence, Q (and hence the word sequence), which is most probable given the the observed data, X:

$$Q = \underset{Q}{\operatorname{argmax}}\, P(Q|X) = \underset{Q}{\operatorname{argmax}}\, P(X|Q)P(Q) \tag{9.6}$$

However, rather than observing the complete data, X, we are now observing a sequence of incomplete feature vectors, X_p. So the problem can be restated as

$$Q = \underset{Q}{\operatorname{argmax}}\, P(Q|X_p) = \underset{Q}{\operatorname{argmax}}\, P(X_p|Q)P(Q) = \underset{Q}{\operatorname{argmax}} \prod_{i=1}^{n} P(x_{pi}|q_i)P(Q) \tag{9.7}$$

where x_{pi} are the n individual feature vectors from which the sequence X_p is composed, and q_i are the states which form the state sequence, Q.

The distribution, $p(x_p|q)$ can be evaluated by integrating over the missing components:

$$p(x_p|q) = \int p(x_p, x_m|q)dx_m = \int p(x|q)dx_m \tag{9.8}$$

The recognition procedure needs very little modification. The HMMs can be trained on clean speech data to obtain the state emission distributions, $p(x|q)$. Then at recognition time each frame is split into missing and present components. For each frame, rather than compute the usual state likelihoods as $p(x|q)$, we instead compute $p(x_p|q)$, which are computed from $p(x|q)$ by integrating over the missing features.

The computation required to evaluate Eq. (9.8) will, of course, depend on the form of the distribution, $p(x|q)$. Missing-data techniques are based on spectral features, which, unlike the cepstral representation favored by most ASR systems, exhibit a high degree of correlation across features. Some of the early missing-data systems attempted to model the spectral data using multivariate Gaussian distributions with full covariance matrices [27]. If the observed feature vector elements have been reordered and segregated as $x = \{x_p, x_m\}$, then the mean and covariance matrix of the Gaussian distribution can be similarly reordered and written as

$$\mu = \{\mu_p, \mu_m\} \tag{9.9}$$

$$C = \begin{Bmatrix} C_{pp} & C_{pmp} \\ C_{mp} & C_{mm} \end{Bmatrix} \tag{9.10}$$

where μ is the reordered mean vector and C is the reordered covariance matrix. The submatrix C_{pp} represents the covariances between pairs of frequency channels in which the data is present (i.e., not masked); C_{mm} is the covariances between pairs of frequency channels in which the data is labeled as missing; and C_{pm} is the covariances between channels marked as present and those marked as missing, and vice versa for C_{mp}. After this reordering, integrating the Gaussian distribution over the missing dimensions produces a marginal distribution of the form

$$p(x_p|q) = \frac{1}{\sqrt{(2\pi)^{|p|}|C_{pp}|}} e^{-\frac{1}{2}(x_p-\mu_p)C_{pp}^{-1}(x_p-\mu_p)^T} \tag{9.11}$$

where $|p|$ is the number of present features.

This distribution is itself a multivariate Gaussian, but of smaller dimension. However, although the form of the marginal distribution is simple there are practical problems that make this distribution unattractive. First, there may be insufficient training data to estimate the parameters of the covariance matrix for each state. Second, there are significant computational overheads. A standard ASR system using a full covariance model will compute the covariance matrices from the training data, and then precompute and store the inverse covariance matrices (known as "precision" matrices) ready for the evaluation of likelihoods at recognition time. For a missing-data system, the precision matrices, C_{pp}^{-1}, depend on which features are present and which are missing, which is obviously not known until recognition time. It is not practical to precompute all possible precision matrices. For example, consider the case in which the spectrum is computed in 64 frequency channels. In this case, there are 2^{64} different possible combinations of missing features and, hence, 2^{64} different partial precision matrices. As the precision matrices cannot be precomputed, they instead have to be computed during decoding. This means that the likelihood computation, which takes place within every state of the acoustic model and is repeated for every time frame, now involves inverting a matrix of typically 64 by 64 elements. This matrix inversion can increase the computational complexity of the likelihood evaluation by an order of magnitude or more, and, for a small- to medium-sized vocabulary task evaluating the state likelihoods is the most time-consuming step of the recognition process.*

An alternative, and more flexible, technique for dealing with the correlation between the spectral features is to employ a Gaussian mixture model (GMM), in which the distribution $p(x|q)$ is modeled as the weighted sum of a number of Gaussian distributions. GMMs used for acoustic modeling typically constrain each

*For large vocabulary systems, the cost of computing likelihoods is secondary to the computational cost of the "decoding" stage, that is, the task of searching for the best state sequence.

Gaussian component to have a diagonal covariance structure. The marginal distribution can now be written as

$$p(x_p|q) = \sum_{\lambda=1}^{M} w_\lambda \frac{1}{\sqrt{(2\pi)^{|p|}|C_{\lambda,pp}|}} e^{-\frac{1}{2}(x_p-\mu_{\lambda p})C_{\lambda,pp}^{-1}(x_p-\mu_{\lambda p})^T} \tag{9.12}$$

where w_λ are the mixture weights of M mixture components. Despite involving an extra summation, unless a very large number of mixtures are employed, the computational cost of Eq. 9.12 is much less than the full covariance case, Eq. 9.11. First, $C_{\lambda,pp}^{-1}$ is simply a diagonal matrix constructed by taking the reciprocal of the elements along the diagonal of $C_{\lambda,pp}$. Second, as the precision matrix is diagonal, the matrix–vector multiplication is of order n rather than of order n^2. So Eq. 9.12 is implemented as

$$p(x_p|q) = \sum_{\lambda=1}^{M} w_\lambda \prod_{i=1}^{n} p(x_i|q, \lambda) \tag{9.13}$$

where $p(x_i|q, \lambda)$ is the univariate Gaussian distribution,

$$p(x_i|q, \lambda) = \frac{1}{\sigma_i\sqrt{2\pi}} e^{-\frac{1}{2}\left(\frac{x_i-\mu_{\lambda,i}}{\sigma_{\lambda,i}}\right)^2} \tag{9.14}$$

and $\sigma_{\lambda,i}$ is the ith element along the diagonal of $C_{\lambda,pp}$.

Note that GMMs can in theory approximate any distribution given that there is sufficient training data to robustly estimate parameters of a large enough number of components. Experiments with the Aurora connected-digit task [84] have shown that good recognition performance can be obtained using mixtures with as few as seven components per state [6].

Exploiting the "Missing-Data Bounds." Figure 9.4 shows an example of a spectral feature vector that might be observed during a noisy speech utterance. Some CASA process has identified the regions of missing data. The frequency channels in which the speech is considered to be masked (i.e., missing) have been labeled m, and the region in which the speech may be considered reliable (i.e., present) has been labeled p. In the standard marginalization approach described in the previous section, the observed values in the region m have absolutely no influence on determining the acoustic model state that is most likely to have produced the observation. However, the region m does tell us something about the speech; we know that the energy that the speech is contributing to the mixture in this region must be *less* than the observed energy. Consider two states, q_1 and q_2, that are modeled by the mean spectra shown in Fig. 9.4. Although they both fit the present regions equally well, only state q_2 is likely to have output energy values that fall beneath the masking energy observed in the region m. We would like this to be reflected in our computation of $p(q|x)$. Whereas before we considered only the influence of the reli-

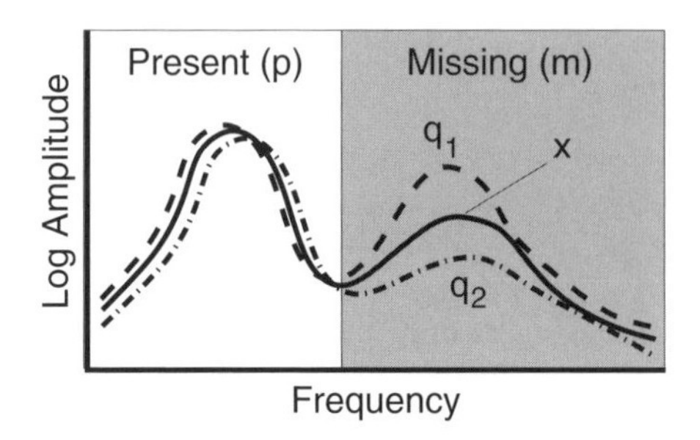

Figure 9.4 An illustration of the benefit of missing-data bounds. x is an observed feature vector. In the low frequencies, the target speech source is known to dominate the acoustic mixture (i.e., the data is labeled as "present"). In the high frequencies, the target is known to be dominated by a masking source (i.e., the data is labeled as "missing"). Consider two potential models of the speech source, q_1 and q_2, the means of which are plotted in the figure. Considering the reliable evidence alone, both models appear to fit the data equally well. Considering x in the missing regions, model q_2 is seen to be a better fit than model q_1, because q_1 is less likely to have produced an observation that could have been masked by the observation x; that is, if q_1 were the explanation of the data we would have expected to have seen evidence of it in the high frequency region.

able region and modeled $p(q|x)$ as $p(q|x_p)$, we now also consider the masked region and evaluate $p(q|x_p = x_p', x_m < x_m')$, where x_p' are the observed features in the unmasked region, and x_m' are the observed features in the masked region.

After application of Bayes' theorem, the distribution we are interested in is

$$f(x) = p(x_p = x_p', x_m < x_m'|q) \tag{9.15}$$

This distribution has been termed a "bounded marginal" [24]. In standard marginalization, we integrate over the missing features from minus to plus infinity. When performing bounded marginalization, we integrate between minus infinity and the observed energy x_m. For the Gaussian mixture model, we make use of the fact that integrating the sum of mixtures over the bounds is the same as summing the integrals of the separate mixtures, and arrive at

$$p(x_p, x_m|q) = \sum_{\lambda=1}^{M}\left(w_\lambda \prod_{i\in p} p(x_i|q, \lambda) \prod_{i\in m} \int_{x=-\infty}^{x_i'} p(x_i|q, \lambda)dx\right) \tag{9.16}$$

The use of bounded marginals has been shown to produce consistent performance improvements over using unbounded marginals [24]. Note that for auditory representations, where x typically represents the cube root of the energy and is therefore always positive, the lower bound of the integration may be effectively taken as 0.

Using Soft Decisions. In all the systems described so far, a binary decision has been made about whether a feature is present or missing. We have justified this by noting that the weaker source makes a negligible contribution to the observed energy (see Section 9.4.1). However, even though each feature may justifiably be labeled as present or missing, in practice we may not be able to reliably estimate this label. The front-end CASA process may only be able to offer a probabilistic judgment; for example, "speech present with 60% confidence." In this case, the spectral–temporal binary mask becomes a "soft mask" containing values in the range 0 to 1. Features no longer belong to either the missing set or the present set; rather, each feature has a probability of belonging to either set. Consequently, $p(x_p, x_m|q)$ is no longer constructed from a product of missing terms and present terms as in Eq. 9.16. Instead, every feature makes a weighted contribution to the missing term and the present term. An implementation of this idea presented by Barker et al. [7] models the likelihood using

$$p(x|q) = \sum_{\lambda=1}^{M} w_\lambda \prod_{i=1}^{d} \left(k_i p(x_i|q, \lambda) + (1 - k_i) \frac{1}{x_i'} \int_{x=-\infty}^{x_i'} p(x_i|q, \lambda)\, dx \right) \quad (9.17)$$

where d is the number of elements in the feature vector, x, and k_i, the probability that the ith feature is present, is taken from the soft mask. The factor $1/x_i'$ seen in the term on the right is a normalizing constant that turns the probability defined by the integral into an average probability density over the allowable range of energy values, 0 to x_i'.

The use of soft masks has been shown to introduce further performance benefits over the standard, discrete mask-bounded marginal system. Unsurprisingly, improvements are greatest in the least stationary noise conditions in which the missing data mask is hardest to estimate reliably [6].

CASA-motivated soft reliability decisions have also been employed in multiband ASR systems (e.g., Glotin et al. [53]). In these systems, the signal is filtered into a small number of overlapping frequency bands (typically 4), and a separate classifier is trained to compute state posterior probabilities, $p(q|x)$, for each band. The classifier outputs for the separate bands are then combined by multiplying weighted versions of the posterior probabilities. The weights are adjusted so that bands judged to be noisy are given less influence. Glotin et al. showed that ITD cues could be used to estimate band weights that produced robust ASR on a task in which simultaneously spoken telephone numbers were recorded with a stereo microphone. Note here, though, that reliability is being judged on the scale of a subband that is typically one-quarter of the full bandwidth of the representation. This contrasts with soft missing-data systems that use a granularity of a single frequency channel. Although multiband systems perform well when the noise is limited in bandwidth and corrupts only one or two frequency bands, they perform less well when the noise covers the full bandwidth of the signal.

9.4.4 Imputation-Based Missing-Data Solutions

Imputation-based approaches to the missing data problem work by filling in the missing values with a "best guess." From a theoretical point of view, this is not an optimal approach. The probability distribution of the missing features given the present features may suggest that there are several very different but equally good ways of reconstructing the data. Choosing any one of these reconstructions does not represent the true uncertainty of the data and may bias the eventual classification. Marginalization is superior in this respect, because it employs the distribution itself rather than one value chosen from it. However, although from an information-theoretic viewpoint marginalization is the more principled approach, there is a good practical reason why imputation may be favorable. Once the missing components of the spectral-feature vector have been reconstructed, all the techniques that are employed in a traditional ASR front end can be employed. For example, the reconstructed features can be transformed into the cepstral domain to reduce their covariance; the energy level can be normalized; cepstral mean normalization can be applied; and delta and double-delta features can be computed. In some circumstances, the performance benefits gained by these techniques may compensate for the bias that the reconstruction introduces.

Imputation techniques fall into two categories. First, "feature compensation" techniques reconstruct the feature vector prior to recognition. Second, "state-based"' imputation techniques form a separate reconstruction of the feature vector within each state of the recognizer.

Feature Compensation. One of the simplest forms of feature compensation is to estimate missing values by interpolating between the closest reliable values. However, such schemes have proven to be largely ineffective [88]. A more sophisticated approach is to reconstruct missing values based on the statistical properties of the spectral vectors. For example, spectral vectors of the clean speech signal can be modeled using a Gaussian mixture model, and a minimum mean-square error (MMSE) estimation of the missing spectral components conditioned on the observation of the reliable components can be formed [43, 90]. This technique is straightforward but does not take into account information represented by the "bounds constraint" discussed in Section 9.4.3.

Raj, Seltzer, and Stern [89] present a correlation-based reconstruction technique that does exploit the "bounds constraint." Spectral vectors of the clean speech signal are considered to be the output of a Gaussian wide-sense stationary (WSS) random process [82]. Under this model the mean spectral vector is constant over time, and the covariance between two spectral–temporal points depends only on the frequency channels of the points and the time difference between them. The mean and covariance fully characterise the process, and can be trivially estimated from clean speech data. Missing features are then reconstructed by using all the present features in the spectrotemporal neighborhood whose correlation with the missing feature is, according to the model, above some fixed threshold. A "bounded" estimate of the missing value can be constructed using an iterative procedure. The initial estimate of each missing value is taken to be the noisy observation. Then each missing element at time t is reestimated, conditioned on the reliable neighbors and the current estimates of the other missing elements at time t. The estimate returned is thresholded so that it does not exceed the masking energy. Each missing element in the current vector is reestimated cyclically until convergence is reached.

By modeling dependencies that span over time, the correlation-based technique can allow the reconstruction of data even when entire frames are missing. However, the simple statistical model that it employs leads to disappointing performance. In the same paper, Raj et al. present a "cluster-based" reconstruction technique that models spectral vectors with a Gaussian mixture model, and again estimate the most probable bounded reconstruction of the missing components given the present data. Despite not modeling interframe dependencies, this technique offers superior performance. They claim that after conversion of the reconstructed spectra into the cepstral domain, recognition performance, measured on the 1000-word vocabulary Resource Management task [85], is better than that achieved with bounded marginalization. A statistical model that combines the cross-temporal modeling of the correlation-based approach and the more detailed mixture modeling of the "cluster-based" scheme would likely lead to further improvements.

Okunu et al. [78] describe an alternative feature-compensation system designed to process speech in the presence of a second speaker. The system first separates the harmonic components of each source by extracting harmonic fragments and then grouping these fragments into two streams (based on fundamental frequency and sound source direction). For each source, the inharmonic regions will be missing. These values are filled in from the "residue," where the residue is the result of spectrally subtracting both harmonic sources from the mixed signal. So essentially the technique separates the harmonic energy but leaves the inharmonic energy unprocessed. Despite this limitation, promising results are reported on a Japanese isolated word recognition task.

Srinivasan et al. [102] compare feature compensation with a marginalization approach. In their system, CASA techniques are employed to estimate local SNR at each time–frequency point, forming a "ratio mask." This mask is then used to scale the noisy spectrum. The resulting clean speech spectrum estimate is transformed into the cepstral domain and applied as input to a conventional speech recognizer. Their experiments employed binaural data in which the source and (nonstationary) masker had a 30 degree separation. For this configuration, they found that their technique was able to produce better results than a missing-data system using a binary mask and marginalization. It should be noted, however, that the sound sources were mixed artificially, using a head-related transfer function, but without simulating room reverberation. This situation leads to relatively reliable ITD and ILD cues for source separation. It is unclear whether the system performance would generalize to situations in which the cues for source separation are more ambiguous and, hence, in which the local SNR estimates are less reliable.

State-Based Imputation. The feature-compensation approaches discussed above form a unique reconstruction of the feature vector. In contrast, state-based imputation schemes work within the HMM-based recognizer and form a separate reconstruction within each HMM state. A naive approach would be to replace the missing values with the unconditional means of each state, but this has been shown to be ineffective [27]. A better approach is to use knowledge of the reliable components. Using the notation introduced in Section 9.3.1, if the observation x is partitioned into its present and missing components, x_p and x_m, we reconstruct x_m in each state q by drawing from the distribution $f(x_m|x_p, q)$. Which value should be drawn from this distribution? Although no one value correctly represents the distribution, some values are better motivated than others. A sensible choice would be to pick the most probable value (i.e., the mode):

$$x'_m = \operatorname*{argmax}_{x_m} p(x_m|x_p, q) \tag{9.18}$$

For a Gaussian distribution, the mode occurs at the mean. For Gaussian mixture models, finding the mode is a difficult problem with no closed-form solution [21]. In particular, the mode of the mixture does not have to occur at the mode of any of the individual components. So, rather than using the mode, previous imputation schemes have usually chosen to impute the expected value [24]:

$$x'_m = E_{x_m|x_p,q} = \int p(x_m|x_p, q)x_m \, dx_m \tag{9.19}$$

Unlike feature compensation, state-based imputation does not directly reconstruct unique spectra for transformation into the cepstral domain. However, the following two-pass strategy can be employed. In the first pass, a spectral representation and state-based imputation are employed. The state sequence associated with the recognition result is recorded. Then with knowledge of this unique state sequence, a state-based reconstruction of the spectral data can be determined. The reconstructed spectrum is then transformed into the cepstral domain for the second recognition pass. In a comparison of imputation techniques, Raj, Seltzer, and Stern were unable to show any performance advantage for state-based schemes over the seemingly more straightforward feature-compensation methods [89].

9.4.5 Estimating the Missing-Data Mask

All the missing-data systems discussed so far require a first stage in which spectrotemporal points are labeled with either a binary present/missing decision (discrete masks), or with a probability value giving a confidence in the reliability of the data (soft masks). With access to the isolated acoustic signals of which an acoustic mixture is composed, it is possible to construct a mask in which all spectrotemporal points with a local SNR greater than 0 dB are marked as present. Using such masks (variously known as a priori masks [24], oracle masks [100], or ideal binary masks [110]), it is possible to achieve reliable recognition performance in the presence of extreme levels of noise (see for example [6]). The challenge for CASA-based ASR systems is to match this level of performance while estimating masks directly from the noisy signal.

Many early missing-data systems have achieved good results using masks derived from very simple noise estimation techniques [109]. The simplest systems form a stationary noise spectrum estimate in regions where there is judged to be no speech activity. Spectrotemporal points are marked as present if the observed energy is greater than the noise estimate. This technique can work even if the noise is not precisely stationary, so long as its dynamic range is not too great. For slowly varying noise, points can be marked as present if the energy exceeds the estimated noise mean by some fixed margin, typically 3 dB. By using such a margin, the reliability of present data can be guaranteed at the expense of marking more data as missing than is strictly necessary. Soft masks can be generated by estimating the local SNR and mapping the estimate onto the range 0 to 1 using a sigmoid function [7]. If a Gaussian noise distribution can be estimated, rather than just a mean noise level, then a true probability can be computed [93]. For small vocabulary tasks, these simple techniques are surprisingly effective [6]. However, for larger vocabulary tasks, or for speech in the presence of highly nonstationary noise sources, more sophisticated techniques are presumed to be necessary.

The attraction of the missing-data approach is that the acoustic pattern-matching stage employs clean speech models and avoids the need for explicitly modeling the noise. However, the mask estimation techniques described above all rely on assumptions about the noise, and only work to the extent that these assumptions remain valid. If the need for noise models has simply been moved from the acoustic match stage to a preliminary speech/background segregation stage, then little has been achieved. Ideally, the missing-data mask should be created using knowledge of the speech source, and with no knowledge specific to the noise present. There have been various CASA-inspired attempts to stay true to this principle. Seltzer, Raj, and Stern pose mask estimation as a Bayesian classification problem using features that they claim make the estimation "free of assumptions about the corrupting noise" [100]. They employ a feature vector based on harmonic comb-filter output, autocorrelation functions, and subband energies, and train separate distributions for the speech and the noise classes. Although a noise model is used, the features chosen ensure a narrow peaky distribution for the speech class, so that the classifier's decision boundary is insensitive to the details of the broader noise class. Hu and Wang [63] describe a mask estimation system based on the use of $F0$ and amplitude modulation (AM) cues that appears to make few noise assumptions. $F0$ and AM are used to define local source fragments that are then grouped across time and frequency on the basis of continuity. The system performs well when evaluated in terms on the SNR improvement of the reconstructed target source, but it has not been evaluated as an ASR front end.

If binaural data is available, then missing data masks can be constructed using localization cues. ITDs and ILDs can be used to estimate the azimuth of origin of the energy at each spectrotemporal point. By looking for clusters in the ILD/ITD estimates taken across all spectrotemporal points, separate sources can be identified (see Section 7.2.2 and Fig. 7.3). Roman, Wang, and Brown present sound-source segregation techniques based on this principle [97]. Harding, Barker, and Brown [59, 58] apply ILD/ITD cues to estimate soft masks for the recognition of speech in the presence of a competing speaker. In this work, the target speaker is assumed to be at an azimuth of 0 degrees and a Bayesian speech/noise decision is made using

ILD and ITD features. Similar work is reported by Palomäki et al. [81]; see Chapter 7, Section 7.6.2 for details. These methods have all been shown to produce very good recognition performance for small vocabulary tasks, provided that the speech and noise sources are separated in azimuth by a few degrees.

Missing-data mask estimation has also been adapted to produce ASR systems that are robust to reverberation. Reverberant energy, arriving from reflections off surfaces in the environment, distorts the direct-path speech spectrum. If the recognizer has been trained in a nonreverberant, condition, this mismatch will lead to poor recognition performance. Not all spectrotemporal regions of the signal will be equally contaminated by reverberant energy. By using metrics that respond to the ratio of direct to reflected energy, a missing-data mask can be constructed such that the recognizer only treats the uncontaminated regions as "present" and ignores the areas dominated by reverberant energy. This approach has been used with some success by Palomäki et al. [80, 79] (see also Chapter 7, Section 7.5.3).

Generally, the missing-data technique allows any CASA system that produces a foreground/ background segregation in the time–frequency domain to be employed as the front end for an ASR system. Although many such CASA systems have been proposed, most have not yet been evaluated in this context.

9.4.6 Difficulties with the Missing-Data Approach

There are two main difficulties with the missing-data approaches described so far. First, there are problems caused by the need for the CASA and ASR components to employ compatible representations. Second, there are serious deficiencies with the bottom-up processing model employed.

In all missing-data systems, the mask acts as an interface between the CASA and ASR components; the CASA front end outputs a spectrotemporal mask, and the ASR back end views the noisy spectrotemporal data through the mask. As the CASA and ASR systems are linked through the mask, they are constrained to operate on compatible spectrotemporal representations. This causes problems for both the ASR and the CASA components.

Spectral representation are not well suited for ASR. As previously mentioned, correlations between neighboring frequency channels and global correlation due to speech-level variation make the data hard to model compactly. Recognition studies employing clean speech routinely demonstrate better performance using cepstral representations that avoid these difficulties. In defense of spectral representations, we might consider that for most applications "clean" conditions are the exception rather than the norm, so it is unclear how much significance to attach to clean speech results. In most situations, it would be better to deploy systems that make fewer errors in noisy conditions, at the expense of the occasional extra error when noise levels are exceptionally low. A more relevant observation is that missing-data techniques employing ideal masks have been shown to perform as well as, or better than, humans on small vocabulary tasks [26]. It remains unclear however, whether these results extend to larger vocabulary tasks. There is some evidence that ideal mask performance deteriorates badly when vocabulary sizes are increased from tens to hundreds of words [102]. More work needs to be done to confirm whether this is a truly general result.

Imputation schemes that allow features to be rerepresented in the cepstral domain may avoid the deficiencies of the spectral representation. However, imputation is a poor way of handling missing data. There have been several attempts to develop missing-data techniques for the cepstral domain that maintain the advantages of marginalization [56, 106, 23, 67]. A naive approach would be to consider the range of possible values for the missing, but bounded, components of a spectral-feature vector, and from this to independently establish the bounds on the elements of the corresponding cepstral vector. This technique does not work well as it produces an unnecessary degree of uncertainty in the cepstral domain. A single missing element in the spectral domain will produce uncertainty in all the cepstral components, but many cepstral vectors that lie within these bounds would not transform back into valid spectral vectors. Häkkinen and Haverinen [56] propose a scheme that effects "spectrum domain marginalization with cepstral features," which operates by transforming the cepstral difference vector (i.e., the observed cepstrum minus the acoustic state mean vector) into the Mel-spectrum domain, weighting the spectral differences so that the influence of corrupt frequency channels is reduced, and then transforming the computation back into the cepstral domain. However, there appears to be little principle behind the choice of weighting terms and it is not possible to employ the "bounds constraint" using this technique. Van Hamme [106] proposes a cepstral technique that is sensitive to the bounds constraint, but it employs a maximum likelihood reconstruction of the spectral vector and, therefore, suffers from the limitations of imputation.

A further difficulty lies in the precise design of the spectral representation. There is a mismatch between the optimum frequency resolution for the CASA front end and that for the ASR back end. CASA systems often use a large number of frequency channels, sometimes with a physiologically unrealistically narrow bandwidth, with the desire to resolve individual harmonics of the sources present. Resolved harmonics are beneficial because (i) they allow sources to be tracked through time (see Section 9.2.2), and (ii) as a source's energy is heavily concentrated in its harmonics, resolving these harmonics allows sources to be detected at unfavourable SNRs (see Fig. 9.5). ASR systems, on the other hand, are usually designed to be "blind" to variation in the harmonic structure of speech. If harmonics are resolved, then variation in $F0$ will lead to variation in the vowel's spectrum. A frequency channel that is centered on a harmonic peak in one vowel example may be centred on a harmonic dip in another. This will cause a huge variance in the observed energy level in any given channel. Furthermore, the regular spacing of the harmonics causes a pattern of correlations between channels that is hard to model statistically without a great deal of training data (i.e., it is necessary to see examples of each speech unit presented across a range of $F0$ values). The result of this conflict in the requirements of ASR and CASA is that most missing-data systems perform CASA with a lower than ideal frequency resolution or, conversely, they perform ASR with a greater resolution than desired. It should be noted that this compromise exists in the temporal domain as well. For example, whereas most ASR systems work best

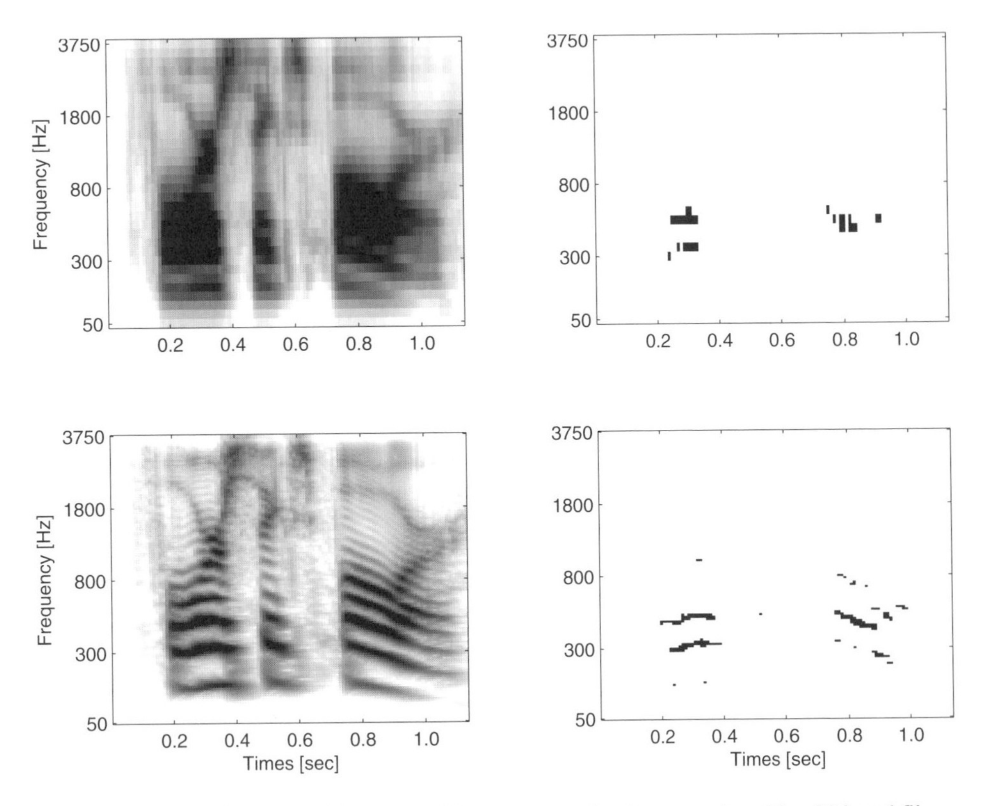

Figure 9.5 Left: Two cochleagrams of the same speech utterance using either 32 broad filter bands (top) or 128 narrower filter bands (bottom). Right: The panels indicate the regions of the cochleagrams that stand clear of the noise floor when the speech is masked with stationary Gaussian white noise added at at an SNR of around –40 dB. In the narrowband representation, where harmonics are better resolved, evidence of the speech source can be found over a larger spectrotemporal region.

with feature vectors that are updated every 10 or 20 ms, source-separation algorithms that exploit temporal fine structure often work on much shorter time scales (e.g. Cooke's "synchrony strands," which operate on a 1 ms time scale [25]).

A more fundamental concern with all the approaches discussed so far is that they assume that speech/background segregation is in some way an "easier" problem than speech recognition. In particular, it is often supposed that missing-data masks can be computed using simple techniques without reference to detailed models of the background noise. The previous section described several techniques that follow this assumption. However, a simple test of these techniques is to present them with a speech signal corrupted by the presence of a second talker. Even the least assuming missing-data systems typically assume that the noise is different from speech. When the noise masker is statistically similar to the target speech source, these systems generally stop working. The system presented by Seltzer et al. [100]

is a case in point; the properties used to distinguish the background from the foreground are all based on a measure of "speechiness." If both the foreground and the background are speech-like, then the system will fail. Other systems simplify the speech-plus-speech problem by making other basic assumptions. For example, [59] tackles conditions in which a target speaker and an interfering speaker are assumed to be spatially separated. This system demonstrates performance that degrades rapidly when localization cues become swamped by reverberation. In fact, evidence suggests that in reverberant conditions listeners solve the source-segregation problem without using spatialization cues even when they are available [32, 31] (see discussion in Chapter 7, Section 7.2.2). Furthermore, listeners are able to understand speech in the presence of a competing speaker in single channel conditions where localization cues are totally absent. Other bottom-up segregation solutions are equally problematic. Harmonic sieving techniques can partially separate simultaneous speakers, but these fail to deal with inharmonic speech segments or with whispered speech mixtures. Clearly, finding a general solution to the foreground mask estimation problem is very challenging.

In the next section, we will argue that the problematic mask estimation step can be avoided by redesign of the standard bottom-up missing data approach. In missing-data systems, the problems of segregation and recognition have been decoupled so that the CASA front end and the ASR back end can be developed independently. This simplification comes at a cost. It is no longer possible to model the mutual constraints that segregation and recognition lend each other. It is these constraints that render both problems soluble. Modeling these constraints entails recoupling the segregation and recognition. The challenge is to recouple these processes in a computationally tractable manner.

9.5 MODEL-BASED CASA AND ASR

The techniques described in the previous section were based on a two-step process: (i) employ primitive CASA in a front end to segregate the spectrotemporal plane into foreground and background regions, and (ii) employ missing-data techniques to recognize the incomplete foreground. However, it is clear that ASA in humans does not, and indeed, cannot work in this way. Primitive grouping processes are insufficient of themselves to suggest an unambiguous segregation. Ambiguity has to be resolved by models of the sound sources present (i.e., Bregman's schemas). Section 9.2.3 reviewed instances in which speech schemas can be clearly demonstrated to override the "default" organization offered by bottom-up cues.

Although the need for a model-driven component to CASA systems has long been realized, the earliest implementations of such systems date from the early 1990s. Ellis' "Prediction-Driven CASA" [45] breaks with the straight bottom-up processing path by taking predictions arising from an internal world model and comparing them against the system input. The results of these comparisons are then used to drive changes to the system's state, that is, the system has certain expectations, and it will search for a way to view the data to try and match these expectations. The initial system was designed to allow sound-source separation and resynthesis, but Ellis has more recently demonstrated how the framework can be extended to facilitate speech recognition [46] (for a more detailed account see Chapter 4). Ellis' system may be compared to another notable model-driven approach, the IPUS system, developed by Lesser et al. [70]. This uses dynamically modifiable front-end signal processing that is able to react to changes in the environment and reanalyze data to resolve ambiguities. The bottom-up and top-down processes are coordinated through a "blackboard architecture," employing a central global workspace that is acted upon by a number of software agents in a manner that resembles human cognition.

Despite the strengths of these early systems, they were not built on statistical foundations and offered no clear route for integration with statistical ASR. More recent efforts have focused on placing bottom-up and top-down processes in a common probabilistic framework. This section will focus on one such technique, known as "speech fragment decoding," which has been inspired by earlier model-driven approaches, but which has been designed from the outset with robust ASR as its goal.

9.5.1 The Speech Fragment Decoding Framework

How may the bottom-up "missing-data"-based model discussed in the previous section be extended to allow schema-driven processes to influence the organization of the acoustic signal? Note that in ASR systems, models of the speech source (i.e. schemas) are already employed for recognition. So one possibility is to allow the primitive processes to suggest multiple interpretations of the acoustic scene, and then to employ the speech models to judge which interpretation produces a speech stream that is most consistent with the set of speech models.

Consider a spectrotemporal representation of a two-second segment of speech that contains 200 (10 ms) frames, where each frame represents the energy in 32 frequency channels. In this example, there are a total of 6400 individual spectrotemporal units. There are therefore 2^{6400} possible foreground/background segregations, that is, each spectrotemporal unit may be independently labeled as belonging to either the foreground or the background. It is clearly impractical to separately consider each of these possible segregations. However, simple data-driven processes (i.e., the application of primitive grouping cues) would show that many of these segregations are unrealistic and do not need to be considered further. We need some system that filters out segmentations that are inconsistent with primitive grouping cues, and passes consistent segmentations forward for consideration by the top-down scoring mechanism.

Consider the cochleagram shown in the upper panel of Fig. 9.6. There are clearly concentrations of energy spanning local spectrotemporal regions that are likely to arise from a single source, for example, noise bursts, harmonics and formant glides. Imagine a bottom-up process that can identify these energy "fragments" and segment the spectrotemporal plane into something resembling the lower panel of Fig. 9.6. If we are confident that each fragment is dominated by the energy of a single source (i.e., that the fragments are "coherent"), then each fragment can be uniquely

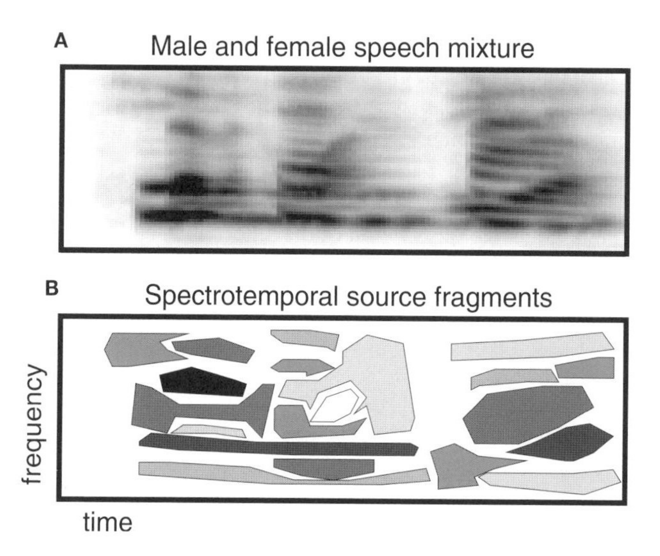

Figure 9.6 (**A**) The cochleagram of the male and female speech mixture seen in Figures 9.2 and 9.3 of this chapter. (**B**) A sketch of a plausible set of spectrotemporal fragments that we would wish an effective primitive CASA system to find in the mixture. Each fragment region is dominated exclusively by either the male or the female source. Note how the fragments relate to speech structures (e.g., harmonics, formants, fricatives) that can be identified in the mixture.

labeled as either part of the foreground or part of the background. The foreground source can now be uniquely described as a subset of the fragments present. In the figure, there are 20 fragments and, hence, there are 2^{20} possible choices of fragment subset that must be considered if we wish to search for the correct description of the foreground. Each of these 2^{20} fragment labelings represents a foreground/background segregation that is consistent with the bottom-up grouping processes.

It is now possible to construct each possible foreground mask and pass the mask to a missing-data speech recognizer of the type described in the previous section. For each hypothesized mask, a likelihood for the best word sequence would be computed. Now, by comparing the likelihoods of the best word sequence across all possible masks the most likely mask/word-sequence pair could be chosen. The winning mask (i.e., foreground/background segregation) would then be a product of bottom-up processes that constrain the possible set of segregations, and top-down processes that jointly search over the space of segregations and word sequences to find the pair that best match the speech models. So in this system, the segregation is not determined solely by the front end, but instead emerges as a product of the recognition process.

In our example, performing the ASR stage 2^{20} times, each time with a different mask, is clearly very computationally expensive. If the utterance is longer, there may easily be 40, 60, or even more sensible fragments, and the computation would again become intractable. However, many pairs of masks will be identical over some period of time and, hence, the decoding will share the same computations.

This observation leads to an efficient implementation in which decodings are arranged in a branching binary tree that evolves over time with complementary mask hypothesis pairs being spawned each time a new fragment commences. The Markov assumptions that underly HMM-based ASR allow mask hypotheses to be merged at the point at which a fragment ends (see Fig. 9.7). Splitting and merging of mask hypotheses occur independently within each state of the standard ASR HMM. For a more detailed description of the algorithm, see Barker et al. [5].

The complete model is illustrated in Fig. 9.8. It is worth stressing again that unlike the systems of the previous section, which employed a "segregate then recog-

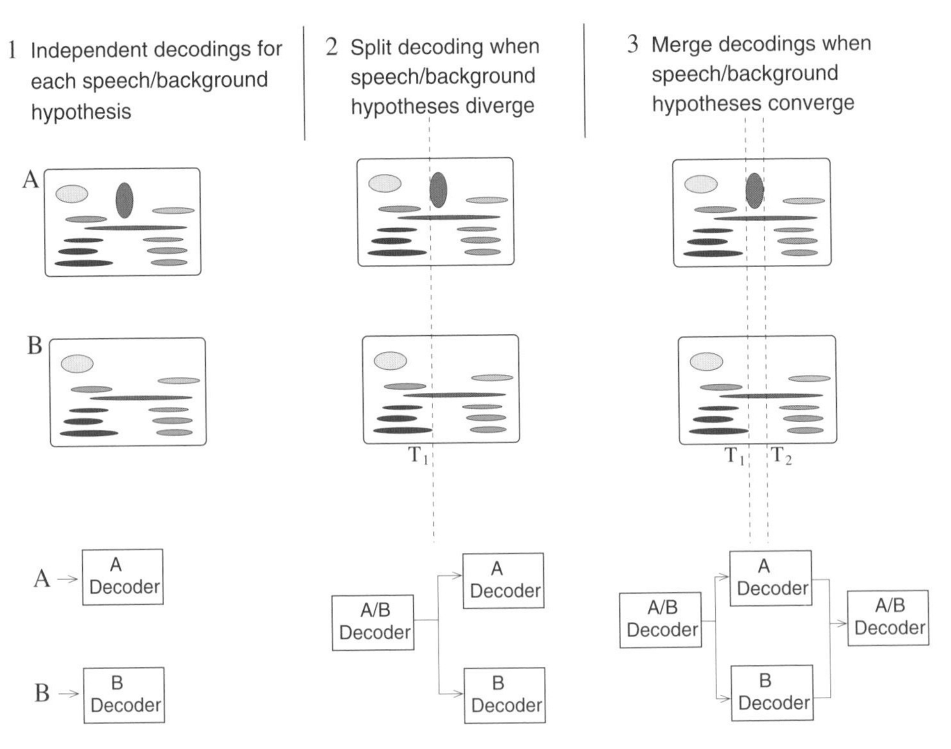

Figure 9.7 A schematic illustrating how the speech fragment decoder efficiently decodes multiple foreground/background hypotheses by only duplicating effort over temporal regions in which the hypotheses differ. In the naive implementation (left) independent word-string decoders are used to process each separate segregation hypothesis. The overall best word-string hypothesis is obtained by comparing the likelihoods of the best-scoring word string suggested by each decoder. In a more efficient implementation (center) segregation hypotheses share a common decoder up to the temporal frame at which they first differ, that is, multiple segregation hypotheses are arranged as a branching binary tree. A further efficiency (right) is realized by merging decodings when segregation hypotheses have identical futures. The word-string HMM decoders are merged on a state-by-state basis; within each state the competing segregation hypotheses are compared and the one with the lower likelihood is rejected.

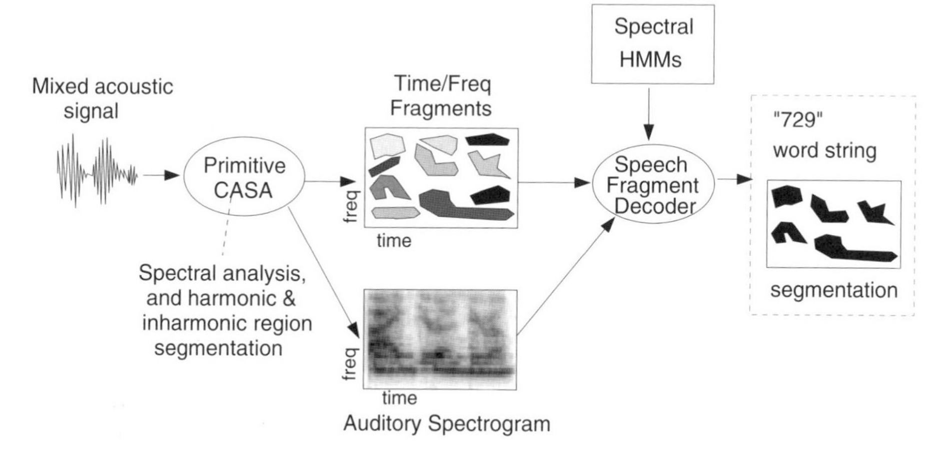

Figure 9.8 An overview of the speech fragment decoding system. Bottom-up processes are employed to locate "coherent fragments" and then a top-down search with access to speech models is used to search for the most likely combination of fragment foreground/background labeling and speech model sequence. Reproduced from [28]. Copyright © 2005 IEEE.

nise" procedure, the segregation now arises as a product of the recognition process. This development is analogous to the earlier development in ASR by which systems progressed from requiring pauses between words to being able to handle continuous speech. In the earlier systems, pause detectors were used to segment the utterance into words, and then the isolated word recognition was applied. In modern continuous speech recognition systems, the position of the word boundaries is something that arises during recognition. If either the acoustic model or language model are changed, then the inferred word boundaries may also change. Likewise, with the speech fragment decoder, if the acoustic model or language model are changed then the foreground/background segregation may change. Barker et al. argues that this is a more accurate model of human behavior in which perceptual organization is under the influence of both speech schemas and primitive grouping processes.

9.5.2 Coupling Source Segregation and Recognition

Most previous CASA-based ASR systems have required a front end that employs primitive processing to perform a complete foreground/background segregation of the spectrotemporal representation [114, 25, 19, 111]. Many of these systems work in two stages. First, primitive processes are employed to identify temporally local fragments of each source (see Chapter 3). Second, these fragments are grouped using primitive sequential grouping cues. Typically, neither stage employs the speech models that are used during speech recognition. Hence, segregation and recognition are essentially decoupled. In contrast, fragment decoding techniques, such as the speech fragment decoder described above and similar techniques such as Shao and

Wang's model-based sequential organization [118], use top-down, model-driven processes to support the sequential grouping of the local fragments. In these systems, segregation and recognition become coupled. In the case of the speech fragment decoder, the sequential grouping is directly built in to the speech recognition process.

The coupled model makes fewer demands of the bottom-up processing component. For example, in the case of the speech fragment decoder, rather than a complete foreground/ background segregation, only local source fragments are required. The fragments themselves may have a relatively small temporal extent (see Fig. 9.6) and the difficult, and often ambiguous, problem of sequential grouping is deferred to the model-based search stage of the system where it is solved by leveraging the power the speech recognition engine.

To illustrate the advantage of the coupled speech fragment decoder, consider the task of extracting and grouping the voiced components of a speech signal. Most previous CASA systems have employed methods for extracting regions of voiced speech from background noise based on tracking of fundamental frequency (for a recent example see [63]). These bottom-up techniques can work well in conditions in which no competing harmonic sound source is present. They mostly operate by first finding an unambiguous $F0$ contour, and then grouping frequency channels that contain energy that is harmonically related to the $F0$ value. If multiple harmonic sources are present, then a multipitch tracking algorithm can be employed to try to simultaneously track each source. Wu et al. [117] present such an algorithm that appears robust over a range of noise conditions. Consider, though, the problem of listening to speech in the presence of a second speaker. If the speakers have similar pitch ranges, then ambiguous situations will arise, such as pitch contours that have overlapping trajectories, or voiced segments that are temporally interlaced and cannot be clearly streamed (see Fig. 9.9). In such cases, no algorithm will be able to produce an unambiguous pitch track for each source without using higher level speech knowledge.

The speech fragment decoder has the potential to deal with the ambiguous situations portrayed in Fig. 9.9 through its use of top-down knowledge. In the cases illustrated in the figure, it is possible to identify unambiguous pitch track segments. For example, in the case of intersecting pitch tracks, the pitch contours would be tracked up to the point of intersection, and again after the intersection, but no assumption would be made about how the segments before and after the intersection are related. This is illustrated in Fig. 9.10. These pitch track segments can be mapped onto spectrotemporal energy fragments. The sequential grouping of these fragments is then disambiguated by the speech models, that is, in the case in the figure both fragment groupings, $\{AB, ab\}$ and $\{Ab, aB\}$, will be separately considered. The correct assignment of fragments to streams will be favoured because this assignment will produce a better fit to the speech models than the incorrect assignment.

Inharmonic energy, such as that which composes the unvoiced speech events that characterize some consonants, is harder to separate. One strategy is simply to deal with the harmonic energy first, and then to fragment the remaining spectrotemporal energy according to the distribution of its peaks. For example, fragments can

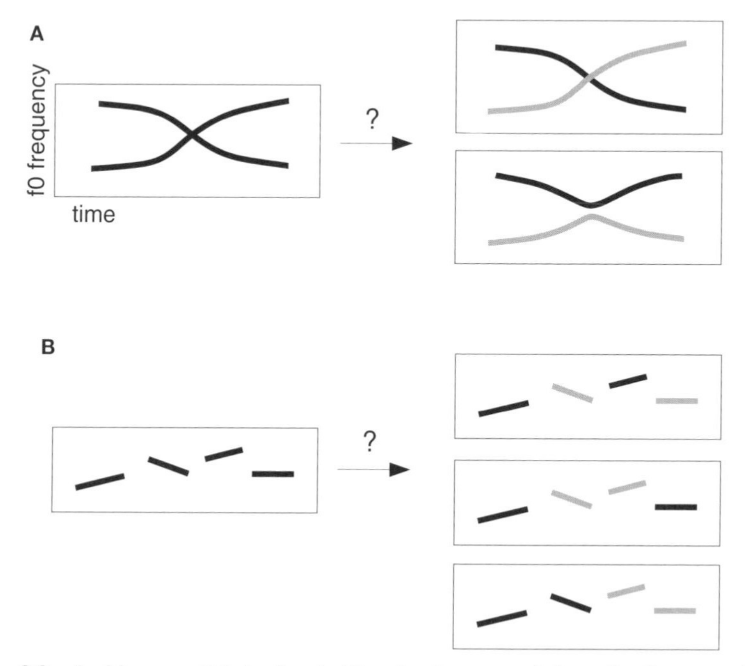

Figure 9.9 Ambiguous pitch tracks. **A**: The simultaneous pitch tracks shown on the left may be interpreted as a rising pitch track and a falling pitch track that cross, or as pitch tracks that move toward each other and then apart again without crossing. **B**: If two speakers are known to be present, the series of pitch track segments on the left may be attributed to the speakers in many different ways.

be defined by locating concentrated regions of energy centered around local energy peaks. One simple implementation that appears to be successful is to apply a watershed segmentation [96] (a popular image-segmentation technique) to a copy of the cochleagram from which the harmonic energy has been removed (for examples of systems employing this technique see [62, 28]). In cases in which there is doubt about where to place fragment boundaries, the bottom-up processes may be tuned to

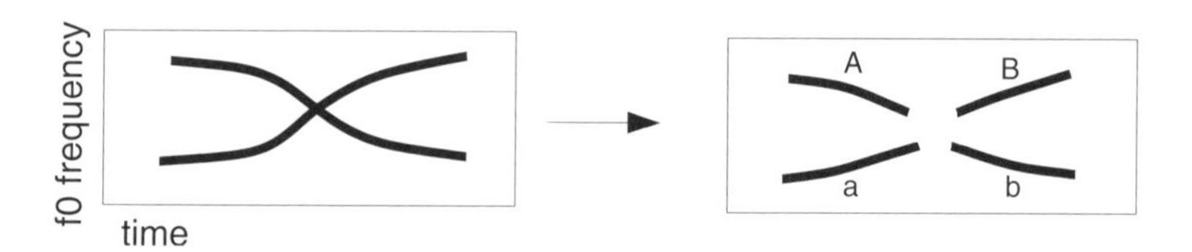

Figure 9.10 The ambiguous pitch tracks shown on the left may be represented as the four pitch track segments shown on the right. The track segments can be formed into corresponding spectrotemporal fragments allowing the alternative streaming $\{AB, \text{ab}\}$ and $\{Ab, aB\}$ to be distinguished by the speech fragment decoder.

err on the side of oversegmentation. Such a tuning leads to smaller fragments that are more likely to be coherent, at the expense of reducing the degree of bottom-up constraint and increasing the size of the top-down search space.

Another technique for coupling bottom-up and top-down processing has been proposed by Srinivasan and Wang [103]. Here, an initial binary mask is constructed using a bottom-up speech separation system. A missing-data recognition system is then used to generate a list of the N most likely recognition hypotheses (an "N-best list") and their associated state sequences. Given the state sequences, the mask can be reexamined and a new state-based decision can be made as to whether each time–frequency unit is present or missing. The "corrected" mask can then be used to recalculate the likelihood of the hypothesis. The highest scoring hypothesis on the rescored N-best list is taken as the output.

In this system, a single bottom-up proposal (i.e., the mask) is being reinterpreted in the light of multiple top-down hypotheses (i.e., state sequences of an N-best list). Clearly, the system requires that the initial bottom-up segregation be sufficiently close that the correct recognition result appears among the N best. If the segmentation is poor, N will have to be made very large. However, increasing N reduces the bottom-up constraint since each "corrected" mask is formed solely on the basis of top-down information (in the extreme in which all hypotheses are considered, the system would become entirely top-down driven). This would be comparable to a speech fragment decoding in which every time-frequency unit is regarded as a separate fragment. In Srinivasan and Wang's system, the degree of bottom-up constraint is a function of the size of the N-best list size, whereas in the speech fragment decoder it is related to the size and number of proposed source fragments. Both approaches to representing the degree of the bottom-up constraint have their advantages and drawbacks. It may be possible to design systems that exploit the benefits of both techniques. For example, one can imagine using a speech fragment decoder to search over a set of masks produced from large fragments (coarse detail), but then employing an N-best list to fine-tune the masks.

9.6 DISCUSSION AND CONCLUSIONS

CASA and ASR have developed from very different perspectives, and bringing the two together presents an enormous challenge. In this chapter, we have shown how CASA has come to be integrated into ASR technology at increasingly deeper levels. The earliest approaches performed signal enhancement and left the ASR back end untouched. In more recent missing-data techniques, CASA is used to label the representation and minor modifications are made to the statistical computations within the speech decoder, but the influence of CASA remains bottom-up. It has long been understood that CASA systems require a top-down, model-based component, and attempts to model such "prediction-based" CASA extend back to the mid-1990s [45]. The recent "speech fragment" approaches to interfacing ASR and CASA have been influenced by these ideas. In these systems, source segregation is solved as part of the recognition problem. As CASA and ASR become more tightly coupled, the potential for human-like ASR increases, but answers to some fundamental questions become more urgent. Two of the more important of these questions are discussed briefly below.

How should we model primitive grouping processes? Listening experiments involving speech seem to suggest there are few grouping "rules" that cannot be stretched under the influence of top-down mechanisms. On this basis, is it realistic to search for immutable spectrotemporal energy fragments of the kind expected by the speech fragment decoding approach? For example, there is much evidence of the influence of "grouping by common onset," but is it safe to build speech fragments on this basis? In "cocktail party" speech there may be many temporal coincidences, where different sources share a common onset. These coincidences could cause severe problems for a rule-driven system, but they appear to cause few problems for listeners. It would seem more prudent that grouping rules be modeled probabilistically. In the case of common onset, elements that start synchronously are just more likely to belong to the same source. How would the output of probabilistic bottom-up grouping be presented to the top-down ASR component? Rather than modeling the segregation as a set of fragments, we would need some softer model. However, a softer model would reduce the influence of bottom-up constraints. If all segregations had some degree of probability, how would we organize a search over the space of segregations? There is a compromize here between hard grouping rules that are unrealistic but conveniently define a narrow set of possibilities, and soft rules that better model reality but introduce computational search problems.

From a modeling perspective, the necessity of a separate layer of bottom-up processes is not as clear as it first seems. Statistical primitive processes may not be clearly distinguishable from schema-driven processes. Some bottom-up grouping cues may even be redundant. Consider common onset again. If a set of formants start synchronously, is there any need to explicitly group them together using a separate bottom-up process? The grouping can be achieved by matching the outcome of alternative grouping hypotheses directly to the speech models. The commonness of the onset is already well represented in the spectral models of the speech signal. Might not other bottom-up organizational cues, such as harmonicity, also be expressed as part of the learned top-down component? Speech models that are trained on speech spectra with resolved harmonics will learn that harmonics are evenly spaced. Foregrounds including evenly spaced harmonics will be more likely to fit to the models, whereas foregrounds including unevenly spaced harmonics from a mixture of sources will not match the speech models.

Despite the possibility of designing a fully model-driven system, such a system appears biologically implausible. An engineer can train models from especially recorded noise-free exemplars of commonly occurring sound sources. However, an organism needs to learn these models from direct experience of the natural environment. In the real world, noise-free examples of a sound source are a rarity. So to fully exploit the environmental data, the sound sources to be modeled need to be segregated from the noise background. However, if segregating the source from the background requires preexisting well-developed models, then an impasse is reached. Employing a hard-wired front end that can partially segregate sounds using rules developed over evolutionary time could solve this dilemma.

What is an appropriate representation for a speech schema? We argued above that by giving more representational power to speech schemas, there is less need to explicitly model bottom-up processes. Current ASR systems employ HMMs to model speech not because they are realistic models of speech acoustics, but because they are simple models from which to make statistical inference. For speech in clean conditions, these have proved to be highly successful. However, in noisy conditions, where large regions of the signal may be masked, they prove inadequate, largely, it is suspected, because they do not present a convenient framework for exploiting acoustic speech cues that operate over longer timescales. For example, listeners are able to take advantage of the fact that attributes such as accent, speech rate, and emotional state are expected to remain reasonably constant over short durations. Getting a clear view of these attributes in a relatively noise-free portion of the signal may constrain the hypotheses that are made during noisier portions. For attributes such as gender, accent, and emotional state that may stay constant over an utterance, it is possible to build separate versions of the speech models for each setting of the attribute in question; hence, we may construct gender-dependent models with a set of models for both male and female speakers, speaker-dependent models with a set of models for each known speaker, speech-rate-dependent models with a set of models for different speaking rates, and so on. However, once we consider the need to simultaneously model all degrees of variation, this approach leads to an unacceptable exponential explosion of the number of parameters required. A better approach is to factor out the influence of each cause of variation in such a way that they can be modeled independently, with each degree of variability being described using a small number of parameters. For example, rather than having ten different models of the word "seven," each at a different pitch, have a single model of the word "seven" and a model of the manner in which pitch influences this base model. Graphical modeling techniques that decompose speech in ways like this are being developed. For example Reyes-Gomez et al. [94] describe a "deformable spectrogram" model that independently models the transition of formants and harmonics (see Chapter 4).

The introduction of a more sophisticated model of speech is clearly a benefit to any ASR system. Typically, improved speech models increase the discriminability between competing word-string hypotheses. However, in the context of the coupled segregation/recognition systems described in the latter part of this chapter, there is a double benefit as the improved models also act to reduce the ambiguity in the speech-background segregation in noisy situations.

However, there is little evidence that human speech perception works in this way. Rather, it appears that speech perception is opportunistic and evidence will be recruited to support a speech hypothesis regardless of whether this evidence is better employed in a competing nonspeech stream. Endogenous attention may play a role here, that is, fragments may be preferentially recruited to the stream that is being actively attended to ([116]). Strong arguments against the use of explicit models of background sources come from common experience. For example, we might ask why it is that people are not able to follow simultaneous conversations, but instead must swap attention between them. Even with short simultaneous digit sequences, listeners are generally unable to transcribe both utterances after hearing the mixture once. However, it is still possible to imagine a model combination approach that mimics human behavior by sharing modeling power unequally between foreground and background sources. A detailed speech model may be employed to represent the foreground, while a less structured "blurry" model with a smaller state space is used to model the background. The level of detail of the models may increase or decrease as attention is switched between streams.

9.7 CONCLUDING REMARKS

In this chapter, we have seen how CASA and ASR have slowly converged. There has been steady progress from the earliest systems, which crudely bolted together preexisting CASA and ASR solutions, to current systems in which CASA and ASR are coupled within a common statistical framework. Although progress has been steady, there is still far to go. The links between speech recognition and auditory scene analysis are profound. Accounting for the robustness of human speech recognition represents a grand challenge for those in the CASA community. It is hard to imagine regarding a CASA model as complete if it does not explain the human ability to understand speech in an unpredictable acoustic environment. Conversely, scene analysis may be posed as the next grand challenge for those in the speech recognition community. It is possible that any system that provides human-like speech recognition performance will need to have simultaneously solved the ASA problem. As both communities seek to address these challenges, it is sure that the boundaries between CASA, with its previous emphasis on bottom-up segregation, and ASR, with its emphasis on top-down search, will become increasingly blurred in the years to come.

REFERENCES

1. J. B. Allen. How do humans process and recognize speech? *IEEE Transactions on Speech and Audio Processing,* 2(4):567–577, 1994.
2. P. F. Assmann and Q. Summerfield. Modeling the perception of concurrent vowels: Vowels with the same fundamental frequency. *Journal of the Acoustical Society of America,* 85(1):327–338, 1989.
3. P. F. Assmann and Q. Summerfield. Modeling the perception of concurrent vowels: Vowels with different fundamental frequencies. *Journal of the Acoustical Society of America,* 88(2):680–697, 1990.
4. J. P. Barker and M. P. Cooke. Is the sine-wave speech cocktail party worth attending? *Speech Communication,* 27:159–174, 1999.

5. J. P. Barker, M. P. Cooke, and D. P. W. Ellis. Decoding speech in the presence of other sources. *Speech Communication,* 45:5–25, 2005.
6. J. P. Barker, M. P. Cooke, and P. D. Green. Robust ASR based on clean speech models: An evaluation of missing data techniques for connected digit recognition in noise. In *Proceedings of Eurospeech,* Aalborg, Denmark, 2001.
7. J. P. Barker, L. Josifovski, M. P. Cooke, and P. Green. Soft decisions in missing data techniques for robust automatic speech recognition. In *Proceeding of the International Conference on Spoken Language Processing,* volume 1, pp. 373–376, Beijing, China, October 2000.
8. L. E. Baum, T. Petrie, G. Soules, and N. Weiss. A maximization technique in the statistical analysis of probabilistic functions of finite state Markov chains. *Annals of Mathematical Statistics,* 41:164–171, 1970.
9. A. J. Bell and T. J. Sejnowski. An information-maximization approach to blind separation and blind deconvolution. *Neural Computation,* 7(6):1004–1034, 1995.
10. J. R. Bellegarda. Statistical techniques for robust ASR: Review and perspectives. In *Proceedings of Eurospeech,* pp. 33–36, Rhodes, Greece, 1997.
11. F. Berthommier and S. Choi. Comparative evaluation of CASA and BSS models for subband cocktail party speech segregation. In *Proceeding of the International Conference on Spoken Language Processing,* Denver, CO, September 2002.
12. *Proceedings of the DARPA Broadcast News Transcription and Understanding Workshop,* Lansdowne, VA, February 1998.
13. S. Boll. Suppression of acoustic noise in speech using spectral subtraction. *IEEE Transactions on Acoustics, Speech and Signal Processing,* 27(2):113–120, 1979.
14. H. Bourlard and S. Dupont. Sub-band based speech recognition. In *Proceedings of the International Conference on Acoustics, Speech and Signal Processing,* pp. 125–127, Munich, Germany, April 1997.
15. A. S. Bregman. Auditory streaming is cumulative. *Journal of Experimental Psychology,* 4(3):380– 387, 1978.
16. A. S. Bregman. *Auditory Scene Analysis.* MIT Press, Cambridge, MA, 1990.
17. A. S. Bregman, J. Abramson, P. Doehring, and C. J. Darwin. Spectral integration based on common amplitude modulation. *Perception and Psychophysics,* 37:483–493, 1985.
18. D. E. Broadbent and P. Ladefoged. On the fusion of sounds reaching different sense organs. *Journal of the Acoustical Society of America,* 29(6):708–710, 1957.
19. G. J. Brown. *Computational Auditory Scene Analysis: A Representational Approach.* PhD thesis, Department of Computer Science, University of Sheffield, UK, 1992.
20. G. J. Brown and M. P. Cooke. Computational auditory scene analysis. *Computer Speech and Language,* 8:297–336, 1994.
21. M. A. Carreira-Perpinan. Mode-finding for mixtures of Gaussian distributions. *IEEE Transactions on Pattern Analysis and Machine Intelligence,* 22(11):1318–1323, 2000.
22. T. D. Carrell and J. M. Opie. The effect of amplitude comodulation on auditory object formation in sentence perception. *Perception and Psychophysics,* 52:437–445, 1992.
23. C. Cerisara. Towards missing data recognition with cepstral features. In *Proceedings of Eurospeech,* Geneva, Switzerland, September 2003.

24. M. Cooke, P. Green, L. Josifovski, and A. Vizinho. Robust automatic speech recognition with missing and unreliable acoustic data. *Speech Communication,* 34(3):267–285, 2001.

25. M. P. Cooke. *Modelling Auditory Processing and Organization.* PhD thesis, Department of Computer Science, University of Sheffield, U.K., 1991.

26. M. P. Cooke. A glimpsing model of speech perception. *Journal of the Acoustical Society of America,* 119:1562–1573.

27. M. P. Cooke, A. C. Morris, and P. D. Green. Missing data techniques for robust speech recognition. In *Proceedings of the International Conference on Acoustics, Speech and Signal Processing,* pp. 863–866, Munich, Germany, April 1997.

28. A. Coy and J. P. Barker. Recognising speech in the presence of a competing speaker using a "speech fragment decoder." In *Proceedings of the International Conference on Acoustics, Speech and Signal Processing,* Philadelphia, March 2005.

29. J. F. Culling and C. J. Darwin. Perceptual separation of simulation vowels: Within and across formant grouping by F0. *Journal of the Acoustical Society of America,* 93(6):3454–3467, 1993.

30. J. F. Culling and C. J. Darwin. Perceptual and computational separation of simultaneous vowels: Cues arising from low-frequency beating. *Journal of the Acoustical Society of America,* 95(3):1559–1569, 1994.

31. J. F. Culling, K. I. Hodder, and C. Y. Toh. Effects of reverberation on perceptual segregation of competing voices. *Journal of the Acoustical Society of America,* 114(5): 2871–2876, 2003.

32. J. F. Culling and Q. Summerfield. Perceptual separation of concurrent speech sounds: Absence of across-frequency grouping by common interaural delay. *Journal of the Acoustical Society of America,* 98(2):785–797, 1995.

33. J. E. Cutting. Auditory and linguistic processes in speech perception: Inferences from six fusions in dichotic listening. *Psychological Review,* 83:114–140, 1976.

34. C. J. Darwin. On the dynamic use of prosody in speech perception. In A. Cohen and S. G. Noteboom, editors, *Structure and Process in Speech Perception,* Institute for Perception Research (IPO), Eindhoven, 1975. Symposium on dynamic aspects of speech perception.

35. C. J. Darwin. Perceptual grouping of speech components differing in fundamental frequency and onset-time. *Quarterly Journal of Experimental Psychology,* 33(A):185–207, 1981.

36. C. J. Darwin and C. E. Bethell-Fox. Pitch continuity and speech source attribution. *Journal of Experimental Psychology: Human Perception and Performance,* 3:665–672, 1977.

37. C. J. Darwin and R. W. Hukin. Effectiveness of spatial cues, prosody, and talker characteristics in selective attention. *Journal of the Acoustical Society of America,* 107(2):970–977, 2000.

38. C. J. Darwin and R. W. Hukin. Effects of reverberation on spatial, prosodic and vocal-tract size cues to selective attention. *Journal of the Acoustical Society of America,* 108(1):335–342, 2000.

39. C. J. Darwin and N. S. Sutherland. Grouping frequency components of vowels: When is a harmonic not a harmonic? *Quarterly Journal of Experimental Psychology,* 36(1):193–208, 1984.

40. A. P. Dempster, N. M. Laird, and D. B. Rubin. Maximum-likelihood from incomplete data via the EM algorithm. *Journal of the Royal Statistical Society Series B,* 39:1–38, 1977.

41. M. F. Dorman, L. J. Raphael, and A. M. Liberman. Some experiments on the sound of silence in phonetic perception. *Journal of the Acoustical Society of America,* 65:1518–1532, 1979.

42. W. J. Dowling. The perception of interleaved melodies. *Cognitive Psychology,* 5:322–337, 1973.

43. S. Dupont. Missing data reconstruction for robust automatic speech recognition in the framework of hybrid HMM/ANN systems. In *Proceedings of the International Conference on Spoken Language Processing,* Sydney, Australia, 1998.

44. S. Dupont and C. Ris. Multiband with contaminated training data. In *Proceedings of Workshop on Consistent and Reliable Acoustic Analysis Cues for Sound Analysis '01,* 2001.

45. D. P. W. Ellis. *Prediction-Driven Computational Auditory Scene Analysis.* PhD thesis, M. I. T, 1996.

46. D. P. W. Ellis. Using knowledge to organize sound: The prediction-driven approach to computational auditory scene analysis and its application to speech/nonspeech mixtures. *Speech Communication,* 27:281–298, 1999.

47. Y. Ephraim, H. Lev-Ari, and W. J. J. Roberts. A brief survey of speech enhancement. In J. C. Whitaker, editor, *The Electronics Handbook,* CRC Press, Boca Raton, FL, 2003.

48. G. Fant. *Acoustic Theory of Speech Production.* Mouton, The Hague, 1960.

49. W. M. Fisher, G. R. Doddington, and K. M. Goudie-Marshall. The DARPA speech recognition research database: Specifications and status. In *Proceedings of the DARPA Workshop on Speech Recognition,* pp. 93–99, February 1986.

50. H. Fletcher. Speech and hearing in communication. In J. B. Allen, editor, *The ASA Edition of Speech and Hearing in Communication.* Acoustical Society of America, New York, 1994.

51. M. J. F. Gales and S. J. Young. HMM recognition in noise using parallel model combination. In *Proceedings of Eurospeech,* Berlin, Germany, September 1993.

52. R. B. Gardner, S. A. Gaskill, and C. J. Darwin. Perceptual grouping of formants with static and dynamic differences in fundamental frequency. *Journal of the Acoustical Society of America,* 85(3):1329–1337, 1989.

53. H. Glotin, F. Berthommier, and E. Tessier. A CASA-labelling model using the localisation cue for robust cocktail-party speech recognition. In *Proceedings of Eurospeech,* pp. 2351–2354, Budapest, Hungary, September 1999.

54. J. J. Godfrey, E. C. Holliman, and J. McDaniel. SWITCHBOARD telephone speech corpus for research and development. In *Proceedings of the International Conference on Acoustics, Speech and Signal Processing,* pp. 517–520, San Francisco, March 1992.

55. Y. Gong. Speech recognition in noisy environments: Asurvey. *Speech Communication,* 16:261– 291, 1995.

56. J. Hakkinen and H. Havarinen. On the use of missing feature theory with cepstral features. In *Proceedings of CRAC '01,* 2001.

57. J. W. Hall, M. P. Haggard, and M. A. Fernandes. Detection in noise by spectro-temporal pattern analysis. *Journal of the Acoustical Society of America,* 76(1):50–56, 1984.

58. S. Harding, J. Barker, and G. J. Brown. Mask estimation based on sound localization for missing data speech recognition. In *Proceedings of the International Conference on Acoustics, Speech and Signal Processing,* Philadelphia, March 2005.

59. S. Harding, J. Barker, and G. J. Brown. Mask estimation for missing data speech recognition based on statistics of binaural interaction. *IEEE Transactions on Speech and Audio Processing,* 14(1):58–67.

60. H. Hermansky and N. Morgan. RASTA processing of speech. *IEEE Transactions on Speech and Audio Processing,* 2(4):578–588, 1994.

61. J. N. Holmes and W. J. Holmes. *Speech Synthesis and Recognition,* 2nd ed. Taylor and Francis, London, 2002.

62. G. Hu and D. L Wang. Auditory segregation based on event detection. In *ISCA Tutorial And Research Workshop on Statistical and Perceptual Audio Processing,* 2004.

63. G. Hu and D. L. Wang. Monaural speech segregation based on pitch tracking and amplitude modulation. *IEEE Transactions on Neural Networks,* 15(5):1135–1150, 2004.

64. A. Janin, J. Ang, S. Bhagat, R. Dhillon, and J. Edwards. The ICSI meeting project: Resources and research. In *In Proceedings of the ICASSP 2004 Meeting Recognition Workshop,* Montreal, May 2004.

65. F. Jelinek. *Statistical Methods for Speech Recognition.* MIT Press, Cambridge, MA, 1998.

66. B. H. Juang. Speech recognition in adverse environments. *Computer Speech and Language,* 5:275–294, 1991.

67. Z. Jun, S. K. Wong, W. Gang, and Q. Hong. Using Mel-frequency cepstral coefficients in missing data technique. *Eurasia Journal of Applied Signal Processing,* 3:340–346, 2004.

68. J.-C. Junqua and J.-P. Haton. *Robustness in Automatic Speech Recognition: Fundamentals and Applications.* Kluwer, Norwell, MA, 1996.

69. P. Ladefoged and D. E. Broadbent. Perception of sequence in auditory events. *Quarterly Journal of Experimental Psychology,* 12:162–170, 1960.

70. V. R. Lesser, S. H. Nawab, and F. I. Klassner. IPUS: An architecture for the integrated processing and understanding of signals. *Artificial Intelligence,* 77:129–171, 1995.

71. W. M. Liberman and I. G. Mattingly. The motor theory of speech perception, revised. *Cognition,* 21:1–36, 1985.

72. J. C. R. Licklider. A duplex theory of pitch perception. In D. Schubert, editor, *Experientia,* volume 7, pp. 128–133. Dowden, Hutchison, and Ross, 1951.

73. P. Lockwood and J. Boudy. Experiments with nonlinear spectral subtractor, Hidden Markov models and the projection for robust speech recognition in cars. *Speech Communication,* 11:215–228, 1992.

74. R. Meddis and M. J. Hewitt. Modeling the identification of concurrent vowels with different fundamental frequencies. *Journal of the Acoustical Society of America,* 91(1):233–245, 1992.

75. G. F. Meyer and F. Berthommier. Vowel segregation with amplitude modulation maps: A reevaluation of place and place-time models. In *ESCA ETRW on The Auditory Basis of Speech Perception,* pp. 212–215, Keele, 1996.

76. S. G. Noteboom, J. P. L. Brokx, and J. J. De Rooij. Contributions of prosody to speech perception. In W. J. M. Levelt and G. B. Flores d'Arcais, editors, *Studies in the Perception of Language.* Wiley, Chichester, 1976.

77. H. G. Okuno, T. Nakatani, and T. Kawabata. Interfacing sound stream segregation to automatic speech recognition —Preliminary results on listening to several sounds simultaneously. In *Proceedings of the Thirteenth National Conference on Artificial Intelligence,* volume 2, pp. 1082–1089, Menlo Park, CA, August 1996.
78. H. G. Okuno, T. Nakatani, and T. Kawabata. Listening to two simultaneous speeches. *Speech Communication,* 27(3–4):299–310, 1999.
79. K. J. Palomäki, G. J. Brown, and J. Barker. Missing data speech recognition in reverberant conditions. In *Proceedings of the International Conference on Acoustics, Speech and Signal Processing,* pp. 65–68, Orlando, May 2002.
80. K. J. Palomäki, G. J. Brown, and J. Barker. Techniques for handling convolutional distortion with "missing data" automatic speech recognition. *Speech Communication,* 43(1–2):123–142, 2004.
81. K. J. Palomäki, G. J. Brown, and D. L. Wang. A binaural processor for missing data speech recognition in the presence of noise and small-room reverberation. *Speech Communication,* 43(4):361–378, 2004.
82. A. Papoulis. *Probability, Random Variables, and Stochastic Processes,* 3rd ed., McGraw-Hill, New York, 1991.
83. D. B. Paul and J. M. Baker. The design of the *Wall Street Journal*-based CSR corpus. In *Proceedings of the DARPA SLS Workshop,* February 1992.
84. D. Pearce and H.-G. Hirsch. The aurora experimental framework for the performance evaluation of speech recognition systems under noisy conditions. In *Proceedings of International Conference on Spoken Language Processing '00,* volume 4, pp. 29–32, Beijing, China, 2000.
85. P. Price, W. M. Fisher, J. Bernstein, and D. S. Pallet. The DARPA 1000-word Resource Management database for continuous speech recognition. In *Proceedings of the International Conference on Acoustics, Speech and Signal Processing,* pp. 651–654, New York, 1988.
86. L. Rabiner and B.-H. Juang. *Fundamentals of Speech Rrecognition.* Signal Processing Series. Prentice-Hall, Englewood Cliffs, NJ, 1990.
87. L. R. Rabiner. A tutorial on hidden Markov models and selected applications. *Proceedings of the IEEE,* 77(2):257–286, 1989.
88. B. Raj. *Reconstruction of Incomplete Spectrograms for Robust Speech Recognition.* PhD thesis, Carnegie Mellon University, 2000.
89. B. Raj, M. L. Seltzer, and R. M. Stern. Reconstruction of missing features for robust speech recognition. *Speech Communication,* 43(4):275–296, 2004.
90. B. Raj, R. Singh, and R. M. Stern. Inference of missing spectrographics features for robust speech recognition. In *Proceedings of the International Conference on Spoken Language Processing,* Sydney, Australia, 1998.
91. R. E. Remez, P. E. Rubin, S. M. Berns, J. S. Pardo, and J. M. Lang. On the perceptual organization of speech. *Psychological Review,* 101(1):129–156, 1994.
92. S. Renals. AMI: Augmented Multiparty Interaction. In *Proceedings of NIST Meeting Transcription Workshop,* Montreal, 2004.
93. P. Renevey and A. Drygajlo. Detection of reliable features for speech recognition in noisy conditions using a statistical criterion. In *Proceedings of CRAC '01,* 2001.
94. M. J. Reyes-Gomez, N. Jojic, and D. P. W. Ellis. Dominant speaker segmentation of sound mixtures with the deformable spectrogram model. *IEEE Transactions on Speech and Audio Processing,* in press.

95. B. Roberts and B. C. J. Moore. The influence of extraneous sounds on the perceptual estimation of first-formant frequency in vowels under conditions of asynchrony. *Journal of the Acoustical Society of America,* 89(6):2922–2932, 1991.

96. J. B. T. M. Roerdink and A. Meijster. The watershed transform: Definitions, algorithms and parallelization strategies. *Fundamenta Informaticae,* 41:187–228, 2001.

97. N. Roman, D. L. Wang, and G. J. Brown. Speech segregation based on sound localization. *Journal of the Acoustical Society of America,* 114(4):2236–2252, 2003.

98. T. Rutkowski, A. Cichocki, and A. K. Barros. Speech enhancement from interfering sounds using CASA techniques and blind source separation. In *Proceedings of the 3rd International Conference on Independent Components Analysis and Blind Source Separation,* pp. 728–733, San Diego, CA, December 2001.

99. M. T. M. Scheffers. *Sifting Vowels: Auditory Pitch Analysis and Sound Segregation.* PhD thesis, Rijksuniversiteit te Groningen, The Netherlands, 1983.

100. M. L. Seltzer, B. Raj, and R. M. Stern. A Bayesian classifier for spectrographic mask estimation for missing feature speech recognition. *Speech Communication,* 43(4):379–394, 2004.

101. M. Slaney and R. F. Lyon. A perceptual pitch detector. In *Proceedings of the International Conference on Acoustics, Speech and Signal Processing,* pp. 357–360, Albuquerque, NM, April 1990.

102. S. Srinivasan, N. Roman, and D. Wang. On binary and ratio time–frequency masks for robust speech recognition. In *Proceedings of the International Conference on Acoustics, Speech and Signal Processing '04,* volume 4, pp. 2541–2544, Jeju Island, Korea, 2004.

103. S. Srinivasan and D. Wang. Robust speech recognition by integrating speech separation and hypothesis testing. In *Proceedings of the International Conference on Acoustics, Speech and Signal Processing,* pp. 89–92, Philadelphia, March 2005.

104. E. Tessier and F. Berthommier. Speech enhancement and segregation based on the localisation cue for cocktail-party processing. In *Proceedings of CRAC '01,* 2001.

105. A. M. Treisman. Contextual cues in selective listening. *Quarterly Journal of Experimental Psychology,* 12:242–248, 1960.

106. H. Van Hamme. Robust speech recognition using missing feature theory in the cepstral or LDA domain. In *Proceedings of Eurospeech,* Geneva, Switzerland, September 2003.

107. A. P. Varga and R. K. Moore. Hidden Markov model decomposition of speech and noise. In *Proceedings of the International Conference on Acoustics, Speech and Signal Processing,* pp. 845–848, Albuquerque, NM, April 1990.

108. R. R. Verbrugge and B. Rakerd. Evidence for talker-independent information for vowels. *Language and Speech,* 29:39–57, 1986.

109. A. Vizinho, P. D. Green, M. P. Cooke, and L. Josifovski. Missing data theory, spectral subtraction and signal-to-noise estimation for robust ASR: An integrated study. In *Proceedings of Eurospeech,* pp. 2407–2410, Budapest, Hungary, September 1999.

110. D. Wang. On ideal binary mask as the computational goal of auditory scene analysis, In *Speech Separation by Humans and Machines,* pp. 181–197. Kluwer Academic, Norwell, MA, 2005.

111. D. L. Wang and G. J. Brown. Separation of speech from interfering sounds based on oscillatory correlation. *IEEE Transactions on Neural Networks,* 20:409–456, 1999.

112. R. M. Warren, C. J. Obusek, R. M. Farmer, and R. P. Warren. Auditory sequence: Confusions of patterns other than speech or music. *Science,* 164:586–587, 1969.

113. R. M. Warren, K. R. Riener, J. A. Bashford, and B. S. Brubaker. Spectral redundancy: Intelligibility of sentences heard through narrow spectral slits. *Perception and Psychophysics,* 57(2):175– 182, 1995.

114. M. Weintraub. *A Theory and Computational Model of Monaural Auditory Sound Separation.* PhD thesis, Department of Electrical Engineering, Stanford University, 1985.

115. P. C. Woodland. Speaker adaptation for continuous density HMMs: A review. In *Proceedings of ITRW on Adaptation Methods for Speech Recognition,* pp. 11–19, Sophia Antipolis, France, 2001.

116. S. N. Wrigley and G. J. Brown. A neural oscillator model of auditory attention. *Lecture Notes in Computer Science,* 2139:1163–1170, 2001.

117. M. Wu, D. L. Wang, and G. J. Brown. A multipitch tracking algorithm for noisy speech. *IEEE Transactions on Speech and Audio Processing,* 11(3):229–241, 2003.

118. S. Yang and D. Wang. Model-based sequential organization in cochannel speech. *IEEE Transactions on Audio, Speech and Language Processing,* 14:289–298, 2006.

119. U. T. Zwicker. Auditory recognition of diotic and dichotic vowel pairs. *Speech Communication,* 3:265–277, 1984.

CHAPTER 10

Neural and Perceptual Modeling

GUY J. BROWN and DELIANG WANG

10.1 INTRODUCTION

Although Bregman's [16] perceptual account of auditory scene analysis (ASA) has been instrumental in stimulating the development of CASA, the body of work presented so far in this book is mostly driven by performance concerns for sound segregation, rather than concerns for modeling the perceptual process of ASA. The latter concerns emphasize the perceptual plausibility of computational models. Moreover, there is little reference in Bregman's account to the neural mechanisms underlying the perceptual organization of sound. Such questions of neural and perceptual plausibility must be addressed before a complete understanding of ASA is possible (see [59]).

Insight into these issues has come from computational models that aim to investigate specific perceptual phenomena and their underlying neural mechanisms, although we know relatively little about the neurobiological substrate of ASA. Models of this nature are inherently related to perceptual and physiological studies of ASA, and can directly contribute to answering the fundamental scientific question of how the auditory system solves the ASA problem. The search for biologically plausible mechanisms may also lead to key constraints for effective representations and algorithms for achieving human-level performance in ASA.

This last chapter describes the CASA effort focused on neural and perceptual modeling. Section 10.2 considers the neural basis of ASA. Section 10.3 explains a number of neuronal models, including that of an auditory neuron. Section 10.4 describes computational models that simulate perceptual phenomena. Section 10.5 presents a neurocomputational framework for ASA, followed by a review of studies on schema-driven grouping in Section 10.6. Finally, in Section 10.7, we discuss several issues that have emerged from the modeling effort.

Computational Auditory Scene Analysis. Edited by DeLiang Wang and Guy J. Brown

10.2 THE NEURAL BASIS OF AUDITORY GROUPING

As described in Chapter 1, Section 1.1, the acoustic input to the auditory system first undergoes frequency analysis within the cochlea. Beyond this peripheral analysis, the input signal is further processed along the central auditory pathway that passes through four nuclei before reaching the auditory cortex. Whereas tonotopic organization (i.e., the systematic arrangement of frequency-specific inputs) appears to be maintained throughout the pathway, various auditory features are extracted at different levels. For example, neurons within the cochlear nucleus exhibit properties akin to onset detection, and cells within the inferior colliculus encode interaural time differences [78]. Although features of the sound are represented in different neural structures, auditory scene analysis that produces organized streams is a holistic process; a stream corresponds to a sound source that shows a variety of auditory features, such as onset, azimuth, and periodicity. How does holistic perception arise from neuronal activity distributed in many brain areas? This is the well-known *binding problem*—how are individual neuronal responses bound together to form percepts? The binding problem occurs because the acoustic input generally contains multiple sound sources, which makes it necessary to resolve the issue of which features should bind with which others to give rise to auditory streams.

So, in neural terms, the ASA problem is a form of the binding problem.

10.2.1 Theoretical Solutions to the Binding Problem

Theoretically speaking, two alternative solutions have been proposed to address the binding problem. The first has been called *hierarchical coding* [9]. The main idea is that individual neurons integrate information in some cortical hierarchy so that cells higher in the hierarchy respond to larger and more specific features of an object. Ultimately, individual objects are represented by individual neurons somewhere in the brain, and for this reason hierarchical coding is also known as the "grandmother cell" or "cardinal cell" representation. This theory is consistent with Hubel and Wiesel's classical observations on the receptive fields of cortical neurons [51], and later findings of face-detecting neurons in the inferior temporal cortex [55].

The hierarchical coding hypothesis has been criticized on both biological and theoretical grounds [45]. A main computational difficulty with the hypothesis, for example, is the apparent requirement for prohibitively many cells to encode all possible combinations of individual features and objects that could occur in a scene [93, 98]. Aware of the limitations of hierarchical coding, von der Malsburg [93] put forward the temporal correlation theory as a major alternative. The main idea is that the temporal structure of neuronal responses provides the neural basis for correlation, which in turn binds features of the same object. In a subsequent paper, von der Malsburg and Schneider [94] demonstrated the utility of the temporal correlation theory in the concrete task of segregating two auditory inputs based on their distinct onset times. Their study introduces the use of neural oscillators as basic units of a neural network, and synchrony and desynchrony between neural oscillations to encode the state of binding among auditory features. This study, along with subsequent neurobiological evidence for coherent oscillations (see Section 10.2.2), led to the formulation of *oscillatory correlation* as a neural mechanism to address the binding problem [86].

The contrast between hierarchical coding and temporal (or oscillatory) correlation is, in a sense, similar to the perennial debate in audition between place coding and temporal coding.

10.2.2 Empirical Results on Binding and ASA

Earlier, we briefly mentioned the biological foundation underlying the hierarchical coding hypothesis. The temporal correlation hypothesis is supported by a growing body of physiological evidence. Oscillations of evoked potentials were observed in early experiments in the olfactory system [42] and the auditory system [43]. Synchronous oscillations in cell recordings were later discovered in the visual cortex [40, 46]. The frequencies of coherent oscillations range from 30 to 70 Hz, compatible with those of electroencephalogram (EEG) gamma rhythms; hence, such oscillations are also called gamma oscillations. Coherent oscillations correlate with perceptual organization for a broad range of stimuli, and are subject to attentional influence. Synchronous firing has been established as a major neural principle for feature binding (for reviews see Singer and Gray [83], Usrey and Reid [88], and Varela et al. [92]).

In the auditory system, gamma oscillations have been observed in localized brain regions both at the cortical level and at the thalamic level, and synchrony between neural oscillations has been found over extended cortical areas [80, 58]. Joliot et al. [53] found that coherent oscillations are correlated with perceptual grouping of click sounds. Synchronous activity was also found from cell recordings in the auditory cortex [39]. Barth and MacDonald [10] found evidence that neural oscillations in the auditory cortex originate within the cortex and synchrony emerges from intracortical interactions. Using multielectrode recordings, Brosch et al. [20] found gamma oscillations in the monkey primary auditory cortex that are correlated with the match between stimulus frequency and the preferred frequency of a cortical unit. In addition, synchronization of gamma oscillations at different recording sites occurs depending on the distance and the match between preferred frequencies of recorded units.

Studies of event-related potentials (ERP) in the brain have found physiological correlates of aspects of ASA. Alain et al. [1, 3] have found, during presentation of harmonic complexes with a single mistuned component, that a feature of the ERP—the object-related negativity (ORN)—is correlated with listener judgments of whether more than one sound source is present, which in turn depends on the degree of mistuning. Another component of the ERP—the late positive wave—is present only during active listening (i.e., when listeners are required to indicate whether they have heard one or two sound sources). They suggest that these results are consistent with Bregman's general division between bottom-up and top-down processes in ASA (see Chapter 1, Section 1.1), with the ORN corresponding to primitive ASA and the late positive wave indicating top-down processing. In addition, Alain and Izenberg [2] reported evidence that simultaneous and sequential grouping processes are differentially influenced by attention. Sussman et al. [85] have investigated the ERP associated with the perception of auditory streaming of tone sequences, in an attempt to determine whether streaming is a result of attentive or preattentive processing. They used an index of preattentive acoustic processing called the mismatch negativity (MMN), and found that when tone repetition is slow, so that listeners hear a single stream, no MMN is observed. However, at faster presentation rates listeners hear two streams, and this is accompanied by the MMN. This finding suggests that auditory streaming is a preattentive process, as claimed by Bregman.

10.3 MODELS OF INDIVIDUAL NEURONS

A fundamental property of a typical neuron is the generation of action potentials or spikes. Inputs from other neurons, also in the form of spikes, trigger changes in the membrane potential of the neuron. When the membrane potential exceeds a threshold, the cell initiates an action potential, which then propagates down the cell's axon with constant amplitude. An action potential is followed by a refractory period during which it is difficult or impossible to generate another spike. The biophysical mechanism underlying action potential generation is very well understood [55], and is best described by the classical Hodgkin–Huxley equations [50]. The Hodgkin–Huxley model is a set of four coupled and nonlinear differential equations that describe changes in membrane potential caused by the flow of ions across the membrane, particularly sodium and potassium. They showed that the conductance of the membrane to sodium and potassium is voltage- and time-dependent, and their equations gave explicit description of such dependencies.

Although the Hodgkin–Huxley equations give an accurate model of the electrophysiological behavior of an individual neuron, they are very complex with many parameters. The complexity of the model, particularly when used as a basic unit of a large neural network, has motivated substantial effort in simplifying the equations. A widely known example is the FitzHugh–Nagumo model [41, 70] which is a two-variable equation. Since a neuronal model typically produces a spike train with a certain frequency when the input is constant, such a model is often referred to as an oscillator. The FitzHugh–Nagumo oscillator may be viewed as an example of a relaxation oscillator. Further simplification led to the commonly used integrate-and-fire oscillator (or spike oscillator) which is a one-variable equation with a discrete operation for spike generation. Relaxation and spike oscillators are described in some detail below, followed by a description of a specific model of an auditory neuron.

10.3.1 Relaxation Oscillators

Relaxation oscillations were first described by van der Pol in 1926 [90] when studying the behavior of a triode circuit. The van der Pol oscillator can be written as a two-variable, first-order differential equation [44, 97]:

$$\frac{dx}{dt} = c\left(y + x - \frac{x^3}{3}\right) \tag{10.1a}$$

$$\frac{dy}{dt} = -\frac{x}{c} \tag{10.1b}$$

The x nullcline (i.e., $dx/dt = 0$) is a cubic curve and the y nullcline is the y axis itself. These two nullclines intersect on the middle branch of the cubic, and the resulting fixed point (where both $dx/dt = 0$ and $dy/dt = 0$) is unstable. As a result, this equation gives rise to an oscillation.

When $c \gg 1$, Eq. 10.1 exhibits two timescales: a fast timescale for the x variable and a slow timescale for the y variable. As a result, the oscillation produced becomes a relaxation oscillation, which is characterized by two timescales.

When studying the synchronization and desynchronization properties of relaxation oscillators, Terman and Wang [27] proposed the following equation:

$$\frac{dx}{dt} = 3x - x^3 + 2 - y + I \tag{10.2a}$$

$$\frac{dy}{dt} = \varepsilon\left\{\alpha\left[1 + \tanh\left(\frac{x}{\beta}\right)\right] - y\right\} \tag{10.2b}$$

where I represents external stimulation to the oscillator. In this equation, the x nullcline is a cubic and the y nullcline is a sigmoid, where α and β are parameters. When $\varepsilon \ll 1$, Eq. 10.2 becomes a relaxation oscillator. When $I > 0$, the two nullclines intersect at a single point on the middle branch of the cubic and the oscillator produces a stable oscillation, illustrated in Fig. 10.1**A**. The periodic orbit alternates between a silent phase (low x values) and an active phase (high x values). Within each of the two phases, the oscillator shows near-steady-state behavior, whereas the transitions between the two phases take place rapidly. Such transitions are referred to as jumps. When $I < 0$, the two nullclines intersect also at a stable fixed point along the left branch of the cubic, illustrated in Fig. 10.1**B**; in this case, no oscillation occurs. The parameter α determines relative times that the oscillator spends in the two phases: a larger α leads to a relatively shorter active phase.

The Terman–Wang oscillator model resembles the van der Pol and FitzHugh–Nagumo oscillators, but provides a dimension of flexibility absent in these models. In neuronal terms, the x variable in Eq. 10.2 corresponds to the membrane potential, and the y variable to the activation state of ion channels. Figure 10.1**C** shows a trace of the x activity when the oscillator is stimulated and, hence, oscillates, akin to a spike train. Network properties of Terman–Wang oscillators will be discussed in Section 10.4.1.

10.3.2 Spike Oscillators

The fact that an action potential is a short pulse of the membrane potential motivates simpler models that produce phenomenologically similar behavior to that of

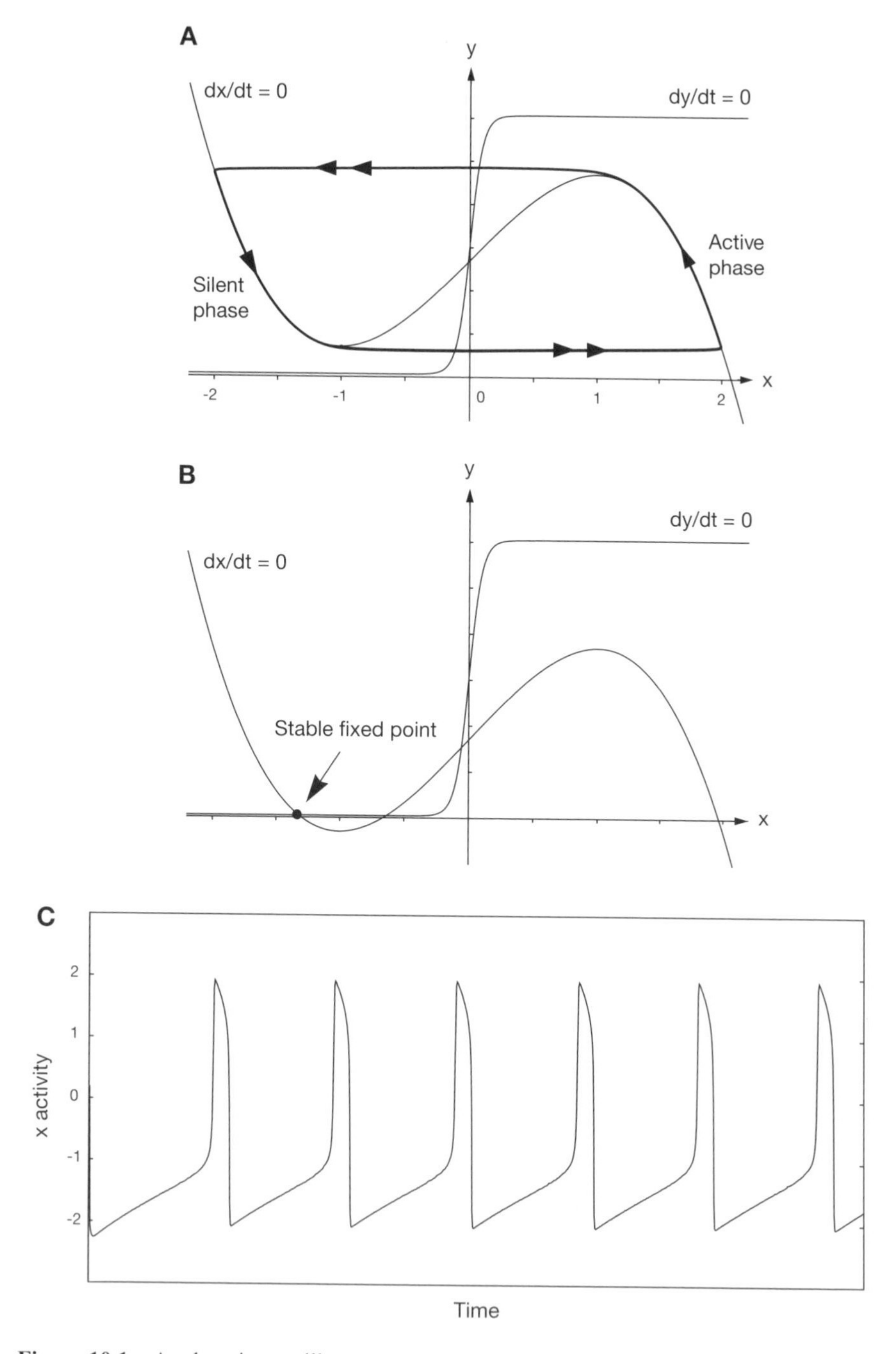

Figure 10.1 A relaxation oscillator. (**A**) With external stimulation, the oscillator has a periodic orbit, shown as the bold curve. The arrows indicate the direction of the orbit, and double arrows indicate jumping. (**B**) Without external stimulation, the oscillator approaches a stable fixed point and does not oscillate. (**C**) The x activity of the oscillator with respect to time. Redrawn from [99]. Copyright © 1999 IEEE.

more complex models. Among such spike oscillators is the widely used integrate-and-fire model [75, 69] on which our following discussion is based. A spike oscillator can be described by one variable corresponding to the membrane potential:

$$\frac{dx}{dt} = -x + I \tag{10.3}$$

where I indicates external stimulation and determines the oscillation period. When the membrane potential reaches a threshold (set to 1) at time t, the oscillator produces a spike and its potential is instantly set to 0. This discrete operation can be mathematically described as

$$x(t^+) = x(t) - 1 \tag{10.4}$$

This algorithmic operation differs from the aforementioned neuronal models in which pulse generation arises from intrinsic dynamics. Nonetheless, in digital simulations this step does not entail much difference.

The above equations can be solved analytically [69]. Let us denote the phase of the oscillator by φ, where $\varphi \in [0, 1]$. We have $x(\varphi) = I[1 - \exp(-\tau\varphi)]$, where $\tau = \log[I/(I - 1)]$ is the period of the oscillation. Figure 10.2**A** shows $x(\varphi)$ and Figure 10.2**B** shows a trace of oscillation.

10.3.3 A Model of a Specific Auditory Neuron

The above models aim to simulate the behavior of a generic neuron. Neurons in the brain vary enormously in terms of their morphology, electrophysiology, and function. As an example of auditory neurons, we describe a model of a stellate neuron in the cochlear nucleus (CN). Innervated by the auditory nerve, the cochlear nucleus is the first major nucleus of the central auditory pathway. The cochlear nucleus is further divided into three regions: the anteroventral CN, posteroventral CN, and dorsal CN. Stellate cells are found primarily inside the anteroventral CN and the posteroventral CN. One type of stellate cell (denoted chop-S) shows physiologically defined "chopper" responses, producing spikes at constant intervals that are independent of stimulus frequency and phase [79]. The stellate model described below was proposed by Hewitt et al. [49] to simulate the chopper response.

The basic neuron model is a variant of the Hodgkin–Huxley equations in which only the potassium conductance is considered. The stellate model includes a dynamic threshold for action potential generation:

$$\tau_\theta \frac{d\theta}{dt} = -[\theta(t) - \theta_0] + cx(t) \tag{10.5}$$

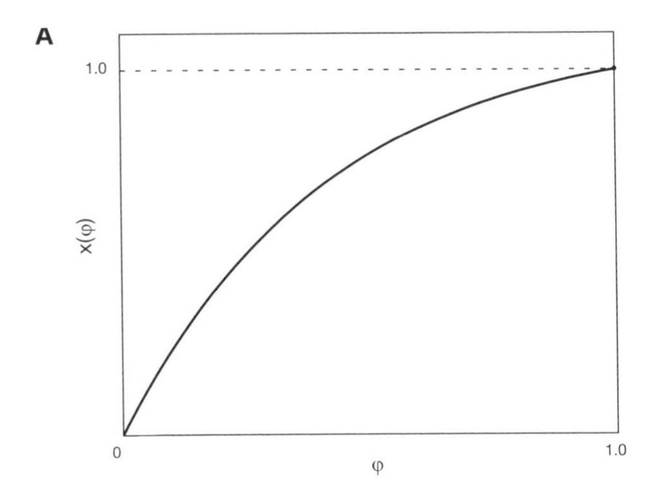

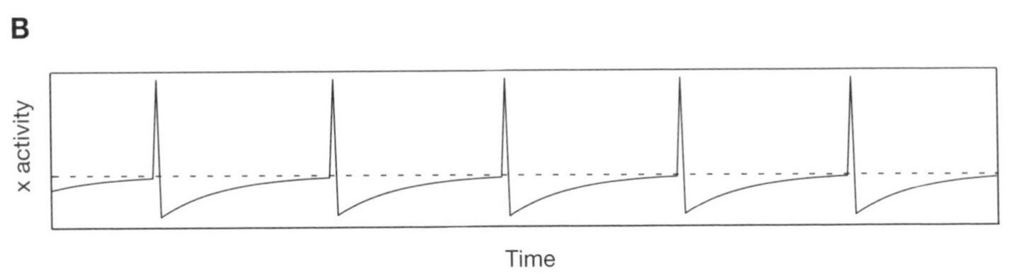

Figure 10.2 A spike oscillator. (**A**) The solution of x, plotted with respect to its phase, φ. The dashed line indicates the oscillator threshold. (**B**) The oscillator activity with respect to time. The spikes shown above indicate the times when the oscillator reaches the threshold. From [98]. Copyright © 1999 IEEE.

where $\theta(t)$ is the time-varying threshold and θ_0 is the resting threshold of the cell model, τ_θ is a time constant for threshold change, x indicates the membrane potential, and c is a constant.

With a fixed set of parameter values, the above single-neuron model generates a spike train in response to a depolarizing (positive) current and its response replicates in vitro recordings of stellate cells. Basically, the membrane potential rises quickly in response to a stimulating current, producing the first spike when the membrane potential reaches the θ value. The parameters of the model are set so that the membrane potential recovers quickly. With a slowly varying θ, the stellate model soon returns to the firing threshold triggering another spike. This response pattern matches that of the chopper response.

In addition, Hewitt et al. [49] couple their stellate neuron model with a cochlear model that simulates cochlear filtering, hair cell transduction, and firing properties of the auditory nerve. They have shown that this composite model exhibits a number of realistic responses to pure tones and amplitude-modulated (AM) stimuli, such as onset responses, the relationship of firing rate to stimulus level, and a bandpass modulation transfer function.

10.4 MODELS OF SPECIFIC PERCEPTUAL PHENOMENA

To what extent are ASA phenomena associated with particular neural mechanisms? Bregman suggests a degree of caution in regard to this issue:

> I am afraid that I am not very hopeful about drawing inferences from function to architecture. . . . [T]hinking in terms of scene analysis may not lead us to find new neural mechanisms. But it may help us to understand the purposes of the ones we know about. ([16], pages 325–326)

In this section, we review a number of studies that have developed computer models of specific ASA phenomena. As said by Bregman, many of these models exploit well-known properties of the auditory system, such as the band-pass nature of cochlear filtering and the randomness apparent in auditory nerve firing patterns. Others propose specific neural circuits associated with pitch analysis and grouping, but must be regarded as speculative in the absence of specific neurophysiological evidence that supports them. However, such models may at least stimulate physiological studies that aim to verify them, and have proven to be productive in the past. A striking example is the neural circuit for extracting interaural time differences hypothesized by Jeffress in 1948 [52], which is strongly supported by many subsequent physiological findings (see Chapter 5).

10.4.1 Perceptual Streaming of Tone Sequences

The phenomenon of auditory streaming has been the subject of a number of computational modeling studies. Auditory streaming can be induced by presenting listeners with a pattern of alternating tones with different frequencies (denoted A and B), giving a continuously repeating sequence of the form ABAB. . . [68, 17, 91] (see also Chapter 1, Section 1.2). When the tone repetition time (TRT) is long, or the frequency difference between the tones is small, listeners perceive the repeating sequence as a musical trill (i.e., a single coherent stream). However, if the TRT is short or the frequency difference between the tones is large, then the A and B tones are perceptually organized into separate streams. In this condition, listeners are able to focus their attention on the stream of A tones or the stream of B tones. Additionally, the two streams separate into figure and ground, so that the attended stream appears to be subjectively louder than the unattended stream [17].

The main challenge for computational models of auditory streaming is to explain the dependence of auditory organization on the frequency difference and TRT. Van Noorden mapped the occurrence of streaming and perceptual coherence as a function of these two parameters, and identified two perceptual boundaries as shown in Fig. 10.3. Below the fission boundary, listeners always hear the sequence of tones as a coherent whole. Above the temporal coherence boundary, the A and B tones are always segregated into different perceptual streams. Between these two boundaries is an ambiguous region, in which listeners may hear streaming or coherence, a change between the two percepts may occur spontaneously, or can be initiated by conscious attentional effort. A further challenge for computational modeling studies

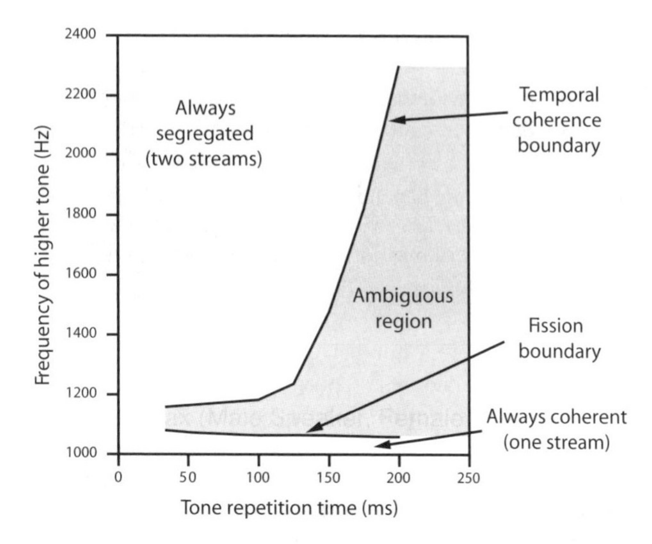

Figure 10.3 Dependence of auditory streaming of ABAB tone sequences on frequency separation and tone repetition time. The frequency of the higher tone is plotted on the ordinate; the frequency of the lower tone was 1000 Hz. The duration of each tone was 40 ms. Data from McAdams and Bregman [60].

is to explain the buildup of streaming over time. Initially, all ABAB tone sequences are perceived as coherent, but the probability of streaming increases over time at a rate that depends upon the TRT [4].

Bregman attributes auditory streaming to sequential grouping heuristics that perceptually group successive tones according to their proximity in time and frequency. A number of computational modeling studies have suggested that such behavior could arise from simple neural circuits, which might be associated with low-level (subcortical) mechanisms of auditory processing. For example, Beauvois and Meddis [13, 14] describe such a model. Their approach is based on two assumptions: that streaming arises from the selective accentuation of peripheral frequency channels, and that spontaneous shifts between foreground and background organization are due to the stochastic nature of auditory nerve firing patterns. The first stage of their scheme is a two-stage cochlear model, which processes the acoustic signal through an ERB filterbank and a model of the inner hair cell–auditory nerve junction [64, 65]. The output of each peripheral frequency channel is then processed by two pathways. In the "temporal integration path," the simulated auditory nerve firing patterns are smoothed by passing them through a leaky integrator with a time constant of 3 ms. The "excitation-level path" adds a cumulative random bias to the output of the temporal integration path, and then smooths it with a longer time constant (70 ms).

Beauvois and Meddis consider the firing output at three auditory filter channels, which are centered at the frequency of the high tone, the frequency of the low tone, and midway between the two. When presented with an acoustic input, the channel with the highest activity in the excitation-level path is selected as the "dominant" (foreground) channel. In the temporal integration path, all channels except the dominant channel are attenuated by a factor of 0.5 (i.e., they become the "background"), and stream segregation is detected on the basis of the relative amplitude levels of the high and low tones in the model's output. If the two tones elicit a different level of activity, then the percept is regarded as streaming; otherwise, perceptual coherence is assumed. The cumulative random bias used in the model has important consequences for its behavior. First, different random walks in each frequency channel cause the dominant channel to fluctuate between the high and low tone, leading to spontaneous changes in organization. Second, the random bias takes time to accumulate, so that the initial percept is coherence, and the likelihood of streaming increases over time. Finally, the model displays a sensitivity to the TRT, because the random bias is proportional to channel activity; when the temporal gaps between successive tones are short, there is little opportunity for a random bias between the two channels to dissipate, and streaming is, therefore, more likely.

McCabe and Denham [61] also describe a model of auditory streaming in which organization arises from competitive interactions between frequency channels. However, unlike the Beauvois and Meddis model, their scheme includes multiple frequency channels with graded inhibitory feedback. The structure of the McCabe and Denham model is shown in Fig. 10.4. It consists of two arrays of neurons, which are presumed to be tonotopically organized. These "foreground" and "background" neural arrays are reciprocally connected by inhibitory inputs, so that activity in a particular frequency region of one array tends to suppress activity in the same frequency region of the other array. Inhibition is graded according to the fre-

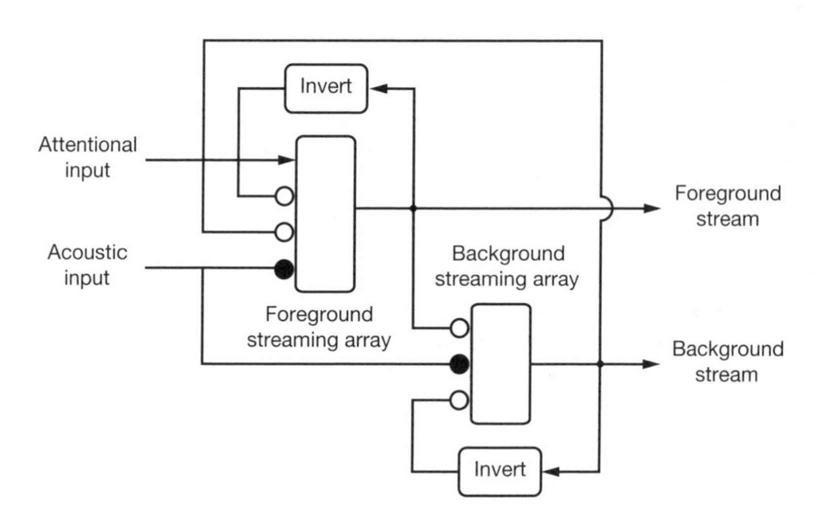

Figure 10.4 Schematic diagram of McCabe and Denham's auditory streaming model. Connections ending in filled circles are excitatory, and connections ending in open circles are inhibitory. Redrawn with permission from [61]. Copyright 1997, Acoustical Society of America.

quency proximity between inputs to the network. In addition, each neural array receives a recurrent inhibitory input that is related to the reciprocal of its own activity. This self-inhibition causes previous activity in the network to suppress differences in subsequent stimuli; it may therefore be regarded as a neural implementation of Bregman's "old plus new" heuristic, which states that listeners tend to perceive a current sound as a continuation of a previous sound, unless there is strong evidence to the contrary (see [16], p. 222). McCabe and Denham show that their model is able to account for the effects of frequency difference and tone repetition rate on the streaming phenomenon, and reproduces the buildup of streaming over time. In addition, interaction between the two neural arrays allows their model to account for the influence of background organization on the perceptual foreground (e.g., see Bregman and Rudnicky [19]).

Both of the above models assume that auditory streaming is encoded in terms of the firing rate across a spatial array of neurons. An alternative explanation is that grouping is signaled by the temporal correlation of neural firings. More specifically, a number of workers have proposed models of auditory streaming in which the grouping of frequency channels is encoded by the temporal synchronization of neural responses (recall Section 10.2.2).

Models of this type have been proposed by Wang [95, 96], Brown and Cooke [23], and Baird [8]. Wang's model [96] is based on the locally excitatory, globally inhibitory oscillator network (LEGION) proposed by Wang and Terman [101]. In their model, each oscillator is defined by an excitatory unit x_i and an inhibitory unit y_i, as described in Eq. 10.2. Oscillators are arranged into a two-dimensional time–frequency grid, with local excitatory connections between neighbors. In addition, all oscillators are coupled with a global inhibitor, which receives excitation from every oscillator and feeds back inhibition. Hence, each oscillator i has the form

$$\frac{dx_i}{dt} = 3x_i - x_i^3 + 2 - y_i + I_i + \rho + S_i \tag{10.6}$$

$$\frac{dy_i}{dt} = \varepsilon\left\{\alpha\left[1 + \tanh\left(\frac{x_i}{\beta}\right)\right] - y_i\right\} \tag{10.7}$$

where S_i represents the combined effects of local excitation and global inhibition. The additional term ρ represents Gaussian noise, which serves to assist the desynchronization of different input patterns. As before, I_i represents the external input to the oscillator.

Input to Wang's model is provided in the form of symbolic binary patterns, for which I_i is set to a small positive value if oscillator i is stimulated, and zero otherwise. The arrangement of oscillators in a time-frequency grid allows the synchronization of oscillators to be influenced by the proximity of acoustic components in time and frequency (it is assumed that the time axis of the oscillator network is created by a system of neural delay lines). More specifically, the strength of excitatory connections between oscillators diminishes according to a Gaussian function with increasing distance in time and frequency. Hence, oscillators that are close in time and frequency tend to synchronize (i.e., they form a stream). In contrast, oscillators that are distant in time and frequency do not receive mutual excitation, and tend to desynchronize because of the action of the global inhibitor. This behavior is illustrated in Fig. 10.5, which shows the response of the oscillator network to tone patterns that induce streaming (panels **A** and **B**) and perceptual coherence (panels **C** and **D**). Wang also demonstrates that his model shows similar behavior for frequency-modulated (FM) tones, and exhibits competition between alternative organizations when a third tone sequence is present.

One criticism of Wang's approach is that oscillatory dynamics proceed very rapidly, so that his model is not able to explain the gradual buildup of auditory

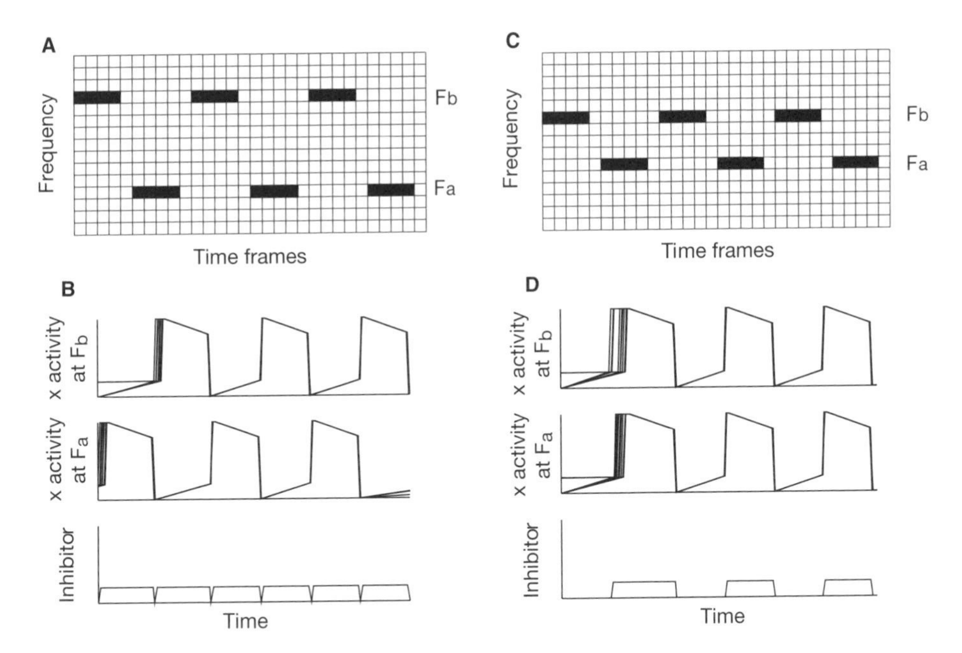

Figure 10.5 Auditory streaming in the neural oscillator model of Wang [96]. (**A**) Time–frequency stimulus pattern corresponding to an alternating tone sequence with a large frequency separation, which is expected to induce streaming. (**B**) Combined x values of all oscillators in response to the pattern of six tones, for the frequency channels excited by the high tone (F_b, top panel) and the low tone (F_a, middle panel). The bottom panel shows the activity of the global inhibitor. Note that the oscillators in the F_b and F_a regions are desynchronized, indicating that streaming has occurred. The global inhibitor receives input from both streams and, therefore, has a frequency twice that of the oscillators in the F_b and F_a regions. (**C**) Stimulus pattern for an alternating tone sequence with a small frequency separation, which is likely to be perceived as coherent. (**D**) Activity for oscillators in the F_b channel (top), F_a channel (middle), and global inhibitor (bottom). The oscillators in the F_b and F_a regions are synchronized, indicating that the tones are grouped to form a single coherent stream. Adapted from Wang [96]. Copyright 1996, Cognitive Science Society.

streaming over time [4]. Indeed, Terman and Wang [86] have shown that a LEGION network requires no more than N oscillator cycles to segregate N objects in the input. Also, Wang's model takes its input from idealized binary patterns rather than sampled acoustic waveforms. Both of these issues are addressed by the oscillator model of Brown and Cooke [23]. Their model subjects the acoustic input to a cochlear frequency analysis (using a gammatone filterbank; see Chapter 1), and then detects onset events at the output of each frequency channel. Grouping is encoded by an array of oscillators (one per frequency channel). Initially, all oscillators are strongly coupled, and show an identical phase response; this is consistent with the finding that listeners initially hear all alternating tone sequences as coherent. During simulation, a simple learning rule reduces the coupling between oscillators whose corresponding onset detectors do not exhibit the same level of activity at the same time. Differences in onset level are greater for frequency channels that are distant in frequency, due to the band-pass nature of auditory filters, and the rate at which coupling is reduced is proportional to the number of onset events that occur. Hence, the network is sensitive to frequency differences between tones in the input, and to the TRT. The gradual decay of coupling between oscillators over time accurately models the buildup of the streaming percept.

On the other hand, Brown and Cooke's model has a number of limitations compared to Wang's approach. First, the former models the phase dynamics of each oscillator using sine circle maps [12], which have a chaotic (rather than periodic) response. This necessitates an expensive cross correlation to assess the state of synchronization in the network, whereas synchronization in Wang's model can be judged by applying a simple threshold to the oscillator activity (see Fig. 10.5). More seriously, time is implicit in Brown and Cooke's model, so that it is not possible to express grouping relationships between acoustic events in different frequency channels at different times. Wang's use of a two-dimensional oscillator grid with an explicit time axis affords a more expressive representation in this respect.

Further extensions of Wang's model have been proposed in order to address its limitations while retaining its advantages. For example, Norris [72] introduces a measure of synchrony for Wang's network, which is calculated as the proportion of time for which two active oscillators are simultaneously active. This allows the performance of the network to be quantified and compared with human data. Norris reports that his version of Wang's model shows reasonable agreement with the shape of the fission boundary and temporal coherence boundary shown in Fig. 10.3, although he does not obtain a close match to human data. Chang [30] obtains better agreement with these two boundaries, using a modification of Wang's network that includes dynamic weights between oscillators. More specifically, Chang uses coupling weights that vary depending upon the TRT, such that the Gaussian distribution of weights is wider along the frequency dimension of the network when the TRT is short. Chang also shows that spontaneous changes between streaming and coherence can be modeled in Wang's network by including a random element in the activity of the global inhibitor.

Norris [72] also extends Wang's model to include grouping by onset synchrony. Onsets are assumed to be detected by a separate process, and are passed to the network as additional external inputs. Coupling is then increased between oscillators that receive onset inputs at approximately the same time, via extra excitatory connections. This approach is only partially successful, however; the modified network groups simultaneous acoustic events more strongly, but does not show the correct behavior for stimulus patterns that place onset cues in competition with temporal and frequency proximity [18]. Norris also adapts Wang's network to include long-range connections and a time-dependent bias in the input to each oscillator, which allow the buildup of streaming over time to be simulated. His model shows the correct qualitative behavior, but does not provide as close a match to human data as the models discussed above [14, 61, 23].

The models of Beauvois and Meddis [14], McCabe and Denham [61], and Brown and Cooke [23] acknowledge the influence of attentional factors in stream segregation, but do not model them explicity. Wang [96] has suggested that the global inhibitor in his model can be regarded as a center for attentional control. He points out that the structure and function of the global inhibitor bears some resemblance to that of the thalamus (specifically the thalamic reticular complex), which has been implicated in selective attention [33].

The notion of selective control of synchronization as a form of attention has been further developed by Baird [8]. His model is motivated by an experiment described by Jones et al. [54], which suggests that rhythmic expectancy plays a role in auditory streaming. Jones et al. replicated and extended an experiment of Bregman and Rudnicky [19] in which listeners were asked to determine the order of presentation of a pair of high-frequency target tones AB or BA, which were embedded in a sequence of low-frequency "captor" tones C. Flanking tones F of intermediate frequency were played directly before the target tones, giving a complete sequence of the form shown in Fig. 10.6. Bregman and Rudnicky found that listeners were better able to judge the order of the A and B tones when the flanking tones were close in frequency to the captor tones. They reasoned that when the flanking tones and target tones were close in frequency, the latter were embedded in a stream FABF so that it was difficult to determine their order. However, Jones et al. noted that the A and B tones were also distinguished from the C and F tones by a difference in rhythm. They found that when the A and B tones had the same rhythm as the C and F tones, listeners found it harder to judge the order of the target tones. Jones et al.

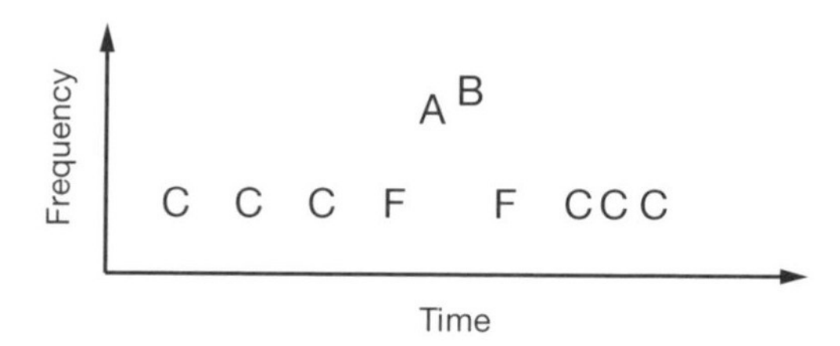

Figure 10.6 The tone sequence used by Bregman and Rudnicky [19]. The task of the listener is to determine the order of the target tones (A and B) in the presence of captor tones (C) and flanking tones (F).

concluded that a difference in rhythm was necessary for listeners to do the task, in addition to the constraints on the frequency of the C and F tones noted by Bregman and Rudnicky.

Following this reasoning, Baird's model of auditory streaming assumes that rhythmic expectancy acts as a sequential grouping cue in stream formation. In his model, neural oscillators establish rhythmic firing patterns that act as a multiscale representation of time. Oscillators synchronize with the salient periodicities in the input signal (i.e., the TRT in the case of alternating tone sequences) in order to generate rhythmic expectations, which prime attention to subsequent points in time (see also [56]). Neural activity patterns that coincide with these expectations are boosted in amplitude and pulled into synchrony with an attentional stream, which models the listener's attentional set. Similarly, activity patterns that deviate from the temporal expectations cause a discrepancy, which leads to the formation of a new stream. Baird suggests that this mechanism might underlie the mismatch negativity response observed in ERP recordings (see Section 10.2.2).

Baird's model suggests a neural basis for the findings of Jones et al. discussed above. In his model, the A and B tones are brought into the attentional foreground because of the mismatch between their rhythm and the expectation generated by the preceding C and F tones; however, frequency differences alone are insufficient to cause the target tones to be segregated when they share the same rhythm as the C and F tones. The representation of time in Baird's model is also of note. Periodic oscillations are used to establish multiple time scales, thus allowing neural activity at two points in time to be sequentially grouped. Hence, his model does not require an explicit time axis as used by Wang [96]; nor does it suffer from the limitations of Brown and Cooke's [23] approach, in which sequential grouping relationships cannot be made between neural activity in the same frequency channel at different times. It is questionable, though, whether Baird's approach will scale to more naturalistic inputs such as speech, in which the periodicities are less regular than those associated with simple tone sequences.

Baird's model considers rhythmic expectancy to be a data-driven attentional process, although his approach also allows voluntary top-down attention in which the attentional stream is synchronized with oscillators at a particular time scale. These bottom-up and top-down attentional mechanisms have been called "exogenous" and "endogenous," respectively (e.g., see [84]). But what exactly is the role of endogenous attention in auditory streaming? Bregman and Rudnicky [19] interpret their findings as evidence that endogenous attention is not required for stream segregation; they note that the F tones in their experiment were "captured" by the C tones to form a stream, even though listeners were instructed to attend to the A and B tones. However, this interpretation has been disputed by Carlyon et al. [29], who point out that subjects may well have been attending to the C tones in Bregman and Rudnicky's experiment, since there was no competing task to draw attention away from them. Accordingly, Carlyon et al. report an auditory streaming experiment in which listener's attention is controlled more carefully. They presented a tone sequence to one ear, and required subjects to perform a competing task in the opposite ear during periods in which attention was to be directed away from the tone sequence. When listeners were instructed to attend to the tone sequence, there was a build-up of streaming consistent with previous studies [4]. However, streaming did not occur when listeners were told to attend to the competing task in the other ear.

Motivated by these findings, Wrigley and Brown [103, 104] have described a neural oscillator model of auditory streaming in which endogenous attention plays a central role. In their model, the influence of endogenous attention is modeled by a Gaussian distribution across a one-dimensional tonotopic array of neural oscillators. This "attentional interest" modulates the coupling between oscillators and an attentional leaky integrator (ALI),which sums and smooths the input from each oscillator. In this model, a group of oscillators is considered to be the subject of attention if their activity is synchronized with the ALI activity. The output from Wrigley and Brown's model is, therefore, an attentional stream, as in Baird's [8] approach. An interesting feature of their model is that sufficiently intense acoustic events can override the endogenous attentional interest, which allows for unconscious (exogenous) redirection of auditory attention by loud sounds. Wrigley and Brown show that a binaural version of their model is able to model the human data of Carlyon et al. [29]. In addition, their model is sufficiently general to explain other auditory grouping phenomena, including the sequential capture of a harmonic from a complex tone [35].

Finally, we note that the above discussion might be taken to suggest a dichotomy between spatial and temporal representations of auditory grouping, as typified by the models of Beauvois and Meddis [14] and Wang [96], respectively. However, this need not be the case. Todd [87] has proposed that mechanisms of rhythm perception and auditory stream segregation are underlain by cortical maps of periodicity-sensitive cells, such that the temporal pattern of the acoustic input is represented spatially in the form of an amplitude modulation (AM) spectrum. His approach therefore represents an interesting middle ground between spatial and temporal coding schemes, and bears some similarity to the multiple-timescale approach of Baird [8]. In Todd's computer model, acoustic events that have highly correlated AM spectra are grouped to form a stream, and those with uncorrelated AM spectra are perceptually segregated. Although his model is functional, Todd hypothesizes a detailed cortical circuit that could perform the correlation of AM information. His computer simulations show qualitative agreement with the effects of TRT and tone frequency on auditory streaming.

10.4.2 Perceptual Segregation of Concurrent Vowels with Different *F*0s

The perceptual segregation of concurrently presented vowels ("double vowels") is another well-known example of ASA that has motivated a number of computer models. Typically, listeners are presented with two (usually synthetic) vowel sounds that start and stop at the same time, and may differ in their fundamental frequency ($F0$). Listeners are able to identify both vowels with a performance above chance level when they have the same $F0$. However, identification performance improves when a small (0.25 semitone) difference in $F0$ ($\Delta F0$) is introduced, and continues to improve with increasing $\Delta F0$ until it asymptotes at about 2 semitones.

Many models of double-vowel segregation assume that the phenomenon is underlain by mechanisms that group acoustic components that share a common $F0$. Among these, the model of Meddis and Hewitt provides the closest match to human data [67]. In their model, grouping is achieved by selection of peripheral frequency channels that are dominated by the same $F0$, as determined by a periodicity analysis of auditory nerve firing patterns. More specifically, a correlogram and summary correlogram are computed for the mixture of two vowels, as described in Chapter 1, Section 1.3.3 (see also Chapter 2, Section 2.4.3). When the vowels have the same $F0$, a single peak is visible in the summary correlogram (see panel **A** of Fig. 10.7).

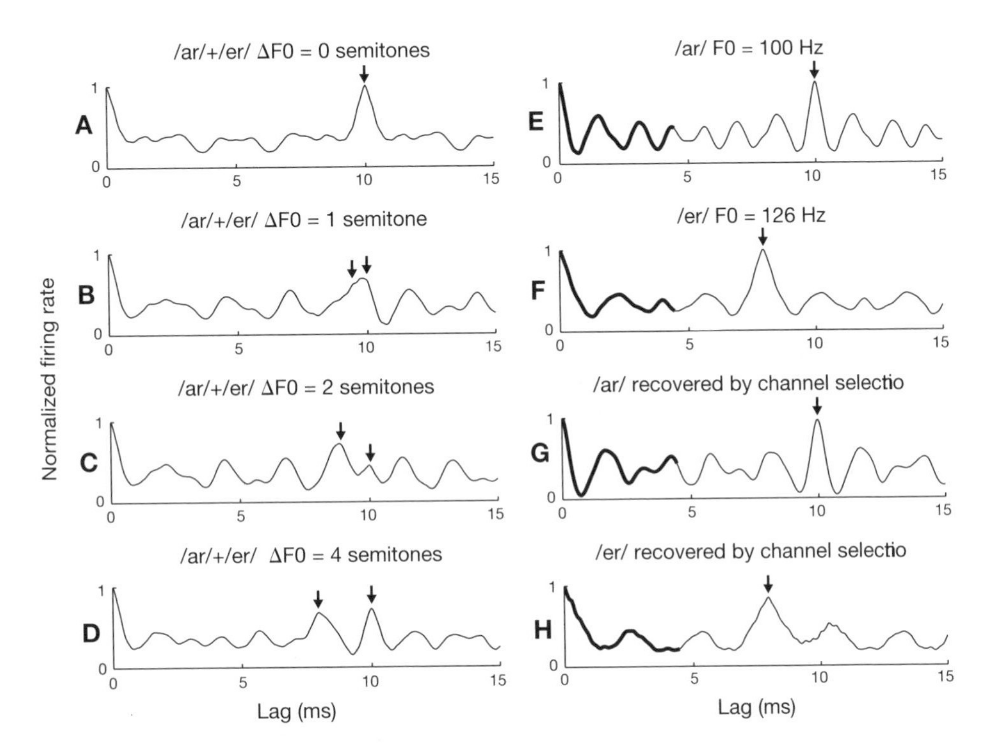

Figure 10.7 Vowel segregation by channel selection. Panels **A–D** show summary correlograms for a mixture of the vowels /ar/ and /er/, for four conditions in which the $F0$ of the vowel /ar/ is held at 100 Hz and the $F0$ of the /er/ is 100 Hz (**A**), 106 Hz (**B**), 112 Hz (**C**), or 126 Hz (**D**). The corresponding differences in fundamental frequency ($\Delta F0$) are 0, 1, 2, and 4 semitones. Arrows indicate the periods of the two vowels. (**E**) Summary correlogram of the isolated vowel /ar/ with $F0 = 100$ Hz. (**F**) Summary correlogram of the isolated vowel /er/ with $F0 = 126$ Hz. (**G**) Summary correlogram of the vowel /ar/ recovered from a mixture of /ar/+/er/ ($\Delta F0 = 4$ semitones). The vowel was recovered by selecting correlogram channels that were dominated by a periodicity of 10 ms. (**H**) Summary correlogram of the vowel /er/, recovered from the mixture of /ar/+/er/ by selecting correlogram channels that were *not* dominated by a periodicity of 10 ms. In panels **E–H**, the bold portion of the summary correlogram (up to 4.5 ms) indicates the "timbre region."

This peak broadens when a small difference in $F0$ ($\Delta F0$) is introduced between the vowels (panel **B**), and splits into a peak at the period of each vowel for $\Delta F0$s of 2 semitones or more (panels **C** and **D**). The strategy employed by the Meddis and Hewitt model is to detect the largest peak in the summary correlogram, and to assume that this corresponds to the period of one of the two vowels. Channels of the correlogram that have a peak at this period are then selected, and used to identify the first vowel; the remaining channels are used to identify the second vowel. Subsequently, a template-matching procedure is used to identify the vowels. The selected correlogram channels that define a vowel are summed across frequency to give a summary correlogram. The short-lag portion of this function (up to 4.5 ms) contains information relating to the spectral shape of the vowel, and is designated the "timbre region" by Meddis and Hewitt. The timbre regions corresponding to each of the two segregated vowels are compared against templates using a Euclidean distance metric, in order to find the best matching vowel identity.

This model explains the role of $\Delta F0$ in determining human performance as follows. When a single $F0$ is present, the frequency channels cannot be partitioned into two sets, and the identity of each vowel must be judged from the timbre region for the vowel mixture. Performance will therefore be mediocre. When a $\Delta F0$ is introduced, channels are segregated according to $F0$, and the timbre region for each segregated vowel bears a closer resemblance to its corresponding template. Performance therefore improves as $\Delta F0$ increases. However, there is little improvement beyond $\Delta F0 = 2$ semitones since the periods of the two vowels are already well separated in the summary correlogram.

The channel selection process is illustrated in panels **D–H** of Fig. 10.7. Panels **E** and **F** show the summary correlograms for two isolated vowels, /ar/ ($F0 = 100$ Hz) and /er/ ($F0 = 126$ Hz). When these two vowels are mixed, they give a $\Delta F0$ of 4 semitones, and the summary correlogram for the resulting mixture is shown in panel **D**. The largest peak in the summary function occurs at 10 ms (the period of /ar/). Selection of correlogram channels dominated by this 10 ms periodicity yields a summary correlogram for the segregated /ar/, shown in panel **G**. Note that the peak at the period of the vowel (10 ms) is more prominent than in the mixture (panel **D**), and that the timbre region of the segregated vowel (bold line in panel **G**) provides a reasonable match to the timbre region of the isolated vowel (bold line in panel **E**). The remaining correlogram channels are combined to give a summary correlogram for the segregated vowel /er/ (panel **H**). Again, comparison with the summary correlogram for the isolated vowel (panel **F**) suggests that the /er/ has been effectively segregated.

An interesting aspect of Meddis and Hewitt's model is that it only uses the $F0$ of one of the vowels to perform segregation. Approaches that attempt to extract the $F0$s of both vowels have proved less successful in modeling the improvement of listener's performance with increasing $\Delta F0$ (e.g., see [7]). Also, the use of the "timbre region" allows a segregated vowel to be matched against a template even when it is described by a subset of frequency channels. This is a novel alternative to the matching of spectral patterns; in the case of the latter, channel selection introduces time–frequency gaps that are usually handled by "missing-data" techniques (see Chapter 9, Section 9.4.2).

Meddis and Hewitt suggest that the physiological basis for their model might take the form of an inhibitory circuit, in which frequency-selective physiological channels inhibit each other if they are dominated by different sound sources (see also [66]). Such a scheme is reminscent of the model of McCabe and Denham [61] discussed in Section 10.4.1. Most simply, inhibition could be employed to suppress the firing of auditory channels that do not encode the "foreground" vowel. Alternatively, inhibition may serve to desynchronize the neural activity elicited by the two vowels. The latter approach has been investigated by Brown and Wang [25], who describe an implementation of Meddis and Hewitt's model within a neural oscillator framework. In their scheme, feedback from a global inhibitor desynchronizes frequency channels that are not dominated by the same $F0$.

In Meddis and Hewitt's model, the firing activity in each frequency channel is assigned to one vowel or the other (a principle of "exclusive allocation"; see Bregman [16], p. 12). In practice, however, some frequency channels may receive substantial input from both vowels (such as those centered on a common harmonic of the two $F0$s). Brown and Wang demonstrate that shared allocation of frequency channels can be implemented within the neural oscillator framework by introducing a slow inhibition mechanism. This enables oscillators corresponding to shared frequency channels to synchronize to two groups (i.e., shared oscillators have a periodicity that is twice that of nonshared oscillators). However, the shared-allocation version of their model does not match human data as closely as the version in which exclusive allocation is enforced.

How else might the neural activity within a single frequency channel be partitioned between two vowels? De Cheveigné [37] suggests that this could be achieved by a principle of cancellation, rather than channel selection. His model is motivated by an experiment in which one constitutent of a double vowel was 10 or 20 dB weaker than the other [38]. For some vowel pairs, all frequency channels were dominated by the stronger vowel. In such cases, the Meddis and Hewitt model predicts no improvement of vowel identification performance when a $\Delta F0$ is introduced; however, a strong effect of $\Delta F0$ was observed.

In order to model this result, de Cheveigné proposes that the output of each peripheral frequency channel is processed by a neural cancellation filter of the form shown in Fig. 10.8**A**. Here, a neuron receives excitation directly from the input, and receives inhibition from a time-delayed version of the input. The neuron fires each time an input spike arrives, unless prevented from doing so by a spike that arrives simultaneously via the inhibitory pathway. This circuit can be approximated by a simple time-domain comb filter of the form

$$y(t) = \max[0, x(t) - x(t - \tau)] \tag{10.8}$$

where $x(t)$ is the input spike probability and τ is the lag associated with the inhibitory pathway. The max operation ensures that the output spike probability $y(t)$ remains nonnegative.

Vowel segregation based on this cancellation filter is illustrated by panels **B** and **C** of Fig. 10.8, for the double vowel /ar/+/er/ with $\Delta F0 = 4$ semitones. The summary correlogram for this mixture is shown in panel **D** of Fig.10.7, and exhibits a clear

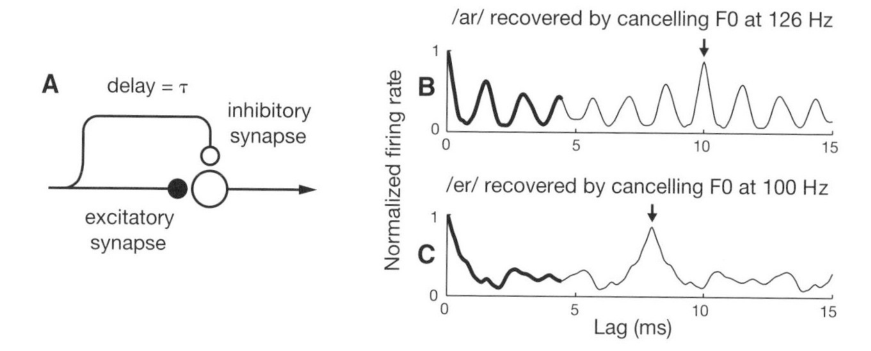

Figure 10.8 Vowel segregation by cancellation. (**A**) Schematic diagram of the neural cancellation filter proposed by de Cheveigné. Redrawn from Fig. 1(a) of [37] with permission. Copyright 1997, Acoustical Society of America. (**B**) Summary correlogram computed after cancellation of the vowel /er/ ($F0$ = 126 Hz), showing that the vowel /ar/ ($F0$ = 100 Hz) has been recovered. (**C**) Summary correlogram computed after cancellation of the vowel /ar/, showing that the vowel /er/ has been recovered. In panels **B** and **C**, the bold portion of the summary correlogram (up to 4.5 ms) indicates the "timbre region," and arrows indicate the period of the vowel. These figures should be compared with panels **E** and **F** of Fig. 10.7, which show the summary correlograms of the isolated vowels.

peak at the period of each vowel (10 ms and 7.9 ms). In the computer model, the output from each auditory filter channel (modeled by a gammatone filter) was half-wave rectified and then passed through the cancellation filter, with the delay parameter τ tuned to the period of one of the vowels. To enable comparison with Fig. 10.7, a summary correlogram was then computed from the filtered output. In panel **B** of Fig. 10.8, the cancellation filter was tuned to the period of the vowel /er/ (7.9 ms); this period has been effectively suppressed, allowing the vowel /ar/ to be recovered. Note that the summary correlogram for the recovered vowel /ar/ bears a very close resemblance to the summary correlogram of the isolated vowel shown in panel **E** of Fig. 10.7. Similarly, tuning τ to the period of the /ar/ (10 ms) allows the vowel /er/ to be recovered, as shown in panel **C** of Fig. 10.8. Again, the summary correlogram for the recovered vowel is very similar to that of the isolated vowel, shown in panel **F** of Fig. 10.7.

Cariani [27, 28] describes a related approach to de Cheveigné's neural cancellation filter, in which the neural circuit is a recurrent network (i.e., the output of each processing unit is fed back to its input). In this scheme, each unit in an array of coincidence detectors receives the same direct input, and has an indirect recurrent input tuned to a particular delay time. The processing rule associated with each unit is given by

$$y(t) = y(t - \tau) + \beta(\tau)[x(t) - y(t - \tau)] \qquad (10.9)$$

where $x(t)$ is the input spike probability, and $\beta(\tau) = \tau/33$ ms determines the rate at which the output $y(t)$ adapts to the input. Note that the inhibitory input in Eq. 10.9 is recurrent, whereas in de Cheveigné's cancellation filter (Eq. 10.8) inhibition arises from a time-delayed version of the direct input. Periodic patterns in the input to the recurrent timing network are amplified by units with matching recurrence times. Similarly, if the input contains a mixture of periodic patterns, they are extracted at different points in the neural array. This behavior allows Cariani's network to simulate the perceptual segregation of double vowels. He finds that for a $\Delta F0$ of one semitone or more, the power spectra of the two vowels can be accurately recovered, and that the quality of the separation improves as the $\Delta F0$ increases. An attractive feature of this scheme is that it does not require the $F0$ of either vowel to be estimated explicitly; rather, attributes such as $F0$ can be computed for periodic input patterns that are already segregated by the neural timing array.

Many other models of double-vowel segregation have been proposed that are variants of the above. For example, the autocorrelation in Meddis and Hewitt's model can be replaced by "strobed" temporal integration, which detects the periodicity in auditory nerve firing patterns whilst maintaining their temporal asymmetry [73, 63]. The latter appears to be significant, since human listeners are able to discriminate sounds in which each period has an exponentially rising ("ramped") envelope from those in which each period is exponentially decaying ("damped") [74]. Models of double-vowel segregation based on amplitude modulation (AM) maps have also been proposed [15], and the Meddis and Hewitt model has been extended to explain the effect of formant transitions that precede or follow a double vowel [6].

Finally, it should be noted that mechanisms other than grouping by common $F0$ may underlie the role of a $\Delta F0$ in the perceptual segregation of double vowels. Culling and Darwin [34] suggest that listeners may exploit fluctuations in the spectral envelope of the vowel mixture that arise from beating between unresolved harmonic components. They describe a computer model based on pattern matching of short-term estimates of the composite vowel spectrum, which provides a good match to listeners performance for small $\Delta F0$s.

10.5 THE OSCILLATORY CORRELATION FRAMEWORK FOR CASA

The neural and perceptual models discussed above have made useful contributions to our understanding of the mechanisms underlying ASA, but by their very nature they are concerned with simple stimuli such as tone sequences or double vowels. Can such models be applied to the processing of complex, real-world sources such as speech, and, therefore, provide a general biologically motivated solution to the ASA problem? Here, we focus on one study that has attempted to answer this question—the neural oscillator-based speech segregation system of Wang and Brown [99].

10.5.1 Speech Segregation Based on Oscillatory Correlation

The neural oscillator architecture for speech segregation described by Wang and Brown effectively melds the earlier systems proposed by the two authors. Wang's model of auditory streaming [96] (see Section 10.4.1) has a biologically plausible architecture, but takes its input from idealized time-frequency patterns. In contrast, Brown's correlogram-based CASA system [21, 22] is able to process digitally sampled audio signals, but represents the auditory scene as a collection of abstract time-frequency elements that do not have a physiological counterpart. Wang and Brown's [99] system combines correlogram-based acoustic analysis with oscillator-based grouping and, therefore, constitutes a biologically plausible architecture that operates on real-world signals.

The structure of the Wang and Brown system is shown schematically in Fig. 10.9. In the first stage, a digitally sampled mixture of speech and noise provides the input to a model of the auditory periphery, which closely follows the description given in Chapter 1, Section 1.3.2. Cochlear frequency analysis is modeled by a bank of 128 gammatone filters, and the output of each filter is processed by the Meddis hair-cell model [64, 65] to give a simulation of auditory nerve firing patterns. In the second stage of the model, "mid-level" representations of the auditory scene are computed from the auditory nerve response. Specifically, a correlogram and summary correlogram are computed at 20 ms intervals, and further features are extracted from the correlogram by a cross-correlation analysis of adjacent channels. The latter operation is motivated by the observation that frequency channels that are dominated by the same harmonic or formant elicit similar patterns of periodicity. Regions of high cross-channel correlation, therefore, indicate the presence of a harmonic or formant, and this information is exploited in the subsequent (segment formation) stage of the model.

The third stage of Wang and Brown's system is a two-layer neural oscillator network, which constitutes the core of their model (Fig. 10.10). The structure of this network mirrors the two conceptual stages of ASA identified by Bregman [16]; the first layer performs segmentation, whereas the second layer performs grouping. The segmentation layer is a LEGION network of the kind discussed in Section 10.4.1: a two-dimensional time–frequency grid of oscillators with a global inhibitor. Each oscillator (i, j) is enabled (receives a positive external input) if the auditory nerve activity computed for frequency channel i and time frame j exceeds an energy threshold. This is easily computed from the correlogram, since the autocorrelation at zero lag corresponds to the energy. Each oscillator (i, j) is connected to its four

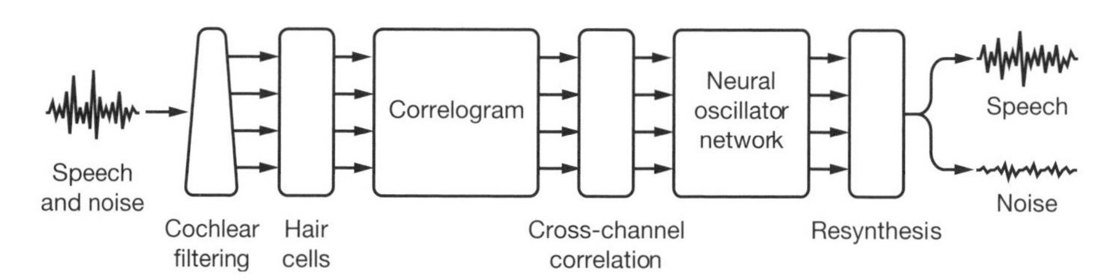

Figure 10.9 Schematic diagram of Wang and Brown's neural oscillator system for segregating speech from noise. Redrawn from [99]. Copyright © 1999 IEEE.

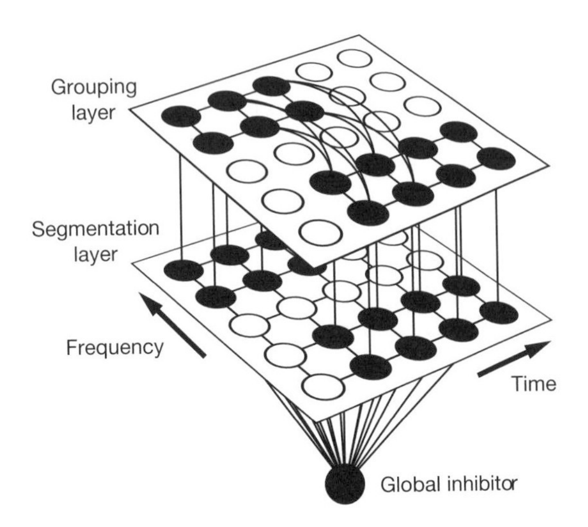

Figure 10.10 Structure of Wang and Brown's two-layer oscillator network. Redrawn from [99]. Copyright © 1999 IEEE.

nearest neighbors in time and frequency. The weight of neighboring connections along the time axis [i.e., between oscillators (i, j) and $(i, j + 1)$] is uniformly set to one. Neighboring connections along the frequency axis [i.e., between oscillators (i, j) and $(i + 1, j)$] are set to one if the cross-channel correlation between channels i and $i + 1$ exceeds a threshold value, otherwise it is set to zero. As a result, segments formed in the oscillator network represent contiguous regions of acoustic energy, such as harmonics and formants of speech. Because of the influence of local excitation and global inhibition, oscillators that encode the same segment are synchronized, and are desynchronized from oscillators that represent different segments. Small segments that do not correspond to significant acoustic events are suppressed by a "lateral potential," which prevents oscillators from receiving an external input unless they belong to a segment that spans at least three time frames.

The second layer of the network consists of a grid of laterally connected oscillators without a global inhibitor, which implements the grouping stage of ASA (see Fig. 10.10). Each oscillator in the grouping layer is enabled only if the corresponding oscillator in the segmentation layer is also enabled. The dynamics of the grouping layer are arranged such that the first oscillators to jump to the active phase are those that correspond to the longest segment in the segmentation layer. This segment may then recruit other oscillators in the grouping layer via lateral connections that link oscillators in different frequency channels, but within the same time frame. The strength of these connections is determined by pitch information, which is derived from the correlogram. More specifically, for each time frame, an $F0$ estimate is derived from the summary correlogram (by selecting the largest peak) and this is used to partition the channels of the correlogram into two sets: those that agree with the $F0$ and those that do not. This process is essentially the same as the channel selection strategy illustrated in Fig. 10.7. Oscillators that belong to the same set (i.e., those that both agree with the $F0$, or both disagree with the $F0$) receive mutual excitatory connections; otherwise, they receive mutual inhibitory connections. The result of this arrangement is to allow the longest segment to recruit other segments to form a group, so long as the two segments have consistent $F0$ information in the majority of overlapping time frames. Furthermore, synchronization is maintained between oscillators within the same segment by a second kind of lateral excitatory connection. As a result, the grouping layer of Wang and Brown's network quickly organizes into two blocks of synchronized oscillators. Each block corresponds to a group of segments that share a common $F0$ (i.e., a stream). Typical output from the grouping layer of the model is shown in Fig. 10.11.

Wang and Brown evaluate their system in terms of a signal-to-noise ratio (SNR) metric, using Cooke's [31] corpus of speech and noise mixtures. In order to compute an SNR after segregation by the system, the oscillator activity in the grouping layer is interpreted as a time–frequency mask (see Fig. 10.11); this mask is then used to weight time–frequency regions during resynthesis, using the approach described by Brown [21, 22]. The authors report that the performance of their model is similar to that of Brown's CASA system [see also [89], which compares the neural oscillator approach with independent component analysis (ICA) techniques for speech segregation]. Brown et al. [26] have also evaluated the neural oscillator approach using an automatic speech recognition (ASR) task. Their study also exploits

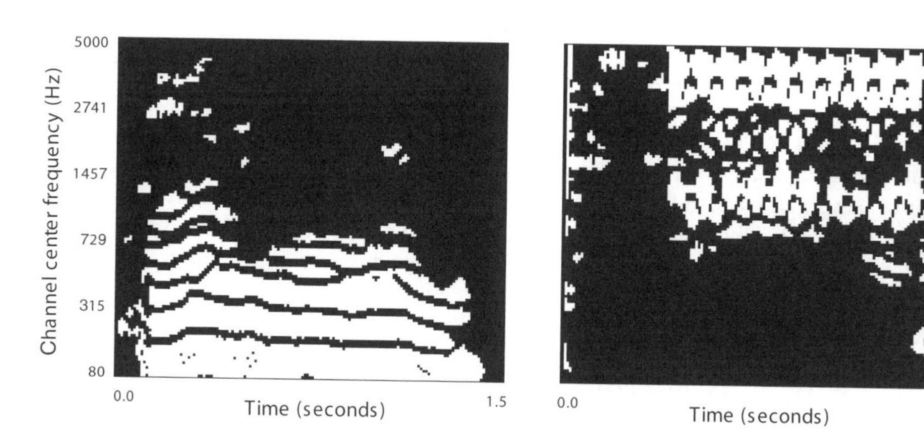

Figure 10.11 Separation of speech from a telephone sound by oscillatory correlation. In response to a mixture of speech and an interfering telephone sound, two synchronized blocks of oscillators (streams) form in the grouping layer of Wang and Brown's model. A snapshot of the network activity taken shortly after the start of simulation shows a block of oscillators corresponding to the speech stream (left panel). At a later time, oscillators corresponding to the telephone sound become active (right panel). These two streams continue to "pop out" in a repeating cycle during the remainder of the simulation. Active oscillators are indicated by white pixels, and inactive (unstimulated) oscillators are shown by black pixels. Redrawn from [99]. Copyright © 1999 IEEE.

the fact that the oscillator activity in the grouping layer can be interpreted as a time–frequency mask and, therefore, provides a suitable input for "missing data" ASR systems (together with suitable acoustic features).

Although the neural oscillator approach has a stronger biological foundation than most CASA systems, it should still be borne in mind that it is a functional model of neural processing rather than a detailed physiological model. Indeed, the Wang and Brown system includes a number of assumptions and simplifications that may not be justified by physiological evidence. For example, the number of segments generated in the segmentation layer is typically larger than the number of synchronized blocks that can be represented in a LEGION network (its "segmentation capacity"). To overcome this,Wang and Brown use an algorithmic version of LEGION that follows the major steps in the dynamical evolution of the network, but is not limited by a segmentation capacity. This is a compromise that is neurally implausible. Another issue is the explicit time axis in the oscillator network, which is presumed to be created by neural delay lines. It is debatable whether such an arrangement of delay lines is physiologically plausible; we return to this point in Section 10.7.

Recently, Pichevar and Rouat [76, 77] have proposed an approach for speech segregation that is also based on a network of Terman–Wang oscillators, but with a different neural architecture. In their scheme, acoustic features are derived from a cochleagram and an amplitude modulation (AM) spectrum. These provide the input to a two-layer oscillator network. The first layer is a two-dimensional LEGION network, similar to the segmentation layer of Wang and Brown's system, in which additional across-frequency connections promote the extraction of harmonic patterns. The second layer of the network is one-dimensional (one oscillator per frequency channel) and identifies temporal correlations between oscillators in the first layer. Synaptic weights between oscillators in the second layer are adapted dynamically, and the output of these oscillators is interpreted as a time-frequency mask. Pichevar and Rouat evaluate their model using Cooke's database of speech and noise mixtures using a log spectral distortion metric. They find that the speech resynthesized after segregation by their model is less distorted than that obtained from Wang and Brown's model, although the latter appears to reject the interfering sounds more effectively.

10.6 SCHEMA-DRIVEN GROUPING

The neural and perceptual models discussed above only address primitive grouping, in which auditory organization emerges from the mechanisms that are assumed to be innate and data-driven. However, it is well known that human listeners are able to apply learned knowledge in ASA—so-called schema-driven grouping. For example, consider the double-vowel paradigm discussed in Section 10.4.2. When the two vowels have the same $F0$, listeners are able to identify both constituents with a performance that is well above chance level; this reflects the role of schema-driven mechanisms, since no bottom-up grouping cues are available. Evidence for schema-driven grouping has come from many other psychophysical experiments, which show that speech and musical sounds can be perceptually integrated even when primitive grouping cues suggest that they should be segregated (for a review, see Chapter 4 of Bregman [16]).

Sagi et al. [81] address the problem of schema-driven ASA within a neural architecture. They focus on the problem of understanding the voice of a single ("attended") speaker in the presence of other speakers. Their approach is based on the cortronic architecture of Hecht-Nielsen [48], which is a type of associative network. Cortronic regions are organized hierarchically, and through training they learn the structural relationships between sounds and linguistic units (words and sentences). Activity in the input region of their network is determined by a short-term wavelet analysis of an acoustic signal. The most active neurons in the input region are selected at each epoch, forming a sparse binary code of the short-term sound input ("sound token"), which is used in the remainder of processing. For simplicity, it is assumed that a word consists of four sound tokens, which are processed in sequence by four sound-processing regions. In response to an input, the network recalls sequences of sound tokens that were learned during training. The authors show that this behavior is robust in the presence of interfering speakers, and that the ability of the network to recover words from a mixture is better than its ability to recover individual sound tokens (the same behavior would be expected of human listeners). However, the model is somewhat abstract and makes a number of restrictive assumptions; in particular, that words must comprise a fixed number of sound tokens. Also, the model requires that word boundaries be detected, and that the extraction of sound tokens from the acoustic input must be synchronized with the processing of those tokens by cortronic regions.

Other workers have suggested that neural oscillations could play a role in schema-driven grouping [94, 24]. This is an attractive idea, because physiological studies suggest an intimate connection between oscillations, temporal synchronization, and memory [11]. More specifically, there is evidence that the convergence of synchronous firing activity on a postsynaptic cell produces optimal conditions for long-term synaptic potentiation. Neural plasticity may, therefore, require temporal synchronization of synaptic inputs (for example, see [82]). Given the putative role of oscillations in solving the binding problem, it is tempting to suggest that oscillations provide a common code that links the mechanisms of memory and sensory-information processing.

Wang et al. [100] have described an oscillatory associative memory that is based on coupled oscillators. Their network is able to segregate concurrently presented patterns that have previously been stored and, therefore, represents an entirely schema-driven approach to segregation. Potentially, such a network could be combined with Wang and Brown's oscillator framework for primitive grouping, in order to model the interaction of top-down and bottom-up grouping mechanisms. In practice, however, the effectiveness of such a model is unclear as issues of invariance and memory capacity are unresolved. Progress has been made recently in improving the capacity of oscillatory associative memories [71].

The model of vowel recognition proposed by Liu et al. [57] may be regarded as an implementation of schema-driven grouping based on an oscillatory associative memory. In their system, acoustic patterns are processed by a three-layer oscillator network, in which each layer (labeled A, B, and C) is assumed to correspond to an area of the auditory cortex. Vowel formants are identified by detecting peaks in the spectrum of the acoustic input, which is obtained by linear prediction. Each formant is then encoded in the A ("feature extracting") layer as a group of synchronized oscillations. Hard-wired connections in the B ("feature linking") layer encode the pattern of formant frequencies associated with different vowels and, therefore, act as a simple associative memory. The C ("categorizing") layer determines the vowel category, by assessing the degree of synchronization in each formant region. When an acoustic input is present, top-down and bottom-up interactions between the A and B layers, together with associative interactions within the B layer, cause a particular vowel pattern to be selected and represented by synchronized oscillations. The authors demonstrate that top-down reinforcement from the B layer is important for robustness in the presence of noise; this is demonstrated by vowel recognition experiments in the presence of multispeaker babble.

Liu et al.'s approach has a number of limitations. First, although it embodies top down and bottom-up information flow, primitive grouping principles are not implemented, and it is not clear how the architecture could be adapted to include them. Second, their system is unable to group sounds that are separated in time (i.e., sequential grouping is not implemented). Finally, Liu et al. do not present any evidence that their system is able to segregate concurrent vowel sounds. More specifically, it is not clear how the network will perform when presented simultaneously with two formant patterns that are both encoded in the B layer.

Clearly, the integration of bottom-up and top-down constraints in neural models remains a challenging problem. Recent progress in this regard is reported by Grossberg et al., who describe a model of auditory stream segregation that is based on the principles of adaptive resonance theory (ART) [47]. This theory claims that learning in neural networks is driven by resonant states. In the first stage of their ART-STREAM model, spectral features are derived by computing the short-term energy within each channel of a gammatone filterbank. These features provide the input to a layered neural network in which spectral and pitch information is represented for each stream. Bottom-up mechanisms detect harmonic patterns in the spectral representation of the input, which in turn leads to the activation of a pitch category. A top-down expectation is then generated, which potentiates harmonic regions associated with the pitch. If the spectral and pitch representations continue to mutually reinforce one another, a resonance develops in the network that corresponds to the formation of a perceptual stream. Grossberg et al. show that the resonant properties of their network allow it to track acoustic sources through interference, as in the auditory continuity illusion [16].

10.7 DISCUSSION

We conclude this chapter with a discussion of key issues raised by the neural and perceptual models discussed above, together with some suggestions for future research.

10.7.1 Temporal or Spatial Coding of Auditory Grouping

The binding problem and the ASA problem are intimately connected, and a fundamental question is how perceptual streams are encoded at the neural level. The oscillatory correlation theory [95, 96] suggests that grouping relationships are encoded by the temporal synchronization of neural responses. Explanations based on selection or suppression of frequency channels represent another plausible mechanism for implementing auditory grouping at the neural level, as exemplified by the model of Meddis and Hewitt [67], but are not incompatible with oscillatory correlation (e.g., see [25]). Alternatively, the models of McCabe and Denham [61] and Grossberg et al. [47] suggest that different streams are represented spatially in arrays of neurons. However, such an approach may lack the generality required to explain the integration of features from different modalities, as in the audio–visual McGurk effect [62].

10.7.2 Physiological Support for Neural Time Delays

Many of the neural models described here require time delays to be generated by neural circuits. This seems reasonable; indeed, systematically arranged neural delay lines have been identified in the auditory midbrain, and are believed to underlie the detection of interaural time differences ([105]; see also Chapter 5). Also, place coding of temporal information is a prevalent principle of auditory information processing. However, neural models that use pitch information require time delays that are considerably longer than those associated with binaural hearing. For example, the cancellation model of de Cheveigné [37] requires time delays that are as long as the longest pitch period to be cancelled (which may be 20 ms or more). There is little physiological evidence to support the existence of such long delay lines; hence, de Cheveigné's model (and others that require long delays, such as Cariani's [28] recurrent timing networks) might be regarded as controversial in this regard. Similarly, there is no strong physiological support for the arrangement of neural time delays employed in Wang and Brown's neural oscillator model.

De Cheveigné's model is phrased in terms of a very specific neural circuit, but other perceptual models are more functional in form and could be implemented at the physiological level in a number of different ways. For example, the autocorrelation-based model of Meddis and Hewitt [67] might be implemented by an arrangement of coincidence detectors and delay lines, and, therefore, appears to require long neural delays in the same manner as de Cheveigné's model. However, other physiological implementations of their model are possible. Using a detailed physiological model similar to that discussed in Section 10.3.3, Wiegrebe and Meddis [102] show that chop-S units located in the ventral cochlear nucleus are able to synchronize selectively to particular pitches. An array of such neurons may, therefore, serve to convert the temporal pitch code in the auditory nerve into a place code in the cochlear nucleus, in much the same way as an autocorrelation analysis. Furthermore, this behavior results from the intrinsic membrane properties of chop-S cells, rather than via neural delay lines.

10.7.3 Convergence of Psychological, Physiological, and Computational Approaches

Many of the neural and perceptual models described in this chapter make predictions that can be tested through psychophysical or physiological experimentation. For example, the neural oscillator model of Wrigley and Brown [104, 103] makes predictions about the "resetting" of auditory attention that have been tested by a psychophysical experiment, and Patterson's auditory image model has motivated many psychophysical experiments that investigate the role of temporal asymmetry in auditory perception [73, 74]. Wang's neural oscillator models [96, 100] also make predictions that could be tested by physiological studies. Similarly, there has been a close link between de Cheveigné's [37] modeling work and his psychophysical studies of double-vowel perception. Futhermore, de Cheveigné has applied his cancellation model to physiological recordings from auditory nerve fibres, which represents an interesting convergence of computational and physiological approaches [36].

Such studies can be viewed as investigating the ASA problem at different levels of Marr's [59] three-level framework for understanding complex information processing systems (computational theory, representation and algorithm, and implementation; see Chapter 1, Section 1.2.2). We expect that computer implementations of neural and perceptual models will play an increasingly important role in ASA research, since they provide a powerful means for integrating physiological and psychophysical data (see [30]).

10.7.4 Neural Models as a Framework for CASA

To date, perhaps the most complete neurobiologically inspired CASA system is the neural oscillator approach proposed by Wang and Brown [99] (see Section 10.5.1). However, the complexity of neural oscillator systems may impede their adoption as a general framework for CASA. The experience of Norris [72] is instructive in this respect; his attempts to integrate common onset and offset cues into Wang's [95] oscillator network met with mixed success, because the oscillatory dynamics associated with existing grouping cues are disturbed when additional cues are introduced. It is desirable for CASA architectures to be flexible and adaptable, and the neural oscillator approach does not score highly in this regard.

On the other hand, a potential benefit of the neural oscillator approach (and of many other neural models) is that it employs a parallel and distributed architecture that is suitable for implementation in hardware. Cosp and Madrenas [32] have described an analog, very large-scale integration (VLSI) implementation of a 16×16 neuron LEGION network. They report that their test device performs segmentation of binary patterns very rapidly, and has a very low complexity and power consumption compared to other hardware and software approaches. Larger devices of this kind could be used as the basis for analog VLSI implementations of oscillator-based CASA architectures, such as those of Wang and Brown [99] and Pichevar and Rouat [76, 77]. Their compact size and low power consumption could make them particularly suitable for application in hearing aids.

10.7.5 The Role of Attention

Relatively few computer models have addressed the role of attention in ASA. The auditory streaming model of Beauvois and Meddis [13, 14] suggests a low-level explanation for some aspects of auditory attention. More comprehensive models have been proposed by Baird [8] (in which auditory attention is viewed as rhythmic expectancy) and Wrigley and Brown [104] (in which endogenous attention is modeled by selectively weighting the activity of units in an oscillator array). An important issue is whether attention is directed to auditory streams formed preattentively, or plays the role of binding auditory features in stream segregation. In other words, does stream segregation precede auditory attention or follow it? This and other related issues concerning the role of attention in ASA are the subject of much debate, and computational models may help to clarify these issues when informed by data from electrophysiological studies of phenomena such as the ORN and MMN (see Section 10.2.2).

10.7.6 Schema-Based Organization

The majority of neural and perceptual models address bottom-up mechanisms of ASA, such as those associated with auditory streaming of tone sequences and $F0$-based sound segregation. Relatively few studies have investigated schema-based organization within neural models, which is perhaps surprising given the substantial literature that exists on associative memory and neural network learning (see [5]). Integration of top-down and bottom-up mechanisms within the same neural framework remains an important problem for future research. It is worth emphasizing that the challenge does not lie in the integration per se, but in uncovering integrative mechanisms whose performance in auditory organization exceeds that achievable by either bottom-up or top-down analysis.

ACKNOWLEDGMENTS

Thanks to Stuart Wrigley for commenting on a draft of this chapter. Brown's research was supported by EPSRC grant GR/R47400/01. Wang's research was supported in part by an AFOSR grant (FA9550-04-01-0117) and an AFRL grant (FA8750-04-1-0093).

REFERENCES

1. C. Alain, S. R. Arnott, and T. W. Picton. Bottom-up and top-down influences on auditory scene analysis: Evidence from event-related brain potentials. *Journal of Experimental Psychology: Human Perception and Performance,* 27(5):1072–1089, 2001.
2. C. Alain and A. Izenberg. Effects of attentional load on auditory scene analysis. *Journal of Cognitive Neuroscience,* 15(7):1063–1073, 2003.

3. C. Alain, B. M. Schuler, and K. L. McDonald. Neural activity associated with distinguishing concurrent auditory objects. *Journal of the Acoustical Society of America,* 111(2):990–995, 2002.
4. S. Anstis and S. Saida. Adaptation to auditory streaming of frequency-modulated tones. *Journal of Experimental Psychology: Human Perception and Performance,* 11(3): 257–271, 1985.
5. M. A. Arbib, editor. *Handbook of Brain Theory and Neural Networks,* 2nd ed. MIT Press, Cambridge, MA, 2003.
6. P. F. Assmann. Modeling the perception of concurrent vowels: Role of formant transitions. *Journal of the Acoustical Society of America,* 100(2):1141–1152, 1996.
7. P. F. Assmann and Q. Summerfield. Modeling the perception of concurrent vowels: Vowels with different fundamental frequencies. *Journal of the Acoustical Society of America,* 88:680–697, 1990.
8. B. Baird. Synchronized auditory and cognitive 40 Hz attentional streams, and the impact of rhythmic expectation on auditory scene analysis. In M. Jordan, M. Kearns, and S. Solla, editors, *Neural Information Processing Systems,* volume 10, pp. 3–10. MIT Press, Cambridge, MA, 1997.
9. H. B. Barlow. Single units and cognition: A neuron doctrine for perceptual psychology. *Perception,* 1:371–394, 1972.
10. D. S. Barth and K. D. MacDonald. Thalamic modulation of high-frequency oscillating potentials in auditory cortex. *Nature,* 383:78–81, 1996.
11. E. Basar, C. Basar-Eroglu, S. Karakas, and M. Schurmann. Brain oscillations in perception and memory. *International Journal of Psychophysiology,* 35(2–3):95–124, 2000.
12. M. Bauer and W. Martienssen. Coupled circle maps as a tool to model synchronization in neural networks. *Network,* 2:345–351, 1991.
13. M. W. Beauvois and R. Meddis. A computer model of auditory stream segregation. *Quarterly Journal of Experimental Psychology,* 43A(3):517–541, 1991.
14. M. W. Beauvois and R. Meddis. Computer simulation of auditory stream segregation in alternating tone sequences. *Journal of the Acoustical Society of America,* 99(4): 2270–2280, 1996.
15. F. Berthommier and G. Meyer. Improving amplitude modulation maps for F0-dependent segregation of harmonic sounds. In *Proceedings of Eurospeech,* pp. 2483–2486, Rhodes, Greece, September 1997.
16. A. S. Bregman. *Auditory Scene Analysis.* MIT Press, Cambridge, MA, 1990.
17. A. S. Bregman and J. Campbell. Primary auditory stream segregation and perception of order in rapid sequences of tones. *Journal of Experimental Psychology,* 89:244–249, 1971.
18. A. S. Bregman and S. Pinker. Auditory streaming and the building of timbre. *Canadian Journal of Psychology,* 32:19–31, 1978.
19. A. S. Bregman and A. I. Rudnicky. Auditory segregation: Stream or streams? *Journal of Experimental Psychology,* 1(3):263–267, 1975.
20. M. Brosch, E. Budinger, and H. Scheich. Stimulus-related gamma oscillations in primate auditory cortex. *Journal of Neurophysiology,* 87:2715–2725, 2002.
21. G. J. Brown. *Computational Auditory Scene Analysis: A Representational Approach.* PhD thesis, University of Sheffield, 1992.

22. G. J. Brown and M. P. Cooke. Computational auditory scene analysis. *Computer Speech and Language,* 8:297–336, 1994.

23. G. J. Brown and M. P. Cooke. Temporal synchronization in a neural oscillator model of primitive auditory stream segregation. In D. F. Rosenthal and H. G. Okuno, editors, *Computational Auditory Scene Analysis,* pp. 87–101. Lawrence Erlbaum, Mahwah, NJ, 1998.

24. G. J. Brown and D. Wang. Timing is of the essence: Neural oscillator models of auditory grouping. In S. Greenberg and W. A. Ainsworth, editors, *Listening to Speech: An Auditory Perspective.* Lawrence Erlbaum, Mahwah, NJ, 2005.

25. G. J. Brown and D. L. Wang. Modelling the perceptual segregation of concurrent vowels with a network of neural oscillators. *Neural Networks,* 10(9):1547–1558, 1997.

26. G. J. Brown, D. L. Wang, and J. Barker. A neural oscillator sound separator for missing data speech recognition. In *Proceedings of the International Joint Conference on Neural Networks,* Washington DC, July 2001.

27. P. Cariani. Neural timing nets for auditory computation. In S. Greenberg and M. Slaney, editors, *Computational Models of Auditory Function,* pp. 235–249. IOS Press, Amsterdam, 2001.

28. P. Cariani. Recurrent timing nets for F0-based speaker separation. In P. Divenyi, editor, *Speech Separation by Humans and Machines,* pp. 31–53. Springer-Verlag, New York, 2005.

29. R. P. Carlyon, R. Cusack, J. M. Foxton, and I. H. Robertson. Effects of attention and unilateral neglect on auditory stream segregation. *Journal of Experimental Psychology: Human Perception and Performance,* 27(1):115–127, 2001.

30. P. Chang. *Exploration of Behavioural, Physiological and Computational Approaches to Auditory Scene Analysis.* Master's thesis, The Ohio State University, Department of Computer Science and Engineering. Available at http://www.cse.ohio-state.edu/pnl/theses.html, 2004.

31. M. P. Cooke. *Modelling Auditory Processing and Organization*. PhD thesis, University of Sheffield, 1991 (also published by Cambridge University Press, Cambridge, UK, 1993).

32. J. Cosp and J. Madrenas. Scene segmentation using neuromorphic oscillatory networks. *IEEE Transactions on Neural Networks,* 14(5):1278–1296, 2003.

33. F. Crick. Function of the thalamic reticular complex: The searchlight hypothesis. *Proceedings of the National Academy of Sciences USA,* 81:4586–4590, 1984.

34. J. F. Culling and C. J. Darwin. Perceptual and computational separation of simultaneous vowels: Cues arising from low-frequency beating. *Journal of the Acoustical Society of America,* 95(3):1559–1569, 1994.

35. C. J. Darwin, R. W. Hukin, and B. Y. Al-Khatib. Grouping in pitch perception: Evidence for sequential constraints. *Journal of the Acoustical Society of America,* 98(2):880–885, 1995.

36. A. de Cheveigné. Separation of concurrent harmonic sounds: Fundamental frequency estimation and a time-domain cancellation model of auditory processing. *Journal of the Acoustical Society of America,* 93(6):3271–3290, 1993.

37. A. de Cheveigné. Concurrent vowel identification. III. A neural model of harmonic interference cancellation. *Journal of the Acoustical Society of America,* 101(5):2857–2865, 1997.

38. A. de Cheveigné, H. Kawahara, M. Tsuzaki, and K. Aikawa. Concurrent vowel identification. I. Effects of relative amplitude and F0 difference. *Journal of the Acoustical Society of America,* 101(5):2839–2847, 1996.

39. R. C. deCharms and M. M. Merzenich. Primary cortical representation of sounds by the coordination of action-potential timing. *Nature,* 381:610–613, 1996.

40. R. Eckhorn, R. Bauer, W. Jordan, M. Brosch, W. Kruse, M. Munk, and H. Reitboeck. Coherent oscillations: A mechanism of feature linking in the visual cortex. *Biological Cybernetics,* 60:121–130, 1988.

41. R. Fitzhugh. Impulses and physiological states in models of nerve membrane. *Biophysical Journal,* 1:445–466, 1961.

42. W. J. Freeman. Spatial properties of an EEG event in the olfactory bulb and cortex. *Electroencephalography and Clinical Neurophysiology,* 44:586–605, 1978.

43. R. Galambos, S. Makeig, and P. J. Talmachoff. A 40-Hz auditory potential recorded from the human scalp. *Proceedings of the National Academy of Sciences USA,* 78:2643–2647, 1981.

44. J. Grassman. *Asymptotic Methods for Relaxation Oscillations and Applications.* Springer-Verlag, New York, 1987.

45. C. M. Gray. The temporal correlation hypothesis of visual feature integration: Still alive and well. *Neuron,* 24:31–47, 1999.

46. C. M. Gray, P. König,A. K. Engel, and W. Singer. Oscillatory responses in cat visual cortex exhibit inter-columnar synchronization which reflects global stimulus properties. *Nature,* 338:334–337, 1989.

47. S. Grossberg, K. K. Govindarajan, L. Wyse, and M. A. Cohen. ARTSTREAM: A neural network model of auditory scene analysis and source segregation. *Neural Networks,* 17:511–536, 2004.

48. R. Hecht-Nielsen. A theory of the cerebral cortex. In *Proceedings of the International Conference on Neural Information Processing (ICONIP),* pp. 1459–1464, Burke, VA, IOS Press, Amsterdam, 1998.

49. M. J. Hewitt, R. Meddis, and T. M. Shackleton. A computer model of a cochlear-nucleus stellate cell: Responses to amplitude-modulation and pure tone stimuli. *Journal of the Acoustical Society of America,* 91(4):2096–2109, 1992.

50. A. L. Hodgkin and A. F. Huxley. A quantitative description of membrane current and its application to conduction and excitation in nerve. *Journal of Physiology (London),* 117:500–544, 1952.

51. D. H. Hubel and T. N. Wiesel. Receptive fields, binocular interaction and functional architecture in the cat's visual cortex. *Journal of Physiology,* 160:106–154, 1962.

52. L. A. Jeffress. A place theory of sound localization. *Journal of Comparative and Physiological Psychology,* 41:35–39, 1948.

53. M. Joliot, U. Ribary, and R. Llinás. Human oscillatory brain activity near to 40 Hz coexists with cognitive temporal binding. *Proceedings of the National Academy of Sciences USA,* 91:11748–11741, 1994.

54. M. R. Jones, G. Kidd, and R. Wetzel. Evidence for rhythmic attention. *Journal of Experimental Psychology: Human Perception and Performance,* 7:1059–1073, 1981.

55. E. R. Kandel, J. H. Schwartz, and T. M. Jessell, *Principles of Neural Science,* 3rd ed,. Elsevier, New York, 1991.

56. E. W. Large and M. R. Jones. The dynamics of attending: How we track time-varying events. *Psychological Review,* 106:119–159, 1999.

57. F. Liu, Y. Yamaguchi, and H. Shimizu. Flexible vowel recognition by the generation of dynamic coherence in oscillator neural networks: Speaker-independent vowel recognition. *Biological Cybernetics,* 71(2):105–114, 1994.

58. R. Llinás and U. Ribary. Coherent 40-Hz oscillation characterizes dream state in humans. *Proceedings of the National Academy of Sciences USA,* 90:2078–2082, 1993.

59. D. Marr. *Vision.* W. H. Freeman, San Francisco, CA, 1982.

60. S. McAdams and A. S. Bregman. Hearing musical streams. *Computer Music Journal,* 3:26–43, 1979.

61. S. L. McCabe and M. J. Denham. A model of auditory streaming. *Journal of the Acoustical Society of America,* 101(3):1611–1621, 1997.

62. H. McGurk and J. McDonald. Hearing lips and seeing voices. *Nature,* 264:746–748, 1976.

63. J. D. McKeown and R. D. Patterson. The time course of auditory segregation: Concurrent vowels that vary in duration. *Journal of the Acoustical Society of America,* 98(4):1866–1877, 1995.

64. R. Meddis. Simulation of mechanical to neural transduction in the auditory receptor. *Journal of the Acoustical Society of America,* 79(3):702–711, 1986.

65. R. Meddis. Simulation of auditory-neural transduction: Further studies. *Journal of the Acoustical Society of America,* 83(3):1056–1063, 1988.

66. R. Meddis. Computer models of possible physiological contributions to low-level auditory scene analysis. *Journal of the Acoustical Society of America,* 115(5):2457, 2004.

67. R. Meddis and M. J. Hewitt. Modeling the identification of concurrent vowels with different fundamental frequencies. *Journal of the Acoustical Society of America,* 91(1):233–245, 1992.

68. G. A. Miller and G. A. Heise. The trill threshold. *Journal of the Acoustical Society of America,* 22:637–638, 1950.

69. R. E. Mirollo and S. H. Strogatz. Synchronization of pulse coupled biological oscillators. *SIAM Journal on Applied Mathematics,* 50:1645–1662, 1990.

70. J. Nagumo, S. Arimoto, and S. Yoshizawa. An active pulse transmission line simulating nerve axon. *Proceedings of the IRE,* 50:2061–2070, 1962.

71. T. Nishikawa, Y. C. Lai, and F. C. Hoppensteadt. Capacity of oscillatory associative-memory networks with error-free retrieval. *Physical Review Letters,* 92(10):108101, 2004.

72. M. Norris. *Assessment and Extension of Wang's Oscillatory Model of Auditory Stream Segregation.* PhD thesis, University of Queensland, School of Information Technology and Electrical Engineering, 2003.

73. R. D. Patterson, M. H. Allerhand, and C. Giguère. Time-domain modeling of peripheral auditory processing: A modular architecture and a software platform. *Journal of the Acoustical Society of America,* 98(4):1890–1894, 1995.

74. R. D. Patterson and T. Irino. Modeling temporal asymmetry in the auditory system. *Journal of the Acoustical Society of America,* 104(5):2967–2979, 1998.

75. C. S. Peskin. *Mathematical Aspects of Heart Physiology.* NYU Courant Institute of Mathematical Sciences, New York, 1975.

76. R. Pichevar and J. Rouat. Cochleotopic/AM topic (CAM) and cochleotopic/spectrotopic (CSM) map based sound separation using relaxation oscillatory neurons. In *Proceedings of the IEEE 13th Workshop on Neural Networks for Signal Processing,* pp. 657–666, September 2003.

77. R. Pichevar and J. Rouat. Monophonic source separation with an unsupervised network of spiking neurones. *Speech Communication,* submitted, 2005.

78. A. N. Popper and R. R. Fay, editors. *The Mammalian Auditory Pathway: Neurophysiology.* Springer-Verlag, New York, 1992.

79. W. S. Rhode and S. Greenberg. Physiology of the cochlear nuclei. In A. N. Popper and R. R. Fay, editors, *The Mammalian Auditory Pathway: Neurophysiology.* Springer-Verlag, New York, 1992.

80. U. Ribary, A. A. Ioannides, K. D. Singh, R. Hasson, J. P. Bolton, F. Lado, A. Mogilner, and R. Llinás. Magnetic field tomography of coherent thalamocortical 40-Hz oscillations in humans. *Proceedings of the National Academy of Sciences USA,* 88(24):11037–11041, 1991.

81. B. Sagi, S. C. Nemat-Nasser, R. Kerr, R. Hayek, C. Downing, and R. Hecht-Nielsen. A biologically motivated solution to the cocktail party problem. *Neural Computation,* 13:1575–1602, 2001.

82. W. Singer. Synchronization of cortical activity and its putative role in information processing and learning. *Annual Review of Physiology,* 55:349–374, 1993.

83. W. Singer and C. M. Gray. Visual feature integration and the temporal correlation hypothesis. *Annual Review of Neuroscience,* 18:555–586, 1995.

84. C. J. Spence and J. Driver. Covert spatial orienting in audition: Exogenous and endogenous mechanisms. *Journal of Experimental Psychology: Human Perception and Performance,* 20(3):555– 574, 1994.

85. E. Sussman, W. Ritter, and H. G. Vaughan. An investigation of the auditory streaming effect using event-related brain potentials. *Psychophysiology,* 36:22–34, 1999.

86. D. Terman and D. L. Wang. Global competition and local cooperation in a network of neural oscillators. *Physica D,* 81:148–176, 1995.

87. N. Todd. An auditory cortical theory of primitive auditory grouping. *Network: Computation in Neural Systems,* 7:349–356, 1996.

88. W. M. Usrey and R. C. Reid. Synchronous activity in the visual system. *Annual Review of Physiology,* 61:435–456, 1999.

89. A. J. W. van der Kouwe, D. L. Wang, and G. J. Brown. A comparison of auditory and blind separation techniques for speech segregation. *IEEE Transactions on Speech and Audio Processing,* 9:189–195, 2001.

90. B. van der Pol. On "relaxation oscillations." *Philosophical Magazine,* 2(11):978–992, 1926.

91. L. P. A. S. van Noorden. *Temporal Coherence in the Perception of Tone Sequences.* PhD thesis, Eindhoven University of Technology, 1975.

92. F. Varela, J. P. Lachaux, E. Rodriguez, and J. Martinerie. The brainweb: Phase synchronization and large-scale integration. *Nature Reviews Neuroscience,* 2:229–239, 2001.

93. C. von der Malsburg. The correlation theory of brain function. Technical Report 81-2, Max-Planck-Institute for Biophysical Chemistry, 1981. (Reprinted in *Models of Neural*

Networks, II, E. Domany, J. L. van Hemmen, and K. Schulten, editors. Springer-Verlag, Berlin, 1994).

94. C. von der Malsburg and W. Schneider. A neural cocktail-party processor. *Biological Cybernetics,* 54:29–40, 1986.
95. D. L. Wang. Auditory stream segregation based on oscillatory correlation. In *Proceedings of the IEEE International Workshop on Neural Networks for Signal Processing,* pp. 624–632, Ermioni, Greece, September 1994.
96. D. L. Wang. Primitive auditory segregation based on oscillatory correlation. *Cognitive Science,* 20:409–456, 1996.
97. D. L. Wang. Relaxation oscillators and networks. In J. G. Webster, editor, *Encyclopedia of Electrical and Electronics Engineering,* pp. 396–405. Wiley, New York, 1999. (Available for download from www.cse.ohio-state.edu/~dwang.)
98. D. L. Wang. The time dimension for scene analysis. *IEEE Transactions on Neural Networks,* 16(6):1401–1426, 2005.
99. D. L. Wang and G. J. Brown. Separation of speech from interfering sounds using oscillatory correlation. *IEEE Transactions on Neural Networks,* 10(3):684–697, 1999.
100. D. L. Wang, J. Buhmann, and C. von der Malsburg. Pattern segmentation in associative memory. *Neural Computation,* 2:94–106, 1990.
101. D. L. Wang and D. Terman. Locally excitatory globally inhibitory oscillator networks. *IEEE Transactions on Neural Networks,* 6(1):283–286, 1995.
102. L. Wiegrebe and R. Meddis. The representation of periodic sounds in simulated sustained chopper units of the ventral cochlear nucleus. *Journal of the Acoustical Society of America,* 115(3):1207–1218, 2004.
103. S. N. Wrigley. *A Theory and Computational Model of Auditory Selective Attention.* PhD thesis, Department of Computer Science, University of Sheffield, 2002.
104. S. N. Wrigley and G. J. Brown. A computational model of auditory selective attention. *IEEE Transactions on Neural Networks,* 15:1151–1163, 2004.
105. T. C. T. Yin and J. C. K. Chan. Interaural time sensitivity in medial superior olive of cat. *Journal of Neurophysiology,* 64(2):465–488, 1990.

Index

Computational Auditory Scene Analysis. Edited by DeLiang Wang and Guy J. Brown

RECEIVED
JAN 12 2007